The Liquid State and Its Electrical Properties

NATO ASI Series

Advanced Science Institutes Series

A series presenting the results of activities sponsored by the NATO Science Committee, which aims at the dissemination of advanced scientific and technological knowledge, with a view to strengthening links between scientific communities.

The series is published by an international board of publishers in conjunction with the NATO Scientific Affairs Division

A	**Life Sciences**	Plenum Publishing Corporation
B	**Physics**	New York and London
C	**Mathematical and Physical Sciences**	Kluwer Academic Publishers Dordrecht, Boston, and London
D	**Behavioral and Social Sciences**	
E	**Applied Sciences**	
F	**Computer and Systems Sciences**	Springer-Verlag
G	**Ecological Sciences**	Berlin, Heidelberg, New York, London,
H	**Cell Biology**	Paris, and Tokyo

Recent Volumes in this Series

Volume 186—Simple Molecular Systems at Very High Density
edited by A. Polian, P. Loubeyre, and N. Boccara

Volume 187—X-Ray Spectroscopy in Atomic and Solid State Physics
edited by J. Gomes Ferreira and M. Teresa Ramos

Volume 188—Reflection High-Energy Electron Diffraction and Reflection Electron Imaging of Surfaces
edited by P. K. Larsen and P. J. Dobson

Volume 189—Band Structure Engineering in Semiconductor Microstructures
edited by R. A. Abram and M. Jaros

Volume 190—Squeezed and Nonclassical Light
edited by P. Tombesi and E. R. Pike

Volume 191—Surface and Interface Characterization by Electron Optical Methods
edited by A. Howie and U. Valdrè

Volume 192—Noise and Nonlinear Phenomena in Nuclear Systems
edited by J. L. Muñoz-Cobo and F. C. Difilippo

Volume 193—The Liquid State and Its Electrical Properties
edited by E. E. Kunhardt, L. G. Christophorou, and L. H. Luessen

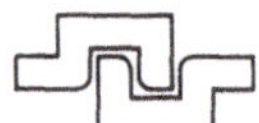

Series B: Physics

The Liquid State and Its Electrical Properties

Edited by

E. E. Kunhardt

Polytechnic University
Weber Research Institute
Farmingdale, New York

L. G. Christophorou

Oak Ridge National Laboratory
Oak Ridge, Tennessee

and

L. H. Luessen

Naval Surface Warfare Center
Dahlgren, Virginia

Plenum Press
New York and London
Published in cooperation with NATO Scientific Affairs Division

Proceedings of a NATO Advanced Study Institute
International Advanced Course on
The Liquid State and Its Electrical Properties,
held July 5-17, 1987,
in Sintra, Portugal

Library of Congress Cataloging in Publication Data

NATO Advanced Study Institute International Advanced Course on the Liquid State and its Electrical Properties (1987: Sintra, Portugal)
The liquid state and its electrical properties / edited by E. E. Kunhardt, L. G. Christophorou, and L. H. Luessen.
p. cm.—(NATO ASI series. Series B, Physics; v. 193)
Bibliography: p.
Includes index.
ISBN-13: 978-1-4684-8025-2 e-ISBN-13: 978-1-4684-8023-8
DOI: 10.1007/978-1-4684-8023-8
1. Liquids—Electric properties—Congresses. 2. Fluids—Electric properties—Congresses. I. Kunhardt, Erich E. II. Christophorou, L. G. III. Luessen, Lawrence H. IV. Title. V. Series.
QC145.4.E45N37 1987 89-3446
530.4′2—dc19 CIP

Softcover reprint of the hardcover 1st edition 1988

A Division of Plenum Publishing Corporation
233 Spring Street, New York, N.Y. 10013

PREFACE

As the various disciplines of science advance, they proliferate and tend to become more esoteric. Barriers of specialized terminologies form, which cause scientists to lose contact with their colleagues, and differences in points-of-view emerge which hinder the unification of knowledge among the various disciplines, and even within a given discipline. As a result, the scientist, and especially the student, is in many instances offered fragmented glimpses of subjects that are fundamentally synthetic and that should be treated in their own right. Such seems to be the case of the liquid state. Unlike the other states of matter -- gases, solids, and plasmas -- the liquid state has not yet received unified treatment, probably because it has been the least explored and remains the least understood state of matter.

Occasionally, events occur which help remove some of the barriers that separate scientists and disciplines alike. Such an event was the ASI on The Liquid State held this past July at the lovely Hotel Tivoli Sintra, in the picturesque town of Sintra, Portugal, approximately 30 km northwest of Lisbon. Since this broad a subject could not be covered in one Institute, the focus of the ASI was on a theme that provided a common thread of understanding for all in attendance -- the Electrical Properties of the Liquid State. Advances in computational power and diagnostic techniques are increasingly leading to fascinating new results on the basic properties of liquids and the physio-chemical interactions occurring in the liquid phase. Such an understanding will undoubtedly open up new technological applications of the liquid state, and the Sintra ASI was just the right stimulation needed for the seventy-odd scientists and engineers in attendance.

Rallying around this theme -- The Liquid State and Its Electrical Properties -- the Institute brought together a remarkable blend of scientists and engineers from the following fields: statistical mechanics, particle kinetics and dynamics, condensed-matter physics, physical chemistry, quantum chemistry, surface chemistry, electro-chemistry, high-voltage engineering, electro-technology, and pulse power. The exposure of each participant to the diverse concepts and research areas of the others made the Institute a stimulating experience for all.

The Institute presented the subject of the liquid state and its electrical properties systematically. The treatment, both theoretical and experimental, started from the microscopic, quantum-statistical foundations of the liquid state and progressed to the macroscopic description of conduction in liquids, and the various applications that arise because of the electrical properties of liquids. Emphasis was placed on fundamental principles and physical points of view, and on achieving a microscopic-macroscopic connection in the description of the liquid state and its electrical properties.

The subject matter was organized into four broad areas: (a) **Theories of Liquid Structures,** (b) **Ionic and Electronic Processes,** (c) **Interfacial Phenomena,** and (d) **Breakdown and Conduction.** These four areas covered the bulk of the Institute. In addition, results of current research were presented in two **Poster Sessions,** and **Future Research Directions** and technological innovations derived from liquid-phase studies were discussed in a special session.

The session on **Theories of Liquid Structures** treated the equilibrium properties of simple liquids from a quantum-statistical and statistical-thermodynamics point-of-view. The tenor in these presentations was the Pair-Correlation Function. The various equations that have been proposed to describe its behavior, and techniques for obtaining their solutions (in special cases) were reviewed. These concepts became more "concrete" with a discussion of electrolytic solutions.

The session on **Ionic and Electronic Processes** principally focused on the nature and dynamical behavior of ions and electrons in liquids. A significant amount of time was devoted to the discussion of electron dynamics in liquids. This was a subject of some controversy that gave rise to a number of special discussion sessions that continued to the end of the ASI. Two points of view were expressed: in one, the liquid was modeled as a fluctuating crystal; in the other, the liquid was viewed as a dense gas. The validity of these models obviously depends on the nature of the liquid, e.g., polar, non-polar, liquid rare gases, etc. The discussions focused on the nature of the localized and extended (free) electron states, and, in particular, on the lowest level of a free electron in the liquid, i.e., the "V_0" level. Accounts were given of the free electron states, their contribution to the electrical behavior, the relaxation of a hot electron into either the V_0 or the localized state, and of the electrical and optical diagnostic techniques which are used in the study to these phenomena.

These accounts carried the Institute over into the area of **Interfacial Phenomena** and the question of how electrons are injected from boundaries, as well as the question of the nature of the boundary layers. Experimental methods for investigating interface phenomena were discussed. As the concentration of charge carriers in the liquid increases, space-charge effects become important. Electro-optic techniques used to investigate these effects and other phenomena associated with charge injection were described. Further increases in charged-carrier concentration due to the presence of large applied electric fields eventually leads to the breakdown of the insulating properties of the liquid. This regime was the subject of the last session.

In this session, on **Breakdown and Conduction**, a review was given of the various theories that have been proposed to explain this phenomena. In highly non-uniform fields, the breakdown channel is observed to progress in the form of a "tree" growing from the point of high field stress. Fractal analysis of this evolution has been attempted and this technique was presented. The concept of streamers in liquids -- the propagating "branches" of the breakdown "tree" -- was introduced and related to that in gases. It was suggested that the theories of liquid breakdown and the interpretation of the observed phenomena are phenomenological in nature, and connection to microscopic processes occurring in the liquid must be sought.

Clearly, the difficulties encountered in the investigations of the liquid state, which lacks the order of the crystal and the large interparticle distances of the dilute gas, have played a role in slowing down the establishment of a knowledge base from which technological applications can grow. The development of new methods, both experimental and theoretical, for the study of the liquid state, and the narrowing of the gap between those who study liquids for their own sake and those who utilize the properties of liquids for applications, would greatly advance liquid state technologies. Areas where significant developments were projected include: (a) the quantum statistical description of electrons in liquids using path-integral methods; (b) the optical micro-probing of the liquid state, the dynamical behavior of charges carriers in liquids, the coupling of microscopic and macroscopic processes and properties, and breakdown; (c) the study of model structures and biological systems (structured liquids) and their interaction with radiation; and (d) liquid crystals and gels.

The comprehensive discussions on the Liquid State and Its Electrical Properties at the Sintra ASI demonstrated that the study of liquids will be a challenging, but rewarding, area for future research and applications.

E. Kunhardt
Polytechnic University
Farmingdale, New York

L. G. Christophorou
Oak Ridge National Laboratory
Oak Ridge, Tennessee

L. H. Luessen
Naval Surface Warfare Center
Dahlgren, Virginia

April 1988

ACKNOWLEDGMENTS

We are grateful to a number of organizations for providing the financial assistance that made the Institute possible. Foremost is the NATO Scientific Affairs Division, which provided the single-most financial contribution for the Institute. In addition, the following US sources made contributions: Naval Surface Warfare Center, Office of Naval Research, Army Electronics Technology and Devices Laboratory, and Los Alamos National Laboratory.

The Editors also wish to acknowledge the following individuals: Mrs. Ann Drury, Polytechnic University, and Mrs. Linda Johnson, Naval Surface Warfare Center, the Institute's Administrative Coordinators for Lectures and Participants, respectively; Ms. Susie M. Anderson, Supervisor for Word Processing at the EG&G Washington Analytical Services Center Office at Dahlgren, Virginia, which had the task of centrally retyping every lecturer's manuscript and producing a camera-ready document for delivery to Plenum Corporation; Mrs. Barbara Kester of the Publications Coordination Office; and finally to the staff and management of the Hotel Tivoli Sintra -- their efforts, along with the natural beauty and charm of Sintra, and the wonderful accommodations and facilities of the Hotel, made this two-week ASI one we will all remember for years to come.

CONTENTS

THEORIES OF LIQUID STRUCTURE

IONIC AND ELECTRONIC PROCESSES

INTERFACIAL PHENOMENA

BREAKDOWN AND CONDUCTION

APPENDICES

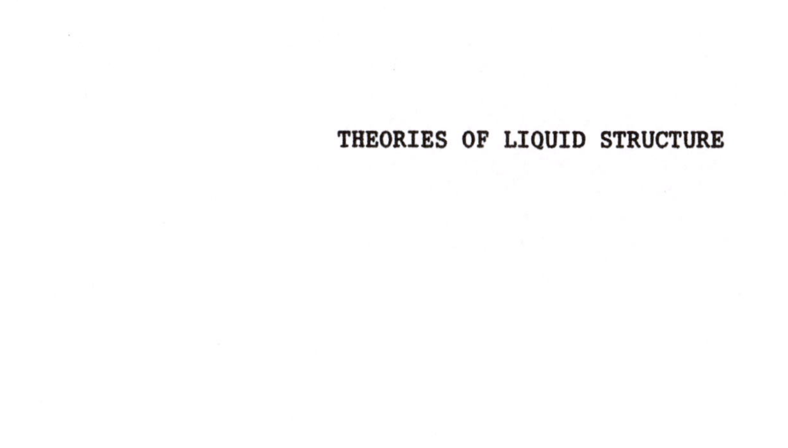

THEORIES OF LIQUID STRUCTURE

FIELD THEORETIC MODELS OF LIQUIDS

David Chandler

Department of Chemistry
University of California
Berkeley, CA 94720 USA

As an alternative to viewing a liquid explicitly a disordered collection of particles, molecular configurations can be described in terms of single particle fields such as the density. This lecture considers this alternative perspective -- how the spatial analog of harmonic oscillator models leads to integral equations (e.g., PY, RISM, ...), and how manageable non-linear treatments have led to an understanding of freezing. Nevertheless, puzzling questions remain concerning symmetry breaking and the formation of glasses.

INTRODUCTION

With standard graphics, a beginning student can literally view the microscopic structure of a liquid as produced with a molecular dynamics or Monte Carlo computer simulation. The relative arrangements of atoms are disordered, fluctuating and, therefore, responding to disturbances or probes of liquid structure. Despite its chaotic nature, however, there are many important regularities, and these are best described with the principles of statistical mechanics and correlation functions.

According to statistical mechanics, our observations of a macroscopic system are the result of sampling all states or fluctuations consistent with the constraints imposed by our measurements. With simulations, assuming trajectories ergodic, one may perform the sampling by averaging over many steps of a trajectory. With realistic intermolecular potentials employed, this procedure is often convenient and accurate. But guidance from analytical approaches are useful too. Indeed, numerically intensive simulations have often proved impractical before the appropriate analytical work illustrated what to expect.

This lecture is concerned with an analytical approach, in particular, a phenomenological description of liquid structure. As we will emphasize, quite a bit of detail can be understood without directly confronting the difficult and system specific problems of intermolecular interactions. Ultimately, of course, quantitative theoretical calculations require accurate models for the interactions between elementary constituents. The field theoretic methods we describe here supplement but not replace such calculations.

CORRELATION FUNCTIONS

Suppose we characterize a bulk system by specifying the volume, V, the temperature $T = (\beta k_B)^{-1}$, and perhaps the chemical potential (if the system is open) or the total number of molecules N (if the system was closed). If these are the only macroscopic characteristics, the average 1-point functions, or <u>fields</u>, seem rather ordinary. For example, consider the density of sites or atoms at position $\underset{\sim}{r}$. At a given instant, it is

$$\rho_\alpha(\underset{\sim}{r}) = \sum_{i=1}^{N} \delta(\underset{\sim}{r}-\underset{\sim}{r}_i^{(\alpha)}) \tag{1}$$

where $\underset{\sim}{r}_i^{(\alpha)}$ is the location of site α of molecule i. See Fig. 1 for an illustration. (For a polyatomic system there are several sites; for a simple fluid like liquid argon, however, there is only one site per molecule, and in that case we can omit the Greek sub- and superscripts.) In the absence of any externally applied spatially dependent field, the average of this density is the bulk density,

$$\langle\rho_\alpha(\underset{\sim}{r})\rangle = \langle \frac{N}{V} \rangle \equiv \rho \quad . \tag{2}$$

It is simply a constant in space and dependent upon macroscopic thermodynamic variables only. Results like this might look boring. The final point of this lecture, however, is that the average behavior of density fields are in fact interesting. To reach that point, it is useful to first consider correlation functions -- simultaneous averages of <u>two</u> 1-point functions -- that are manifestly interesting even for a macroscopically homogeneous system.

At any instant, the density will deviate from its average, and the deviation of one position $\underset{\sim}{r}$ will affect that at another position $\underset{\sim}{r}'$. These correlations between activities at different points in space are characterized by correlation functions

$$\chi_{\alpha\gamma}(\underset{\sim}{r},\underset{\sim}{r}') = \langle\Delta\rho_\alpha(\underset{\sim}{r})\Delta\rho_\gamma(\underset{\sim}{r}')\rangle \tag{3}$$

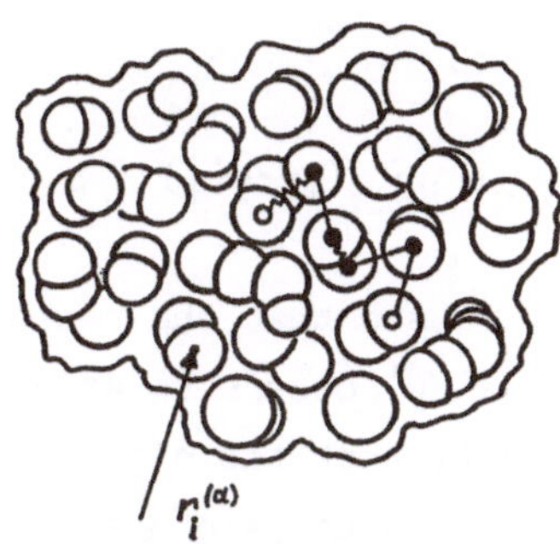

Fig. 1. Instantaneous configuration of molecules in a liquid. The point $\underset{\sim}{r}_i^{(\alpha)}$ refers to the location of the αth atom or interaction site of molecule i. The cluster diagram superimposed on the drawing illustrates the physical meaning of a typical term in the integral equation described in the INTEGRAL EQUATION section. In particular, the springs denote intramolecular pair distribution functions. These are the off-diagonal (i.e., $\alpha \neq \gamma$) components of the intramolecular response function, $\omega_{\alpha\gamma}(|r-r'|)$. The diagonal components of the response function are delta functions which are depicted diagrammatically as simple circles.

where $\Delta\rho_\alpha(\underset{\sim}{r}) = \rho_\alpha(\underset{\sim}{r}) - \langle\rho_\gamma(\underset{\sim}{r})\rangle$. Owing to its role in linear response or the functional derivative relationship

$$\delta\langle\rho_\alpha(\underset{\sim}{r})\rangle/\delta\phi_\gamma(\underset{\sim}{r}') = -k_B T\ \chi_{\alpha\gamma}(\underset{\sim}{r},\underset{\sim}{r}') \quad , \tag{4}$$

the density-density correlation function is also properly viewed as a response function. [Here, $\phi_\alpha(\underset{\sim}{r})$ is a potential field acting on site γ.] The derivation of (4) follows directly from the standard equilibrium ensemble distribution laws of statistical mechanics. See, for example, Chandler (1982). According to Eq. (4), large spontaneous fluctuations (the right-hand side) imply large responses to applied disturbances (the left-hand side).

On substituting Eq. (1) into Eq. (4), the response function naturally breaks into two parts. One involves the <u>intra</u>molecular response or pair correlations:

$$\omega_{\alpha\gamma}(|\underset{\sim}{r}-\underset{\sim}{r}'|) = \langle\delta(\underset{\sim}{r} - \underset{\sim}{r}' - \underset{\sim}{r}_1^{(\alpha)} + \underset{\sim}{r}_1^{(\gamma)})\rangle \quad . \tag{5}$$

The other part involves the <u>inter</u>molecular pair distribution:

$$\rho g_{\alpha\gamma}(|\underset{\sim}{r}-\underset{\sim}{r}'|) = \frac{1}{\rho}\langle N\delta(\underset{\sim}{r}_1^{(\alpha)} - \underset{\sim}{r})(N-1)\delta(\underset{\sim}{r}_2^{(\gamma)} - \underset{\sim}{r})\rangle$$

$$\equiv \rho[h_{\alpha\gamma}(|\underset{\sim}{r}-\underset{\sim}{r}'|) + 1] \quad . \tag{6}$$

In particular, for a uniform one-component molecular fluid, by partitioning of sums and accounting for the equivalency of different molecules, you find

$$\chi_{\alpha\gamma}(\underset{\sim}{r}, \underset{\sim}{r}') = \rho\omega_{\alpha\gamma}(|\underset{\sim}{r}-\underset{\sim}{r}'|) + \rho^2 h_{\alpha\gamma}(|\underset{\sim}{r}-\underset{\sim}{r}'|) \tag{7}$$

In light of Eqs. (4) - (7), we see that the response of sites α at $\underset{\sim}{r}$ due to a disturbance to sites γ at $\underset{\sim}{r}'$ occurs to an extent proportional to the conditional probability density that the sites α are present at $\underset{\sim}{r}$ given that sites γ are found at $\underset{\sim}{r}'$. These pair distributions are measured by neutron and x-ray scattering. This fact is true because the scattering of x-rays and neutrons are due to interference of radiation coming from pairs of scattering centers.

INTEGRAL EQUATIONS

A central problem in liquid state theory is the calculation of these pair correlation functions. One phenomenological way to address the problem is based on a mechanistic view which pictures the pair correlations between atoms on different molecules as arising from all possible simple linear sequences of intramolecular and direct intermolecular pair correlations. We will use the notation $c_{\alpha\gamma}(|\underset{\sim}{r}-\underset{\sim}{r}'|)$ to denote this direct correlation or coupling function. With diagrams, the idea is expressed as

$\rho^2 h_{\alpha\gamma}(|\underset{\sim}{r}-\underset{\sim}{r}'|) =$

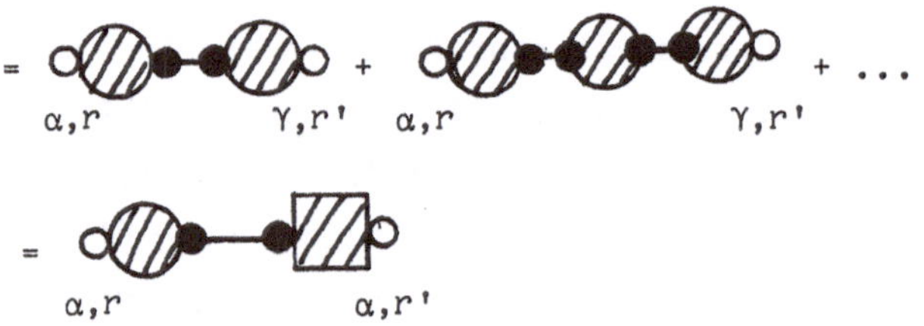

$$\equiv \sum_{\eta,\lambda} \int d\underset{\sim}{r}'' \int d\underset{\sim}{r}''' \; \rho\omega_{\alpha\eta}(|\underset{\sim}{r}-\underset{\sim}{r}'|)\; c_{\eta\lambda}(|\underset{\sim}{r}''-\underset{\sim}{r}'''|) \times \chi_{\lambda\gamma}(|\underset{\sim}{r}'''-\underset{\sim}{r}'|) \tag{8}$$

where the last equality makes clear the meaning of the diagrammatic notation (Chandler, 1982; Landanyi and Chandler, 1975; Chandler and Pratt, 1976; Pratt and Chandler, 1977). The second to last equality is the result of summing all the chains depicted in the first equality. The reader can verify this fact by expanding the second equality via iterative solution employing

$\chi_{\alpha\gamma} =$  (9)

as the starting point.

Equation (8) is a type of Dyson equation known as the Chandler-Andersen (or RISM -- "reference interaction site method") equation (Chandler, 1982; Chandler and Andersen, 1972). It is a generalization of the Ornstein-Zernike integral equation for simple atomic fluids (Hansen and McDonald, 1986). It is an integral equation which relates the unknown $h_{\alpha\gamma}(|\underset{\sim}{r}-\underset{\sim}{r}'|)$ to the equally unknown $c_{\alpha\gamma}(|\underset{\sim}{r}-\underset{\sim}{r}'|)$. Indeed, Eq. (8) is essentially a <u>definition</u> of $c_{\alpha\gamma}(r)$. Another relationship is required to close the equation, and to construct a closure a field theoretic perspective can be suggestive. We turn to such a perspective now.

EQUIVALENT GAUSSIAN FIELD THEORY

Here, we focus on one molecule in the liquid, say molecule 1, and regard all the rest as a bath. See Figure 2. To describe the configurational state of the bath or solvent, we use the dynamical density fields, $\rho_s(\underset{\sim}{r})$. [Roman subscripts rather than Greek labels are used to help distinguish bath sites from those of the solute -- the tagged molecule. In the end, of course, they will all be the same stuff.] In general, the statistical properties of these fields are quite complicated. At the very least, however, they are characterized by their average behavior and their variance in the absence of the solute, $\chi_{ss'}(|\underset{\sim}{r}-\underset{\sim}{r}'|)$. With only this information, we can construct the probability functional

$$P[\rho_s(\underset{\sim}{r})] \propto \exp\{ \sum_{\alpha,s} \int d\underset{\sim}{r}\, c_{\alpha s}(|\underset{\sim}{r}_1^{(\alpha)} - \underset{\sim}{r}|)\rho_s(\underset{\sim}{r}) - \frac{1}{2} \sum_{s,s'} \int d\underset{\sim}{r} \int d\underset{\sim}{r}'\, \Delta\rho_s(\underset{\sim}{r}) \chi^{-1}_{ss'}(\underset{\sim}{r}, \underset{\sim}{r}') \Delta\rho_{s'}(\underset{\sim}{r}')\} \qquad (10)$$

where $\chi^{-1}_{ss'}(\underset{\sim}{r}, \underset{\sim}{r}')$ is the functional inverse of the solvent response function; i.e.,

$$\sum_{s''} \int d\underset{\sim}{r}''\, \chi^{-1}_{ss''}(\underset{\sim}{r}, \underset{\sim}{r}'') \chi_{s''s'}(\underset{\sim}{r}'', \underset{\sim}{r}') = \delta_{ss'} \delta(\underset{\sim}{r}-\underset{\sim}{r}') \ . \qquad (11)$$

The second part to the right-hand side of Eq. (10) provides a Gaussian distribution for the solvent, and the first part is a linear coupling between the solvent and solute sites.

In the linear coupling term, the coupling function, $c_{\alpha s}(r)$, is in fact the very same function introduced in Eq. (8). To understand this fact, we can use Eq. (10) to perform the appropriate averages that

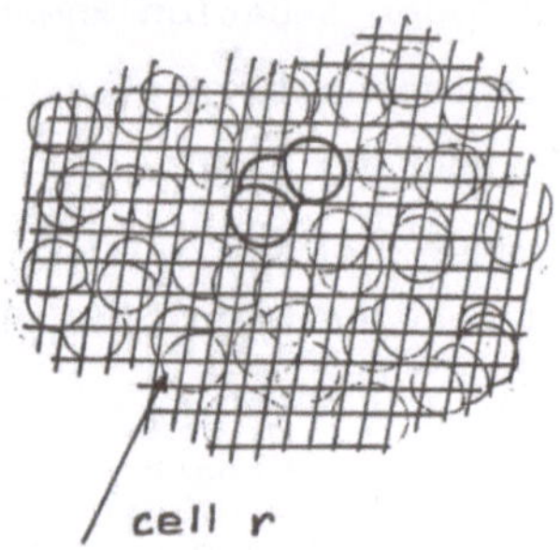

Fig. 2. One molecule in a bath. The instantaneous configuration of the bath can be specified by listing the particle coordinates, $\{\underset{\sim}{r}_i^{(s)}\}$, or equivalently the occupation numbers for positions in space -- the infinitesimal cells illustrated schematically with the grid. To within a factor of the volume element, the occupation numbers are the density fields, $\rho_s(\underset{\sim}{r})$.

determine $h_{\alpha s}(r)$. First note

$$\rho h_{\alpha s}(|\underset{\sim}{r}-\underset{\sim}{r}'|) = \langle\delta(\underset{\sim}{r}-\underset{\sim}{r}_1^{(\alpha)})[\rho_s(\underset{\sim}{r})-\rho]\rangle \tag{12}$$

where we make no distinction between $\langle\rho_s(\underset{\sim}{r})\rangle$ and ρ. [There would be a difference in a multicomponent system -- a detail the reader can study as an exercise.] Thus

$$\rho^2 h_{\alpha s}(|\underset{\sim}{r}-\underset{\sim}{r}'|) = \int s(\underset{\sim}{r}_1^{(1)},\ldots,\underset{\sim}{r}_1^{(n)}) \prod_{\alpha=1}^{n} d\underset{\sim}{r}_1^{(\alpha)} \; \square \; \mathcal{D}\rho_s(\underset{\sim}{r})\; P[\rho_s(\underset{\sim}{r})]$$

$$\times\; \delta(\underset{\sim}{r}-\underset{\sim}{r}_1^{(\alpha)})\; [\rho_s(\underset{\sim}{r}')-\rho]\rho\;, \tag{13}$$

where $s(\underset{\sim}{r}^{(1)},\ldots)$ is the intramolecular distribution for the tagged molecule. The functional integration over the density fields means that one is to integrate $\rho_s(\underset{\sim}{r})$ over all possible values of density for each point in space. Here, for visualization, one can imagine dividing space into a grid of cells as is illustrated in Fig. 2; $\underset{\sim}{r}$ is then the cell label; and the density, $\rho_s(\underset{\sim}{r})$, for each cell is to be integrated. In the continuum limit of small cells, this multivariable integration is a functional integration.

The functional integration in Eq. (13) can be performed because $P[\rho_s(\underset{\sim}{r})]$ is Gaussian. The result of all the integrations is

$$\rho^2 h_{\alpha s}(|\underset{\sim}{r}-\underset{\sim}{r}'|) = \;\circ\!\!\oslash\!\!\bullet\!\!-\!\!-\!\!-\!\!\bullet\!\!\boxslash\!\!\circ \quad {}_{\alpha,r}\qquad {}_{s,r'} \tag{14}$$

Now remember that the solvent is made up of molecules just like the solute. Thus, the χ-function on the right-hand side (i.e., the response function represented by the rectangle) is the χ-function determined by $h_{\alpha s}(|\underset{\sim}{r}-\underset{\sim}{r}'|)$, Eq. (7). Hence, Eq. (14) is the Chandler-Andersen equation (8).

With this model, one may also compute the solvation energy, or more precisely, the excess chemical potential $\Delta\mu$ for the tagged solute. The calculation is left as an exercise to the reader. The result is

$$-\beta\Delta\mu = \sum_{\alpha,s} \rho \int d\underset{\sim}{r}\; c_{\alpha s}(r) + \frac{1}{\langle N \rangle}$$

$$= \sum_{\alpha,s} \rho \int d\underset{\sim}{r}\; c_{\alpha s}(r)$$

$$+ \frac{1}{2} \sum_{\alpha,\gamma} \sum_{s,s'} \int d\underset{\sim}{r} \int d\underset{\sim}{r}' \int d(\underset{\sim}{r}_1^{(\alpha)} - \underset{\sim}{r}_1^{(\gamma)})\, \omega_{\alpha\gamma}(|\underset{\sim}{r}_1^{(\alpha)} - \underset{\sim}{r}_1^{(\gamma)}|)$$

$$\times\; c_{\alpha s}(|\underset{\sim}{r}_1^{(\alpha)} - \underset{\sim}{r}|)\; \chi_{ss'}(|\underset{\sim}{r}-\underset{\sim}{r}'|)\; c_{s'\gamma}(|\underset{\sim}{r}'-\underset{\sim}{r}^{(\gamma)}|) \tag{15}$$

where the second equality makes explicit the diagrammatic meaning of the first. As the diagrammatic depiction indicates, the result is that of a "reaction field" theory. In particular, the first term is a mean field contribution, and the second arises from processes in which the solute pushes on the solvent, and the solvent reacts back (i.e., responds) and pushes the solute. There are no further contributions because the solvent is modeled as a Gaussian bath, and Gaussian (or harmonic) baths exhibit no nonlinear responses.

CLOSURES

The discussion presented above suggests that the coupling function is like an effective pair potential (in units of $-k_BT$). It's a rather special potential since its use in a Gaussian field theory is supposed to yield pair correlation functions appropriate to liquids where the actual potentials of interactions are very strong and nonlinear. It is perhaps a surprising fact that simple statements about this function can lead to rather satisfactory theories of liquid structure. For example, when considering nonassociated liquids (e.g., carbon tetrachloride, benzene, carbon disulfide, ...), it is reasonable to consider only the effects of packing forces when examining the liquid structure (Chandler et al., 1983). For this purpose, one often adopts hard core models in which the liquid molecules are conceived of as overlapping fused hard spheres. In that case

$$g_{\alpha\gamma}(r) = 0\,, \quad r < d_{\alpha\gamma}\,. \tag{16}$$

where $d_{\alpha\gamma}$ is the distance of closest approach (as prescribed by molecular geometry and van der Waals radii). Further, since the range of direct interaction between neighboring pairs is also determined by this same set of lengths, one is led to the closure

$$c_{\alpha\gamma}(r) \approx 0 \ , \quad r > d_{\alpha\gamma} \ . \tag{17}$$

The task of determining the liquid structure then reduces to finding the $c_{\alpha\gamma}(r)$ for $r < d_{\alpha\gamma}$ which makes $g_{\alpha\gamma}(r) = 0$ for $r < d_{\alpha\gamma}$. Once found, the pair correlations are determined for all distances by inverse Fourier transform of

$$\hat{h}_{\alpha\gamma}(k) = \left(\hat{\underset{\sim}{\omega}}(k)\hat{\underset{\sim}{c}}(k)[\underset{\sim}{1} - \rho\hat{\underset{\sim}{\omega}}(k)\hat{\underset{\sim}{c}}(k)]^{-1}\hat{\underset{\sim}{\omega}}(k)\right)_{\alpha\gamma} \tag{18}$$

where the carots denote Fourier transformation. Equation (18) is, as the reader can verify, another way of writing the Chandler-Andersen equation (8).

This approach is called the RISM theory. Tens of applications of this theory for many different liquids have been quite successful. Representative applications by several workers are found in Trans. Faraday Soc. Discussion (1978). Chandler (1978 and 1982) has reviewed some of this work. Specific examples are Lowden and Chandler (1974), Hsu et al. (1976) and Hsu and Chandler (1978). In its simplest version, that is, when it is applied to single site spherical particles, the RISM theory for hard core molecules reduces to the Percus-Yevick theory for the hard sphere fluid.

Extensions to polar and associated liquids employ different closures. These theories are called "extended RISM." The hypernetted chain closure (HNC) is most often applied (Rossky, 1985), and with reasonable success. See, for example, Pettitt and Rossky (1982), Chiles and Rossky (1984), and Pettitt and Rossky (1986). But this choice of closure is perhaps unnecessarily adhoc, and others have been suggested (Chandler et al., 1986b).

DENSITY FUNCTIONAL THEORY AND SYMMETRY BREAKING

The pair correlations between atoms or sites in a molecular fluid pertain to the microscopic spontaneous fluctuations that occur in a macroscopically homogeneous fluid. Under certain circumstances, these fluctuations conspire collectively or in concert to form ordered phases such as crystals. The description of these transformations of phase is beyond the scope of the linear (i.e., Gaussian) theory we have outlined thus far. The incorporation of nonlinearities is the subject we turn to now. The methodology we consider is known as density functional theory.

For simple atomic fluids and solids, this remarkable formulation of equilibrium statistical mechanics was pioneered more than 20 years ago. See, for example, Morita and Hiroike (1961), Stillinger and Buff (1962) and Lebowitz and Percus (1963). The formulation rests on a variational principle enunciated by Morita and Hiroike (1961):

> There exists a free energy functional, $F[\rho(\underset{\sim}{r})]$, such that $\delta^n F/\delta\rho(\underset{\sim}{r}) \ldots \delta\rho(\underset{\sim}{r}')$ generates all correlation functions, and $F[\langle\rho(\underset{\sim}{r})\rangle]$ is the minimum of $F[\rho(\underset{\sim}{r})]$ on the surface $\int d\underset{\sim}{r}\,\rho(\underset{\sim}{r}) = \langle N\rangle$.

This principle can be viewed as a microscopic version of thermodynamics. It says that the computation of correlation functions is equivalent to identifying $F[\rho(\underset{\sim}{r})]$ and determining the density field which minimizes that functional. Notice too that it says all statistical information of the equilibrium state (i.e., all multipoint correlation functions) is contained in the reversible work functional for the 1-point density field. It's such an important result, it should cause all but the most casual reader to pause.

It is convenient to separate this free energy functional into two parts:

$$F[\rho(\underset{\sim}{r})] = F_0[\rho(\underset{\sim}{r})] + F_I[\rho(\underset{\sim}{r})] \tag{19}$$

where $F_0[\rho(\underset{\sim}{r})]$ is the free energy functional for the case where different particles did not interact with one another, and $F_I[\rho(\underset{\sim}{r})]$ is the correction to that ideal gas expression. One may show that (Morita and Hiroike, 1961)

$$F_0[\rho(\underset{\sim}{r})] = \int d\underset{\sim}{r}\; \rho(\underset{\sim}{r})[\ln\rho(\underset{\sim}{r})-1] \;. \tag{20}$$

which has the physical interpretation of an entropy of mixing. Further, the interaction contribution can be expressed formally as (Morita and Hiroike, 1961; Stillinger and Buff, 1962),

$$F_I[\rho(\underset{\sim}{r})] = -\frac{1}{2}\int d\underset{\sim}{r}\int d\underset{\sim}{r}'\; \bar{c}[\underset{\sim}{r},\underset{\sim}{r}';\rho(\underset{\sim}{r})]\; \Delta\rho(\underset{\sim}{r})\Delta\rho(\underset{\sim}{r}') \;, \tag{21}$$

where $\Delta\rho(\underset{\sim}{r})$ is the deviation in density from that of a convenient reference state. If we assume the free energy is analytic, the coupling functional, $\bar{c}[\underset{\sim}{r},\underset{\sim}{r}';\rho(\underset{\sim}{r})]$, can be expanded about that of the reference state. To lowest order when the reference state is a homogeneous fluid of density ρ, one finds

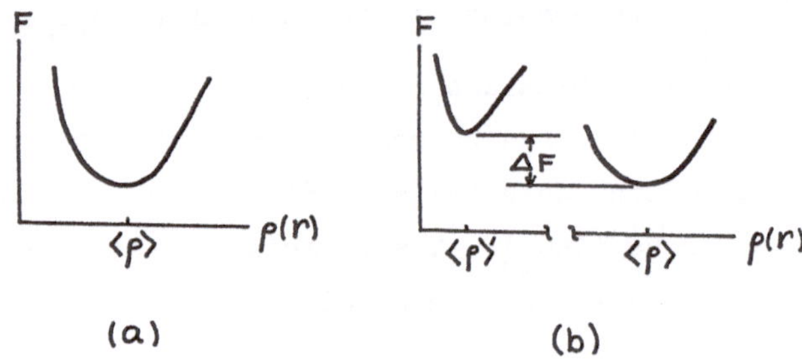

Fig. 3. Free energy functionals of density fields. Case (a) is monostable, i.e. $\delta F/\delta\rho(\underset{\sim}{r})$ possesses only one solution, and a linear theory can describe the region close to the minimum. Case (b) shows two stable or metastable phases characterized by different average density fields. When two phases are in equilibrium, the value of ΔF in Case (b) coincides with equality of the grand canonical free energy for the two stable states.

$$\bar{c}[\underset{\sim}{r},\underset{\sim}{r}';\ \rho(\underset{\sim}{r})] \approx c(|\underset{\sim}{r}-\underset{\sim}{r}'|;\ \rho) \tag{22}$$

where $c(r)$ is the direct correlation function. In other words,

$$\begin{aligned}[\delta^2 F/\delta\rho(\underset{\sim}{r})\delta\rho(\underset{\sim}{r}')]_{\text{uniform}} &= \frac{1}{\rho}\,\delta(\underset{\sim}{r}-\underset{\sim}{r}') - c(|\underset{\sim}{r}-\underset{\sim}{r}'|) \\ &= \chi^{-1}(|\underset{\sim}{r}-\underset{\sim}{r}'|)\ .\end{aligned} \tag{23}$$

In the second equality we are identifying the connection between the direct correlation function and the functional inverse of the response function for a uniform simple atomic fluid. The reader can verify this result using the Ornstein-Zernike equation [i.e., the Chandler-Andersen equation (8) for an atomic or single site fluid] and Eq. (11). As $\rho(\underset{\sim}{r}) \to \rho$, Eq. (22) becomes exact. Its use for all $\rho(\underset{\sim}{r})$, however, is an approximation -- the hypernetted chain approximation.

In a harmonic approximation, $F[\rho(\underset{\sim}{r})]$ is a quadratic functional with its minimum at $\rho(\underset{\sim}{r}) = \rho$ and its curvature given by $\chi^{-1}(|\underset{\sim}{r}-\underset{\sim}{r}'|)$. The minimization principle, $\delta F/\delta\rho(\underset{\sim}{r}) = 0$, is a linear equation with only one solution, $\rho(\underset{\sim}{r}) = \rho$. See Fig. 3. One way to construct a nonlinear density field theory where $F[\rho(\underset{\sim}{r})]$ may have more than one minimum is to use without approximation the $\rho\ell n\rho$ term, Eq. (20), but to approximate the coupling functional as done in Eq. (22). This approximation for the free energy functional has been successfully applied as a theory for freezing of simple fluids (Ramakrishnan and Yussouff, 1979). That is, with Eqs. (19)-(22), $\delta F/\delta\rho(\underset{\sim}{r}) = 0$ is a nonlinear equation with multiple solutions; and the crystalline solutions of broken symmetry accurately describe the freezing of atomic fluids (Haymet, 1987). Since this theory relies on a truncated Taylor expansion to estimate F_I, the theory is a type of "Landau theory" or "mean field theory." In other words, one assumes analyticity of the free energy functional in the vicinity of a phase transition.

In implementing the theory, one requires information about the interparticle potentials only as it appears implicitly through $c(|\underset{\sim}{r}-\underset{\sim}{r}'|)$ or equivalently the pair correlation function of the dense homogeneous fluid. The fact that this simple approach which requires only information about the homogeneous system actually succeeds at explaining the freezing of, for example, liquid argon deserves some comment. In particular, one should note that $F[\rho(\underset{\sim}{r})]$ is the reversible work function for changing density fields. By employing Eq. (22), one assumes a relatively small change in the interparticle interaction contribution to the work when changing $\rho(\underset{\sim}{r})$ from that of a dense fluid to that of a solid. This assumption seems reasonable since the nearest neighbor environment of a particle in a dense fluid is rather similar to that in a solid. Nevertheless, subtle changes in that environment can have a more pronounced and collective effect on the entropy term. For this reason, it is important to retain the nonlinear structure of Eq. (20).

DENSITY FUNCTIONAL THEORY OF POLYATOMICS

Nonsimple molecular fluids and solids can also be analyzed with density functional theory. Here, the trick is to regard molecules as objects composed of atoms. (An alternative view in which molecules are made up of spherical harmonics is possible, but then the density fields involve orientational variables as well as points in space.) With the atomistic view of molecules, as we have illustrated in the first part of this lecture, one can use the site or atomic density fields, $\{\rho_\alpha(\underset{\sim}{r})\}$, to describe configurational states. Furthermore, the following generalization of atomic density functional theory has been derived (Chandler et al., 1986a):

> There exists a free energy density functional, $F[\{\rho_\alpha(\underset{\sim}{r})\}]$ such that $\delta^n F/\delta\rho_\alpha(\underset{\sim}{r}) \ldots \delta\rho_\alpha(\underset{\sim}{r}')$ generates all correlation functions, and $F[\{\langle\rho_\alpha(\underset{\sim}{r})\rangle\}]$ is the minimum of $F[\{\rho_\alpha(\underset{\sim}{r})\}]$ on the surface $\langle N\rangle = \int d\underset{\sim}{r}\ \rho_\alpha(\underset{\sim}{r})$.

Notice again the remarkable nature of this result. It implies that all information is contained in the 1-point density fields and the variation of the free energy with those fields. As with the atomic case, one may write

$$F[\{\rho_\alpha(\underset{\sim}{r})\} = F_0[\{\rho_\alpha(r)\}] + F_I[\{\rho_\alpha(\underset{\sim}{r})\}] \tag{24}$$

where

$$F_I[\{\rho_\alpha(\underset{\sim}{r})\}] = -\frac{1}{2}\sum_{\alpha,\gamma}\int d\underset{\sim}{r}\int d\underset{\sim}{r}'\ \bar{c}_{\alpha\gamma}[\underset{\sim}{r},\underset{\sim}{r}';\{\rho_\alpha(\underset{\sim}{r})\}]\Delta\rho_\alpha(\underset{\sim}{r})\Delta\rho_\gamma(\underset{\sim}{r}') \tag{25}$$

and when expanded about the uniform fluid reference state,

$$\bar{c}_{\alpha\gamma}[\underset{\sim}{r}, \underset{\sim}{r}';\{\rho_\alpha(\underset{\sim}{r})\}] \approx \bar{c}_{\alpha\gamma}[\underset{\sim}{r}, \underset{\sim}{r}';\rho] = c_{\alpha\gamma}(|\underset{\sim}{r}-\underset{\sim}{r}'|,\rho) \quad . \tag{26}$$

The entropy of mixing functional, however, does not have standard $\rho\ell n\rho$ form. Indeed,

$$F_0[\{\rho_\alpha(\underset{\sim}{r})\}] \neq \sum_\alpha \int d\underset{\sim}{r}\; \rho_\alpha(\underset{\sim}{r})[\ell n\; \rho_\alpha(\underset{\sim}{r})-1] \quad . \tag{27}$$

For this equality to hold, we would have to imagine that atoms within the same molecule were uncorrelated. The different atoms are, however, bonded; and the configurational constraints imposed by the bonding reduce the mixing entropy from that estimated with the right-hand side of (27). Further, for the physical reasons discussed in the preceding section, we require an accurate treatment of the entropy contribution to understand transformation of phases, spontaneous assemblies and many other symmetry breaking phenomena.

The determination of the entropy of mixing functional with bonding constraints is obtained by introducing an additional set of fields -- "fugacity fields" -- akin to the procedures of Legrange multipliers in multivariable calculus. The result of this analysis is the set of variational equations

$$0 = \delta F_0/\delta z_\alpha(\underset{\sim}{r}) \tag{28a}$$

and

$$0 = \delta[F_0 + F_I]/\delta\rho_\alpha(\underset{\sim}{r}) \tag{28b}$$

where

$$F_0[\{\rho_\alpha(\underset{\sim}{r})\},\{z_\alpha(\underset{\sim}{r})\}] = -\sum_{\alpha=1} \int d\underset{\sim}{r}\; \rho_\alpha(\underset{\sim}{r})\ell n\; z_\alpha(r)$$
$$+ \int s(r^{(1)},\ldots,r^{(n)}) \prod_{\alpha=1}^{n} z_\alpha(\underset{\sim}{r}^{(\alpha)})d\underset{\sim}{r}^{(\alpha)} \tag{29}$$

and $F_I = F_I[\{\rho_\alpha(\underset{\sim}{r})\}]$ is given by (25) and presumably well approximated with (26). It is evident, as the reader can verify, that the solutions of Eq. (28a) satisfy the bonding constraints as specified by the intramolecular distribution function, $s(\underset{\sim}{r}^{(1)},\ldots,\underset{\sim}{r}^{(n)})$.

Equations (25), (26), (28) and (29) have been applied to liquid water to describe the freezing of that system to form ice Ih (Ding et al., 1987). The calculations required as input the pair correlation functions of the liquid. These were taken from experiment. The theory predicts a freezing temperature of -5°C (the experimental result is, of course, 0°C), and a density of the solid as 0.030 molecules/Å^3 (experiment is the same to that number of figures). The theory therefore predicts that ice floats on water; it also predicts a positive coefficient of thermal expansion of the solid in accord with experiment.

THE FUTURE

There are still a number of important problems where the field theoretic approaches we have outlined have yet to be applied successfully. These include the solvation of strongly bound complexes, the formation of glasses and also the phenomenon of liquid crystallinity. In the first of these, the strength of *inter*molecular bonding may invalidate approximations like Eqs. (22) or (26), and it is not yet clear how to progress beyond that stage. In the second example, one is faced with the phenomenon of nonergodicity, perhaps described by a reversible work functional with a chaotic multiple of deep metastable minima. It is not at all clear that the free energy relevant to *stable* equilibrium states considered herein is the appropriate functional for the glassy case too. The issues here are also pertinent to fundamental questions about spontaneous symmetry breaking. They are perhaps highlighted by the example of liquid crystallinity. For nematic liquid crystals, translational symmetry of the equilibrium state requires that $\langle\rho_\alpha(\underset{\sim}{r})\rangle = \rho$. Yet, the fundamental principle of density functional theory implies that the set of $\langle\rho_\alpha(\underset{\sim}{r})\rangle$'s contain all information concerning the equilibrium state. Yet the order parameter characterizing a liquid crystal seems absent from the 1-point density fields. The resolution of this apparent paradox resides in the demonstration that the orientational order and broken symmetry of a liquid crystal is contained in objects like $\langle\int d\underset{\sim}{r}\, r^2[\rho_\alpha(\underset{\sim}{r}) - \rho_\gamma(\underset{\sim}{r})]\rangle$, where α and γ are nonequivalent sites. The demonstration is left as an exercise for the reader.

ACKNOWLEDGMENTS

My recent research in this area was supported by grants from the U.S. N.S.F. and N.I.H.. I am grateful for this support and indebted to my students and collaborators, especially K. Ding, J.D. McCoy and S.J. Singer for sharing their ideas and work on this topic.

REFERENCES

Chandler, D., and Andersen, H.C., 1972, Optimized cluster expansions for classical fluids. II. Theory of Molecular Liquids, *J. Chem. Phys.*, 57:1930.

Chandler, D. and Pratt, L.R., 1976, Statistical mechanics of chemical equilibria and intramolecular structures of non-rigid molecules in condensed phases, J. Chem. Phys., 65:2925.

Chandler, D., 1978, Structures of molecular liquids, Ann. Rev. Phys. Chem., 29:441.

Chandler, D., 1982, Equilibrium theory of polyatomic fluids, Studies in Statistical Mechanics, VIII, ed. by E.W. Montroll, J.L. Lebowitz (North Holland, Amsterdam), p. 275.

Chandler, D., Weeks, J.D. and Andersen, H.C., 1983, The van der Waals picture of liquids, solids and phase transformations, Science, 220:787.

Chandler, D., McCoy, J.D. and Singer, S.J., 1986a, Density functional theory of nonuniform polyatomic systems. I. General formulation, J. Chem. Phys., 85:5971.

Chandler, D., McCoy, J.D. and Singer, S.J., 1986b, Density functional theory of nonuniform polyatomic systems. II. Rational closures for integral equations, J. Chem. Phys., 85:5977.

Chiles, R.A. and Rossky, P.J., 1984, Evaluation of reaction free energy surfaces in aqueous solution: an integral equation approach, J. Am. Chem. Soc., 106:6867.

Ding, K., Chandler, D., Smithline, S.J. and Haymet, A.D.J., 1987, Density functional theory for the freezing of water, 1987, Phys. Rev. Lett. 59:1698.

Faraday Disc. Chem. Soc., 1978, No. 66, "Structure and Motion in Molecular Liquids."

Hansen, J.P. and McDonald, I.R., 1986, "Theory of Simple Liquids, 2nd Ed.," Academic, New York.

Haymet, A.D.J., 1987, Freezing, Science, 236:1076.

Hsu, C.S., Chandler, D. and Lowden, L.J., 1976, Application of the RISM equation to diatomic fluids: The liquids nitrogen, oxygen and bromine, Chem. Phys., 14:213.

Hsu, C.S. and Chandler, D., 1978, RISM calculation of the structure of liquid acetonitrile, Mol. Phys., 36:215.

Landanyi, B.M. and Chandler, D., 1975, New type of cluster theory for molecular fluids: interaction site cluster expansion, J. Chem. Phys., 62:4308.

Lebowitz, J.L. and Percus, J.K., 1963, Statistical thermodynamics of nonuniform fluids; Asymptotic behavior of the radial distribution function, J. Math. Phys., 4:116, 248.

Lowden, L.J. and Chandler, D., 1974, Theory of intermolecular pair correlations for molecular liquids. Applications to the liquids carbon tetrachloride, carbon disulfide, carbon diselenide and benzene, J. Chem. Phys., 61:5228.

Morita, T. and Hiroike, K., 1961, A new approach to the theory of classical fluids. III. General treatment of classical systems, Progr. Theor. Phys., 25:537.

Pettitt, B.M. and Rossky, P.J., 1982, Integral equation predictions of liquid state structure for waterlike intermolecular potentials, J. Chem. Phys., 77:1451.

Pettitt, B.M. and Rossky, P.J., 1986, Alkali halides in water: ion-solvent correlations and ion-ion potentials of mean force at infinite dilution, J. Chem. Phys., 84:5836.

Pratt, L.R. and Chandler, D., 1977, Interaction site cluster series for the Helmholz free energy and variational principle for chemical equilibria and intramolecular structures, J. Chem. Phys., 66:147.

Ramakrishnan, T.V. and Yussouff, M., 1979, First-principles of order-parameter theory of freezing, Phys. Rev. B., 19:2775.

Rossky, P.J., 1985, The structure of polar molecular liquids, Ann. Rev. Phys. Chem., 36:321.

Stillinger, F.H. and Buff, F.P., 1962, Equilibrium statistical mechanics of inhomogeneous fluids, J. Chem. Phys., 37:1.

THE STRUCTURE OF SIMPLE FLUIDS

Donald A. McQuarrie

University of California
Department of Chemistry
Davis, CA 95616 USA

INTRODUCTION

In this chapter, we shall discuss the structure of simple fluids through the radial distribution function, which is the central quantity of most statistical mechanical theories of fluids (McQuarrie, 1976; Hansen and McDonald, 1986). Although all the equations and techniques presented here are applicable to polyatomic fluids (Gray and Gubbins, 1984), for simplicity we shall consider only simple fluids, that is, those that interact by way of a spherically-symmetric, angle-independent intermolecular potential. This chapter is meant to be tutorial in nature, and so in section IMPERFECT GASES, we review the virial expansion of imperfect gases and then in section DISTRIBUTION FUNCTIONS AND LIQUIDS, we introduce several of the types of distribution functions that are used to describe the structures of liquids. A central quantity of this section is the radial distribution function. In section THERMODYNAMIC PROPERTIES OF LIQUIDS, we express the thermodynamic properties of a fluid as functionals of the radial distribution function and in section INTEGRAL EQUATIONS FOR THE RADIAL DISTRIBUTION FUNCTION, we discuss several integral equations that give the radial distribution function in terms of the intermolecular potential. Three important quantities that are introduced in this section are the potential of mean force, the direct correlation function and the Ornstein-Zernike equation. Section SOME NUMERICAL RESULTS OF THE FLUID-THEORY INTEGRAL EQUATIONS consists of a brief comparison of the numerical results of the various integral equations to computer simulations for a fluid of hard spheres and a Lennard-Jones fluid. In section SOLUTIONS OF STRONG ELECTROLYTES, we apply the integral equation formalism to a dilute solution of charged particles and show how the Debye-Hückel theory can be obtained as a limiting law and in section STATISTICAL MECHANICAL PERTURBATION THEORY, we introduce statistical mechanical perturbation theory and use it to derive the van der Waals equation and then discuss its modern extensions. Finally, in section TRANSPORT IN LIQUIDS, we briefly discuss the time-correlation function formalism of transport processes in fluids and derive an expression for the self-diffusion coefficient as an integral of the velocity time-correlation function.

IMPERFECT GASES

The thermodynamic properties of fluids that are sufficiently dilute can be expressed as a power series in the number density, ρ. For example, the pressure of a dense gas is given by

$$\beta p = \rho + B_2(T)\,\rho^2 + B_3(T)\,\rho^3 + \dots , \tag{1}$$

where $\beta = 1/kT$ and the coefficient of ρ^j, $B_j(T)$, is called the jth virial coefficient. Statistical mechanics gives us an explicit relation for each virial coefficient as a functional of the intermolecular potential. The simplest and most useful of these relations is that of the second virial coefficient, $B_2(T)$, which is given by

$$B_2\,(T) = -2\pi \int_0^\infty [e^{-\beta\, u(r)} - 1]r^2\, dr\ . \tag{2}$$

In this equation, u(r) is the intermolecular potential between two isolated molecules.

To find virial expansions for thermodynamic quantities other than the pressure, start with

$$p = -\left(\frac{\partial A}{\partial V}\right)_{N,T} = -\left(\frac{\partial A}{\partial \rho}\right)_{N,T}\left(\frac{\partial \rho}{\partial V}\right)_{N,T} = \rho^2\left(\frac{\partial (A/N)}{\partial \rho}\right)_{N,T} . \tag{3}$$

If we combine Eqs. (1) and (3), then we obtain

$$\left(\frac{\partial(\beta\, A/N)}{\partial \rho}\right)_{N,T} = \frac{\beta p}{\rho^2} = \frac{\rho + B_2(T)\,\rho^2 + \dots}{\rho^2} ,$$

which upon integration between the limits of a dilute density where the gas is ideal and an arbitrary density gives

$$\frac{\beta\, A}{N} = \frac{\beta\, A_{ideal}}{N} + B_2(T)\,\rho + \frac{1}{2}\,B_3(T)\,\rho^2 + \dots\ . \tag{4}$$

Now use the Gibbs-Helmholtz equation

$$U = \left(\frac{\partial(\beta\, A)}{\partial \beta}\right)_{N,V} , \tag{5}$$

and Eq. (4) to obtain a virial expression for the thermodynamic energy, U:

$$\frac{\beta U}{N} = \frac{\beta U_{ideal}}{N} - T \sum_{n=1}^{\infty} \frac{1}{n} \frac{dB_{n+1}(T)}{dT} \rho^n . \tag{6}$$

Finally, using the thermodynamic relation A = U - TS and Equations (4) and (6) gives a virial expansion of the entropy

$$\frac{S}{NK} = - \left(\frac{\partial(A/Nk)}{\partial T} \right)_{N,V} = \frac{S_{id}}{Nk} - \sum_{n=1}^{\infty} \frac{1}{n} \frac{d(TB_{n+1})}{dT} \rho^n . \tag{7}$$

As attractive and convenient as virial expansions are, the problem is that they do not converge at liquid densities. It turns out that a many-body approach is required for a theoretical study of liquids, and we shall develop the standard statistical mechanical n-body distribution function formalism in the next section.

DISTRIBUTION FUNCTIONS AND LIQUIDS

We first introduce a function $P^{(N)}(r_1,r_2,\ldots,r_N)$ such that $P^{(N)}(r_1,r_2,\ldots,r_N)dr_1dr_2\ldots dr_N$ is the probability that particle 1 is in the volume element dr_1 at r_1, that particle 2 is in dr_2 at r_2, and so on. This probability is given by

$$P^{(N)}(r_1,\ldots,r_N)dr_1..dr_N = \frac{e^{-\beta U_N(r_1,\ldots,r_N)} dr_1..dr_N}{Z_N(V,T)} , \tag{8}$$

where $U_N(r_1,r_2,\ldots,r_N)$ is the total potential energy of the N-body system and where $Z_N(V,T)$, called the configuration integral, is given by

$$Z_N(V,T) = \int\!\!..\!\!\int dr_1..dr_N \, e^{-\beta U_N(r_1,\ldots,r_N)} . \tag{9}$$

We can integrate $P^{(N)}(r_1,\ldots,r_N)$ over the configurations of some set of particles, say n + 1 to N, to obtain the reduced distribution function $P^{(n)}(r_1,\ldots,r_n)$

$$P^{(n)}(r_1,\ldots,r_n) = \int\!\!..\!\!\int dr_{n+1}..dr_N \, P^{(N)}(r_1,\ldots,r_N) . \tag{10}$$

This reduced distribution function is the probability that particle 1 is in dr_1 at r_1,.., that particle n is in dr_n at r_n, irrespective of the location of the remaining particles.

Generally we are not interested in whether particle 1 is located at r_1 and particle 2 is located at r_2 and so on, but in whether any particle is in dr_1 at r_1 and some other particle is in dr_2 at r_2 and so on. Thus we introduce another distribution function $\rho^{(n)}(r_1, r_2, .., r_n)$ called an n-particle density function, which is given in terms of $P^{(n)}(r_1, r_2, .., r_n)$ by the relation

$$\rho^{(n)}(r_1, .., r_n)\, dr_1 .. dr_n = \frac{N!}{(N-n)!} P^{(n)}(r_1, .., r_n)\, dr_1 .. dr_n \; . \qquad (11)$$

The combinatorial factor in front of $P^{(n)}(r_1, ..., r_n)$ arises from the fact that there are N choices for the first particle, N-1 for the second particle, and so on. Equation (11) gives the "probability" that <u>any</u> particle is in dr_1 at $r_1, ...,$ that <u>any</u> particle is in dr_n at r_n, irrespective of the configuration or location of the other N-n particles. The integration of $\rho^{(n)}(r_1, ..., r_n)$ over the configurations of its n variables is

$$\int .. \int \rho^{(n)}(r_1, .., r_n)\, dr_1 .. dr_n = \frac{N!}{(N-n)!} \; . \qquad (12)$$

For a homogeneous fluid, $\rho^{(1)}(r_1)$ is independent of r_1, and so we have that $\rho^{(1)}(r_1) = \rho^{(1)}$ = constant. We can determine the form of $\rho^{(1)}$ by

$$\int \rho^{(1)}(r_1) dr_1 = N = \rho^{(1)} \int dr_1 = \rho^{(1)} V \; , \qquad (13)$$

or

$$\rho^{(1)} = \rho = \frac{N}{V} \; . \qquad (14)$$

The first step of Eq. (13) follows from Eq. (11). If there are no interparticle interactions, then from Eq. (8) we have

$$\rho^{(n)}(r_1, .., r_n) = \rho^n \qquad \text{(ideal fluid)} \; . \qquad (15)$$

Generally, however, Eq. (15) does not hold, and we introduce an n-body correlation function, $g^{(n)}(r_1, r_2, ..., r_n)$, through the relation

$$\rho^{(n)}(r_1, .., r_n) = \rho^n \, g^{(n)}(r_1, .., r_n) \; , \qquad (16)$$

which serves to define $g^{(n)}(r_1, .., r_n)$. For a homogeneous fluid, we have

$$g^{(1)}(r_1) = 1,$$

and more generally

$$g^{(2)}(r_1, r_2) = \frac{V^2}{N^2} N(N-1) \frac{\int .. \int dr_3 .. dr_N e^{-\beta U_N(r_1, .., r_N)}}{Z_N(V,T)}$$

$$= V^2\left[1 + O\left(\frac{1}{N}\right)\right] \frac{\int\cdots\int dr_3 \cdots dr_N \, e^{-\beta U_N(r_1,\ldots,r_N)}}{Z_N(V,T)} . \quad (17)$$

In a homogeneous fluid, $g^{(2)}(r_1,r_2)$ depends upon only the distance between particles 1 and 2, and so we have

$$g^{(2)}(r_1,r_2) = g^{(2)}(|r_2-r_1|) = g^{(2)}(r_{12}) . \quad (18)$$

We shall see below that all the thermodynamic properties of a dense fluid can be expressed as functionals of $g^{(2)}(r_{12})$ if, as is usually done, we assume that the total intermolecular potential is pair-wise additive, that is, we assume that

$$U_N(r_1,r_2,\ldots,r_N) \simeq \sum_{i<j} u(r_{ij}) , \quad (19)$$

where $u(r_{ij})$ is the intermolecular potential of two isolated molecules i and j. Note that there are N(N-1)/2 terms in the right-hand side of Eq. (19). It turns out that $g^{(2)}(r_{12})$ has a nice physical meaning which can be obtained as follows:

Equation (12) for n = 2 gives

$$\iint dr_1 dr_2 \, \rho^{(2)}(r_1,r_2) = N(N-1) .$$

We now convert to center-of-mass and relative coordinates

$$\iint d\mathbf{R}_{CM} \, dr_{12} \, \rho^2 \, g^{(2)}(r_{12}) = N(N-1) ,$$

and then integrate over $d\mathbf{R}_{CM}$ to obtain

$$N\int dr_{12} \, \rho g^{(2)}(r_{12}) = N(N-1) .$$

Suppressing the superscript on $g^{(2)}(r_{12})$ and introducing spherical coordinates gives

$$\int_0^\infty \rho \, g(r) \, 4\pi r^2 dr = N-1 . \quad (20)$$

Thus we see that $\rho g(r)$ is the number of particles between r and r+dr about a particle fixed at the origin, or that the quantity $\rho g(r)$ is a local density. The function g(r) is called a radial distribution function, and Fig. 1 shows the radial distribution function of a fluid of

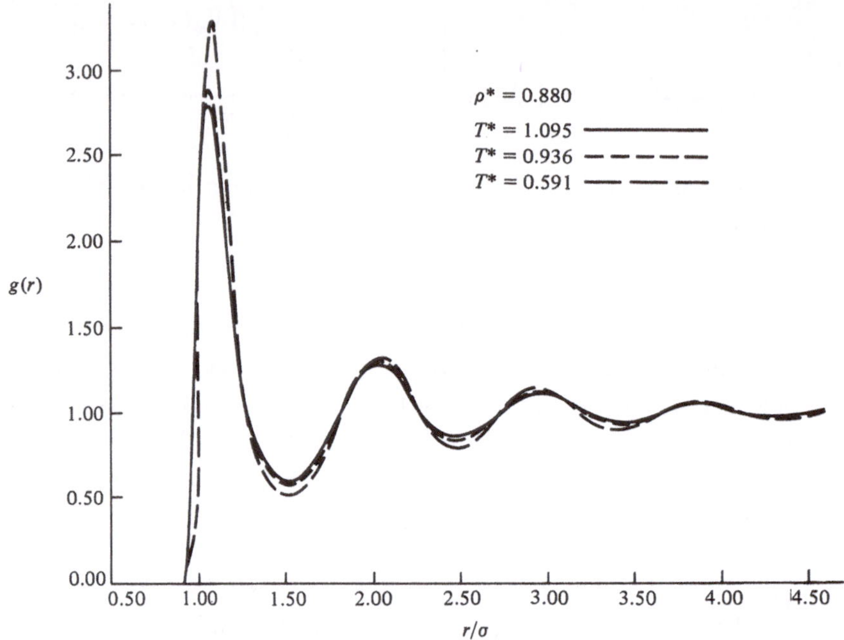

Fig. 1. The radial distribution function of a fluid of molecules obeying a Lennard-Jones 6-12 potential from molecular dynamics calculations. The reduced temperature, $T^* = kT/\varepsilon$ and the reduced density, $\rho^* = \rho\sigma^3$. (McQuarrie, 1976)

molecules obeying a Lennard-Jones 6-12 potential from molecular dynamics calculations (McQuarrie, 1976).

The radial distribution function of a fluid can be determined by neutron or x-ray scattering. For example, it can be shown that the differential cross-section for neutron scattering, $d\sigma/d\Omega$, is given by (Hansen and McDonald, 1986)

$$\frac{d\sigma}{d\Omega} = Nb^2 S(k) \ , \tag{21}$$

where

$$|k| = \frac{4\pi}{\lambda} \sin \frac{\theta}{2} \ ,$$

and

$$S(k) = 1 + \rho \int e^{-i\mathbf{k}\cdot\mathbf{r}} \, [g(r) - 1] d\mathbf{r} \ . \tag{22}$$

Thus a measurement of $d\sigma/d\Omega$ gives $g(r)$ by Fourier inversion.

Because of relations like Eq. (22), it is common to introduce the notation

$$h(r) = g(r) - 1 \ . \tag{23}$$

The function h(r) is called the indirect correlation function and its Fourier transform

$$\hat{h}(k) = \rho \int e^{-ik\cdot r}\, h(r)dr \ , \tag{24}$$

is called the structure factor of the fluid.

THERMODYNAMIC PROPERTIES OF LIQUIDS

As we stated earlier, it is possible to express all the thermodynamic properties of a liquid in terms of g(r), if we assume that the intermolecular potential is pair-wise additive. The thermodynamic energy is easily expressed in terms of g(r). We shall do this first by a physical argument and then by a formal approach. For a monatomic fluid, the total energy is given by

$$E = \frac{3}{2} NkT + \langle U\rangle \ , \tag{25}$$

where ⟨U⟩ is the average potential energy of the fluid. We can express ⟨U⟩ in terms of g(r) by the following physical argument: $\rho g(r)4\pi r^2 dr$ is the number of particles located in a spherical shell of radius r and thickness dr about a central molecule located at the origin. Now $u(r)\ \rho g(r)\ 4\pi r^2 dr$ is the interaction energy of the molecules in the spherical shell with the central molecule, and so the integral in Eq. (26) represents the total interaction of the central molecule with the rest of the molecules of the fluid. The factor of N in front of the integral arises because any of the N molecules of the fluid can be designated as the central molecule and the factor of 1/2 arises because a central molecule in one case will be in the spherical shell in another case, and thus we would overcount by a factor of 2 in the average potential energy otherwise. Thus ⟨U⟩ is given

$$\langle U\rangle = \frac{N}{2}\int_0^\infty u(r)\ \rho g(r)4\pi r^2 dr \ . \tag{26}$$

We can obtain the same result formally in the following way: the average potential energy is given by

$$\langle U\rangle = \frac{\int\!..\!\int dr_1..dr_N\ U_N(r_1,..,r_N)e^{-\beta\, U_N(r_1,..,r_N)}}{Z_N(V,T)} \ . \tag{27}$$

We now assume that $U_N(r_1,...,r_N)$ is pair-wise additive

$$U_N(r_1,..,r_N) = \sum_{i<j} U(r_{ij}) \ . \tag{28}$$

All N(N-1)/2 terms in Eq. (28) contribute the same result to $\langle U \rangle$ and so we can write Eq. (27) as

$$\langle U \rangle = \frac{N(N-1)}{2} \frac{\int \cdots \int dr_1 \cdots dr_N \, u(r_{12}) e^{-\beta U_N(r_1,\ldots,r_N)}}{Z_N(V,T)} .$$

We integrate over the coordinates of particles 3 to N and use Eq. (11) for the definition of $\rho^{(2)}(r_1,r_2)$ to obtain

$$= \frac{1}{2} \iint dr_1 dr_2 \, u(r_{12}) \, \rho^{(2)}(r_1,r_2) .$$

If we convert to center-of-mass and relative coordinates as we did in deriving Eq. (20), then we obtain

$$= \frac{N}{2} \int dr_{12} \, u(r_{12}) \, \rho g(r_{12}) = \langle U \rangle , \tag{29}$$

in agreement with Eq. (26).

Other thermodynamic properties can be expressed in terms of g(r) (McQuarrie, 1976; Hill, 1956). For example, one equation for the pressure in terms of g(r) is

$$\beta p = \rho - \frac{\beta \rho^2}{6} \int_0^\infty r \frac{du}{dr} g(r) 4\pi r^2 dr . \tag{30}$$

INTEGRAL EQUATIONS FOR THE RADIAL DISTRIBUTION FUNCTION

To complete our theoretical treatment of liquids, we need a procedure to calculate g(r). It so happens that there is no exact equation for g(r), but there are several accurate approximate equations. Four equations that have had some success for fluids are the Kirkwood equation, the Born-Green-Yvon equation, the Percus-Yevick equation and the Hypernetted chain equation (McQuarrie, 1976). All four of these equations are integral equations for g(r) in terms of u(r).

The Born-Green-Yvon equation can be derived from a simple force-balance argument. First we write

$$F(r_1,r_2) = f(r_1,r_2) + \int dr_3 f(r_1,r_3) \, \rho^{[3]} (r_3 | r_1,r_2) , \tag{31}$$

where $F(r_1,r_2)$ is the total force on a particle at r_1 given that there is another particle at r_2, $f(r_1,r_2)$ is the direct force on the particle at

r_1 due to the particle at r_2 [i.e., $f(r_1,r_2) = -\nabla_1 u(r_{12})$], and the integral term in Eq. (31) is the force on the particle at r_1 due to all the other particles in the fluid (i.e. particles 3 to N). The quantity $\rho^{[3]}(r_3|r_1,r_2)$ is the conditional density of particles at r_3 given that there is a particle at r_1 and another at r_2. By analogy with Bayes theorem of probability theory, this conditional density is given by

$$\rho^{(3)}(r_1,r_2,r_3) = \rho^{[3]}(r_3|r_1,r_2)\rho^{(2)}(r_1,r_2) \ . \tag{32}$$

To cast this exact force-balance expression into a usable form, we need to define a quantity $w^{(n)}(r_1,..,r_n)$ by

$$g^{(n)}(r_1,..,r_n) \equiv \exp\left\{-\beta\, w^{(n)}\ (r_1,..,r_n)\right\}$$

$$= V^n \frac{\int..\int dr_{n+1}..dr_N\ e^{-\beta\, U\ (r_1,..,r_N)}}{Z_N(V,T)} \ . \tag{33}$$

By taking logarithms we obtain

$$-\beta\, w^{(n)}(r_1,..,r_n) = \ln\left[\frac{V^N}{Z_N}\right] + \ln\left[\int..\int dr_{n+1}..dr_N\ e^{-\beta\, U_N}\right] .$$

Now take the gradient with respect to the coordinates of particle 1 to obtain

$$-\nabla_1 w^{(n)}(r_1,..,r_n) = \frac{\int..\int dr_{n+1}..dr_N(-\nabla_1 U_N)e^{-\beta\, U_N(r_1,..,r_N)}}{\int..\int dr_{n+1}..dr_N\ e^{-\beta\, U_N(r_1,..,r_N)}} \ . \tag{34}$$

The right-hand side of Eq. (34) is the force acting on particle 1, with particles 1 through n fixed at positions r_1 through r_n and with all the N-n other particles averaged over all configurations, weighted by the Boltzmann factor $e^{-\beta\, U_N}$. The quantity $w^{(n)}(r_1,..,r_n)$ is the potential whose gradient gives this average force. Thus $w^{(n)}(r_1,..,r_n)$ is called the potential of mean force.

Recall that the force-balance expression, Eq. (31) is

$$F(r_1,r_2) = f(r_1,r_2) + \int dr_3\ f(r_1,r_3)\ \frac{\rho^{(3)}(r_1,r_2,r_3)}{\rho^{(2)}(r_1,r_2)} \ .$$

If we introduce $w^{(2)}(r_1,r_2)$ and change the notation slightly, then we obtain

$$-\nabla_1 w^{(2)}(r_{12}) = -\nabla_1 u(r_{12}) - \rho \int dr_3 \nabla_1 u(r_{13}) \frac{g^{(3)}(r_1,r_2,r_3)}{g^{(2)}(r_1,r_2)} . \tag{35}$$

The integrand in Eq. (35) contains the quantity $w^{(3)}(r_1,r_2,r_3)$ through the relation

$$g^{(3)}(r_1,r_2,r_3) = e^{-\beta W^{(3)}(r_1,r_2,r_3)} .$$

We now assume that the potential of mean force is pair-wise additive

$$w^{(3)}(r_1,r_2,r_3) = w^{(2)}(r_1,r_2) + w^{(2)}(r_1,r_3) + w^{(2)}(r_2,r_3) , \tag{36}$$

which implies that

$$g^{(3)}(r_1,r_2,r_3) = g^{(2)}(r_1,r_2) g^{(2)}(r_1,r_3) g^{(2)}(r_2,r_3) . \tag{37}$$

Equation (36) or (37) is known as the superposition approximation. It should be noted that the superposition approximation is a more severe approximation than assuming that the total intermolecular potential is pair-wise additive. One way to look at it is to interpret Eq. (37) by saying that the probability that there are particles at positions r_1, r_2 and r_3 is the product of the pair-wise probabilities. Figure 2 shows the potential of mean force for a fluid of molecules that interact via a Lennard-Jones potential. Note that the definition of $w^{(2)}(r)$ in terms of $g(r)$ implies that $w^{(2)}(r)$ has minima (maxima) where $g(r)$ has maxima (minima). As Fig. 4, the radial distribution function of a hard-sphere fluid, shows, even a dense fluid of hard spheres has an effective attractive potential.

Under the superposition approximation, the above force-balance expression becomes the Born-Green-Yvon (BGY) equation

$$-kT\nabla_1 \ln g(r_{12}) = \nabla_1 u(r_{12}) + \rho \int dr_3 \nabla_1 u(r_{13}) g(r_{13}) g(r_{23}) , \tag{38}$$

which is a closed integral equation for $g(r)$ in terms of $u(r)$. Two other integral equations for $g(r)$ are the Percus-Yevick (PY) equation

$$e^{\beta u(r_{12})} g(r_{12}) = 1 + \rho \int dr_3 \left[1 - e^{\beta u(r_{13})} \right] g(r_{13}) h(r_{23}) , \tag{39}$$

and the hypernetted chain equation (HNC)

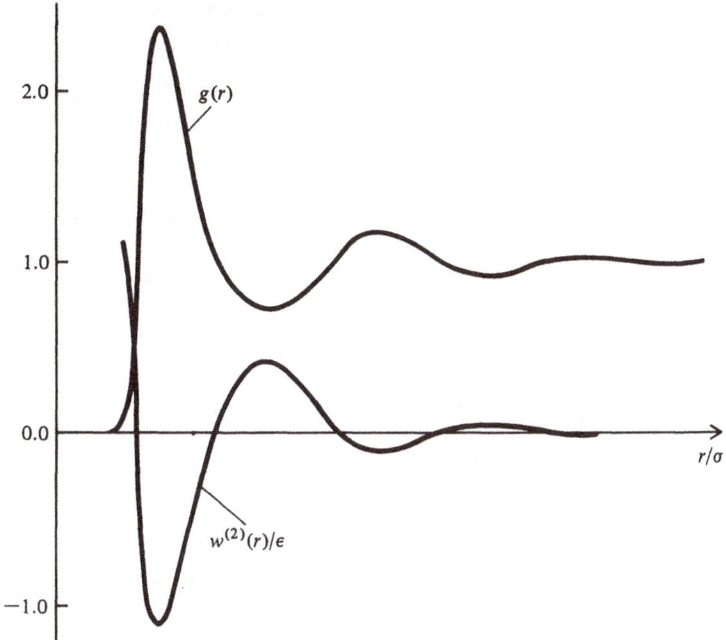

Fig. 2. The radial distribution function g(r) and the corresponding potential of mean force $w^{(2)}(r)$ for a dense fluid. Note that $w^{(2)}(r)$ has minima where g(r) has maxima and vice versa (McQuarrie, 1976).

$$\ln g(r_{12}) + \beta\, u(r_{12}) = \rho \int dr_3 [h(r_{13}) - \ln g(r_{13}) - \beta\, u(r_{13})] h(r23) \ . \tag{40}$$

The Percus-Yevick and hypernetted chain equations can be "derived" in a similar manner. To do this, we must introduce one last function, the so-called direct correlation function, c(r). The quantity $h(r_{12}) = g(r_{12}) - 1$ is a measure of the total influence of a particle at $\mathbf{r}_1$ on a particle at $\mathbf{r}_2$. Ornstein and Zernike proposed a division of $h(r_{12})$ into two parts, a direct part and an indirect part:

$$h(r_{12}) = c(r_{12}) + \rho \int dr_3\, c(r_{13}) h(r_{23}) \ . \tag{41}$$

The term $c(r_{12})$ represents the direct part and the integral term represents the influence of the particle at $\mathbf{r}_2$ on that at $\mathbf{r}_1$ as mediated by a third particle. Equation (41) is called the Ornstein-Zernike equation and is really a definition of the direct correlation function c(r). Note that the Fourier transform of the Ornstein-Zernike equation gives

$$\hat{h}(k) = \hat{c}(k) + \rho \hat{c}(k) \hat{h}(k) \ . \tag{42}$$

Although the direct correlation function does not seem to lend itself to the nice physical interpretation of the radial distribution function, g(r), or the total correlation function h(r) = g(r) -1, it has the theoretical advantage that it has a shorter range than the radial distribution function and its structure is simpler (Fig. 3).

To derive the Percus-Yevick and hypernetted chain equations, we write (McQuarrie, 1976)

$$\begin{aligned} c(r) &= g_{total}(r) - g_{indirect}(r) \\ &= g(r) - g_{indirect}(r) \ . \end{aligned} \tag{43}$$

If we assume that $g_{indirect} = e^{-\beta[w(r)-u(r)]}$, then

$$c_{PY}(r) = g(r)[1 - e^{\beta u(r)}] \ . \tag{44}$$

If we substitute this approximation into the Ornstein-Zernike equation, then we obtain the Percus-Yevick equation, Eq. (39).

The HNC equation can be obtained by linearizing $g_{indirect}$ in terms of β, giving

$$g_{indirect} = 1 - \beta[w(r) - u(r)] \ ,$$

and (45)

$$c_{HNC}(r) = g(r) - 1 + \beta[w(r) - u(r)] \ .$$

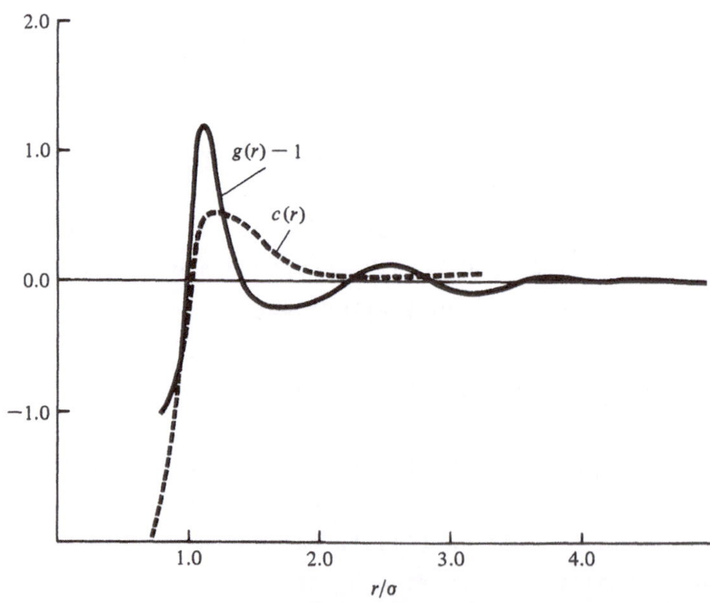

Fig. 3. A comparison of the behavior of the direct correlation function and the radial distribution function. Note that the direct correlation function has a much simpler structure and a much shorter range (McQuarrie, 1976).

The substitution of this approximation into the Ornstein-Zernike equation yields the hypernetted chain equation, Eq. (40). Notice that in both cases

$$c(r) \longrightarrow -\beta\, u(r) \text{ for large } r\ . \tag{46}$$

SOME NUMERICAL RESULTS OF THE FLUID-THEORY INTEGRAL EQUATIONS

In this section we shall present some of the many numerical results that have been obtained from the statistical-mechanical theory of fluids. We shall concentrate on two intermolecular potentials, the hard sphere potential and the Lennard-Jones potential. Although certainly lacking in realistic detail, the hard-sphere potential

$$u(r) = \begin{matrix} \infty & r < \sigma \\ 0 & r > \sigma \end{matrix}$$

has the advantage of mathematical simplicity, since many results have been obtained analytically. Physically, it attributes the molecules a non-zero size, and it turns out that many structural features of liquids can be approximated by hard-sphere fluids. Figure 4 shows the radial distribution function of a hard-sphere fluid as calculated from the

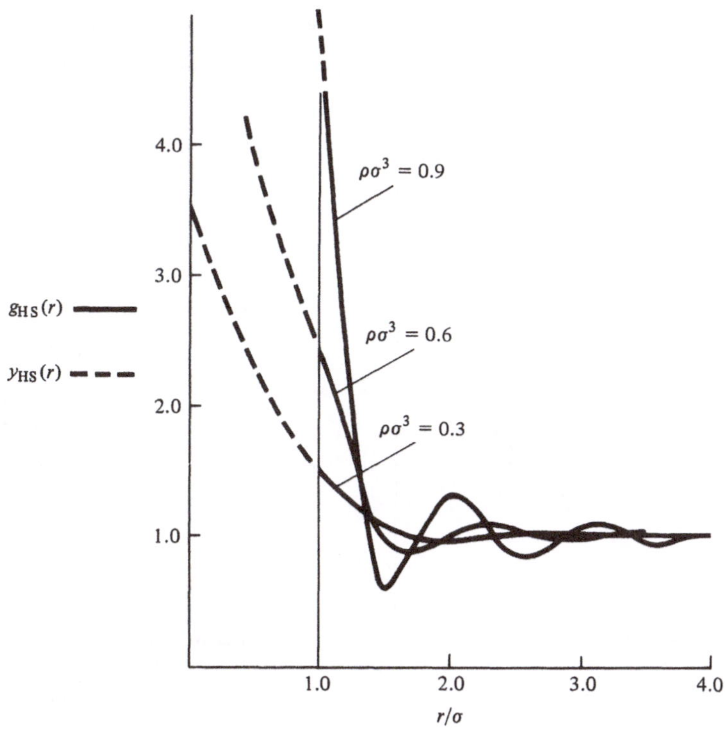

Fig. 4. The radial distribution for a hard-sphere fluid as calculated from the Percus-Yevick equation by Throop and Bearman (1965). The dashed lines represent the function $y(r) = \exp\,[u(r)]g(r)$. Notice that $y(r)$ is a continuous function at $r = \sigma$. (McQuarrie, 1976)

Percus-Yevick equation [Eq. (39)]. Note that the radial distribution function of a hard-sphere fluid (Fig. 4) and that of a Lennard-Jones fluid (Fig. 1) have very similar short-range, or local, structure.

The Percus-Yevick equation has the unique property that it is possible to solve the hard-sphere equation of state analytically (Thiele, 1963; Wertheim, 1963)

$$\frac{\beta p}{\rho} = \frac{1 + 2y + 3y^3}{(1 - y)^2} \text{ (pressure equation) ,} \tag{47a}$$

$$\frac{\beta p}{\rho} = \frac{1 + y + y^2}{(1 - y)^3} \text{ (compressibility equation) ,} \tag{47b}$$

where y is a reduced density given by

$$y = \frac{\pi\rho\sigma^3}{6} = \frac{\sqrt{2}\,\pi v_o}{6v} = 0.74\,\frac{v_o}{v} , \tag{48}$$

and where v_o is the closest-packed density, $v_o = \sigma^3/\sqrt{2}$. There are two equations of state presented above because the Percus-Yevick equation is not exact and so the resulting equation of state depends upon which thermodynamic formulas are used to obtain it from the radial distribution function. Equation (30) is only one of several equations that can be used. Figure 5 compares the analytical expressions given by Eqs. (47)

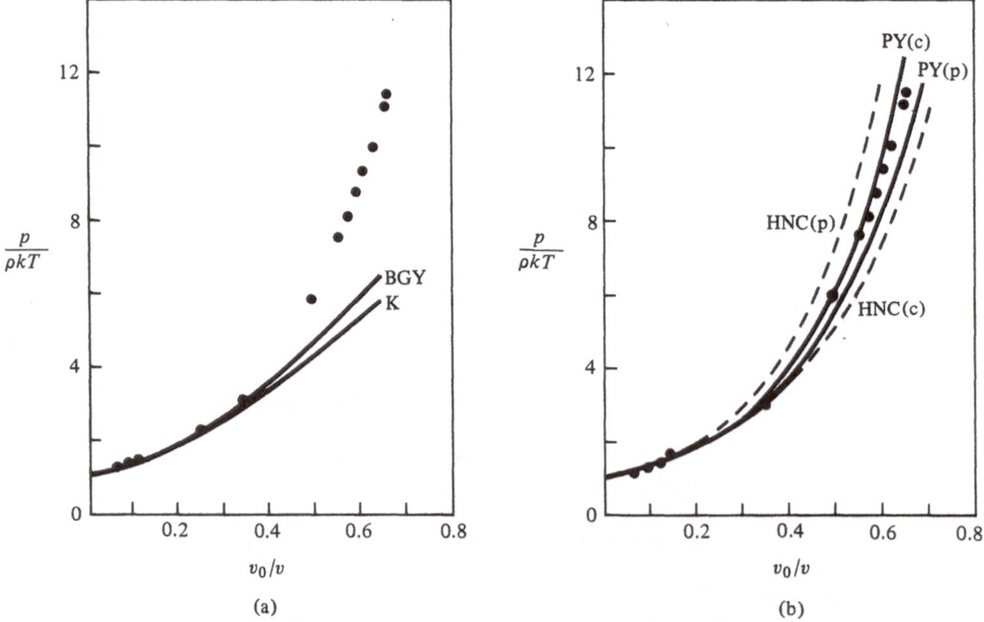

Fig. 5. A comparison of the equation of state of hard spheres obtained analytically from the Percus-Yevick equation and numerically from the hypernetted chain equation with computer simulation results (McQuarrie, 1976).

and the numerical results from the hypernetted chain equation [Eq. (40)] with computer simulation results.

A hard-sphere equation of state that is in almost exact agreement with computer simulations is the so-called Carnahan-Starling equation of state (Carnahan and Starling, 1969)

$$\frac{\beta p}{\rho} = \frac{1 + y + y^2 - y^3}{(1 - y)^3} . \tag{49}$$

The Carnahan-Starling equation of state is obtained from the virial expansion of a hard sphere fluid by curve-fitting and extrapolation. Table 1 compares values of the compressibility factor, Z, and the excess entropy, S^{ex}, of a hard sphere fluid obtained from the Percus-Yevick equation and the Carnahan-Starling equation.

Figure 6 shows the radial distribution function calculated from the Percus-Yevick equation for the Lennard-Jones 6-12 potential

$$u(r) = 4\varepsilon \left[\left(\frac{\sigma}{r}\right)^{12} - \left(\frac{\sigma}{r}\right)^{6} \right] ,$$

and Table 2 compares the pressure and excess energy of a Lennard-Jones fluid as calculated by means of the Percus-Yevick equation, the hypernetted chain equation and computer simulation (Monte Carlo). In Table 2, the reduced density $\rho*$ is given by $\rho\sigma^3$ and the reduced temperature T^* is given by kT/ε. The experimental reduced critical parameters are $\rho* = 0.32$ and $T^* = 1.28$. As can be seen from Table 2 the Percus-Yevick equation compares more favorably than the hypernetted chain equation with the Monte Carlo calculations for a short-ranged potential such as the Lennard-Jones potential.

SOLUTIONS OF STRONG ELECTROLYTES

The classical low concentration theory of solutions of electrolytes is the Debye-Hückel theory. This theory models the positive and negative ions of charges z_+ and z_- as point ions dissolved in a solvent that is modeled as a dielectric continuum of dielectric constant ε. The inter-action between two ions is given by Coulomb's law

$$u_{ij} = \frac{z_i z_j}{\varepsilon r} . \tag{50}$$

In a dilute solution, g(r) = 1, and in particular we write

$$g_{ij}(r) \simeq 1 - \beta w_{ij}(r) . \tag{51}$$

All the four integral equations discussed in section THERMODYNAMIC PROPERTIES IN LIQUIDS become identical in this limit and can be solved

Table 1. Comparative values of the compressibility factor Z and the excess entropy for a hard-sphere fluid (Carnahan and Starling, 1970).

density		PY(c)		PY(p)		CS	
y	v/v_o	Z	S^{ex}	Z	S^{ex}	Z	S^{ex}
0.0884	8.38	1.447	0.028	1.444	0.029	1.446	0.028
0.1562	4.74	1.965	0.101	1.946	0.105	1.959	0.102
1.2128	3.48	2.579	0.212	2.520	0.219	2.559	0.215
0.2616	2.83	3.304	0.360	3.170	0.365	3.260	0.361
0.3060	2.42	4.187	0.548	3.930	0.546	4.101	0.547
0.3444	2.15	5.192	0.765	4.757	0.747	5.047	0.759
0.3817	1.94	6.461	1.04	5.757	0.992	6.226	1.00
0.4150	1.78	7.978	1.36	6.857	1.25	7.617	1.33
0.4515	1.64	10.03	1.78	8.358	1.61	9.474	1.72
0.4840	1.53	12.50	2.27	10.031	2.00	11.68	2.17
0.5142	1.44	15.52	2.84	11.596	2.46	14.33	2.69
0.5405	1.37	18.89	3.44	14.007	2.86	17.26	3.24
0.5700	1.30	23.76	4.27	16.845	3.87	21.44	3.98
0.5972	1.24	29.89	5.26	20.119	4.07	26.63	4.85

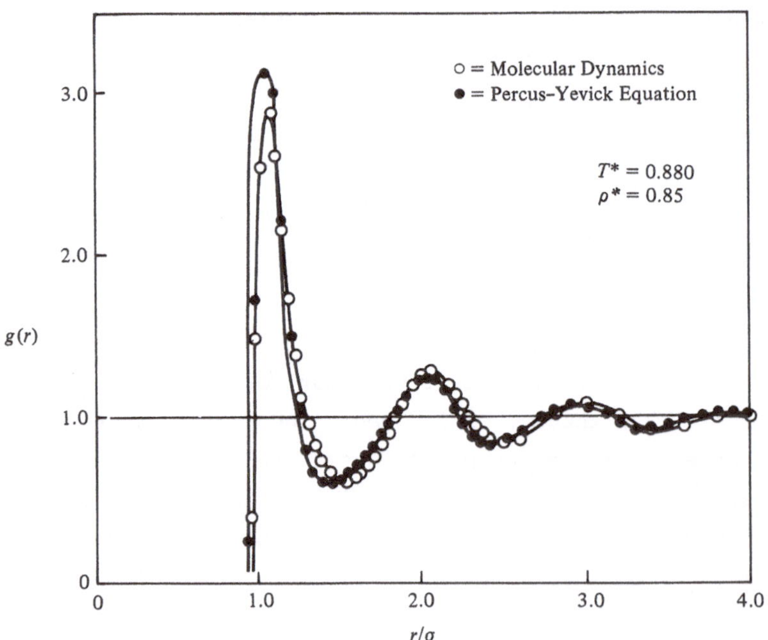

Fig. 6. A comparison of the radial distribution function of a Lennard-Jones 6-12 fluid calculated from the Percus-Yevick equation (solid circles) with computer simulation (molecular dynamics) results (open circles). The reduced temperature $T^* = kT/\varepsilon = 0.880$ and the reduced density $\rho^* = \rho\sigma^3 = 0.85$ (Mandel et al., 1970).

Table 2. A comparison between Monte Carlo results and the predictions of integral-equation theories of the thermodynamic properties of a Lennard-Jones fluid. MC = Monte Carlo results of Hansen and Verlet (1969); PY and HNC results are from Levesque (1966) and Verlet and Levesque (1967). c = from compressibility equation; p = from pressure equation. (Hansen and McDonald, 1986)

$T^* = 2.74$								
	MC		HNC			PY		
ρ^*	$\frac{\beta p}{\rho}$	$\frac{U^{ex}}{N\varepsilon}$	$\frac{\beta p^c}{\rho}$	$\frac{\beta p^v}{\rho}$	$\frac{U^{ex}}{N\varepsilon}$	$\frac{\beta p^c}{\rho}$	$\frac{\beta p^v}{\rho}$	$\frac{U^{ex}}{N\varepsilon}$
0.3	1.040	-1.783	1.050	1.083	-1.787	1.056	1.070	-1.791
0.4	1.199	-2.371	1.177	1.279	-2.351	1.186	1.234	-2.368
0.55	1.653	-3.207	1.542	1.901	-3.127	1.614	1.722	-3.238
0.7	2.641	-3.902	2.213	3.160	-3.693	2.437	2.646	-3.931
0.8	3.604	-4.281	2.895	4.528	-3.852	3.345	3.603	
1.0	7.388	-4.180	5.095	9.113	-3.261	6.808	6.670	-4.576

$T^* = 1.35$						
	MC		PY		HNC	
ρ^*	$\frac{\beta p}{\rho}$	$\frac{U^{ex}}{N\varepsilon}$	$\frac{\beta p^v}{\rho}$	$\frac{U^{ex}}{N\varepsilon}$	$\frac{\beta p^v}{\rho}$	$\frac{U^{ex}}{N\varepsilon}$
0.3	0.352	-2.090	0.396	-2.18		
0.35	0.298	-2.405	0.376	-2.48		
0.4	0.272	-2.470	0.386	-2.77	0.388	-2.79
0.45	0.280	-3.030	0.434	-3.07	0.442	-3.08
0.5	0.303	-3.372	0.532	-3.38	0.557	-3.37
0.55	0.415	-3.704	0.692	-3.69	0.755	-3.68
0.65	0.850	-4.343	1.256	-4.32	1.492	-4.26
0.7	1.166	-4.684	1.689	-4.65	2.086	-4.52

analytically. For example, the hypernetted chain equation for a multicomponent solution is

$$\ln g_{ij}(r_{12}) = -\beta\, u_{ij}(r_{12}) + \sum_{k=1}^{s} \rho_k \int dr_3 \left[h_{ik}(r_{13}) - \ln g_{ik}(r_{13}) - \beta\, u_{ik}(r_{13}) \right] h_{kj}(r_{23}) \,. \tag{52}$$

If we introduce Eq. (51) and neglect products of distribution functions, then this equation becomes

$$w_{ij}(r_{12}) = u_{ij}(r_{12}) - \beta \sum_{k=1}^{s} \rho_k \int dr_3\, u_{ik}(r_{13}) w_{kj}(r_{23}) \,. \tag{53}$$

If we now let

$$w_{ij}(r) = \frac{z_i z_j}{\varepsilon}\, w(r) \,,$$

and use Eq. (50), then Eq. (53) becomes

$$w(r) = \frac{1}{r} - \frac{\kappa^2}{4\pi} \int dr_3\, \frac{w(r_{23})}{r_{13}} \,, \tag{54}$$

where κ^2 is given by

$$\kappa^2 = \frac{4\pi\beta}{\varepsilon} \sum_{j=1}^{s} \rho_j z_j^2 \,. \tag{55}$$

By taking the Fourier transform of Eq. (55), we obtain

$$\tilde{W}(k) = \tilde{U}(k) - \frac{\kappa^2}{4\pi} \tilde{W}(k)\tilde{U}(k) \,, \tag{56}$$

where

$$\tilde{W}(k) = \frac{1}{k^2 + \kappa^2} \,. \tag{57}$$

The inversion of this quantity gives

$$w(r) = \frac{e^{-\kappa r}}{r} \,, \tag{58}$$

or

$$w_{ij}(r) = \frac{z_i z_j}{\varepsilon} \frac{e^{-\kappa r}}{r} \,, \tag{59}$$

as the potential of mean force of the Debye-Hückel theory. This form for the potential of mean force is called a screened coulombic potential. As we show below, the coulombic potential about a central ion is screened by an atmosphere of ions of opposite sign. The radial distribution function of the Debye-Hückel theory is

$$g_{ij}(r) = 1 - \frac{\beta z_i z_j}{\varepsilon} \frac{e^{-\kappa r}}{r} . \tag{60}$$

The charge density, $c_i(r)$, about a central ion of type i is given by

$$c_i(r) = \sum_{j=1}^{s} z_j \rho_j g_{ij}(r) , \tag{61}$$

$$= - \frac{z_i \kappa^2}{4\pi} \frac{e^{-\kappa r}}{r} ,$$

$$= - \frac{z_i \kappa^2}{4\pi} \frac{e^{-\kappa r}}{r} , \tag{62}$$

where the summation vanishes because of electroneutrality. An alternative expression of electroneutrality is

$$\int_0^\infty c_i(r) 4\pi r^2 dr = -z_i , \tag{63}$$

which says that the total charge about a central ion of charge z_i is equal in magnitude but of opposite sign. The integrand in Eq. (63)

$$p_i(r)dr \equiv c_i(r) 4\pi r^2 dr$$

$$= -z_i \kappa^2 r \, e^{-\kappa r} \, dr . \tag{64}$$

The quantity $p_i(r)$, the fraction of charge between the spherical shells of radius r and r + dr, is plotted against r in Fig. 7. We can see from this figure that a central ion is surrounded by an atmosphere of ions of opposite sign, and that the thickness of this ionic atmosphere is $1/\kappa$, the reciprocal of the Debye-Hückel parameter.

Using the Debye-Hückel radial distribution functions given by Eq. (60), we can calculate all the thermodynamic properties analytically. For example, if we substitute Eq. (60) into the equation for the energy of a multicomponent system

$$E = \frac{3}{2} NkT + 2\pi N \rho \sum_{i,j} x_i x_j \int_0^\infty u_{ij}(r) g_{ij}(r) r^2 dr , \tag{65}$$

we obtain

$$\frac{\beta E^{ex}}{V} = - \frac{\kappa^3}{8\pi} , \tag{66}$$

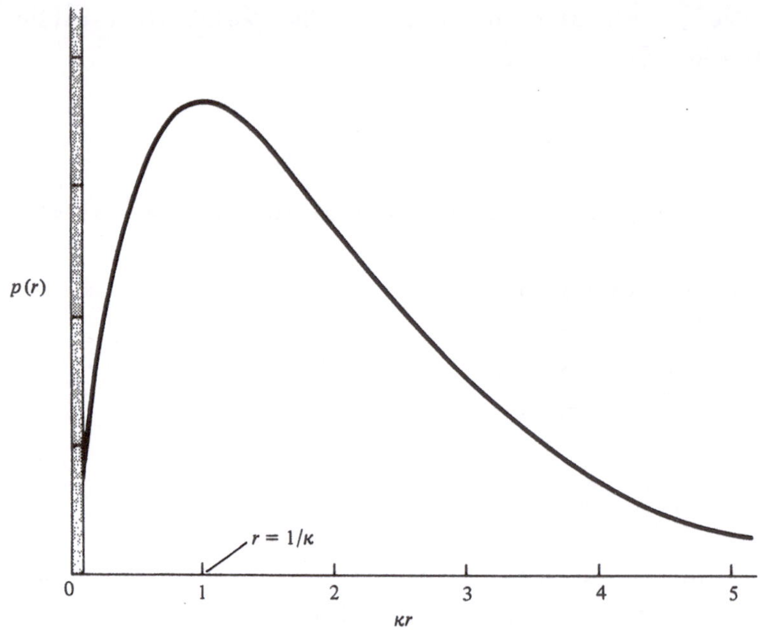

Fig. 7. The fraction of charge between the spherical shells of radius r and r + dr. This plot illustrates the idea of an ionic atmosphere in an electrolyte solution (McQuarrie, 1976).

where $E^{ex} = E - E^{ideal}$. If we substitute this expression into the Gibbs-Helmholtz equation,

$$\frac{E}{V} = \left(\frac{\partial(\beta A/V)}{\partial\beta} \right)_{N,V} ,$$

then we can integrate with respect to β to obtain

$$\frac{\beta A^{ex}}{V} = -\frac{\kappa^3}{12\pi} . \tag{67}$$

The chemical potential of the jth species is given by

$$\mu_j = \left(\frac{\partial A}{\partial N_j} \right)_{T,V,N_{k=j}} , \tag{68}$$

and eventually we obtain

$$\mu_j^{ex} = kT\ln\rho_j - \frac{\kappa z_j^2}{2\varepsilon} . \tag{69}$$

We can obtain a relation for the activity coefficient, γ_j, of the jth ionic species by comparing this result to the thermodynamic formula

$$\mu_j^{ex} = kT \ln\rho_j\gamma_j ,$$

giving

$$\ln \gamma_j = -\frac{\kappa z_j^2}{2\varepsilon kT} . \tag{70}$$

For a salt whose chemical formula is $C_{\nu_+}A_{\nu_-}$, the experimentally observed quantity is the mean ionic activity coefficient $\gamma_\pm = (\gamma_+^{\nu_+}\gamma_-^{\nu_-})^{1/\nu_+ + \nu_-}$. After some amount of manipulation, one obtains the result

$$\ln \gamma_\pm = -|z_+z_-| \frac{\kappa}{2\varepsilon kT} . \tag{71}$$

This is one of the central results of the Debye-Hückel theory. This result, being exact in the limit of low concentrations, is called the Debye-Hückel limiting law. Equation (71) predicts that a plot of ln $\gamma_\pm$ versus the square root of the concentration [see Eq. (55)] should be linear and that the slope of the straight line should be proportional to the product of the valences of the cations and anions. Figure 8 shows some experimental data for 1-1, 1-2 and 1-3 salts, in agreement with the Debye-Hückel limiting law. The data in Fig. 8 are for very low salt concentrations (<0.01 molar) and are expected to agree with the Debye-Hückel theory since it is, indeed, a limiting law. For higher

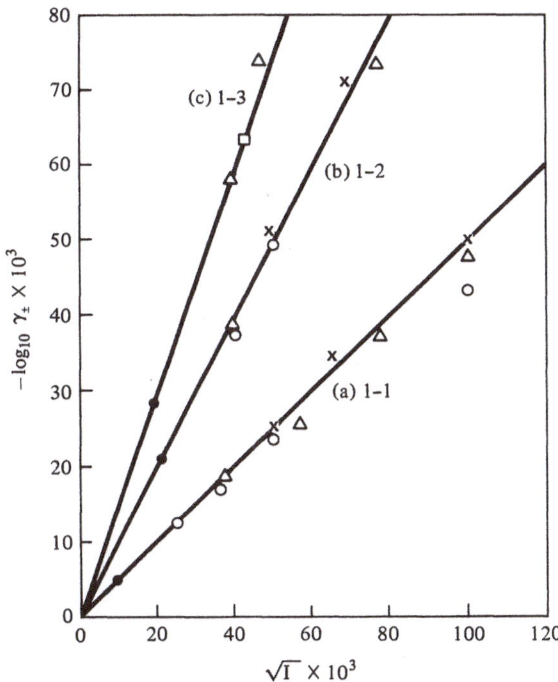

Fig. 8. The logarithms of the mean ionic activity coefficients of some 1-1 salts (such as NaCl), 2-1 salts (such as $CaCl_2$) and 3-1 salts (such as $LaCl_3$) versus the ionic strength $I = \frac{1}{2} \sum_j \rho_j Z_j^2$, which is proportional to $\rho^{1/2}$.

concentrations we should expect to observe deviations from Eq. (71) and Fig. 9 shows some typical experimental data. Notice that not only does one observe deviations from the linear behavior predicted by the limiting law, but that different salts display their own individual characteristics, reflecting their different ionic sizes. As will be seen in Professor Rasaiah's chapter, more general curves like those observed in Fig. 9. can be obtained by solving the full integral equations [see, for example, Eq. (52)] rather than linearized, low concentration limiting forms [see, for example, Eq. (53)].

STATISTICAL MECHANICAL PERTURBATION THEORY

Much experimental evidence indicates that the structure of a liquid is determined primarily by the short-ranged repulsive forces and that the relatively long-ranged attractive part of the potential provides a net force that is seen as a somewhat uniform attractive potential. Thus, in a sense we picture the repulsive part of the potential as determining the structure of the liquid and the attractive part as holding the molecules together at some specified density. This physical picture suggests that

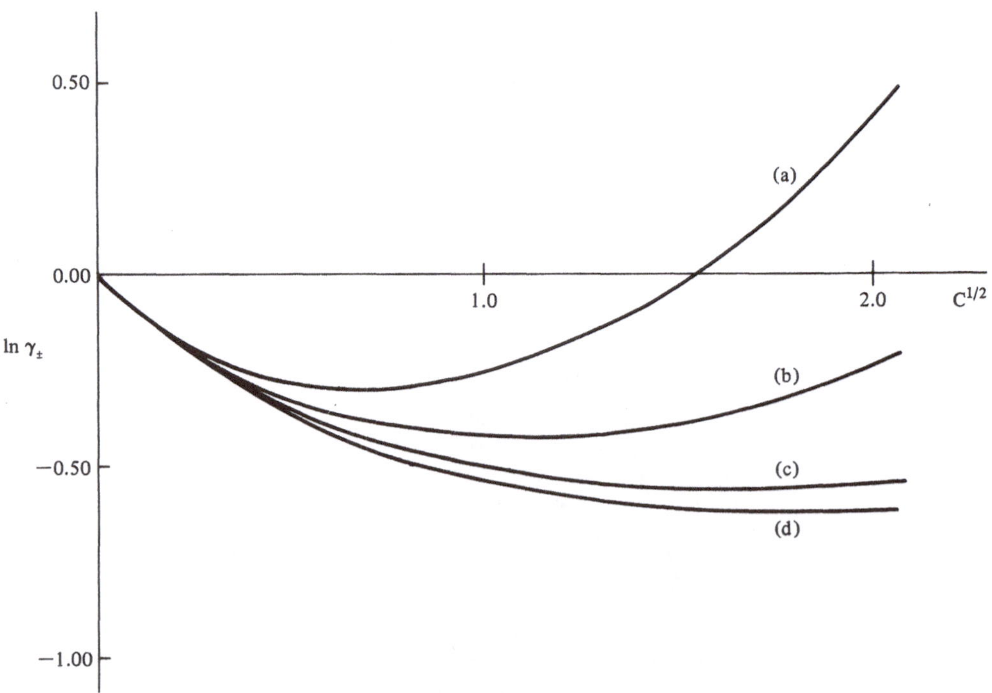

Fig. 9. Logarithm of the mean ionic activity coefficient of aqueous solutions of alkali halides at 25°C. (a) LiCl, (b) NaCl, (c) KCl, (d) RbCl (Robinson and Stokes, 1965).

we treat a fluid as a system of molecules that are governed primarily by the repulsive potential and that the attractive potential be treated as a perturbation. The unperturbed system, i.e. the system of purely repulsive molecules, is often taken to be a system of hard spheres since this system has been fairly well studied.

We let the total potential energy of the system be separated into two parts

$$U_N = U_N^{(0)} + U_N^{(1)} , \tag{72}$$

where $U_N^{(0)}$ is the potential energy of the unperturbed (reference) system and $U_N^{(1)}$ is a perturbation term. The configuration integral [Eq. (9)] for this potential is

$$Z_N(V,T) = \int..\int dr_1..dr_N\, e^{-\beta[U_N^{(0)} + U_N^{(1)}]} .$$

We now multiply and divide $Z_N(V,T)$ by

$$\int..\int dr_1...dr_N e^{-\beta\, U_N^{(0)}} ,$$

to obtain

$$Z_N(V,T) = \int..\int dr_1..dr_N e^{-\beta\, U_N^{(0)}} \times \frac{\int..\int dr_1..dr_N\, e^{-\beta[U_N^{(0)} + U_N^{(1)}]}}{\int..\int dr_1..dr_N\, e^{-\beta\, U_N^{(0)}}} . \tag{73}$$

Note that the second factor here can be considered to be the average of $\exp(-\beta\, U_N^{(1)})$ over the unperturbed, or reference, system. Thus we can write Eq. (73) in the form

$$= Z_N^{(0)} \left\langle e^{-\beta\, U_N^{(1)}} \right\rangle_0 , \tag{74}$$

where $Z_N^{(0)}$ is the configuration integral of the unperturbed system and $\langle\ \rangle_0$ represents a canonical average in the unperturbed system. Using Eq. (74), the Helmholtz free energy of the system is

$$A = A_o + A^{(1)} , \tag{75}$$

where A_o is the Helmholtz free energy of the unperturbed system and $A^{(1)}$ is given by

$$A^{(1)} = -kT \ln \left\langle e^{-\beta U_N^{(1)}} \right\rangle_0 . \tag{76}$$

If we expand in powers of β, then we obtain

$$A^{(1)} = \langle U_N^{(1)} \rangle_0 - \beta \left[\frac{\langle (U_N^{(1)})^2 \rangle_0 - \langle U_N^{(1)} \rangle_0^2}{2} \right] + O(\beta^2) . \tag{77}$$

As usual, we assume that the intermolecular potential is pair-wise additive [Eq. (19)], and so $\langle U_N^{(1)} \rangle_0$ becomes

$$\begin{aligned} \langle U_N^{(1)} \rangle_0 &= \left\langle \sum_{i<j} u^{(1)'}(r_{ij}) \right\rangle_0 = \frac{N(N-1)}{2} \langle u^{(1)}(r_{12}) \rangle_0 \\ &= \frac{N(N-1)}{2} \frac{\int \cdots \int dr_1 \cdots dr_N \, u^{(1)}(r_{12}) e^{-\beta U_N^{(0)}(r_1,\ldots,r_N)}}{Z_N^{(0)}} \\ &= \frac{1}{2} \int\int dr_1 dr_2 \, u^{(1)}(r_{12}) \rho_0(r_1,r_2) \\ &= \frac{N\rho}{2} \int dr \, u^{(1)}(r) g_0(r) . \end{aligned} \tag{78}$$

The zero subscript in the last two terms indicates that the distribution function is that of the reference system.

Perhaps the simplest application of the statistical mechanical perturbation theory of fluids is a derivation of the van der Waals equation. To derive the van der Waals equation, we first write the two-body intermolecular potential as the summation of a hard sphere part $u_{HS}(r)$ and an attractive part $u^{(1)}(r)$,

$$u(r) = u_{HS}(r) + u^{(1)}(r) . \tag{79}$$

We also assume that the perturbation term is small enough that we can

$$\begin{aligned} \left\langle e^{-\beta U_N^{(1)}} \right\rangle_0 &\simeq 1 - \beta \langle U_N^{(1)} \rangle_0 \\ &\simeq e^{-\beta \langle U_N^{(1)} \rangle_0} , \end{aligned} \tag{80}$$

where $\langle U_N^{(1)} \rangle_0$ is given by Eq. (78). Certainly $g_0(r)$, the radial distribution function of a hard sphere fluid, was not available to van der Waals, and he effectively approximated $g_0(r)$ by

$$g_0(r) = 0 \qquad r < \sigma$$
$$\qquad = 1 \qquad r > \sigma , \tag{81}$$

where σ is the diameter of the hard spheres. Using this form for $g_0(r)$, $\langle U_N^{(1)} \rangle_0$ becomes

$$\left\langle U_N^{(1)} \right\rangle_0 = 2\pi N\rho \int_0^\infty u^{(1)}(r)\, r^2 dr = -aN\rho , \tag{82}$$

where

$$a = -2\pi \int_0^\infty u^{(1)}(r)\, r^2 dr . \tag{83}$$

Note that a is a positive quantity because $u^{(1)}(r)$, being attractive, is everywhere negative. So far, then we have

$$Z_N(V,T) = Z_N^{(0)}(V,T) e^{\beta a\rho N} , \tag{84}$$

and so the pressure is given by

$$p\beta = \left(\frac{\partial \ln Z_N^{(0)}}{\partial V} \right)_{N,T} - \frac{a\rho^2}{kT} . \tag{85}$$

For an ideal gas $U_N = 0$ and the configuration integral is equal to V^N.

The final approximation of the van der Waals theory is to assume that the hard-sphere configuration integral is the form V_{eff}^N, where the effective volume is the volume available to a molecule in the fluid. By a geometric argument, V_{eff} is given by

$$V_{eff} = V - \frac{2\pi N\sigma^3}{3} , \tag{86}$$

and so

$$Z_N^{(0)} = V_{eff}^N = (V - Nb)^N , \tag{86}$$

where

$$b = \frac{2\pi\sigma^3}{3} . \tag{87}$$

If we substitute Eq. (87) into Eq. (85), then we obtain the van der Waals equation

$$\beta p = \frac{\rho}{1 - b\rho} - \beta a\rho^2 , \tag{88}$$

where the constants a and b have their usual meanings.

The perturbation theoretical derivation of the van der Waals equation suggests a number of obvious improvements. The most obvious ones are:

1. Use a better form for $g_0(r)$;
2. Use a better form for p_0, or $Z_N^{(0)}$;
3. Use a more realistic reference system;
4. Consider higher order terms in β.

Points 1 and 2 can be incorporated straightforwardly using the ideas presented earlier in this chapter. For example, we could use the analytic Percus-Yevick equations of state for hard spheres (Eqs. 47a and b) or the Carnahan-Starling equation of state (Eq. 49) for p_0. Furthermore, we could use the hard-sphere radial distribution function obtained numerically from one of the integral equations or even that calculated from computer simulation. Points 3 and 4 are less straightforward and represent contributions that were made around 1970 by Barker and Henderson (1976) and by Weeks, Chandler and Andersen (1971). The results of these two approaches are comparable and are illustrated in Figs. 10 and 11 and Table 3.

TRANSPORT IN LIQUIDS

In this final section, we shall introduce the basic idea of a time-correlation function and the relation of time-correlation functions

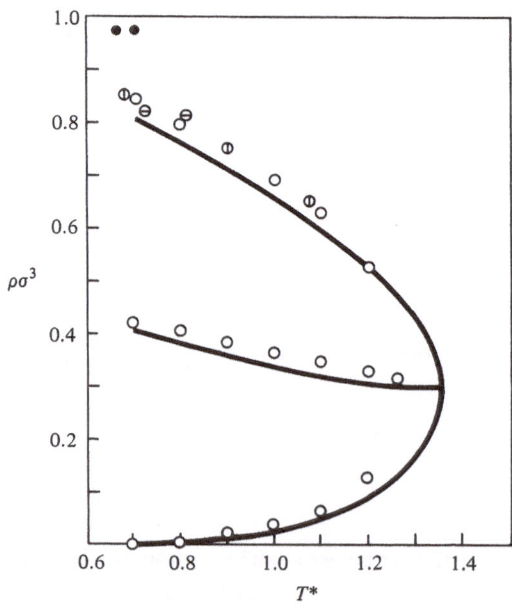

Fig. 10. Densities of coexisting phases for a fluid obeying the 6-12 potential according to a statistical mechanical perturbation theory developed by Barker and Henderson. The points are a mixture of machine calculations and actual experimental data (from Barker and Henderson, 1967).

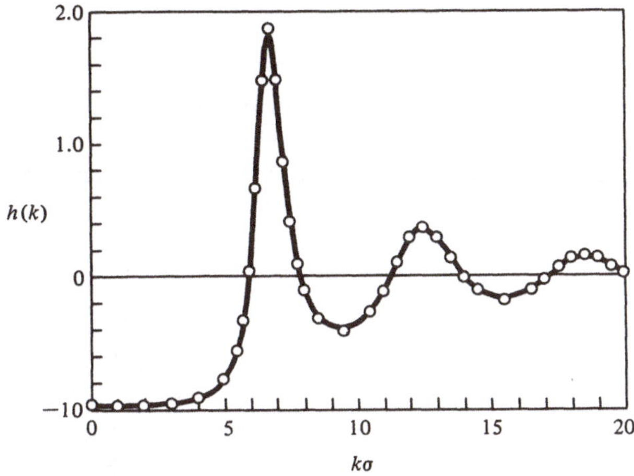

Fig. 11. Plot of the structure factor h(k) for $\rho^* = \rho\sigma^3 = 0.844$, $T^* = kT/\varepsilon = 0.723$. The line is calculated according to a statistical mechanical perturbation theory developed by Weeks et al. (1971), and the circles are molecular dynamics results.

to transport coefficients in preparation of the chapter by Dr. Hubbard. Let A $\{p(t),q(t)\}$ be a function of the phase space coordinates where q(t) represents all the generalized coordinates that are necessary to describe the system and p(t) represents all the conjugate momenta. Generally we can express quantities as

$$p(t) = p(p,q;t) \quad \text{and} \quad q(t) = q(p,q;t) \ ,$$

where p and q represent initial values. Thus we can write $A\{p(t),q(t)\}$ as

$$A\{p(t),q(t)\} = A(p,q,;t) = A(t) \ .$$

The (classical) time-correlation function of A(t) is

$$\langle A(t)A(0)\rangle = \int\cdots\int dpdq\ f(p,q)\ A(p,q;t)A(p,q;0) \ ,$$

where f(p,q) is the equilibrium phase space distribution function. In most cases of interest the correlation of A(t) with A(0) will decrease with increasing time and $\langle A(t)A(0)\rangle$ will decay as a function of time.

The time-correlation function formalism of transport processes provides expressions for the various microscopic transport coefficients as functionals of time-correlation functions, each transport coefficient being a functional of a different time-correlation function. There are many ways to derive these expressions, but here we shall present a method first developed in general by Helfand (1961). For simplicity we shall consider only the case of self-diffusion.

Table 3. A comparison between the perturbation theory and Monte Carlo results for the excess free energy of a Lennard-Jones fluid on the $T^* = 0.75$, 1.15, and 1.35 isotherms (from Weeks et al, 1971).

		$-\beta A^{ex}/N$		
T^*	ρ	CWA	Baker-Henderson	Monte Carlo
0.75	0.1	0.55	0.57	0.81
	0.2[a]	1.15	1.16	1.48
	0.3[a]	1.78	1.77	2.11
	0.4[a]	2.42	2.38	2.68
	0.5[a]	3.06	2.96	3.23
	0.6[a]	3.65	3.48	3.74
	0.7[a]	4.14	3.90	4.17
	0.8[a]	4.46	4.16	4.47
	0.84	4.51	4.20	4.54
1.15	0.1	0.29	0.30	0.39
	0.2	0.60	0.61	0.73
	0.3	0.92	0.92	1.05
	0.4[a]	1.23	1.20	1.33
	0.5[a]	1.51	1.46	1.59
	0.55[a]	1.63	1.56	1.69
	0.6[a]	1.74	1.65	1.79
	0.65	1.82	1.72	1.84
	0.75	1.88	1.76	1.88
	0.85	1.77	1.63	1.78
1.35	0.1	0.22	0.23	0.30
	0.2	0.45	0.46	0.56
	0.3	0.68	0.69	0.80
	0.4	0.90	0.89	1.00
	0.5	1.09	1.05	1.16
	0.6	1.22	1.16	1.26
	0.7	1.26	1.18	1.29
	0.8	1.16	1.07	1.19

[a] Corresponds to one-phase metastable states in the liquid-gas two-phase region.

We start with the diffusion equation

$$\frac{\partial G(r,t)}{\partial t} = D\nabla^2 G(r,t) , \tag{89}$$

with the initial condition

$$G(r,0) = \delta(\mathbf{r}) . \tag{90}$$

If $\hat{G}(k,t)$ is the Fourier transform of $G(r,t)$, then Fourier transformation of Eq. (89) yields

$$\frac{\partial \hat{G}(k,t)}{\partial t} = - k^2 D\hat{G}(k,t) , \tag{91}$$

whose solution is

$$\hat{G}(k,t) = e^{-k^2 Dt} . \tag{92}$$

According to this result

$$\left(\frac{\partial^2 \hat{G}(k,t)}{\partial k^2} \right)_{k=0} = - 2Dt . \tag{93}$$

We can derive another expression for this derivative by differentiating

$$\hat{G}(k,t) = \int d\mathbf{r}\, e^{i\mathbf{k}\cdot\mathbf{r}} G(r,t) ,$$

directly, giving

$$\left(\frac{\partial^2 \hat{G}(k,t)}{\partial k^2} \right)_{k=0} = - \frac{1}{3} \int_0^\infty dr\, 4\pi r^2\, G(r,t)$$

$$= - \frac{1}{3} \langle r^2(t) \rangle . \tag{94}$$

By comparing Eqs. (93) and (94) we find that

$$\langle r^2(t) \rangle = 6Dt , \tag{95}$$

an equation that was first derived by Einstein in his early work on Brownian motion. If the diffusing particle had started at the point r_0 rather than at the origin, then Eq. (95) would be

$$\langle |\mathbf{r}(t)-\mathbf{r}(0)|^2 \rangle = 6Dt . \tag{96}$$

But now we can write

$$\mathbf{r}(t)-\mathbf{r}(0) = \int_0^t \mathbf{v}(t')dt' ,$$

and

$$|\mathbf{r}(t)-\mathbf{r}(0)|^2 = \int_0^t dt' \int_0^t dt''\, \mathbf{v}(t')\cdot\mathbf{v}(t'') .$$

If we take the ensemble average of both sides of this equation, then we obtain

$$\langle |\mathbf{r}(t)-\mathbf{r}(0)|^2 \rangle = \int_0^t dt' \int_0^t dt'' \; \langle \mathbf{v}(t')\cdot\mathbf{v}(t'')\rangle \; . \tag{97}$$

Using the condition of the stationarity of an equilibrium ensemble average and the time reversibility of the classical equations of motion, we can write the integrand in Eq. (97) as

$$\langle \mathbf{v}(t')\cdot\mathbf{v}(t'')\rangle \underset{\substack{\uparrow \\ \text{stationarity}}}{=} \langle \mathbf{v}(t'-t'') \cdot \mathbf{v}(0)\rangle \underset{\substack{\uparrow \\ \text{time reversibility}}}{=} \langle \mathbf{v}(t''-t') \cdot \mathbf{v}(0)\rangle \; . \tag{98}$$

Thus Eq. (97) becomes

$$\langle |\mathbf{r}(t)-\mathbf{r}(0)|^2 \rangle = \int_0^t dt' \int_0^t dt'' \; \langle \mathbf{v}(t''-t')\cdot\mathbf{v}(0)\rangle \; .$$

If we now let $\tau = t''-t'$, change the order of integration, and carry out the one obvious integration, then we get

$$\langle |\mathbf{r}(t)-\mathbf{r}(0)|^2 \rangle = 6Dt = 2t \int_0^t \left(1 - \frac{\tau}{t}\right) \langle \mathbf{v}(\tau)\cdot\mathbf{v}(0)\rangle \, d\tau \; . \tag{99}$$

If the quantity $\langle \mathbf{v}(\tau)\cdot\mathbf{v}(0)\rangle$ decays sufficiently rapidly and t is chosen to be sufficiently large, then one can neglect the term τ/t in the integrand and finally obtain

$$D = \frac{1}{3} \int_0^\infty \langle \mathbf{v}(\tau)\cdot\mathbf{v}(0)\rangle \, d\tau \; , \tag{100}$$

which is our desired expression for D. It should be clear from the derivation that Eq. (100) is limited to processes for which the diffusion equation is a suitable description, which means processes that occur slowly with respect to microscopic times and for which spatial variations are smooth.

Other macroscopic transport equations can be used to derive relations for other transport coefficients as functionals of time-correlation functions. For example, the Navier-Stokes equation can be used to derive a time-correlation function expression for the coefficient of shear viscosity.

$$\eta = \frac{1}{VkT} \int_0^\infty dt \; \langle J(0)J(t)\rangle \; ,$$

where

$$J = \sum_{j=1}^{N} \left(\frac{p_{xj}p_{zj}}{m} + z_j F_{jx} \right) ,$$

and where F_{jx} is the x-component of the intermolecular forces acting upon particle j.

REFERENCES

Barker, J.A., and Henderson, D., 1967, Perturbation Theory and Equation of State for Fluid. II. A Successful Theory of Liquids, J. Chem. Phys., 47:4714.

Barker, J.A., and Henderson, 1976, What is a "liquid"? Understanding the states of matter, Rev. Mod. Phys., 48:587.

Carnahan, N.F. and Starling, K.E., 1969, Equation of state for nonattracting rigid spheres, J. Chem. Phys., 51:635.

Carnahan, N.F. and Starling, K.E., 1970, Thermodynamic properties of a rigid-sphere fluid, J. Chem. Phys., 53:600.

Gray, C.G. and Gubbins, K.E., 1984, "Theory of Molecular Liquids", Clarendon Press, Cambridge.

Hansen, J.P. and McDonald, I.R., 1986, "Theory of Simple Fluids", Academic Press, New York.

Hansen, J.P. and Verlet, L., 1969, Phase transitions of the Lennard-Jones system, Phys. Rev., 184:151.

Helfand, E., 1961, Theory of the molecular friction constant, Phys. Fluids,, 4:681.

Hill, T.L., 1956, "Statistical Mechanics", McGraw-Hill, New York.

Levesque, D., 1966, Study of the Percus-Yevick, hypernetted chain, and Born-Green equations for classical fluids, Physica, 32:1985.

Mandel, F., Bearman, F.J. and Bearman, M.Y., 1970, Numerical solutions of the Percus-Yevick equation for the Lennard-Jones (6-12) and hard-sphere potentials, J. Chem. Phts., 52:3315.

McQuarrie, D.A., 1976, "Statistical Mechanics", Harper and Row, New York.

Robinson, R.A., and Stokes, R.H., 1965, "Electrolyte Solutions", 2nd Ed., Butterworths, London.

Thiele, E., 1963, Equation of state for hard spheres, J. Chem. Phys., 39:474.

Throop, G.J., and Bearman, R.J., 1965, Numerical Solution of the Percus-Yevick Equation for the Hard-Sphere Potentials, J. Chem. Phys., 42:2408.

Verlet, L. and Devesque, D., 1967, Theory of classical fluids, Physics, 36:254.

Weeks, J.D., Chandler, D., and Andersen, H.C., 1971, Role of repulsive forces in determining the equilibrium structure of simple liquids, J. Chem. Phys., 54:5237.

Werthein, M.S., 1963, Exact solution of the Percus-Yevick integral equation for hard spheres, Phys. Rev. Letters, 10:321.

DYNAMIC PROCESSES IN LIQUIDS AND SELECTED TOPICS RELATED TO THE DYNAMICS OF IONS AND ELECTRONS IN LIQUIDS

Joseph B. Hubbard

National Bureau of Standards
Center for Chemical Engineering
Thermophysics Division
Gaithersburg, MD 20899 USA

PRELIMINARIES

It has been a rather difficult task formulating these lectures to meet the needs of such a diverse audience. Interests here range over an incredibly broad spectrum -- from general theories of disordered media and systems far from equilibrium to detailed analyses of charge transfer experiments, electrical breakdown phenomena, along with the rich field of physics and chemistry of solvated electrons.

Therefore, I have attempted, perhaps unwisely, to bridge the gap between formal theoretical considerations and phenomenological models in two seminars. The first deals with non-equilibrium statistical mechanics as applied to liquids and, save for a brief introduction to the theory of activated processes in condensed phases, consists of classical, graduate-level, textbook theory. The second lecture is a bit more ambitious, insofar as I feel obliged to derive the limiting law for electrical conductance and discuss classical charge recombination dynamics, as well as introduce some relatively modern ideas on multi-state transport processes. Finally, I hope to have time to talk about such things as Wiener integrals, Kac's formula, the connection between Brownian motion and quantum mechanics, as well as an application of these ideas to the localization of an electron in a disordered system of repulsive scatterers.

Out of necessity and, I admit, personal preference, my style will be mosaic and somewhat impressionistic rather than classical-linear, and so I apologize beforehand for the cavalier derivations, the terse notation, and the huge omissions, such as the expression of the transport coefficients (thermal conductivity, viscosity, etc.) in terms of time-correlation functions. Again, there is so much ground to be covered in so short a time that, with a scrupulous approach, one remains immersed in formalism and never gets to the "real stuff". I have no doubt that at the end of the second lecture most of you will still be asking for the "real stuff", and I have great confidence that your intellectual needs will be at least partially satisfied by the other lectures, as well as through the informal discussions we had during the two weeks of the ASI.

From a strictly logical viewpoint, one should approach liquids from quantum statistical mechanics; however, with a few important exceptions, the collective chemical physics aspects of liquids is described quite well by many-body classical mechanics, and this is the approach taken here.

First, imagine a Euclidian space of 2n dimensions where the axes are molecular coordinates q_i and momenta p_i ($i = 1 \ldots n$). With a classical Hamiltonian **H**, the equations of motion read (Landau and Lifshitz, 1969; Tolman, 1938; McQuarrie, 1976; Hansen and McDonald, 1976)

$$q_j = \frac{\partial H}{\partial p_j} \qquad p_j = - \frac{-\partial H}{\partial q_j} \,. \tag{1}$$

Next, consider a very large number **N** of such systems having identical macroscopic variables (density, volume, energy ...), and define a number density $f(p, q, t)$ in this 2 n dimensional phase space such that f dp dq is the number of systems whose phase points lie in dp dq at time t. Note that f is normalized by

$$\int f(p, q, t)\, dp\, dq = \mathbf{N} \,. \tag{2}$$

The ensemble average of any phase space function ψ can now be defined by the weighted average

$$\langle\psi\rangle = \mathbf{N}^{-1} \int \psi(p, q)\, f(p, q, t)\, dp\, dq \,, \tag{3}$$

and this is usually identified with some thermodynamic function.

There are several interesting points to note about the phase space density function. For one, the time dependence of f is not arbitrary, but is governed strictly by the laws of classical mechanics. Observe that since **N** is conserved, we have the continuity expression (Tolman, 1938)

$$\frac{\partial f}{\partial t} = - \sum_{i=1}^{n} \left[\frac{\partial f}{\partial q}\dot{q}_i + \frac{\partial f}{\partial p}\dot{p}_i\right] , \tag{4}$$

or equivalently

$$\frac{\partial f}{\partial t} = - \sum_{i=1}^{n} \left[\frac{\partial H}{\partial p_i} \frac{\partial f}{\partial q_i} - \frac{\partial H}{\partial q_i} \frac{\partial f}{\partial p_i} \right] , \tag{5}$$

which is the Liouville equation and is at the core of just about all classical statistical mechanics. Equation (4) says that the motion of an ensemble of phase points is identical to that of an "ideal" incompressible 2 n dimensional fluid. In addition, the phase space density, as well as the phase space volume element, is displacement invariant:

$$f(p_0, q_0, t_0) \longrightarrow f(p, q, t)$$
$$\partial p_0 \; \delta q_0 \longrightarrow \delta p \; \delta q \qquad (t_0 \longrightarrow t) \; . \tag{6}$$

Phase space is a useful notion because it provides for the introduction of geometrical/topological concepts in statistical mechanics, and this provides for close linking between intuitive and formal reasoning.

Let's analyze the relation between a so-called phase space correlation function and the conjugate response (susceptibility), which is a measure of the system's reaction to an externally imposed perturbation. Suppose one takes a system at equilibrium with Hamiltonian H_0 and applies a time dependent external force starting at t=0, so that (McQuarrie, 1976; Hansen and McDonald, 1976)

$$H = H_0 + H_1 = H_0 - A(p, q) \; F(t) \; , \tag{7}$$

where A is some function of the coordinates and momenta of the unperturbed system at time t. If we assume linear response plus causality, then we can express the response to F(t) as the phase average

$$\langle B(p, q) \rangle \equiv \langle B(t) \rangle = \int_0^t d\tau \; \phi \; (t - \tau) \; F(t) \; , \tag{8}$$

where ϕ is called the response function. The frequency dependent susceptibility $\chi(\omega)$ is usually defined as the Fourier component of $\phi(t)$ which, via the convolution theorem of Fourier transforms, is related to the response by

$$\langle B(\omega) \rangle = \chi(\omega) \; F(\omega) \; . \tag{9}$$

The problem is how to relate ϕ to A and B. The solution consists in determining the phase space distribution f(p, q, t) to first order in F(t), and then calculating the ensemble average

$$\langle B(t) \rangle = \int dp \; dq \; B(p, q) \; f(p, q, t) \; . \tag{10}$$

First, decompose $f = f_0 + \delta f$ and use the Liouville equation to give, to first order in δf,

$$\frac{\partial\ \delta f}{\partial t} = \left(H_0,\ \delta f\right) - F(t)\left(A,\ f_0\right) . \tag{11}$$

$$\left(A,\ B\right) = \sum_i \left[\frac{\partial A}{\partial q_i}\frac{\partial B}{\partial p_i} - \frac{\partial A}{\partial p_i}\frac{\partial B}{\partial q_i}\right] \qquad \text{(Poisson Bracket)}$$

In terms of the Liouville operator L, defined by

$$L\psi = i\left(H,\ \psi\right) \qquad i = \sqrt{-1} , \tag{12}$$

one can formally write

$$\frac{\partial\ \delta f}{\partial t} = -\ iL_0\ \delta f - F(t)\left(A,\ f_0\right) , \tag{13}$$

where "0" refers to the unperturbed system. This is a first-order differential equation with the solution

$$\delta f(t) = -\int_0^t e^{-i(t\ -\ \tau)L_0}\left(A,\ f_0\right) F(\tau)\ d\tau \tag{14}$$

$$\delta f(0) = 0 .$$

Since we are free to choose $B(t) = 0\ (t < 0)$,

$$\langle B(t)\rangle = -\int_0^t d\tau \int dp\ dq\ B(p,\ q)e^{-i(t\ -\ \tau)L_0}\left(A(\tau),\ f_0(\tau)\right)F(\tau)$$

$$\langle B(t)\rangle = -\int_0^t d\tau \int dp\ dq\ e^{i(t\ -\ \tau)L_0}\ B\left(A(\tau),\ f_0(\tau)\right)F(\tau)$$

$$= -\int_0^t d\tau\ F(\tau) \int dp\ dq \left(A(\tau),\ f_0(\tau)\right)\ B(t\ -\ \tau) , \tag{15}$$

where use has been made of the easily derived fact that $\exp\ (iL_0t)$ is unitary (adjoint equals inverse). It is readily demonstrated that Eq. (15) can be written in the form Eq. (8) with $\phi(t)$ given by

$$\phi(t) = \int dp\ dq \left(f_0,\ A(0)\right)\ B(t)$$

$$= \int dp\ dq\ f_0\left(A(0),\ B(t)\right)$$

$$= \langle\left(A(0),\ B(t)\right)\rangle . \tag{16}$$

Suppose that f_0 is canonical; i.e.,

$$f_0 = \exp\ (-\beta\ H_0)\ /\ \int dp\ dq\ \exp(-\beta\ H_0) . \tag{17}$$

Then

$$\left(f_0, A\right) = \beta f_0 \frac{dA}{dt} = \beta \dot{A} f_0, \tag{18}$$

and

$$\phi(t) = \beta \langle \dot{A}(0) B(t) \rangle . \tag{19}$$

Moreover, since we are averaging over an equilibrium canonical ensemble, $\langle A(0) B(t) \rangle$ is stationary, which means

$$\frac{d}{ds} \langle A(s) B(t + s) \rangle = 0 \qquad \text{(time displacement invariance)} , \tag{20}$$

and so an equivalent form for $\phi(t)$ is

$$\phi(t) = -\beta \langle A(0) \dot{B}(t) \rangle . \tag{21}$$

This says that the response function of the system to a weak but otherwise arbitrary external force is given by a two-point correlation function comprised of the ensemble averaged product of the unperturbed variable when the field is first turned on, and the rate of change of the conjugate response measured at the same time as the desired response function.

Examples of linear response functions (susceptibilities) include the frequency dependent electrical conductivity (the Fourier transform of an equilibrium current autocorrelation function), dielectric susceptibility, which is the transform of a dipole moment autocorrelation function, along with stress, heat flux, and an assortment of velocity correlation functions.

By this time, you may have guessed that a time correlation function is a far more difficult object to calculate than an ordinary equilibrium partition function, and insofar as the latter is virtually impossible to evaluate exactly for realistic liquids, one might well wonder where we go from here.

THE MEMORY FUNCTION

One useful trick is to represent time correlation functions in terms of auxiliary or "memory" functions, and hope that good approximations to memory functions will be easy to obtain.

Suppose we have a normalized autocorrelation function of a phase space function $A(p, q)$ where, as before, A implicitly depends on time through p and q (McQuarrie, 1976; Hansen and McDonald, 1976):

$$\psi(t) = \langle A(0) A(t) \rangle \qquad \psi(0) = 1 . \tag{22}$$

Without loss of generality one can also choose

$$\langle A(t) \rangle = 0 . \tag{23}$$

Since the time evolution of A is given by

$$\frac{dA}{dt} = iLA, \quad \exp(itL)\, \dot{A}(0) = \dot{A}(t) , \tag{24}$$

one can differentiate ψ twice with respect to time and use the easily demonstrated Hermitian (self adjoint) property of L to arrive at

$$\begin{aligned} \frac{d^2\psi}{dt^2} &= - \langle iLA \exp(itL)\, iLA \rangle \\ &= - \langle \dot{A}(0)\, \dot{A}(t) \rangle , \end{aligned} \tag{25}$$

with the initial condition (derivable from stationarity)

$$\langle A(0)\, \dot{A}(0) \rangle = 0 . \tag{26}$$

Laplace transformation (variable s) of Eq. (25) yields

$$\begin{aligned} s^2\, \tilde{\psi}(s) - s &= - \tilde{\gamma}(s) \\ \gamma(t) &= \langle \dot{A}(0)\, \dot{A}(t) \rangle , \end{aligned} \tag{27}$$

or equivalently

$$s\, \tilde{\psi}(s) - 1 = - \left[1 - \frac{\tilde{\gamma}(s)}{s}\right]^{-1} \tilde{\gamma}(s)\, \tilde{\psi}(s) . \tag{28}$$

Upon Laplace transform inversion via the convolution theorem, one obtains

$$\begin{aligned} \frac{d\psi}{dt} &= - \int_0^t d\tau\, K(\tau)\, \psi(t - \tau) \\ \tilde{K}(s) &= - \left[1 - \frac{\tilde{\gamma}(s)}{s}\right]^{-1} \tilde{\gamma}(s) , \end{aligned} \tag{29}$$

which looks impressive, except that the memory function K is expressed in terms of an unknown correlation function which is itself closely related to the function we are trying to determine. Nonetheless, Eq. (29) turns out to be quite useful, partly because non-trivial constraints can be placed on K(t), and partly because simple ansatzes about the form of K appear to be at least qualitatively valid.

Though it is not really essential, at this point I would like to discuss the properties of K using the algebra of projection operators (Hansen and McDonald, 1976; Hynes and Deutch, 1975), which was introduced into statistical mechanics by R. Zwanzig back in 1960. It should be emphasized that the utility of projection operators lies in the approximations suggested, just as Dirac's bra-ket notation in quantum mechanics can be provocative as well as economical. In many instances the price paid for a "streamlined" mathematical derivation is blurred physical insight, and so the extra effort of alternative approaches can be worth while. I hope to have the time to present an example of this later on.

First, note that an alternative form for the "intermediate" memory function $\gamma(t)$ is, from Eq. (24)

$$\gamma(t) = \langle \dot{A}(0) \exp(i\, tL)\dot{A}(0)\rangle \, , \tag{30}$$

and so its Laplace transform is formally

$$\tilde{\gamma}(s) = \langle \dot{A}(0)\, (s - iL)^{-1}\, \dot{A}(0)\rangle \, . \tag{31}$$

Now define a projection operator **P** so that **P** projects any phase space function $G(p, q)$ onto $A(p, q)$:

$$\begin{aligned} \mathbf{P}\, G &= A(p, q) \left[Q^{-1} \exp(-\beta\, H)\right] \int dp'\, dq'\, A(p', q')\, G(p', q') \\ &= f_0\, A(p, q) \int dp'\, dq'\, A(p', q')\, G(p', q') \, , \end{aligned} \tag{32}$$

where f_0 is the equilibrium canonical distribution function. By use of a general operator identity and the Hermitian property of the Liouville operator $\langle a^* Lb\rangle = i\, \langle \dot{a}\, b\rangle = \langle (La)^* b\rangle$, $\tilde{\gamma}(s)$ may be written

$$\begin{aligned} \tilde{\gamma}(s) &= \langle \dot{A} \left[s - i(1 - \mathbf{P})L\right]^{-1} \dot{A}\rangle \\ &\quad - \langle \dot{A}(s - iL)A\rangle\, \langle \dot{A} \left[s - i(1 - \mathbf{P})L\right]^{-1} \dot{A}\rangle \\ \tilde{\gamma}(s) &= \langle \dot{A} \left[s - i(1 - \mathbf{P})L\right]^{-1} \dot{A}\rangle \\ &\quad - s^{-1}\, \tilde{\gamma}(s)\, \langle \dot{A} \left[s - i(1 - \mathbf{P})L\right]^{-1} \dot{A}\rangle \, , \end{aligned} \tag{33}$$

which implies that the transform of $K(t)$ may be represented as

$$\tilde{K}(s) = \langle \dot{A} \left[s - i(1 - \mathbf{P})L\right]^{-1} \dot{A}\rangle \, , \tag{34}$$

and therefore

$$K(t) = \langle \dot{A} \exp \left[it\, (1 - \mathbf{P})L\right] \dot{A}\rangle \, . \tag{35}$$

The operator $(1 - \mathbf{P})$ is usually called the orthogonal projector since $\mathbf{P}(1 - \mathbf{P})$ annihilates a phase-space function. Equation (35) is useful because it expresses the memory function explicitly as a projected correlation function of the phase-space variables.

It turns out that Eq. (35) can be used to derive time correlation function expressions for thermal transport coefficients, but here we will only have time to show how this representation can be manipulated as to yield constraints on K(t).

First, recall that because of the stationarity of an equilibrium time correlation function

$$\langle A(0)\, A(t)\rangle = \langle A(\tau)\, A(t+\tau)\rangle ,$$

$$\frac{d}{d\tau}\langle A(\tau)\, A(t+\tau)\rangle = 0 = \langle \dot{A}(\tau)\, A(t+\tau)\rangle + \langle A(\tau)\, \dot{A}(t+\tau)\rangle , \text{ and}$$

$$\langle \dot{A}(0)\, A(0)\rangle = 0$$

$$\langle \ddot{A}(0)\, \dot{A}(0)\rangle = 0 , \text{ etc.} \tag{36}$$

Now expand the exponential in Eq. (35) and consider just the first two terms:

$$\begin{aligned} K(t) &= \sum_{n=0}^{\infty} \frac{t^n}{n!} \langle (iLA) \left[i(1-P)L \right]^n (iLA)\rangle \\ &= \langle \dot{A}(t)\, \dot{A}(t)\rangle + t\langle (iLA) \left[i(1-P)L \right] iLA\rangle + 0(t^2) \\ &= \langle \dot{A}(0)\, \dot{A}(0)\rangle + t \left[\langle (iLA)\, (iL)\, (iLA)\rangle \right. \\ &\qquad\qquad \left. - \langle (iLA)\, (iPL)\, (iLA)\rangle \right] + 0(t^2) \\ &= \langle \dot{A}\, \dot{A}\rangle + t \left[\langle \dot{A}(t)\, \ddot{A}(t)\rangle - \langle \dot{A}(t)\, A(t)\rangle \langle A(0)\, \ddot{A}(t)\rangle \right] + 0(t^2) \\ &= \langle \dot{A}\, \dot{A}\rangle + \text{zero} + 0(t^2) , \end{aligned} \tag{37}$$

and you see that the term linear in t vanishes identically. This procedure can be readily extended to show that all odd powers of t vanish, and so K(t) must be an even (therefore real) function of time. Using the Schwartz inequality, $\int f^* g \leq [\int f^* f]^{1/2}\, [\int g^* g]^{1/2}$, it is also easy to prove that the memory function is bounded by

$$- \langle \dot{A}(0)\, \dot{A}(0)\rangle \leq K(t) \leq \langle \dot{A}(0)\, \dot{A}(0)\rangle . \tag{38}$$

Such conditions can be quite useful in the interpretation and scrutiny of molecular dynamics computer simulations as well as in the construction of theories.

Rigor aside, the most useful aspect of the memory function approach is that non-trivial, physically realistic results emerge from absurdly simple suppositions about K(t). For example, go back to Eq. (29) and interpret ψ as the normalized velocity autocorrelation function of a Brownian particle:

$$\frac{d\psi}{dt} = - \int_0^t d\tau\, K(\tau)\, \psi(t-\tau) . \tag{39}$$

Choose an environmental "collision" model with a flat power spectral density, which means $K(\tau) = \gamma\delta(t)$, where $\delta(t)$ is a Dirac delta function, and ψ decays as a simple exponential. Even though this choice of K violates the rigorous criteria just established, it turns out that a very useful disordered medium dynamics model (constant friction Langevin), gives precisely this result for single-particle motion. If we had taken K(t) to be a simple exponential (again, this is not strictly legitimate), then it is easy to show by way of the convolution theorem of Laplace transforms that ψ decays as the sum of two oscillating exponentials, and this does a remarkably good job of describing the intermediate time structure of the velocity correlation functions "observed" in molecular dynamics simulations of simple liquids.

LANGEVIN AND FOKKER-PLANCK DYNAMICS

The memory function approach is formally exact and mathematically convenient, and this is because it completely avoids detailed consideration of the interactions. A restricted but more direct approach is to take a guess at an equation of motion for, say, a tagged molecule or particle in an isotropic medium, where the collisional effects are prescribed by some stochastic model; this is the spirit of Langevin dynamics.

Consider the following expression for the velocity **u** of such a particle (McQuarrie, 1976; Hansen and McDonald, 1976)

$$\frac{d\mathbf{u}}{dt} = - \xi \mathbf{u} + \mathbf{F}(t) , \tag{40}$$

where ξ is some constant friction coefficient and **F**(t) is a random function of time endowed with the following properties:

$$\langle \mathbf{F}(t) \rangle = 0 , \qquad \langle \mathbf{u}(t) \cdot \mathbf{F}(t) \rangle = 0 , \text{ and}$$

$$\langle \mathbf{F}(t) \cdot \mathbf{F}(s) \rangle = \phi(|t - s|) . \tag{41}$$

The ensemble average of the random force therefore vanishes, it is statistically independent of the velocity, and this force is time-stationary. In addition, ϕ is assumed to be appreciable only in a small region around $|t - s| = 0$, so that the dynamic influence of individual "collisions" are short-lived.

What is meant by solving an equation such as (40) is that one obtains the probability that the solution is **u**(t) at time t, given that, with probability one, the solution was **u**(0) at t = 0. If we call this probability density $\psi(\mathbf{u}, t; \mathbf{u}_0)$, then the initial condition is

$$\psi \longrightarrow \delta(\mathbf{u} - \mathbf{u}_0) \text{ as } t \longrightarrow 0 . \tag{42}$$

Additionally, as $t \longrightarrow \infty$, ψ must become Maxwellian regardless of the choice of u_0, and so

$$\psi \longrightarrow \left(\frac{m}{2\pi k_B T}\right)^{3/2} \exp\left(\frac{-mu^2}{2k_B T}\right) \text{ as } t \longrightarrow \infty , \tag{43}$$

where m is the particle mass.

Equation (40) is readily solved to give

$$u(t) = u_0 e^{-\xi t} + e^{-\xi t} \int_0^t dse^{+\xi s} F(s)\, ds , \tag{44}$$

and since $\langle F(t)\rangle = 0$,

$$\langle u(t)\rangle = u_0 e^{-\xi t} . \tag{45}$$

The mean square velocity is also easily obtained:

$$\langle u^2\rangle = u_0^2 e^{-2\xi t} + e^{-2\xi t} \int_0^t d\tau \int_0^t dse^{+\xi(\tau+s)} \langle F(\tau) \cdot F(s)\rangle$$

$$= u_0^2 e^{-2\xi t} + \frac{\gamma}{2\xi}(1 - e^{-2\xi t}) \text{ , and}$$

$$\gamma = \int_{-\infty}^{+\infty} dx\ \phi(x) = \frac{6\ \xi\ k_B T}{m} , \tag{46}$$

where use has been made of the rapid decay of the random force correlation function, and equipartition has been invoked to obtain the constant γ. I won't go into detail, but it is a straightforward enough tedious matter to show that the quantity $[u - u_0 \exp(-\xi t)]$ has a Gaussian distribution, i.e.,

$$\psi(u,\ t;\ u_0) = \left[\frac{\beta\, m}{2\pi\alpha(t)}\right]^{3/2} \exp\left[\frac{-\beta\, m\left(u(t) - u_0 e^{-\xi t}\right)^2}{2\alpha(t)}\right] \text{, where}$$

$$\alpha(t) = 1 - \exp(-2\xi t),\ \beta = (k_B T)^{-1} , \tag{47}$$

and so you see that ψ indeed evolves to the Maxwell form regardless of the initial velocity. Also, it is not too difficult to show that the displacement $\left[r - r_0 - \xi^{-1} u_0 \left(1 - \exp(-2\xi t)\right)\right]$ is Gaussianly distributed, which means that the displacement probability density is given by Eq. (47) with $\alpha(t) = \xi^{-2}\left(2\xi t - 3 + 4\exp(-\xi t) - \exp(-2\xi t)\right)$ and $\left[u(t) - u_0 \exp(-\xi t)\right]^2$ replaced by $\xi^2 \left(r - r_0 - \xi^{-1} u_0[1 - \exp(-\xi t)]\right)^2$. This immediately gives the Einstein limits

$$\langle(r - r_0)^2\rangle \longrightarrow u_0^2\, t^2 \qquad (t \longrightarrow 0)$$

$$\langle(\mathbf{r} - \mathbf{r}_0)^2\rangle = \frac{6k_BT}{m\xi} t = 6\, \dot{D}t \qquad (t \longrightarrow \infty) . \tag{48}$$

It happens that the connection between random velocity and random displacement is a special case of a more general property of random variables having a Gaussian distribution: if A(t) is a Gaussian random process (GRP), then so is $\int dt\, \Gamma(t)\, A(t)$, where Γ is some nonrandom square integrable function (Fox, 1978).

The "constant ξ" Langevin equation is restricted to random forces with a very short (δ in fact) correlation time, and so it is legitimate to inquire about the possibility of a more general Langevin theory. The equation (Hansen and McDonald, 1976; Hynes and Deutch, 1975)

$$\frac{d\mathbf{u}(t)}{dt} = - \int_0^t d\tau\, K(t - \tau)\, \mathbf{u}(\tau) + \mathbf{F}(t) \tag{49}$$

is capable of providing an exact dynamical description, which should not be surprising, considering the similarity with the memory function formalism. Here the random force and memory kernel K are related by

$$\langle\mathbf{F}(t - \tau) \cdot \mathbf{F}(\tau)\rangle = \frac{3k_BT}{m} K(t) , \tag{50}$$

while $\mathbf{F}(t)$ is again uncorrelated with the Langevin velocity

$$\langle\mathbf{F}(t) \cdot \mathbf{u}(t)\rangle = 0 . \tag{51}$$

In some cases working with a Langevin equation can prove to be conceptually vague or crude, and it may be preferable to deal directly with the conditional probability density $\psi(\mathbf{u}, t; \mathbf{u}_0)$. According to Eq. (47), ψ is the solution to some sort of diffusion equation in velocity space, and it is of interest to obtain this equation in a more direct manner than by computing the moments via the Langevin equation, and thereby conclude that ψ must be Gaussian. In fact, consider

$$\psi(\mathbf{u}, t + \Delta t) = \int \psi(\mathbf{u} - \Delta\mathbf{u}, t)\, K(\mathbf{u} - \Delta\mathbf{u}; \Delta\mathbf{u}) d\, \Delta\mathbf{u} \text{ , and}$$

$$\Delta\mathbf{u} = \mathbf{u} - \mathbf{u}_0 , \tag{52}$$

where K is now defined as the transition probability from state ($\mathbf{u} - \Delta\mathbf{u}$) to $\mathbf{u}$ during the time interval Δt. This is an example of a "Chapman-Kolmogorov" equation, which is at the heart of the theory of Markov processes. Expand the LHS in a Taylor series about $\Delta t = 0$, do the same for the RHS about $\Delta\mathbf{u} = 0$, and retain those terms linear in Δt and quadratic in $\Delta\mathbf{u}$:

$$\psi(\mathbf{u}, t) + \frac{\partial\psi}{\partial t} \Delta t = \int\Bigg[\psi(\mathbf{u}, t) - \frac{\partial\psi}{\partial u_j} \Delta u_j$$

$$+ \frac{1}{2}\left[\frac{\partial^2 \psi}{\partial u_j^2}\right] \Delta u_j^{\,2} + \frac{\partial^2 \psi}{\partial u_i \, \partial u_j} \Delta u_i \, \Delta u_j \Bigg] \cdot \Bigg[K(u;\, \Delta u)$$

$$- \frac{\partial K}{\partial u_j} \Delta u_j + \frac{1}{2} \frac{\partial^2 K}{\partial u_j^2} \Delta u_j^{\,2} + \frac{\partial^2 K}{\partial u_i \, \partial u_j} \Delta u_i \, \Delta u_j \Bigg] d\Delta u \; . \tag{53}$$

Here indices labelling the Cartesian components are summed over with the restriction $i < j$. This equation can be written in a more compact form

$$\frac{\partial \psi}{\partial t} \Delta t = - \frac{\partial}{\partial u_j} \left[\psi \langle \Delta u_j \rangle\right] + \frac{1}{2} \frac{\partial^2}{\partial u_j^2} \left[\psi \langle \Delta u_j^{\,2} \rangle\right]$$

$$+ \frac{\partial^2}{\partial u_i \, \partial u_j} \left[\psi \langle \Delta u_i \, \Delta u_j \rangle\right] , \tag{54}$$

where $\langle \; \rangle$ represents the velocity ensemble averaging operator

$$\int K(u;\, \Delta u) \; d \, \Delta \, u \; . \tag{55}$$

Equation (54) is a slightly generalized form of the Fokker-Planck (FP) equation which reverts to the more familiar velocity diffusion

$$\frac{\partial \psi}{\partial t} = \xi \, \nabla_u \cdot (\psi \, u) + \frac{\xi \, k_B T}{m} \nabla_u^{\,2} \, \psi \, , \tag{56}$$

under the assumptions of spatial isotropy, energy equipartition, and $\Delta t \longrightarrow 0$; i.e.,

$$\langle \Delta u_j \rangle = - \xi \, u_j \, \Delta t + \ldots$$

$$\langle \Delta u_j^{\,2} \rangle = \frac{2 \xi \, k_B T}{m} \Delta t + \ldots$$

$$\langle \Delta u_i \, \Delta u_j \rangle = 0 \qquad (i \neq j) \; . \tag{57}$$

The solution to Eq. (56) with the initial condition $\psi(t = 0) = \delta(u - u_0)$ is given by Eq. (47).

Needless to say, an FP equation can incorporate variables other than a Langevin velocity; for example, Kramers' equation is a bivariate FP which involves both spatial coordinates and velocity. In a slightly more abstract sense, one might consider the time evolution of a probability density $P(y, t)$ where, for simplicity, y is taken to be a continuous scalar on the real line. It turns out that a "legitimate" FP equation can be written as (Fox, 1978; Van Kampen, 1981)

$$\frac{\partial P(y, t)}{\partial t} = - \frac{\partial}{\partial y} \left[A(y) \; P\right] + \frac{1}{2} \frac{\partial^2}{\partial y^2} \left[\, B(y) \; P\right] , \tag{58}$$

where A and B are any two differentiable real functions with B always positive. Equation (58) can also be expressed as the derivative of a probability flux:

$$\frac{\partial P(y, t)}{\partial t} = - \frac{\partial}{\partial y} J(y, t) \text{ and}$$

$$J(y, t) = A(y)\, P - \frac{1}{2} \frac{\partial}{\partial y} \left[B(y)\, P\right] , \tag{59}$$

which is convenient for formulating boundary conditions.

Returning to the FP velocity Eq. (56); note that the coefficients must be obtained from the appropriate Langevin equation. This, of course, carries over to the more general Eq. (59), and in fact, the FP formulation is mathematically equivalent to the Langevin approach.

One final comment on Langevin equations: consider the form

$$\frac{dy(t)}{dt} = A(y) + F(t) , \tag{60}$$

where now A is some nonlinear function of y and F has the "strict" or white-noise Langevin property

$$\langle F(t)\rangle = 0 \text{ and } \langle F(t)\, F(t')\rangle = \gamma\, \delta(t - t') \text{ (Gaussian white noise).} \tag{61}$$

It is not difficult to show that the corresponding FP equation is

$$\frac{\partial P(y, t)}{\partial t} = - \frac{\partial}{\partial y} \left[A(y)\, P\right] + \frac{\gamma}{2} \frac{\partial^2 P}{\partial y^2} . \tag{62}$$

However, this simple correspondence no longer holds for

$$\frac{d\, y(t)}{dt} = A(y) + D(y)\, F(t) , \tag{63}$$

and this is because of the pathological nature of F(t): Gaussian white noise looks like a random string of Dirac delta functions but D(y) is smoother, and Eq. (63) has nothing to say about whether, for a particular value of t, we want D[y(t)] before, after, or during a random kick via F(t). In the jargon of mode-coupling, we don't know if D(y) F(t) is convective, dissipative, or some combination of the two. This is not to say that nonlinear Langevin equations cannot be used, but that clear thinking about the underlying physics is absolutely essential.

Before going on to other topics, I should present the phase-space analogue of the FP equation (constant external field) which can be derived in the same way as the velocity space FP (Wax, 1954):

$$\frac{\partial \psi}{\partial t} + (\mathbf{u} \cdot \nabla_r)\, \psi + (F_{ext} \cdot \nabla_u)\, \psi = \xi\, \nabla_u \cdot (\psi\, \mathbf{u}) + \frac{\xi\, k_B T}{m} \nabla_u^2\, \psi . \tag{64}$$

This is known as Chandrasekhar's equation. For those of you familiar with the kinetic theory of gases, note the identity of the LHS with the streaming part of the Liouville equation for the reduced single particle distribution function, whereas the RHS can be viewed as the Brownian dynamics analogue of the collision terms in the Boltzmann equation.

ACTIVATED PROCESSES

This first lecture will be concluded with a brief discussion of theories of activated processes in condensed phases, which will hopefully serve as an introduction to the topics explored by the other lecturers. The word "activated" used here refers to some process, usually chosen to be stochastic, by which a particle or system acquires sufficient energy to make a transition from one state to other states. In order to accomplish this, the system must pass over, or tunnel through, a set of potential energy barriers in some configuration space, and the rate at which this occurs is usually calculated from a phase-space density flux evaluated at, or integrated over, some judiciously chosen locations on the energy surface. In condensed systems, a key role is played by the rate at which the particle exchanges energy with its surroundings (bath), and this is usually characterized by a scalar coupling constant γ. If tunneling is ignored, then for low friction or small γ, the rate determining step is the activation of the particle to an energy higher than the barrier (one-dimensional reaction coordinate). For high damping (large friction) the rate is controlled by collisions with the bath when the particle is near the barrier maximum. Somewhere in between large and small γ, the reaction rate should be maximal: this is the regime of transition state theory (TS) which says that the rate is identical with the equilibrium probability flux over the barrier. The well-known TS theory will be considered in detail by other lecturers. The point is that for TS to be a good approximation, γ must be large enough so that local thermal equilibrium holds, yet small enough so that a particle, having just crossed the barrier, will not immediately experience another collision and recross. Of course, it is impossible for both of these conditions to be valid simultaneously and so TS actually gives a maximum reaction velocity. One question is "How good (or bad) is TS, and over what range of γ's does TS hold?"

A novel approach (proposed by Skinner and Wolynes, 1978) for non-diffusive particle motion starts with an "irreversible" Liouville equation for a reduced phase space distribution function (single particle)

$$\frac{\partial f(p, q, t)}{\partial t} = - L^* f(p, q, t) \text{ and } \int f \, dp \, dq = 1 , \tag{65}$$

where

$$f(p, q, t) = f_0 + \delta f(p, q, t) \text{ and}$$

$$f_0 = Q^{-1} \exp \left[- \beta \left(\frac{p^2}{2m} + V(q) \right) \right] . \tag{66}$$

Here one assumes thermal equilibrium in each of two wells (a and b), save for a fraction of particles δf not equilibrated at $t = 0$. Let the fraction in well "a" be given by the restricted integral

$$\delta f_a = \int_a dp \, dq \, f(p, q, t) , \tag{67}$$

and define an adjoint to L^* which operates through the equilibrium distribution on a phase function:

$$f_0 \, LA(p, q) = L^* f_0 \, A(p, q) . \tag{68}$$

Now a relaxation or escape rate can be defined as the $s \longrightarrow 0$ limit of the Laplace transform of the restricted autocorrelation of δf:

$$R_E = \lim_{s \to 0} \tilde{C}_a(s)$$

$$C_a(t) = \langle \delta f_{a,0}^{\,2} \rangle^{-1} \left[\langle \delta f_a (0) \exp \left[- Lt \right] \delta f_a(0) \rangle \right]$$

$$\tilde{C}_a(s) = \langle \delta f_{a,0}^{\,2} \rangle^{-1} \left[\langle \delta f (0) \left[s + L \right]^{-1} \delta f_a(0) \rangle \right] , \tag{69}$$

where $\langle \; \rangle$ denotes a full phase space integration weighted with the equilibrium distribution. Next, split the dynamical operator L into a convective (streaming) part and a collision part, i.e.,

$$L = L_{st} + \gamma L_{col}$$

$$L_{st} = u \frac{\partial}{\partial q} - m^{-1} \frac{\partial V(q)}{\partial q} \frac{\partial}{\partial u} , \quad (p = mu) ,$$

and perform a perturbation expansion in γ. The result is an expansion of the resolvent as

$$\left[s + L \right]^{-1} = \left[s + L_{st} \right]^{-1} + \sum_{m=1}^{\infty} \left[-\gamma \left[s + L_{st} \right]^{-1} L_{col} \right]^m \left[s + L_{st} \right]^{-1} . \tag{71}$$

A convenient choice for L_{col} is one that simply randomizes the velocity (fixed position) of a particle after each collision

$$L_{col} = 1 - \alpha \int du \exp \left[- \frac{\beta \, mu^2}{2} \right] \text{ and}$$

$$\alpha = \left(\frac{m\beta}{2\pi}\right)^{1/2} . \tag{72}$$

Using this "sparse but strong collision" formulation, Skinner and Wolynes (1978) were able to evaluate the first terms in the resolvent expansion of Eq. (69) (for a single square well, the entire series can be summed), and thereby calculate reaction rates as a function of γ and a few potential energy parameters. Their most important conclusion is that impacts near the barrier maximum give a rate significantly lower than that predicted by TS, i.e., reverse crossing is an important effect in condensed systems.

Kramers' approach to the barrier crossing problem was to begin with a (q, p)-FP equation, which corresponds to choosing the "dense but weak collision" operator (Kramers, 1940):

$$L_{col} = \frac{\partial}{\partial u}\left[D_u \frac{\partial}{\partial u} + \gamma u\right], \text{ where}$$

$$D_u = \gamma/m\beta \text{ and } u = m^{-1}p . \tag{73}$$

For the high friction case (γ large) one assumes local thermal equilibrium, which means that the phase-space distribution has the form

$$f(p, q, t) = f(q, t) \exp\left[-\frac{\beta p^2}{2m}\right], \tag{74}$$

and f satisfies the Smoluchowski equation

$$\frac{\partial f}{\partial t} = \gamma^{-1} \frac{\partial}{\partial q}\left[\frac{dV(q)}{dq} + \frac{1}{\beta}\frac{\partial}{\partial q}\right] f . \tag{75}$$

The associated stationary flux is

$$J_0 = -\gamma^{-1}\left[\frac{dV(q)}{dq} + \frac{1}{\beta}\frac{\partial}{\partial q}\right] f , \tag{76}$$

or equivalently

$$J_0 = -(\beta\gamma)^{-1}\left[\exp\left(-\beta V(q)\right)\frac{\partial}{\partial q}\left(f \exp\left[\beta V(q)\right]\right)\right]$$

$$J_0 = \frac{(\beta\gamma)^{-1} f(q) \exp\left(\beta V(q)\right)\Big|_{q_a}^{q_b}}{\int_{q_a}^{q_b} dq \exp\left((\beta V(q)\right)}, \tag{77}$$

where q_a and q_b are chosen as the minima in the two wells (q_b is taken as large if the RH portion of the barrier has no minimum). Kramers' approach was to ignore back scattering from the potential to the right of q_b and assume harmonic oscillators forms for V at q_a and the barrier maximum at q_c, i.e., $V(q_a) = 1/2(2\pi\omega_a)^2(q - q_a)^2$ (ω_a = frequency). This produces the escape rate

$$R_E = \frac{2\pi\omega_a\omega_c}{\gamma}\exp(-\beta\, h_c) \qquad (\gamma \text{ large}) , \tag{78}$$

where h_c is the potential energy difference between q_c and q_a.

For small γ, the motion should resemble that of a weakly damped oscillator, and so it is the particle energy, rather than position, that changes in small increments. In this instance, it can be shown a diffusion equation in "energy/action space" is a good description, i.e.,

$$\frac{\partial \bar{f}(E,\, t)}{\partial t} = \gamma \frac{\partial}{\partial I}\left(I\,\bar{f} + \beta^{-1}\, I\, \frac{\partial \bar{f}}{\partial E}\right)$$

$$I(E) = \int p\, dq \qquad \bar{f} = \int f\, dI/I \tag{79}$$

or

$$\frac{\partial \bar{f}}{\partial t} = \gamma\left[1 + \left(\frac{dE}{dI}\right)\frac{\partial}{\partial E}\right]\left[\bar{f} + \beta^{-1}\frac{\partial \bar{f}}{\partial E}\right] . \tag{80}$$

$I(E)$ is the action integral taken over one period (the area inside a phase-space curve of constant energy) and $\bar{f}$ is a coarse-grained f in "action space". The steady flux is now

$$J_0 = \frac{\gamma\beta^{-1}\, \bar{f}(E)\, \exp(\beta\, E)\Big|_{E_a}^{E_b}}{\displaystyle\int_{E_a}^{E_b} dE\, I^{-1}\, \exp(BE)} . \tag{81}$$

The energy integrals here are a bit trickier to deal with than q-integrals, but an important point can be gleaned directly from the fluxes: for large γ, the escape rate varies as γ^{-1}, while for small γ, R_E is proportional to γ, and this implies that R_E has a maximum at some intermediate γ. With some simplifying harmonic assumptions about $V(q)$, Kramers calculates the low friction escape rate

$$R_E = \gamma\beta\, h_c\, \exp(-\beta\, h_c) \qquad (\gamma \text{ small}) , \tag{82}$$

so the Arrhenius factor appears in both extremes. While a great deal of effort has gone into improving and generalizing Kramers' theory, his original work continues to occupy a central position in statistical physics.

At the risk of sounding like a displaced medieval scholar, I begin the second lecture with a discussion (mercifully compact) of the Debye-Falkenhagen-Onsager (DFO) theory of electrolytic conductance (Falkenhagen, 1934; Fuoss and Accascina, 1959). Although vestiges of this once famous theory lie strewn about in statistical physics, it is not generally considered an appropriate research topic for a modern theorist, and this is due mainly to its asymptotic and precocious nature -- though exact in the low concentration limit, it has proven exceedingly difficult to extend DFO to higher concentrations or to weak electrolytes. Moreover, an exact derivation of even the first term (limiting law) in the language of modern statistical mechanics (van Hove time correlations, linear response theory) does not yet exist, though Professor Friedman and his group may soon rectify this situation. The reason for this "gap" is that DFO incorporates many-body correlations and self-consistent dynamics from the outset, while a full microscopic theory requires a formal introduction of these essentials in subtle ways (Resibois, 1968).

Define a time-dependent two point correlation function g_{ab} so that $n_a\, g_{ab}(r_a, r_b)\, dV_a$ is the number of ions of species a in a volume dV_a about the point r_a, given that, with probability one, there is an ion of type b located at r_b. The ions a and b may be of the same or different species. It should be obvious that

$$g_{ab}(r_a, r_b) = g_{ba}(r_b, r_a) \text{ and} \tag{83}$$

$$g_{ab} \longrightarrow 1 \quad \text{as} \quad |r_b - r_a| \longrightarrow \infty \,.$$

The correlation functions satisfy a continuity equation in the configuration space of two particles (six dimensions) [1].

$$\frac{\partial g_{ab}}{\partial t} + \mathrm{div}_a\, J_a + \mathrm{div}_b\, J_b = 0 \,, \tag{84}$$

where $J_a(b)$ are the probability fluxes for the a(b) particles. The flux J_a is given by:

$$J_a = \omega_a^o\, z_a\, g_{ab}\, (E - \nabla_a\, \phi_b) - \beta^{-1}\, \omega_a^o\, \nabla_a\, g_{ab}$$

$$(\text{electron charge} \equiv 1\ ,\ z = \text{valence}) \,, \tag{85}$$

---

[1] In order to obtain the exact limiting law for any ionic transport property, it is sufficient to consider a two-point probability flux conservation law along with the assumption of Brownian motion for the ions.

with J_b the same with a and b interchanged, where ω^o is the single ion mobility (infinite dilution) [2], $\mathbf{E}$ is the external field, and ϕ_b is the potential due to a "b" ion located at $\mathbf{r}_b$. If the solution is so dilute that g differs only slightly from unity, then to first order in $h = (g - 1)$ we can write Poisson's equation as

$$\nabla_a^2 \phi_b(\mathbf{r}_a, \mathbf{r}_b) = -\frac{4\pi}{\varepsilon_0}\left[\sum_c a_c n_c h_{cb} + z_b \delta(\mathbf{r}_a - \mathbf{r}_b)\right] \text{ with}$$

$$h_{ab} = g_{ab} - 1 , \qquad \sum_c a_c n_c = 0 , \text{ and} \tag{86}$$

$$\varepsilon_0 = \text{dielectric constant} ,$$

where the summation constraint is electrical neutrality. The associated probability flux is

$$J_a = \omega_a^o \left[-\beta^{-1} \nabla_a h_{ab} + z_a (1 + h_{ab}) \mathbf{E} - z_a \nabla_a \phi_b\right] . \tag{87}$$

Choose $\mathbf{E}$ to be constant, use the symmetry $\nabla_a h_{ab} = - \nabla_b h_{ab}$, and substitute Eqs. (86) and (87) into the continuity condition Eq. (84) with $\partial h_{ab}/\partial t = 0$ to arrive at

$$\beta^{-1} (\omega_a^o + \omega_b^o) \nabla^2 h_{ab}(r) + z_a \omega_a^o \nabla^2 \phi_b(r) + z_b \omega_b^o \nabla^2 \phi_a(-r)$$

$$= (z_a \omega_a^o - z_b \omega_b^o) \mathbf{E} \cdot \nabla h_{ab}(r) , \tag{88}$$

where derivatives are with respect to $\mathbf{r} = \mathbf{r}_b - \mathbf{r}_a$. Set $\mathbf{E} = 0$ and since ϕ^o must be an even function of r, and since h and $\phi \longrightarrow 0$ as $r \longrightarrow \infty$, one recovers

$$\beta^{-1} (\omega_a^o + \omega_b^o) h_{ab}^o + (z_a \omega_a^o \phi_b^o + z_b \omega_b^o \phi_a^o) = 0 . \tag{89}$$

By requiring solutions of Eq. (89) of the form

$$h_{ab}^o = z_a z_b h^o , \qquad \phi_a^o = -\beta^{-1} z_a h^o , \tag{90}$$

we see that Eq. (89) becomes an identity and Eq. (86) reverts back to the equilibrium-linearized Poisson Boltzmann equation, whose solution may be written in terms of h^o as

$$h^o(r) = -\frac{\beta}{\varepsilon_0} \frac{e^{-kr}}{r} , \quad \kappa = \text{Debye Kappa}, \ \kappa^2 = \frac{4\pi\beta}{\varepsilon_0} \sum_i z_i^2 n_i . \tag{91}$$

---

[2] The ion mobility at infinite dilution (ω^o) is a phenomenological parameter in this theory. Although there have been several recent attempts to calculate ω^o from a fundamental consideration of ion-solvent dynamics, these treatments do not possess the degree of rigor (asymptotically exact) associated with the DFO theory.

Since $\mathbf{E}$ is assumed to be small, one writes

$$\phi_a = \phi_a^o + \delta\phi_a \qquad h_{ab} = h_{ab}^o + \delta h_{ab} , \tag{92}$$

where the perturbations are now odd functions of r. Focus on a binary electrolyte, so we need consider only one function $\delta h_{12}(\mathbf{r}) = -\delta h_{21}(\mathbf{r})$, and Poisson's Eq. (86) becomes

$$\nabla^2 \delta\phi_2(\mathbf{r}) = -\frac{4\pi z_1}{\varepsilon_0} n_1 \delta h_{12}(\mathbf{r}) , \text{ with}$$

$$\delta\phi_1(\mathbf{r}) = \delta\phi_2(\mathbf{r}) . \tag{93}$$

To lowest order in the perturbations, Eq. (88) then becomes

$$\beta^{-1}(\omega_1^o + \omega_2^o) \nabla^2 \delta h_{12}(\mathbf{r}) + (z_1 \omega_1^o - z_2 \omega_2^o) \nabla^2 \delta\phi_2(\mathbf{r})$$
$$= (z_1 \omega_1^o - z_2 \omega_2^o) z_1 z_2 \mathbf{E} \cdot \nabla h^o(\mathbf{r}) , \tag{94}$$

which can be solved by Fourier transforms. Substitute Eq. (93) into (94), noting that

$$h^o(k) = -\left[\frac{\beta}{\varepsilon_0}\right] \frac{4\pi}{k^2 + \kappa^2} \text{ (wavevector } \mathbf{k}) , \tag{95}$$

and it is not hard to show that the Fourier components of $\delta\phi$ satisfy

$$\delta\phi_2(\mathbf{k}) = 4\pi z_1 z_2 \beta \kappa^2 q \varepsilon_0^{-1} \left[\frac{i \mathbf{k} \cdot \mathbf{E}}{k^2(k^2 + \kappa^2)(k^2 + q \kappa^2)}\right] , \tag{96}$$

where

$$q = \frac{\omega_1^0 z_1 - \omega_2^0 z_2}{(z_1 - z_2)(\omega_1^0 + \omega_2^0)} , \quad 0 < q < 1 . \tag{97}$$

What we want is the perturbed electric field acting on ion 2, which is given by

$$\delta\mathbf{E}_2(\mathbf{r})\Big|_{r=0} = -\nabla \delta\phi_2(\mathbf{r})\Big|_{r=0}$$
$$\delta\mathbf{E}_2(\mathbf{k}) = -i \mathbf{k} \delta\phi_2(\mathbf{k}) . \tag{98}$$

The inverse Fourier transform of $\delta\mathbf{E}_2(\mathbf{k})$ with $r = 0$ then gives us

$$\delta\mathbf{E}_2(0) = \frac{i}{(2\pi)^3} \int \mathbf{k} \delta\phi_2(\mathbf{k}) d^3 k , \tag{99}$$

and so from Eq. (96) we require the integral

$$I = \frac{1}{(2\pi)^3} \int \frac{\mathbf{k}(\mathbf{k} \cdot \mathbf{E}) d^3 k}{k^2(k^2 + \kappa^2)(k^2 + q \kappa^2)} . \tag{100}$$

The angular average replaces $\mathbf{k}(\mathbf{k} \cdot \mathbf{E})$ by $(1/3)\,(k^2\,\mathbf{E})$, and the integral over $|\mathbf{k}|$ is performed via the calculus of residues (poles at $i\kappa$, $iq^{1/2}\kappa$):

$$I = \frac{1}{12\pi(1 + q^{1/2})\kappa}\,\mathbf{E}\ . \tag{101}$$

The total electric field acting on ion 2 is then

$$\mathbf{E} + \delta\mathbf{E}_2(0) = \left[1 - \frac{|z_1\ z_2|\ q\ \beta\ \kappa}{3\varepsilon_0(1 + q^{1/2})}\right]\mathbf{E}\ , \tag{102}$$

which is called the ion-atmosphere relaxation effect. Note that an identical result is obtained for $\delta\mathbf{E}_1(0)$, regardless of the valences. Since $\omega\, z\mathbf{E}$ is the velocity of an ion of mobility ω in the external field, Eq. (102) can be interpreted as an effective reduction in mobility:

$$\delta\omega = -\frac{\omega^0\ |z_1\ z_2|\ q\ \beta\ \kappa}{3\varepsilon_0(1 + q^{1/2})}\ ,\quad \kappa^2 = \frac{4\pi\beta}{\varepsilon_0}\sum_i z_i^2\, n_i\ , \tag{103}$$

where $\delta\omega$ is proportional to the square root of concentration through the κ-factor.

The simplest way of calculating the electrophoretic contribution to $\delta\omega$ is to focus on a particular ion (location $r = 0$) in solution together with its atmosphere, the electrical charge density of which is

$$\delta\rho = \Sigma\, z_a\ \delta n_a\ ,\quad \delta n_a = n_a - n_a^0\ , \tag{104}$$

where δn_a is the difference between the local and bulk density of ion species a. In the presence of the external field $\mathbf{E}$, this charge density experiences a force $f = \mathbf{E}\delta\rho$ which induces motion in the surrounding liquid, and this motion is then transmitted to the central ion via hydrodynamic (viscous) interactions. The central ion therefore migrates in a counter-current produced by the interaction of its atmosphere with the external field. δn_a is related to the local potential ϕ:

$$\delta n_a = -\,z_a\ \phi\ n_a^0\beta\ ,\quad \phi = \frac{z_b e^{-\kappa r}}{r}\ , \tag{105}$$

where the exponential in the Boltzmann factor has been linearized and we have assumed spherical symmetry since the field is weak (these arguments can be rigorously justified). The associated atmospheric charge density may be written

$$\delta\rho = \frac{z_b}{4\pi}\,\kappa^2\,\frac{e^{-\kappa r}}{r}\ , \tag{106}$$

and with the further assumption (easily justified) of slow, steady, incompressible hydrodynamic flow, we can formulate an augmented Navier-Stokes equation

$$\eta \nabla^2 \mathbf{v} - \nabla p + \mathbf{E}\, \delta\rho = 0 , \tag{107}$$

where p is the hydrodynamic pressure and η is the solvent shear viscosity. Taking Fourier components in Eq. (107), one arrives at

$$- \eta\, k^2 \mathbf{v}(\mathbf{k}) - i\mathbf{k}\, p(\mathbf{k}) + \mathbf{E}\, \delta\rho(\mathbf{k}) = 0 , \tag{108}$$

with the solution

$$\mathbf{v}(\mathbf{k}) = \frac{\delta\rho(\mathbf{k})}{\eta} \left[\frac{k^2\, \mathbf{E} - \mathbf{k}\,(\mathbf{k} \cdot \mathbf{E})}{k^4}\right] \text{ and}$$

$$\mathbf{k} \cdot \mathbf{v}(\mathbf{k}) = 0 \quad \text{(incompressibility)} , \tag{109}$$

with $\delta\rho(\mathbf{k})$ given by

$$\delta\rho(\mathbf{k}) = \frac{-z_b}{\kappa^{-2} k^2 + 1} . \tag{110}$$

What we require is the velocity at the origin $\mathbf{v}(0)$, which is given by the inverse Fourier transform of Eq. (109) with $\mathbf{r} = 0$:

$$\begin{aligned} \mathbf{v}(0) &= \frac{1}{(2\pi)^3} \int \mathbf{v}(\mathbf{k})\, d^3\mathbf{k} \\ &= \frac{-z_b}{(2\pi)^3 \eta} \left[\frac{8\pi}{3}\right] \mathbf{E} \int_0^\infty \frac{dk}{[\kappa^{-2} k^2 + 1]} , \end{aligned} \tag{111}$$

where we have again integrated over the directions of $\mathbf{k}$. The integral in Eq. (111) is easily evaluated to give

$$\mathbf{v}(0) = \frac{-z_b\, \kappa}{6\pi\, \eta}\, \mathbf{E} , \tag{112}$$

and, as in the relaxation effect, this can be expressed in terms of an electrophoretic correction to the single ion mobility

$$\delta\omega = \frac{-\kappa}{6\pi\eta} , \tag{113}$$

which is the same for all ions. The total correction to the mobility is then given by the sum

$$\delta\omega_i = \frac{- \omega_i^0\, q\beta\kappa\, |z_1 z_2|}{3\varepsilon_0(1 + q^{1/2})} - \frac{\kappa}{6\pi\eta} \quad (i = 1, 2) , \tag{114}$$

which is called the "limiting law for electrical conductance". Its derivation and experimental verification was a major scientific achievement.

CHARGE RECOMBINATION

Charge recombination dynamics a la Smoluchowski is a topic both illuminating and tractable, inasmuch as it deals with transient two-particle Brownian motion, non-uniform fields, and boundary conditions. Imagine a pair of ions, or an ion and an electron, initially separated in a viscous dielectric continuum (ε_0) by a given distance r_0. In fact, imagine that an ensemble of such pairs are initially prepared, and ask for how the two-body probability density $P(r, t; r_0)$ evolves in time. In particular, zero separation (consider point charges) with $r = 0$ will imply irreversible recombination, while infinite separation at long times means escape forever (Onsager, 1934, 1935, and 1938). If one assumes the existence of a mutual diffusion coefficient and posits the validity of the Stokes-Einstein relation (at this point, the introduction of so-called "Levy flights" to model electron migration would be of interest) then an accurate calculation of P is not too difficult. In short, let P satisfy:

$$\frac{\partial P(r, t; r_0)}{\partial t} = D \nabla^2 P - \xi^{-1} (F \cdot \nabla)P$$

$$F = -q^2 r/\varepsilon_0 r^2 \qquad P(r, 0; r_0) = \delta(r - r_0) , \qquad (115)$$

where q is the magnitude of the ion or "electron" charge and ξ^{-1} is a friction coefficient (ratio of F to drift velocity). Assume further that $D = (\beta \xi)^{-1}$ holds along with spatial isotropy, and one obtains

$$\frac{\partial P}{\partial t} = D \left[\frac{1}{r^2} \frac{\partial}{\partial r} \left(r^2 \frac{\partial P}{\partial r}\right) + \frac{\alpha}{r^2} \frac{\partial P}{\partial r}\right] \text{ with}$$

$$\alpha = \beta q^2/\varepsilon_0 \text{ and } P(r, o; r_0) = (4\pi r^2)^{-1} \delta(r-r_0) , \qquad (116)$$

where α is called the Onsager length. What we want is the probability that an ion pair initially separated by r_0 will still be apart at time t, and this is given by

$$W(r_0, t) = \int d^3r \, P(r, t; r_0) \quad , \quad W(r_0, 0) = 1 , \qquad (117)$$

where the integral is over all space. The poor man's approach to this problem, which turns out to be an excellent approximation, especially at long times, is to make a "prescribed diffusion" ansatz (Mozumder, 1968, 1974; Hong and Noolandi, 1978):

$$P(r, t; r_0) = F(t)\,(4\pi Dt)^{-3/2} \exp\left[-(r - r_0)^2/4Dt\right], \tag{118}$$

where F(t) is self-consistently determined. Spatial integration as in Eq. (117) produces

$$\frac{dF(t)}{dt} = -4\pi D\alpha \left[F(t)/(4\pi Dt)^{3/2}\right] \exp\left[-r_0^2/4Dt\right],$$

$$F(0) = 1 \quad , \tag{119}$$

which is easily solved. The results are

$$F(t) = \exp\left(\frac{-\alpha}{(4\pi D)^{1/2}} \int_{t^{-1}}^{\infty} ds\; s^{-1/2} \exp\left[-r_0^2 s/4D\right]\right)$$

$$W(r_0, t) = \exp\left\{-\frac{\alpha}{r_0} \operatorname{erfc}\left[r_0/(4Dt)^{1/2}\right]\right\}, \tag{120}$$

where the complementary error function is indicated. At short times, W drops off very fast as

$$W(r_0, t) \simeq 1 - \frac{\alpha}{r_0^2}\left[\frac{4Dt}{\pi}\right]^{1/2} \exp\left[-r_0^2/4Dt\right] \text{ (t small)}, \tag{121}$$

and at long times the net survival probability goes to

$$W(r_0, t) = \left[1 + \frac{\alpha}{(\pi Dt)^{1/2}}\right] \exp\left[-\alpha/r_0\right] \text{ (t large)}, \tag{122}$$

which happens to be exact. There are a few interesting points about this expression: the infinite time escape probability is independent of the diffusion coefficient, the rate at which this limit is attained does depend on D, and the decay at long times is very slow ($t^{-1/2}$); in fact, the sluggish dynamics strongly resembles critical point fluctuations.

Very similar effects arise in a large class of nonlinear reaction-diffusion problems, where the initial condition is a special inhomogeneity and the system is unbounded. Needless to say, even a qualitative mathematical analysis of such systems is a formidable task. One interesting example is

$$\frac{\partial C(r, t)}{\partial t} = D\,\nabla^2 C - k\,C^2$$

$$C(r, 0) = \frac{N_0}{(\pi r_0^2)^{3/2}} \exp\left[-r^2/r_0^2\right], \quad (N_0 >> 1), \tag{123}$$

which is the simplest description of free radical recombination [concentration C(r, t)] in pulse radiolysis. The total number of unreacted radicals remaining at time t is

$$N(t) = \int d^3r\, C(r, t) , \tag{124}$$

and with a prescribed diffusion ansatz, Eq. (123) is easily converted into the homogeneous Riccati form

$$\frac{d\,W(\tau)}{d\tau} = - \frac{\alpha^* W^2}{(1 + \tau)^{3/2}} \quad \text{with} \quad W(\tau) = N_0\, N(\tau) , \quad \tau = 4Dt/r_0^2 ,$$

$$W(0) = 1 , \text{ and } \alpha^* = (2\pi)^{-3/2} (kN_0/4Dr_0) . \tag{125}$$

Here τ is a dimensionless diffusion time and α^* is essentially the ratio of a characteristic diffusion time to a recombination time. Equation (125) is readily solved to give

$$W(\tau) = \frac{1}{1 - 2\alpha^*(1 + \tau)^{-1/2} + 2\alpha^*}$$

$$W(\infty) = \frac{1}{1 + 2\alpha^*} . \tag{126}$$

In this instance, unlike ion-pair recombination, the fraction that escapes to infinity depends on the diffusion coefficient. It's easy to generalize this analysis to describe inhomogeneous bimolecular recombination in d-dimensions:

$$W(\tau) = \frac{d - 2}{(d - 2) - 2\, \alpha_d^*(1 + \tau)^{(2-d)/2} + 2\alpha_d^*} \qquad (d \neq 2)$$

$$W(\tau) = \frac{1}{1 + \alpha_2^* \ln(1 + \tau)} \qquad (d = 2) . \tag{127}$$

Note that nothing survives at infinite time for d = 2 or less and that the decay is always algebraic ($d \neq 2$). It turns out that even for the low dimensional cases and higher reaction orders (3 and 4) prescribed diffusion provides a surprisingly accurate description of $W(\tau)$ over the entire time history; however, the derived concentration profiles $C(r, t)$ aren't very reliable.

Go back to the ion-pair recombination problem and add a uniform external electric field **E**, so that the survival probability becomes a function of the field strength (Onsager, 1934, 1935, 1938). One still assumes Smoluchowski dynamics as in Eq. (115), but now the force is given by

$$\mathbf{F} = - \frac{q^2 \mathbf{r}}{\varepsilon_0 r^2} - q\, \mathbf{E} , \tag{128}$$

and the prescribed diffusion probability density takes the form (Mozumder, 1968, 1974; Hong and Noolandi, 1978)

$$P(r,\ t;\ r_0,\ E) = F(t)(4\pi\ Dt)^{-3/2}\ \exp\left[\frac{(x-x_0 + \mu Et)^2 + (y-y_0)^2 + (z-z_0)^2}{4Dt}\right]$$

$$\text{with}\quad \mu = q\beta\, D = q\xi^{-1} \quad \text{and} \quad E = E\,\hat{x}\ , \tag{129}$$

where μ is the ion (electron) mobility. Spatial integration produces an equation for $F(t)$ similar to Eq. (119) which is readily solved to give

$$-\ln F(t) = \frac{2\alpha}{\sqrt{\pi}} \exp\left[\frac{q\beta\ Ex_0}{2}\right] \int_k^\infty dk\ \exp\left[-\ r_0^2\ k^2 - \frac{\mu^2\ E^2}{16D^2k^2}\right] \quad \text{with}$$

$$k = (4Dt)^{-1/2}\ , \quad \alpha = \text{Onsager length}\ , \tag{130}$$

and the escape probability ($t \longrightarrow \infty$) becomes

$$W(r_0,\ E,\ \infty) = \exp\left\{\left[-\ \frac{\alpha}{r_0}\right] \exp\left[-\ \frac{q\beta\ Er_0}{2}(1 - \cos\Theta)\right]\right\}, \tag{131}$$

where Θ is the angle between r_0 and E. If the field is small enough so that the exponential can be linearized one obtains

$$W(r_0,\ E,\ \infty) = \left[1 + \frac{\alpha\ q\beta\ E}{2}(1 - \cos\Theta)\right] \exp\left[-\ \alpha\ /r_0\right] \quad (E\ \text{small})\ , \tag{132}$$

so the escape probability, relative to the zero-field case, is enhanced by a factor $\sim \alpha\, q\beta\, E$, i.e., it is proportional to β^2 and $1/\varepsilon_0$. Again it happens that Eq. (132) is exact. Also, observe that neither the diffusion coefficient nor the mobility appear in the expression for $W(\infty)$.

MULTI-STATE DYNAMICS AND TRANSPORT

One should bear in mind that the observed linearity of transport phenomena (electrical conductance, for example) does not arise from linear behavior at the microscopic level; it is a consequence of macroscopic averaging. In short, just because one observes that current density and field are related by $J = \sigma E$ and that σ is independent of E over a considerable range, it cannot be concluded that the individual charge carriers are actually gliding along with the appropriate constant drift velocity. Indeed, if that were the case we would not be having this Institute. A more realistic picture would be a highly irregular, more or less wildly fluctuating trajectory with a very small field-induced directional bias, and it is the long time average over many such trajectories that constitutes a macroscopic observation.

Now imagine a "massive" or Brownian particle, and suppose that in addition to the constant Langevin friction generated by the rapidly fluctuating random force, there exists a more slowly decaying dissipative force, or, on the other hand, assume that the coupling with an external field (via mobility, valence, etc.) fluctuates in time, which is an example of a convective process. In the case of fluctuations internal to the system, one anticipates an enhanced friction coefficient as the slow force correlation time diminishes, while in the convective case, a short correlation time for mobility fluctuations should lead to an increase in that transport coefficient.

More generally, however, the time correlation functions of tagged particle dynamical variables will be significantly altered as a consequence of these fluctuations, and as it is becoming possible to obtain information on such correlations from experiment or simulation, we shall focus on the structure of these functions rather than just their integrals. So as to avoid physical ambiguity, white noise for the "auxiliary stochastic process" is assumed only as a limiting case, and while this introduces considerable mathematical complexity (non-linear Langevin equations are replaced by more general stochastic equations), one is comforted by the awareness that the physics has been built in at the outset rather than tortuously extracted or divined at the end of a calculation (Fox, 1978; Van Kampen, 1981; van der Ziel, 1954).

As a preliminary to studying multi-state transport processes, it is advisible to at least become acquainted with the so-called "master equation", which is essentially a Chapman-Kolmogorov form specialized for time-stationary Markov (short-memory) processes (Fox, 1978; Van Kampen, 1981):

$$\frac{\partial P(y,\ t)}{\partial t} = \int \Big[K(y;\ y')\ P(y',\ t) - K(y';\ y)\ P(y,\ t)\Big]\ dy' \ . \tag{133}$$

Here $K(y;\ y')$ is defined as the transition probability per unit time for going from state y' to y ($y' \longrightarrow y$), and $P(y,\ t)$ is the probability distribution for observing state y at time t. For discrete states, the analog of Eq. (133) is

$$\frac{d\ P_n(t)}{dt} = \sum_{l} \Big[K_{nl}P_l(t) - K_{ln}P_n(t)\Big] \ , \tag{134}$$

and so the master equation is just a birth-death or gain-loss expression for the time evolution of the probability of each state.

Solving a master equation typically involves determining $P(y,\ t)$ given the initial condition $P(y,\ 0)$ and some assumptions about the form of the transition probability K. For instance, take $K(y;\ y') = k(y)$, which implies that the transition probability depends only on one state, and Eq. (133) becomes

$$\frac{\partial P(y, t)}{\partial t} = \int \left[k(y)\, P(y', t) - k(y')\, P(y, t)\right] dy' , \tag{135}$$

which is sometimes called a "Kubo-Anderson process". Since P is normalized over y, this equation is easily solved to give

$$P(y, t) = P(y, 0)\, e^{-at} + \frac{k(y)}{a}\left[1 - e^{-at}\right] , \quad a = \int k(y)\, dy , \tag{136}$$

which says that the initial distribution decays exponentially to k(y)/a with a rate given by the integrated state-transition probability.

As an example of a discrete state master equation, consider the class of one-step processes. Introduce a raising (lowering) operator $R(R^{-1})$ which acts on some f(n) defined on the set of integers by

$$R\, f(n) = f(n + 1) \text{ and } R^{-1}\, f(n) = f(n - 1) , \tag{137}$$

and Eq. (134) may be rewritten

$$\frac{d\, P_n(t)}{dt} = (R - 1)\, D_n\, P_n + (R^{-1} - 1)\, B_n\, P_n , \tag{138}$$

where the birth and death rates (B_n, D_n) cannot be negative, but either may be zero.

To arrive at a Fokker-Planck type equation from here, treat n as a continuous variable, expand $R(R^{-1})$ in a Taylor series as

$$R = 1 + \frac{\partial}{\partial n} + \frac{1}{2}\frac{\partial^2}{\partial n^2} + \ldots$$

$$R^{-1} = 1 - \frac{\partial}{\partial n} + \frac{1}{2}\frac{\partial^2}{\partial n^2} + \ldots , \tag{139}$$

and the FP approximation to the one-step master equation becomes

$$\frac{\partial\, P(n, t)}{\partial t} = \frac{\partial}{\partial n}\left\{D(n) - B(n)\right\} P + \frac{1}{2}\frac{\partial^2}{\partial n^2}\left\{D(n) + B(n)\right\} P , \tag{140}$$

where D and B are now continuous non-negative functions of the state variable n. The behavior in the vicinity of a stationary state n_s is easily deduced by writing $n = n_s + \delta n$, expanding in δn, and retaining only the low-order terms:

$$\frac{\partial P(n, t)}{\delta t} = \left\{D'(n_s) - B'(n_s)\right\} \frac{\partial}{\partial \delta n}\left[\delta n\; P\right] + \left\{\frac{D(n_s) + B(n_s)}{2}\right\} \frac{\partial^2\, P}{\partial \delta n^2} ,$$

$$D'(n_s) = \left.\frac{dD}{dn}\right|_{n_s} \quad \text{and} \quad D(n_s) = B(n_s) . \tag{141}$$

It is then a simple matter to calculate mean-square fluctuations and time correlation functions in the stationary state as

$$\langle \delta n^2 \rangle_s = \frac{1}{2} \left[\frac{D(n_s) + B(n_s)}{D'(n_s) - B'(n_s)} \right] , \text{ where}$$

$$\langle \delta n(0) \; \delta n(t) \rangle_s = \langle \delta n^2 \rangle_s \; e^{-at} \quad \text{and}$$

$$a = D'(n_s) - B'(n_s) , \tag{142}$$

so it's hard to observe anything except exponential decay and Lorentzian lineshapes if you hang around stable stationary states and small-step dynamics.

Macroscopic "laws" are extracted from the master equation as follows: multiply Eq. (133) by the state variable y and integrate over y to produce

$$\frac{d\langle y \rangle}{dt} = \int \left[y \, \frac{\partial P(y, t)}{\partial t} \right] dy = \iint (y' - y) \, K(y'; y) \, P(y, t) \, dy \, dy'$$

$$= \int m_1(y) \, P(y, t) \, dy = \langle m_1(y) \rangle , \text{ where}$$

$$m_1(y) = \int (y' - y) \, K(y';y) \, dy' . \tag{143}$$

If the first-order "jump" moment m_1 happens to be a linear function of y (say $C_1 y$), then

$$\frac{d\langle y \rangle}{dt} = C_1 \langle y \rangle , \tag{144}$$

and one recovers a linear macroscopic form. It's not difficult to obtain a similar expression for the variance of $y(\sigma^2 = \langle y^2 \rangle - \langle y \rangle^2)$:

$$\frac{d\sigma^2}{dt} = C_2 \langle y \rangle + 2 \, C_1 \sigma^2 \langle y \rangle , \tag{145}$$

where the constant C_2 is obtained from a second-order jump moment $m_2(y)$.

Anything resembling a scrupulous analytical approach to variable-state transport processes would require the services of a coven of mathematicians, and so, lacking the necessary conjuring power, I must resort to inference, plausibility arguments, and, hopefully, common sense.

Consider the following stochastic equation (Fox, 1978; Van Kampen, 1981; Kubo, 1963):

$$\frac{d\psi}{dt} + \xi(t)\psi = F(t) \qquad (t > 0) , \text{ where}$$

$$\psi(t) = \langle \psi(t) \rangle + \delta\psi(t) ,$$

$$\xi(t) = \xi_0 + \delta\xi(t) \qquad (\xi_0 > 0) ,$$

$$\langle\delta\xi(t)\rangle = 0, \quad \langle F(t)\rangle = 0, \quad \langle\delta\psi(t)\rangle = 0 \,,$$

$$\langle\delta\xi(t')\, F(t)\rangle = 0, \quad \langle\psi(0)\rangle = 1 \,, \text{ and}$$

$$\langle\psi(t')\, F(t)\rangle = 0 \,, \tag{146}$$

so that $\langle\psi\rangle$ might be a normalized velocity autocorrelation function of a tagged particle in a Langevin medium with a fluctuating collision frequency or resistance $\delta\xi(t)$. $\delta\xi$ will be unspecified except we insist that if $\delta\xi$ has a sufficiently short correlation memory or begins to resemble white noise, the renormalized friction coefficient is increased, i.e., $\xi_0 \longrightarrow \xi_0 + \Delta\xi$ $(\Delta\xi > 0)$. This is what one expects if $\delta\xi$ arises from dissipative (internal) fluctuations in the medium surrounding the particle, and, in fact, there is a very close connection between Eq. (146) and the generalized Langevin equation discussed earlier, but we won't go into that here. The additive noise $F(t)$ is included for achieving and sustaining equilibrium; it does not play a significant role in the following discussion.

Noting that dissipative terms change their sign under time reversal $(t \longrightarrow -t)$, one writes the time-reversed analogue of Eq. (146):

$$\dot{\psi} - \xi(t)\,\psi = +\,F(t) \qquad (t < 0) \,, \tag{147}$$

and this immediately gives

$$\delta\dot{\psi} - \xi_0\,\delta\psi = \delta\xi(t)\,\langle\psi(t)\rangle + \Big\{\delta\xi(t)\,\delta\psi(t) - \langle\delta\xi(t)\,\delta\psi(t)\rangle\Big\} + F(t) \qquad (t < 0) \,. \tag{148}$$

We need $\delta\psi$ for negative times because this determines the initial condition $\psi(0)$, which, for this example, is stochastic. Temporally speaking, the correlation function $\langle\psi(t)\rangle$ depends on $\langle\psi(s)\rangle$, $\delta\psi(s)$, $\delta\xi(s)$, $s < t$, and, in turn, the fluctuation $\delta\psi(s)$ depends on $\langle\psi(\tau)\rangle$, $\delta\psi(\tau)$, and $\delta\xi(\tau)$ at previous times $\tau < s$. Equation (148) is not solvable in general, so we invoke a closure ansatz called "first-order smoothing" which amounts to throwing away the bracketed fluctuation term in Eq. (148), and then $\delta\psi$ becomes (Bourret, 1964, 1965, 1966; Zwanzig, 1964)

$$\delta\psi(t) = e^{+\xi_0 t}\int_{-\infty}^{t} e^{-\xi_0 \tau}\Big[\delta\xi(\tau)\,\langle\psi(\tau)\rangle + F(\tau)\Big]d\tau \qquad (t \gtrless 0) \,, \tag{149}$$

which implies

$$\langle\dot{\psi}\rangle + \xi_0 \langle\psi\rangle = -\int_0^t d\tau\, e^{+\xi_0 (t-\tau)} \langle\delta\xi(t)\ \delta\xi(\tau)\rangle \langle\psi(\tau)\rangle \quad (t > 0) . \tag{150}$$

Since we are dealing with equilibrium fluctuations, we have

$$\langle\delta\xi(t)\ \delta\xi(\tau)\rangle = \langle\delta\xi(0)\ \delta\xi(t - \tau)\rangle , \tag{151}$$

and so Eq. (150) can be solved by Laplace transform (variable s):

$$\langle\tilde{\psi}(s)\rangle = \frac{1}{(s + \xi_0) + \tilde{\Gamma}(s - \xi_0)} ,$$

$$\tilde{\Gamma}(s) = \int_0^\infty dt\ e^{-st} \langle\delta\xi(0)\ \delta\xi(t)\rangle . \tag{152}$$

Choose $\Gamma(t) \sim \delta(t)$, recover a renormalized (enhanced) friction coefficient, and $\langle\psi\rangle$ decays as a simple exponential. To model "real stuff", take

$$\Gamma(t) = ae^{-bt} \qquad (a,\ b > 0) , \tag{153}$$

and

$$\langle\tilde{\psi}(s)\rangle = \frac{(s - \xi_0 + b)}{(s - \xi_0 + b)(s + \xi_0) + a} . \tag{154}$$

The roots of the denominator are

$$\begin{matrix} s_+ \\ s_- \end{matrix} = \left\{ -\frac{b}{2} \left[1 \pm \left(1 - \frac{4\gamma}{b^2} \right)^{1/2} \right] \right.$$

$$\gamma = \xi_0 (-\xi_0 + b) + a , \tag{155}$$

and therefore $\langle\psi(t)\rangle$ decays as the sum of two oscillating exponentials. The steady diffusion coefficient is given by $\langle\tilde{\psi}(0)\rangle$, so in this rather crude approximation, the particle can become "trapped" or "localized" under several conditions; for instance, when the correlation frequency b equals the steady collision frequency ξ_0. Also, even if ξ_0 vanishes the diffusion coefficient remains finite (~b/a). I won't go into the details, but it turns out that "first-order smoothing" becomes exact for a class of stationary two-state Markov processes with equal transition probabilities--dichotomic Markov processes--so that the preceding calculation has some physical relevance.

As a variation of Eq. (146) consider Langevin dynamics with a randomly modulated oscillator frequency $\delta\omega$ (real):

$$\dot{\psi} + \xi_0\ \psi + i\ \delta\omega\ (t)\ \psi = F(t) ,\ \langle\delta\omega\ (t)\rangle = 0 ,\ i = \sqrt{-1} ,\ \xi_0 > 0 , \tag{156}$$

where restrictions analogous to those imposed in Eq. (146) are assumed. The velocity correlation function is then (in this case, $\psi(0)$ can be considered non-random)

$$\langle\psi\rangle = e^{-\xi_0 t} \langle \exp(-i \int_0^t \delta\omega\,(\tau)\; d\tau)\rangle \;, \tag{157}$$

and if the expanded exponential is truncated at the quadratic level, one has

$$\langle\psi\rangle = e^{-\xi_0 t} \left[1 - \frac{1}{2} \int_0^t d\tau_1 \int_0^t d\tau_2 \langle \delta\omega\,(\tau_1)\;\delta\omega\,(\tau_2)\rangle + \ldots\right] . \tag{158}$$

As before, take

$$\frac{d}{d\tau} \langle \delta\omega\,(\tau)\;\;\delta\omega\,(t + \tau)\rangle = 0 \;, \tag{159}$$

and again supposing that

$$\Gamma(t) = \langle \delta\omega\,(0)\;\;\delta\omega\,(t)\rangle = ae^{-bt} \qquad (a,\; b > 0) \;, \tag{160}$$

one arrives at

$$\langle\psi\rangle = e^{-\xi_0 t} \left[1 - \frac{a}{b^2} (bt + e^{-bt} - 1)\right] . \tag{161}$$

For short correlation times b is large, and

$$\langle\psi\rangle = e^{-\xi_0 t} \left[1 - \frac{a}{b}\, t + \ldots\right] \qquad \text{(b large)} \;,$$

$$\int_0^\infty \langle\psi\rangle\; dt = \xi_0^{-1} \;\; \left[1 - \frac{a}{b}\, \xi_0^{-1} + \ldots\right] , \tag{162}$$

and so the diffusion coefficient is reduced by introducing random oscillations. For long correlation times, one has the "frozen" fluctuation result:

$$\langle\psi\rangle = e^{-\xi_0 t} \left[1 - \frac{1}{2}\, at^2 + \ldots\right] \qquad \text{(b small)} \;, \tag{163}$$

and diffusion vanishes for $a\xi_0^{-2} \geq 1$.

Of course, if $\delta\omega$ is taken to be a stationary Gaussian random process, then Eq. (157) takes the familiar form (Fox, 1978; Van Kampen, 1981):

$$\langle\psi\rangle = e^{-\xi_0 t} \;\exp \left[- \frac{1}{2} \int_0^t d\tau_1 \int_0^t d\tau_2\; \Gamma(\tau_1 - \tau_2)\right] , \tag{164}$$

which becomes, for an exponential Γ,

$$\langle\psi\rangle = e^{-\xi_0 t} \exp\left[-\left(\frac{a}{b^2}\left[bt + e^{-bt} - 1\right]\right)\right], \text{ where}$$

$$\Gamma(\tau_2 - \tau_1) = a \exp -b|\tau_2 - \tau_1| , \tag{165}$$

and this has the limiting behavior

$$\langle\psi\rangle = \exp\left[-\left(\xi_0 + \frac{a}{b}\right) t\right] \qquad \text{(b large) and}$$

$$\langle\psi\rangle = e^{-\xi_0 t} \exp\left[-\frac{1}{2} at^2\right] \qquad \text{(b small) .} \tag{166}$$

For this so-called "Uhlenbeck-Ornstein" process, one therefore obtains correlation functions whose decay can range from simple exponential to Gaussian.

Starting at the other end from a Brownian oscillator, one can easily devise a class of quasi-tractable Kubo-type models incorporating random damping and random frequency fluctuations (Kubo, 1963).

One should not conclude from this discussion that we are somehow limited to single-particle dynamics and ordinary, though stochastic, differential equations. One can also deal rather easily with many-body dynamics in some subset of phase space, where a portion of the time evolution operator is itself stochastic, provided everything remains linear with respect to the distribution. Consider the following random operator equation:

$$\frac{\partial\psi(\Omega, t)}{\partial t} = L_0(\Omega, t)\,\psi + L_1(\Omega, t)\psi , \tag{167}$$

where Ω denotes a subset of phase-space variables, L_0 is a "projected" Liouville operator, and L_1 is a "centered" random operator whose ensemble or statistical average vanishes, i.e., $\langle L_1\rangle = 0$. Switch to the integral form of Eq. (167) which is formally

$$\psi(\Omega, t) = \exp(L_0 t)\,\psi(\Omega,0) + \int_0^t d\tau \exp\left[L_0(t - \tau)\right] L_1(\tau)\,\psi(\tau) , \tag{168}$$

and re-introduce the projection operator P via the usual relations

$$P L_0 = L_0 P \qquad P L_1 P = 0$$

$$P\psi(\Omega,0) = \psi(\Omega,0) \qquad P\psi = \langle\psi\rangle . \tag{169}$$

The projector in this case is just an ensemble-averaging operator, the initial distribution is taken here to be non-random, and the fluctuating part of the distribution is given by the orthogonal projection

$$\delta\psi = (1 - P)\,\psi(\Omega, t) . \tag{170}$$

A few straightforward iterations yield a closed expression for the correlation function

$$\langle\psi(\Omega, t)\rangle = \exp(L_0 t)\ \psi(\Omega, 0) - G_0 D\langle\psi(\Omega, t)\rangle$$

where

$$G_0 f = \int_0^t d\tau \exp\left[L_0(t - \tau)\right] f(\tau) \quad \text{and}$$

$$D = - PL_1 \sum_{n=1}^{\infty} \left\{G_0\left[(1 - P)L_1\right]\right\}^n . \tag{171}$$

Equation (171) comprises a Dyson expansion and D is usually called the Dyson or "mass" operator. A diagrammatic expansion is the form in which the expression usually appears in the literature (Dyson, 1949).

The "first-order smoothing" closure discussed previously can now be viewed as the summation of an infinite subseries in a Dyson expansion:

$$\delta\psi = G_0\ L_1\langle\psi\rangle$$

$$\langle\psi\rangle = \exp\ (L_0 t)\ \psi(\Omega, 0) + G_0\ PL_1\ G_0\ L_1\ \langle\psi\rangle\ , \tag{172}$$

which says that the fluctuation $\delta\psi(t)$ depends only on L_1 and $\langle\psi\rangle$ at previous times. When expressed in terms of the initial distribution $\psi(\Omega, 0)$, Eq. (172) becomes

$$\langle\psi(\Omega, t\rangle = \sum_{n=0}^{\infty} \left[G_0\ PL_1\ G_0\ L_1\right]^n \exp(L_0 t)\ \psi(\Omega, 0)\ , \tag{173}$$

and so the smoothed Markovian evolution operator associated with Eq. (173) involves sums of products of two-point correlation functions associated with the stochastic operator L_1.

Dyson-type equations have been used extensively in quantum electrodynamics, quantum field theory, statistical mechanics, hydrodynamic instability and turbulent diffusion studies, and in investigations of electromagnetic wave propagation in a medium having a random refractive index (Tatarski, 1961). Also, this technique has recently been employed to study laser light scattering from a macromolecular solution in an electric field.

It would appear that a judicious combination of Dyson expansion and renormalization group methods might be a powerful tool for a variety of problems in non-equilibrium statistical physics.

FUNCTIONAL INTEGRATION, BROWNIAN MOTION, AND QUANTUM MECHANICS (Simon, 1979, 1985)

As many of you are aware, there exists an intimate relationship between time-dependent stochastic processes in classical systems and stationary-state quantum mechanics, and I would like to briefly discuss this connection using M. Kac's famous formula for Wiener/Feynmann path integrals, along with an application to the problem of an electron interacting with a large number of static random impurities via a short-range repulsive pseudopotential (Kac and Luttinger, 1974). This is one example where topological disorder has a profound influence on the electronic density-of-states function $g(\varepsilon)$; moreover, the exact form of $g(\varepsilon)$ can be calculated in the low-energy limit and it can be shown that the $\varepsilon \longrightarrow 0$ wavefunctions correspond to "localized states". Although the random scatterer problem was originally formulated for dilute systems, it is not difficult to extend many of the results to dense, strongly interacting "impurities" such as liquids, so our theoretical ruminations will have some relevance to "real stuff".

As a classic example of the relation between functional integrals (integrals over some space of functions) and stochastic differential equations, consider the case of Eq. (167) with

$$\Omega \equiv \mathbf{r},\ L_0 = \nabla^2,\ L_1 = \xi(\mathbf{r})$$

$$\Big[\xi(\mathbf{r})\Big]_\xi = 0, \text{ and } \quad \psi(\mathbf{r},\,0) = \delta(\mathbf{r}) \ , \tag{174}$$

so that the ξ-ensemble average of $\xi(\mathbf{r})$ vanishes, the projected operator L_0 describes ordinary diffusion in configuration space, and the initial condition on the distribution is that it collapse to a Dirac delta function. We, therefore, have a heat or diffusion equation with a spatially random sink or source term. For any prescribed function $\xi(\mathbf{r})$, ψ may be written as a Wiener/Kac functional integral

$$\psi(\mathbf{r},\,t) = \left[\delta(\rho(t) - \mathbf{r}) \exp\left\{\int_0^t d\tau\ \xi\Big(\rho(\tau)\Big)\ d\tau\right\}\right]_\rho \ , \tag{175}$$

where the brackets with subscript ρ refer to an integration over a so-called "Wiener measure"; in other words, the average is over a set $\{\rho(t)\}$ of continuous Brownian paths which begin at $\mathbf{r} = 0$ ($t = 0$) and terminate at position $\mathbf{r}$ at time t. The "self correlation function" is now derived from the distribution by performing an ensemble average over all "permissible" functions $\xi(\mathbf{r})$ (ξ does not have to be continuous):

$$\left[\psi(r, t)\right]_\xi = \left[\delta\left(\rho(t) - r\right) \exp\left\{\frac{1}{2}\int_0^t \int_0^t \Gamma\left(\rho(\tau); \rho(\tau')\right) d\tau' \, d\tau\right\}\right]_\rho ,$$

$$\Gamma\left(\xi(r); \xi(r')\right) \equiv \left[\xi(r) \, \xi(r')\right]_\xi , \tag{176}$$

where the Wiener/Kac integration and ξ-averages have been interchanged, and we have assumed, for explicitness, that $\xi(r)$ is a Gaussian random variable. Equation (176) is therefore a functional integral representative of the solution of a special class of Dyson equation. Note that Γ is just a two-point spatial correlation function of ξ. Another interesting feature of Eq. (176) is that it happens to be closely related to the solution of a certain non-random partial differential equation, i.e.,

$$\frac{\partial \psi(r, t)}{\partial t} = \nabla^2 \psi - \Gamma(r) \psi , \quad \Gamma(r) \geq 0$$

$$\psi(r, 0) = \delta(r) . \tag{177}$$

This is the Kac formula which we shall now discuss. Let $T(t_n - t_{n-1}; x_n | x_{n-1})$ be a (one-dimensional) transition probability for a Brownian process, so that T is the probability a particle located at x_{n-1} at time t_{n-1} will be at x_n at time t_n:

$$T = \left[4\pi(t_n - t_{n-1})\right]^{-1/2} \exp\left\{- \frac{(x_n - x_{n-1})^2}{4(t_n - t_{n-1})}\right\} . \tag{178}$$

The probability of realizing a particular Brownian trajectory is then

$$\int_{a_1}^{b_1} dx_1 \cdots \int_{a_N}^{b_N} dx_N \prod_{n=1}^{N} T(t_n - t_{n-1}; x_n | x_{n-1}) . \tag{179}$$

Imagine that this integral defines a "Wiener measure" of the set of functions x(t):

$$a_1 < x(t_1) < b_1, \ldots, a_N < x(t_N) < b_N . \tag{180}$$

If N is finite and x(t) is constant within each interval, then the associated Wiener integral of some functional $\alpha\left[x(t)\right] = \alpha(x_1, \ldots, x_N)$ looks like an ordinary multiple integral, and in fact this expression can be evaluated exactly whenever α is the exponential of some linear or bilinear function of the x's. For the more general case of infinite N, Kac proved that if $\Gamma(x)$ is a real, positive, continuous function of x, then

$$\left[\exp\left\{- \int_0^t \Gamma\left(x(\tau)\right) d\tau\right\}\right]_x = \lim_{N\to\infty} \left[\exp - \frac{t}{N} \sum_{n=1}^{N} \Gamma\left[x\left(\frac{nt}{N}\right)\right]\right]_x , \tag{181}$$

and moreover,

$$\psi(x,\ t) = \left[\delta\left(x(t) - x\right) \exp\left\{- \int_0^t \Gamma\left(x(\tau)\right)\, d\tau\right\}\right]_{x(t)} \tag{182}$$

is the solution of

$$\frac{\partial\psi}{\partial t} = \frac{\partial^2 \psi}{\partial x^2} - \Gamma(x)\psi, \qquad \Gamma(x) \geq 0 \quad \text{and} \quad \psi(x,\ 0) = \delta(x). \tag{183}$$

Although such functional-integral representations are not particularly useful for analytical work, they have proven to be remarkably well-adapted to the capabilities of modern computers, as you have heard from Professor Chandler.

Since linear reaction-diffusion equations such as Eq. (183) are relatively easy to solve, a good deal of effort has gone into establishing connections between these and diffusive-type stochastic processes. For example, consider diffusion in a medium containing a random distribution of static traps, so that a particle contacting the surface of a trap is immediately and irreversibly transformed or annihilated and the trap is unaffected. Given some initial distribution of Brownian particles, how does this distribution evolve in time and what fraction survive? This problem would be almost trivial if the trap distribution were periodic and the initial particle distribution were uniform, for in that case the diffusive flux across the surface of each unit cell vanishes, and the many-body problem is converted into a few- or even one-body problem involving but a single trap. For random traps one would expect significant differences from the periodic trap case (same density) at short times, while at long times a diffusing particle must avoid a large number of traps and hence sample many trap configurations. In fact, at long times, one anticipates the validity of

$$\frac{\partial\psi}{\partial t} = D\, \nabla^2\, \psi - K\, \psi\ , \tag{184}$$

where the trap-averaged absorption coefficient K should be proportional to the macroscopic or coarse-grained trap density, and indeed this is exactly what occurs for periodic traps. A curious and quite surprising feature of the random trap problem is that Eq. (184) is not valid at long times, which implies that there is a profound difference between the coarse-grained trap versions of the two problems. In other words, a long-lived trap-avoiding particle samples, on the average, very different topologies in configuration space, depending on whether the traps are periodic or random.

The connection between these meanderings and electrons in simple liquids is that if one can somehow calculate the "conditional Wiener integral" (Simon, 1979, 1985; Kac and Luttinger, 1974)

$$\psi(0,\ t) = \left[\delta\Big(x(t) - 0\Big) \exp\left\{- \int_0^t \Gamma\Big(x(\tau);\ 0\Big)d\tau\right\}\right]_{x(t)}$$

$$\psi(x,\ 0) = \delta(x) \qquad (x_N = x_0 = 0) \quad (N \longrightarrow \infty)\ , \tag{185}$$

then by an inverse Laplace transform of ψ we have the density of states (per unit volume) of an electron in some pseudopotential comprised of a random configuration of scatterers. In Eq. (185), the "0" in the delta function and in Γ signify that we must average only over the subset of Brownian paths that begin and terminate at the origin $x = 0$. These closed paths are the continuum limit of the "necklesses" discussed by Prof. Chandler earlier. Of course, the implicit assumption here is that the electronic motion is adiabatic. It so happens that for a hard sphere pseudopotential, for which Γ is either zero or infinity, the functional integral Eq. (185) can be computed analytically in the limit $t \longrightarrow \infty$, which, via a standard Tauberian theorem of Laplace transforms, gives the exact asymptotic low-energy form for $g(\varepsilon)$ (density of states) [3]. This remarkable and powerful result was first derived by Donsker and Varadhan (1975) in a 40-page mathematical tour de force entitled "Asymptotics of the Wiener sausage." Independently of the mathematicians, R.F. Kayser and I (1983) obtained a nearly identical result employing a similar but much more succinct "method of bounds" which I shall now sketch.

Consider a trap configuration selected from some arbitrary probability distribution. Since the addition of traps can only decrease the number density ψ everywhere, we can introduce a perfectly absorbing spherical surface of radius R centered at the origin ($r = 0$) and write

---

3 This Tauberian theorem is as follows. Suppose that

$$\lim_{t \to \infty} t^{-\alpha} \ln\psi(t) = -B$$

where B is a positive constant and $0 < \alpha < 1$. Then the Laplace transform of $\psi(t)$, $g(\varepsilon)$, has the asymptotic form

$$\lim_{\varepsilon \to 0} \varepsilon^{\beta} \ln g(\varepsilon) = -A \quad \text{where}$$

$$\beta = \frac{\alpha}{1 - \alpha}$$

and

$$\ln A = \ln(1 - \alpha) + \frac{\alpha}{1 - \alpha} \ln \alpha - (1 - \alpha) \ln B$$

$$\psi(0,\ t) = \sum_{\Lambda} \psi_{\Lambda}(0,\ t)\ P_{\Lambda} > \sum_{\Lambda} \psi'_{\Lambda}(0,\ t)\ P_{\Lambda}\ , \tag{186}$$

where the prime refers to the "absorbing-shell" modified configuration, Λ labels a single configuration of an infinite number of traps, and P_{Λ} is the probability of realizing that configuration. Both Λ -summations are intractable; however, since each term is inherently positive, one can sum over that subclass $\Lambda\ (0)$ for which there are no traps inside this sphere of radius R and still ensure that

$$\psi(0,\ t) > \sum_{\Lambda\ (0)} \psi'_{\Lambda}(0,\ t)\ P_{\Lambda}\ . \tag{187}$$

The point is that since we have a perfectly absorbing surface, ψ' is independent of $\Lambda\ (0)$, and in fact $\psi'_{\Lambda}(0,\ t) \equiv \psi^*(0,\ t)$, where ψ^* is the solution to

$$\frac{\partial\psi^*}{\partial t} = D\ \nabla^2\ \psi^*\ , \quad \psi^*(r,\ t) = 0 \text{ at } |r| = R\ , \text{ and}$$

$$\psi^*(r,\ 0) = \delta(r) \quad \text{(Point Source)}\ . \tag{188}$$

The long-time solution of Eq. (188) is of the form

$$\psi^*(0,\ t) = a^{-d} \exp\left[-ADt/R^2\right]\ , \tag{189}$$

where a is the trap radius (spherical traps), d is the spatial dimension, and A is an easily determined dimensionless constant. We may now conclude that

$$a^d\ \psi(0,\ t) > P(0) \exp\left[-ADt/R^2\right]\ , \tag{190}$$

where $P(0) = \sum_{\Lambda(0)} P_{\Lambda}$ is the probability that this sphere of radius R contains no traps. For spatially uncorrelated traps, for example, one has the Poisson result

$$P(0) = \exp\ (-n_T BR^d)\ , \tag{191}$$

where n_T is the trap density and BR^d is the volume of the d-sphere. To obtain a greatest lower bound, combine Eqs. (191) and (190), maximize the product with respect to R with t fixed, and arrive at

$$a^d\ \psi(0,\ t) > \exp\left[-\ C(n_T^{2/d}\ Dt)^{d/(d+2)}\right] \quad \text{(g.l.b.), where}$$

$$C = B(2A/dB)^{2/(d+2)}\ . \tag{192}$$

This result was also obtained independently by Grassberger and Procaccia (1982).

A much more difficult task is to derive a good rigorous least upper bound on $\psi(0,\ t)$; however, if one is clever enough it is possible to construct (l.u.b.) $\equiv$ (g.l.b.), so that Eq. (192) is actually an equality.

This was formally known as the "Kac-Lifshitz conjecture" (Lifshitz, 1965, 1968). Consider a random configuration of traps and imagine constructing a finite size lattice of d-hypercubes, each cube of edge ξR, where R is the lattice spacing. These cubes are large enough so that the average trap density in a cube is well-defined, and the linear extent of this lattice is NR with N a large number. The cellular lattice or template is therefore characterized by

$$\xi << 1, \qquad \xi R >> a \quad \text{(trap radius)}$$
$$N >> 1, \qquad n_T(\xi R)^d >> 1 \,. \tag{193}$$

The trap-averaged density at the origin is now split into two configuration summations as

$$\psi(0, t) = \sum_{\Lambda_1} \psi_{\Lambda_1}(0, t)\, P_{\Lambda_1} + \sum_{\Lambda_2} \psi_{\Lambda_2}(0, t)\, P_{\Lambda_2}, \tag{194}$$

where Λ_1 designates those trap configurations with at least M cubic cells completely devoid of traps, and Λ_2 refers to configurations with fewer than M cells empty. For a random trap distribution, elementary combinatorial reasoning yields an upper bound on the Λ_1 summation:

$$\sum_{\Lambda_1} < (4\pi Dt)^{-d/2} \sum_{m=M}^{N^d} \binom{N^d}{m} \exp\left[-mn_T(\xi R)^d\right] \left\{1 - \exp\left[-n_T(\xi R)^d\right]\right\}^{N^d-m}, \tag{195}$$

which, via Stirling's asymptotic formula, simplifies to [4]:

$$\sum_{\Lambda_1} < (4\pi Dt)^{-d/2} N^d \exp\left[-M\, n_T(\xi R)^d + M \ln (N^d/M)\right] . \tag{196}$$

Also, it is not difficult to show that the Λ_2 summation is bounded above as

$$\sum_{\Lambda_2} < a^{-d} \exp(-K\, T) = a^{-d} \exp(-K_0 a^{d-2}\, Dt/R^d) \,, \tag{197}$$

where K_0 is a dimensionless constant. In order to obtain a least upper bound for $\psi(0, t)$, combine Eqs. (194), (196), (197), fix the time t at some arbitrarily large value, fix $\xi << 1$, and then vary R, N, and M so that the bound is minimized [5]. A few lines of algebra produces

---

[4] In order to use Stirling's formula we further assume that

$$N_d \exp\left[-n_T(\xi R)^d\right] << 1 \text{ and } N^d >> M >> 1$$

These assumptions are consistent with the derived least upper bound.

[5] For the upper bound, note that the fraction of cells completely devoid of traps, MN^{-d}, vanishes as $t \rightarrow \infty$, and that the number of traps in a cell is, by Eq. (193), large and increasing with time. However, the total number of vacant cells M grows with time and the density at the origin will be greatest when these cells are centered about the origin in a sphere of radius $M^{1/d}R$.

$$R = a(K_0/A')^{1/d}\ (Dt/a^2)^{2/d(d+2)}$$

$$M = N^{d(1-\gamma)}, \qquad |\gamma| < 1$$

$$N \sim t^{|\beta|}, \qquad \beta = \frac{d-2}{d(d+2)(1-\gamma)}, \tag{198}$$

where A′ is a dimensionless constant, and this immediately leads to

$$a^d\ \psi(0,\ t) < \exp\left[-A'\left(Dt/a^2\right)^{d/(d+2)}\right] \quad (\text{l.u.b}), \tag{199}$$

which means that the time exponent d/(d+2) is exact. The elegant and more penetrating analysis of Donsker and Varadhan (1979) proves that, in fact, A′ is such that Eqs. (199) and (192) are identical, so we now have the exact asymptotic solution to a many-body reaction-diffusion problem, as well as the analytic evaluation of a non-trivial conditional Wiener integral. Incidentally, if the traps are allowed to interact with one another via an arbitrary potential, it is not too difficult to prove that the preceding analysis remains valid, provided the trap density n_T is replaced by the "quenched trap equation-of-state" (p_T/k_BT) (Kayser and Hubbard, 1983, 1984). This, of course, implies that the density-of-states function has some interesting features at a phase transition of the scattering medium. Also, note that the (g.l.b) "hole radius" R increases with time as $t^{1/(d+2)}$, which means that the growth rate of the dominant class of density fluctuations is sub-diffusive. This translates into low energy localized states for the quantum Lorentz gas, which are also known as the "fluctuation levels" of Lifshitz. For instance, with d = 3, g(ε) takes the form (Lifshitz, 1965, 1968; Friedberg and Luttinger, 1975; Luttinger and Tao, 1983)

$$\underset{\varepsilon\to\varepsilon_0}{g(\varepsilon)} \sim \exp\left(-E/(\varepsilon-\varepsilon_0)^{3/2}\right), \tag{200}$$

where E is an easily determined number (positive) and ε_0 is a spatially uniform background potential. This is in marked contrast to

$$\underset{\varepsilon\to\varepsilon_0}{g(\varepsilon)} \sim (\varepsilon-\varepsilon_0)^{1/2}, \tag{201}$$

which is valid for any periodic "repulsive scatterer" distribution, and which arises from delocalized electron states. The exact form of g(ε) for the random impurity problem over the entire range of ε is still an unsolved problem (Friedberg and Luttinger, 1975; Luttinger and Tao, 1983). However, the behavior of g(ε) in the immediate vicinity of a metal-insulator transition or mobility edge is analytically discernible via instanton, field-theoretic, and renormalization group techniques.

These are very recent developments which lie well beyond the scope of this lecture.

REFERENCES

Bourret, R.C., 1965, Can. J. Phys., 43:619; 1966, Can. J. Phys., 44:2519; 1964, Phys. Lett., 12:323.

Donsker, M.D., and Varadhan, S.R.S., 1975, Commun. Pure Appl. Math., 28:525; 1979, Commun. Pure Appl. Math., 32:721.

Dyson, F.J., 1949, Phys. Rev., 75:486 and 1736.

Falkenhagen, H., 1934, "Electrolytes," Oxford, New York.

Fox, R.F., 1978, Physics Reports, No. 3, 48:181.

Friedberg, R., and Luttinger, J. M., 1975, Phys. Rev. B. 12:4460; 1983, J. M. Luttinger and R. Tao, Ann. Phys., 145:185.

Fuoss, R.M., and Accascina, 1959, "Electrolytic Conductance," Interscience, New York.

Grassberger, P., and Procaccia, I., 1982, J. Chem. Phys., 77:6281.

Hansen, J.P., and McDonald, I.R., 1976, "Theory of Simple Liquids," Academic, New York.

Hong, K.M., and Noolandi, J., 1978, J. Chem. Phys., 68:5163; 1978, J. Chem. Phys., 69:5026.

Hynes, J.T., and Deutch, J.M., 1975, in: "Physical Chemistry an Advanced Treatise" XIB, Eyring, Henderson and Jost, eds., Academic, New York, Chap. 11.

Kac, M., and Luttinger, J.M., 1974, J. Math Phys., 15:183.

Kayser, R.F., and Hubbard, J.B., 1983, Phys. Rev. Lett. 51:79: 1984, J. Chem. Phys. 80:1127.

Kramers, H.A., 1940, Physica (Utrecht), 7:284.

Kubo, R., 1963, J. Math Phys., 4:174.

Landau, L.D., and Lifshitz, E.M., 1969, "Statistical Physics," Addison-Wesley, Reading, MA.

Lifshitz, I.M., 1968, Soviet Phys. J.E.T.P. 26:462; 1965, Soviet Phys. U.S.P. 7:549.

McQuarrie, D.A., 1976, "Statistical Mechanics," Harper and Row, New York.

Mozumder, A., 1968, J. Chem. Phys., 48:1659; 1974, J. Chem. Phys., 61:780.

Onsager, L., 1935, Ph.D. thesis, Yale University; 1934, J. Chem. Phys., 2:599; 1938, Phys. Rev., 54:554.

Resibois, P.M.V., 1968, "Electrolyte Theory," Harper and Row, New York.

Simon, B., 1979, "Functional Integration and Quantum Physics," Academic, New York; 1985, J. Stat. Phys., 38:65.

Skinner, J.L., and Wolynes, P.G., 1978, J. Chem. Phys., 69:2143.

Tatarski, V.I., 1961, "Wave Propagation in a Turbulent Medium," McGraw-Hill, New York.

Tolman, R.C., 1938, "The Principles of Statistical Mechanics," Oxford, London.

van der Ziel, A., 1954, "Noise," Prentice, Englewood Cliffs, N.J.

Van Kampen, N.G., 1981, "Stochastic Processes in Physics and Chemistry," North-Holland, New York.

Wax, N., 1954, "Selected Papers on Noise and Stochastic Processes," Dover, New York.

Zwanzig, R., 1964, Physica, 30:1109.

IONIC AND ELECTRONIC PROCESSES

THEORIES OF ELECTROLYTE SOLUTIONS

Jayendran C. Rasaiah

Department of Chemistry
University of Maine
Orono, ME 04469

INTRODUCTION

The systematic study of electrolyte solutions by van't Hoff and Arrhenius (1887) established physical chemistry as a scientific discipline. Electrolytes are important not only in solution chemistry and in the chemistry of electrode processes, but also in geochemistry and oceanography and in many areas of biophysics and biochemistry. "Salt solutions", as they are often referred to by chemists, are ubiquitous; they are easily handled in the laboratory and their properties accurately measured. Reams of experimental data on these solutions have been collected but their explanation and correlation in terms of the molecular (or microscopic) properties of the system continue to present major challenges to theoretical chemists. This has also lead to the formulation of sophisticated new techniques in statistical mechanics and the extensive use of computer simulation to study ionic fluids.

In this lecture, I will discuss a few of the theories of electrolytes developed since the classic work of Debye and Huckel (1923). Rapid progress in this field came after it was realized that many of the methods used since 1960 to study fluids could also be applied, with some modifications for the long-range coulomb forces, to ionic solutions. This lecture is not a catalog or review of this progress but a tutorial which concentrates on a few of the important steps in this chronicle and naturally reflects my own tastes and experience. The presentation is at an introductory level but is comprehensive, except for the omission of the very successful semi-empirical theory of electrolytes developed by Pitzer (1973) and his collaborators Mayorga (Pitzer and Mayorga, 1973) and Kim (Pitzer and Kim, 1974). Reviews of the subject have appeared periodically (Rasaiah, 1973; Watts, 1973; Friedman and Dale, 1977;

Olivares and McQuarrie, 1975; Outhwaite, 1975; Blum, 1980, Baus and Hansen, 1980; Hafskjold and Stell, 1982; and Friedman, 1981 and 1987) and the reader is referred to them for some of the details. The monographs and texts by Friedman (1962, 1985), McQuarrie (1976), Watts and McGee (1976) and Berry et al. (1980) also contain accounts of the modern theory. The review by Baus and Hansen (1980) contains an excellent discussion of the one-component plasma.

An electrolyte solution contains at least two components; a solvent and a solute which dissociates into at least two oppositely charged ions present in proportions that must satisfy the electroneutrality condition:

$$\sum_{i=1}^{\sigma} c_i e_i = 0 \ . \tag{1}$$

Here c_i and e_i are the concentration and charge, respectively, of the species i, and σ is the number of components in solution. In the usual preparative concentration range (0.1 to 2 molar) the reduced densities (or concentrations) of solute and solvent are very different. In a 2M NaCl solution for example, the reduced densities of solute and solvent are 0.02 and 0.8, respectively, in units of $\rho\sigma^3$, where ρ is the density and σ is the diameter of the species (solute or solvent molecules) under consideration. Although the solute-reduced densities are relatively small, the long-range coulomb forces between the ions have a profound effect on the equilibrium properties of the system. These properties, which are averages over the configurations of the system in phase space, can be calculated in principle using the methods of equilibrium statistical mechanics from the intermolecular potentials. Examples of these are the heats of dilution, the activity and osmotic coefficients, and the partial molar volumes, which are all experimentally measurable, and have been extensively tabulated (Harned and Owen, 1950; Robinson and Stokes, 1959, Hovarth, 1985). The difficulty with a comprehensive molecular theory of ionic solutions is that the solute potentials for the ions (e.g., 1.0 m NaCl in aqueous solution) are not completely known. Hence, model potentials are often used and the credibility of a theory is determined by means of self-consistency tests or by comparison with the essentially exact results obtained by computer simulation of the same model system. The main conclusions of these studies are that, with some exceptions, several theoretical methods are now available to predict the equilibrium properties of model electrolytes. The exceptions are the properties of electrolytes where ion-association is appreciable; for instance, the behavior of higher-valence electrolytes at low concentrations at room temperature, or aqueous 1-1 electrolytes near the critical point of water. The modern theory of weak electrolytes is also just beginning to be developed (Lee et al., 1985). Apart from these drawbacks, the accuracy of the methods now available allows us to conclude that it is not so much the theories that need improvement but the model potentials themselves, and it is possible that this will be a major area of study in the future.

There are generally two ways to proceed in determining the equilibrium properties of a solution. The configurational averages can be carried out simultaneously over the solute and solvent species, or they may be carried out successively over the solvent, at fixed positions of the solute, and then over the coordinates of the solute species. The first method is due to Kirkwood and Buff (1951) and has been applied most successfully, until recently, to simple nonelectrolyte solutions and mixtures. The second is the McMillan-Mayer theory which has been more widely used in the study of electrolytes, polymers, and proteins. Our discussion of electrolytes will be in terms of the McMillan-Mayer (MM) formalism. For recent applications of the Kirkwood Buff theory see Perry et al. (1988).

MODEL POTENTIALS

The solvent-averaged solute potentials used in the MM theory are really free energies and depend not only on the separation between the ions but also on the temperature T and the density ρ_s of the solvent molecules. Assuming pairwise additivity, the potential of N ions at locations $r_1, r_2, \ldots r_N$ is given by

$$U_N\,(r_1,\ \ldots\ r_N,\ T,\ \rho_s) = \Sigma\, u_{ij}(r;T,\rho)\ , \tag{2}$$

where the pair "potential" has the form

$$u_{ij}\,(r_{ij};T,\ \rho_s) = u_{ij}^{*}(r) + e_i e_j/\varepsilon_o r\ , \tag{3}$$

where r is the distance between the ions i and j, ε_o is the dielectric constant of the solvent, and u_{ij}^{*} (r) is the short-range potential which is generally unknown. The second term in this equation is the long-range coulomb potential which was the source of the many difficulties encountered in earlier theories of electrolytes applied to solutions at moderately high concentrations; most of the difficulties in the theory now lie in the characterization of the short-range part of the potential $u_{ij}^{*}(r;T,\rho_s)$ which includes:

(a) the repulsive part of the interaction between two ions,

(b) the effect of solvent granularity,

(c) dielectric effects not included in the coulomb term,

(d) effects associated with the detailed molecular structure of the solvent around an ion, and

(e) van der Waals and chemical interactions, e.g., dissociation effects in a weak acid.

Figure 1 shows the result of an early Monte Carlo simulation [McDonald and Rasaiah (1975)] of the force between two ions in a Stockmayer fluid. The oscillations in the potential and the force are due to the solvent granularity modified by the ion-dipole and dipole-dipole interactions in the solution. The potential energy of interaction between pairs of ions in a real solution have some of the features found for the simple system shown in Fig. 1, but the details are known only for a few cases which have been investigated recently. Interesting examples are (Figs. 2 and 3) the potentials of average force between two chloride ions, and between the alkali metal (Li, Na, and K) and chloride (Cl) ions calculated recently by Pettitt and Rossky (1986). Distinct minima corresponding to contact- and solvent-separated ion-pairs are found for oppositely charged ions, while the minimum in the Cl^{-}-Cl^{-} interaction is apparently related to bridging by water molecules (Dang and Pettitt, 1987). Fascinating accounts of computer simulations of the structure and dynamics of a pair of Na^{+} and Cl^{-} ions in water are given by Berkowitz et al. (1984), Karim and McCammon (1986a), and Belch et al. (1986). From these accounts it appears unlikely, at least to me, that the potential of average force for these ions is pairwise additive, and it may become expedient to use an effective concentration-dependent pair potential in future applications. A theory proposed by Adelman (1976a, 1976b) shows how this might be implemented. One consequence of this is that the dielectric constant in Eq. (3) should be replaced by that of the solution (Friedman, 1982).

The lack of detailed information on the potential of average force, particularly in early work, led to the use of model potentials, the simplest of which is the restricted primitive model (RPM) in which the short-range potential is that for hard spheres of diameter σ:

$$\begin{aligned} u_{ij}^{*}(r) &= \infty \qquad (r < \sigma) \\ &= 0 \qquad (r > \sigma) \,. \end{aligned} \tag{4}$$

The model treats the solution as a mixture of charged hard spheres of diameter σ in a structureless medium of dielectric constant ε_0. Other models for strong electrolytes have also been reported quite extensively in the literature. They are the square-mound (or well) model (Rasaiah and Friedman, 1968a; Rasaiah, 1970a) in which the hydration spheres around the ions are represented by a barrier of finite height (or depth) and width equal to the diameter of a water molecule, and the co-sphere

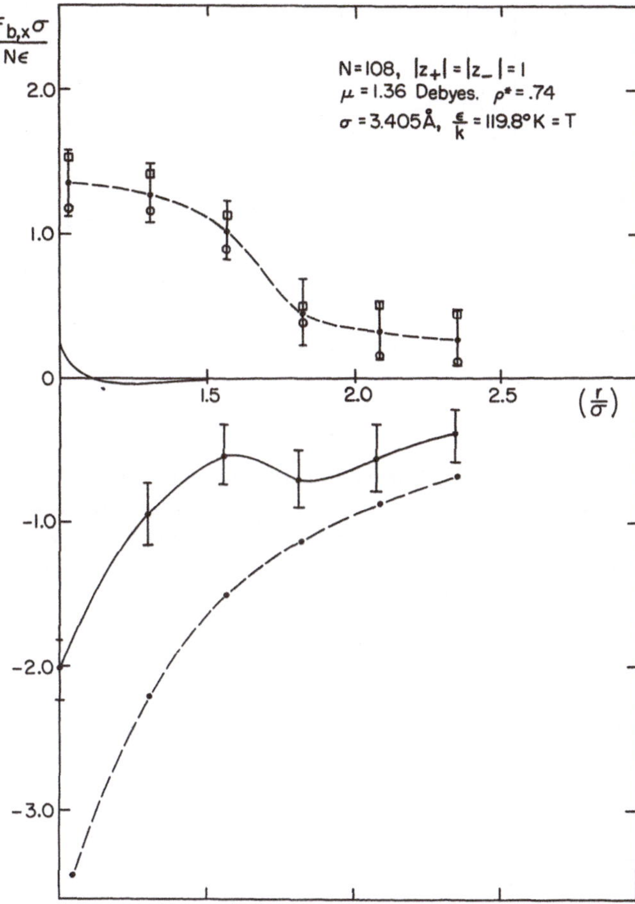

Fig. 1. The average force (in reduced units) on an ion b as a function of the distance from an ion of opposite sign. The curves from top to bottom are the solvent force, $< F_{b,x}^{solv} > \sigma/N\varepsilon$; the Lennard-Jones force, $F_{b,x}^{LJ} \sigma/N\varepsilon$; the total force, $< F_{b,x} > \sigma/N\varepsilon$; and the Coulomb force, $< F_{b,x}^{coul} > \sigma/N\varepsilon$. The total force is the sum of the solvent force, the Lennard-Jones force, and the coulomb force (from McDonald and Rasaiah, 1975).

overlap model, proposed by Ramanathan and Friedman (1971), in which the short-range potential includes a softer core repulsion term (COR_{ij}) varying as r^{-9}, a dielectric repulsion term (CAV_{ij}) varying as r^{-4}, and a Gurney term (GUR_{ij}) which represents contributions from the overlap of co-spheres:

$$u_{ij}^*(r) = COR_{ij} + CAV_{ij} + GUR_{ij} \ . \tag{5}$$

The first two terms are repulsive while the Gurney term may be attractive or repulsive. The cavity term is small and the only adjustable parameters are the coefficients A_{ij} in the Gurney term:

$$GUR_{ij} = A_{ij} \ V_{mu}(r)/V_w \ , \tag{6}$$

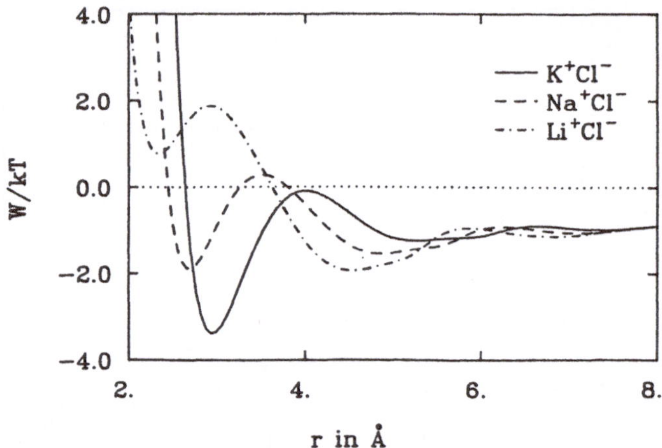

Fig. 2. Potentials of mean force for K^+ Cl^- (solid line), Na^+ Cl^- (dashed line), and Li^+ Cl^- (dot-dash line) in water at 25°C. The potential for water is a modification of TIPS and the interionic potentials are of the Huggins-Mayer form (from Pettitt and Rossky, 1986).

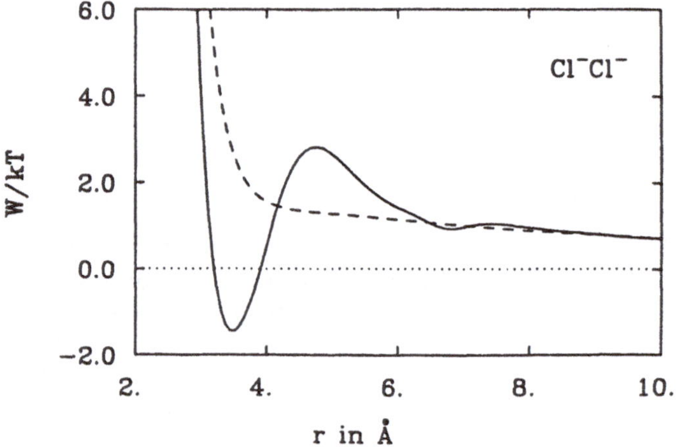

Fig. 3. Potentials of mean force for Cl^--Cl^- (solid line) in TIPS water at 25°C (see caption of Fig. 2). The dashed line shows the curve for the primitive model (from Pettitt and Rossky, 1986).

where $V_{mu}(r)$ is the mutual volume of overlap of the co-spheres around i and j when the distance between them is r, and V_W is the molar volume of water. Further refinements to this model, which include the effects of solvent granularity on the potential of average force, have been introduced by Tembe et al. (1982) in their study of the Fe^{2+}-Fe^{3+} electron exchange in water. Either of these models can be used with an accurate theory to fit the experimental thermodynamic data for strong electrolytes with parameters varying from one species to another in a physically plausible way. Paradoxically this shows that the models used to fit these thermodynamic data are by no means unique and other probes [e.g., x-ray and neutron diffraction or the distinct self-diffusion coefficient; Friedman (1986, 1988), Zhong and Friedman (1988)] have to be used to further characterize and distinguish between them. Deviations from the primitive model have been discussed by Stell (1973).

Another model has been studied recently in the investigation of weak electrolytes (Lee et al., 1985). This is the sticky electrolyte model (SEM) which is identical to the RPM except for a delta function interaction between oppositely charged ions a distance $L<\sigma$. It mimics chemical bonding between positive and negative ions and was first introduced by Cummings and Stell (1984) in their study of chemical reactions in non-ideal systems. The model is a modification of Baxter's model of adhesive hard spheres (1968a), the main difference being that bonding or stickiness takes place within the core rather than on the surface. All of these Hamiltonian models have a repulsive potential at short interionic distances and a coulombic interaction with the bulk dielectric constant in the denominator which is asymptotically correct at large separations. The RPM is the simplest of the models with both these important features.

THE MCMILLAN-MAYER THEORY

McMillan and Mayer (1945) showed that the same statistical methods used in the study of gases can be applied to the solute particles in solution, provided the solute potentials are averaged over the configurations of the solvent molecules (see Figs. 1, 2 and 3) and the solution is under an additional pressure Π, called the osmotic pressure. The osmotic pressure ensures that the chemical potential of the solvent in the two phases is the same. A literal description of this is to imagine that the two phases are separated by a semi-permeable membrane which permits only the solvent molecules to diffuse from one side to the other. The additional pressure on the solution required to maintain equilibrium between the two phases at the same temperature is the osmotic pressure.

By using solvent-averaged potentials the emphasis in the McMillan-Mayer theory is on the solute molecules. Van't Hoff suggested in 1907 that the osmotic coefficient ϕ, in the McMillan-Mayer system, measures the deviation of the osmotic pressure of the solution from the ideal gas behavior and is defined by

$$\phi = \Pi/ckT \; . \tag{7}$$

Using standard thermodynamic arguments, the excess Helmholtz free energy per unit volume, A^{ex}, is obtained by integration over the solute concentration c:

$$A^{ex} = ckT \int_0^c (\phi - 1) d\ln c' \; , \tag{8}$$

and the mean ionic activity coefficients γ_{+-} can be obtained by differentiation of the Helmholtz free energy with respect to the concentration, i.e.,

$$dA^{ex}/dc = kT\ln\gamma_{+-} \;. \tag{9}$$

It should come as no surprise that the osmotic pressure, like the pressure of a gas, has a virial expansion:

$$\Pi/ckT = \Phi = 1 + B_2(T)c + B_3(T)\,c^2 + \ldots \;, \tag{10}$$

where the "virial coefficients" are determined by the solvent-averaged potentials. Defining the Mayer f-function by

$$f_{12}(r) = \exp[-\beta\, u_{12}(r)]-1 \;, \tag{11}$$

the second and third virial coefficients are given by,

$$B_2(T) = -1/2 \int f_{12}(r_{12})\, d\vec{r}_2 \text{ and} \tag{12a}$$

$$B_3(T) = -1/3 \int f_{12}(r_{12}) f_{13}(r_{13}) f_{32}(r_{32}) d\vec{r}_2\, d\vec{r}_3 \;. \tag{12b}$$

If we represent the f-bond by a line with two open circles to denote the coordinates of particles 1 and 2, then the first three virial coefficients can be depicted graphically as

```
B2(T) = o---●                                   (13)

           ●
          / \
B3(T) =  o---●                                  (14)

         ●---●     ●---●     ●---●
         |   |     |\  |     |\ /|
         |   |     | \ |     | X |
B4(T) =  o---●  +  o---●  +  o---● .            (15)
```

Here blackening a circle implies integration over the coordinates of the particle represented by the circle. Further generalizations of this notation to include the symmetry number of each graph and the concentration or density of the particles whose coordinates are being integrated over is a powerful shorthand; it is described in texts and monographs (Friedman, 1985; Hansen and McDonald, 1986) to which the interested reader is referred to for additional details.

The McMillan-Mayer theory provides the theoretical foundation for van't Hoff's supposition of ideal gas behavior for the osmotic pressure of the solute; it also shows how deviations from this behavior, which are observed at higher solute concentrations, can be studied with the same theory. The assumption of pair-wise additivity of the solvent-averaged solute potentials, which we have assumed to simplify our discussion, is not completely justified, but even this can be taken into account in principle in the virial expansion when the non-additive terms are known (Friedman, 1962).

Another approach to the thermodynamic properties of solutions is to calculate them from the solute-solute distribution functions rather than from the virial coefficients. Approximations to these functions, which correspond to the summation of a certain class of terms in the virial series to all orders in the solute concentration (or density), have already been worked out for simple fluids, and the McMillan-Mayer theory states that the same approximations may be applied to the solute particles in solution provided the solvent-averaged potentials are used to determine the solute distribution functions. Examples of these approximations are the Percus-Yevick (PY) (1958), Hypernetted-Chain (HNC), mean-spherical (MS), and Born-Green-Yvon (BGY) theories. Before discussing them we will review some of the properties of distribution functions and their relationship to the observed thermodynamic variables.

The distribution function $g_{ij}(\vec{r}_1,\vec{r}_2)$ is related to the probability of finding species i and j at $\vec{r}_1$ and $\vec{r}_2$, respectively; more precisely

$$c_j g_{ij}(\vec{r}_1,\vec{r}_2) = \text{Probability that species } j \text{ is at } \vec{r}_2 \text{ assuming } i \text{ is at } \vec{r}_1. \tag{16}$$

The function $g_{ij}(\vec{r}_1,\vec{r}_2)$ is the ensemble average of the product of two delta functions, one at $\vec{r}_1$ and the other at $\vec{r}_2$, i.e.,

$$g_{ij}(\vec{r}_1,\vec{r}_2) = V^2 < \delta(\vec{r}_1-\vec{r})\delta(\vec{r}_2-\vec{r}\,')> . \tag{17}$$

For an isotropic fluid, $g_{ij}(\vec{r}_1,\vec{r}_2) = g_{ij}(r)$, where $r = |\vec{r}_1 - \vec{r}_2|$. In the limit of low concentrations, the distribution function is approximated by the Boltzmann exponential formula

$$g_{ij}(r,T) \rightarrow \exp(-\beta\, u_{ij}(r)) \qquad \text{as } c \rightarrow 0 , \tag{18}$$

where $u_{ij}(r)$ is the pair potential. For charged systems, however, the potential function multiplied by β appearing in the exponential factor is the screened coulomb potential defined by

$$q_{ij}(r) = -\beta\, e_i e_j \exp(-\kappa r)/\varepsilon_o r , \tag{19}$$

where κ is the inverse Debye length defined by

$$\kappa^2 = \frac{4\pi e^2}{\varepsilon_o kT} \sum_{i=1}^{\sigma} c_i z_i^2 . \tag{20}$$

As $r \rightarrow \infty$, $g_{ij}(r) \rightarrow 1$, signifying no correlation between the particles. Also as $r \rightarrow 0$ the repulsion between two particles ensures that $g_{ij}(r) \rightarrow 0$. The charge density at a distance r away from an ion is related to the distribution function by

$$\rho_i(r) = \sum_{j=1}^{\sigma} c_j e_j g_{ij}(r) \ . \qquad (21)$$

The thermodynamic properties are obtained using standard relations of statistical mechanics, keeping in mind that the potentials of average force are free energies [Rasaiah and Friedman (1968b), Rasaiah (1970b), and that the solution is electrically neutral (Friedman and Ramanathan, 1970)]. The following relations provide directly the experimental property shown in parenthesis:

(a) The virial equation (the osmotic coefficient from isopiestic measurements)

$$\phi_v = 1 - \frac{1}{6ckT} \sum_{i=1}^{\sigma} \sum_{j=1}^{\sigma} c_i c_j \int r \frac{\partial u_{ij}(r)}{\partial r} g_{ij}(r) \, d\vec{r} \ ; \qquad (22)$$

(b) The compressibility equation (the activity coefficient from cell EMF's)

$$\frac{\partial \ln \gamma_\pm}{\partial \ln c} = \frac{1}{cG_{+-}} - 1 \ , \qquad (23)$$

where $$G_{+-} = \int (g_{+-}(r)-1) d\vec{r} \ ; \qquad (24)$$

(c) The energy equation (the heat of dilution from calorimetric measurements)

$$E^{ex} = \frac{1}{2} \sum_{i=1}^{\sigma} \sum_{j=1}^{\sigma} c_i c_j \int \frac{[\partial \beta \, u_{ij}(r)]}{\partial \beta} g_{ij}(r) \, d\vec{r} \ , \qquad (25)$$

where for an ionic solution

$$\frac{\partial[\beta \, u_{ij}(r)]}{\partial \beta} = \frac{e_i e_j}{\varepsilon_o r} \left[1 + \frac{d\ln \varepsilon_o}{d\ln T} \right] + \frac{\partial[\beta \, u^*_{ij}(r)]}{\partial \beta} \ , \qquad (26)$$

and $d\ln\varepsilon_o/d\ln T = -1.3679$ for water at 25 C;

(d) The volume equation (the partial molar volume of the solute from density measurements)

$$V^{ex} = \frac{1}{2} \sum_{i=1}^{\sigma} \sum_{j=1}^{\sigma} c_i c_j \int \frac{\partial u_{ij}(r)}{\partial P_o} g_{ij}(r) \, d\vec{r} \ , \qquad (27)$$

where

$$\frac{\partial u_{ij}(r)}{\partial P_o} = - \frac{e_i e_j}{\varepsilon_o r} \left[\frac{d\ln \varepsilon_o}{d\ln P_o} \right] + \frac{\partial u_{ij}^*(r)}{\partial P_o} \ , \qquad (28)$$

and $d\ln\varepsilon_o/d\ln P_o = 47.1 \times 10^{-6}$ for water at 25 C.

The equilibrium properties obtained in the McMillan-Mayer system are for a fixed temperature T, chemical potential of the solvent μ_1, and concentration of a solute c. They can be converted by thermodynamic manipulation to the corresponding properties at the usual experimental conditions of constant T, P_o, and c; the corrections are usually small, but significant at concentrations above 0.1 molar (Friedman, 1972).

THE DIRECT-CORRELATION-FUNCTION AND INTEGRAL-EQUATION APPROXIMATIONS

The behavior of fluids and solutions would be more easily understood if the distribution functions could be calculated exactly for a simple but realistic model potential at a moderately high concentration (e.g., 1.0M). This is unfortunately not the case, and it is necessary to devise accurate but computationally convenient approximations for the distribution functions. They are more easily formulated in terms of the direct correlation function, $c_{ij}(r)$, which is defined by the Ornstein-Zernike (OZ) equation (1914):

$$h_{ij}(r_{12}) = c_{ij}(r_{12}) + \sum_{k=1}^{\sigma} c_k \int h_{ik}(r_{13}) c_{kj}(r_{32})\, d\vec{r}_3 \ , \tag{29}$$

where $h_{ij}(r) = g_{ij}(r)-1$. The second term is a convolution integral and represents the (indirect) contribution to the h-function beyond what is already in the c-function. The direct correlation function $c_{ij}(r)$ completely determines the distribution function $g_{ij}(r)$; this is easily seen by taking the fourier transform of the OZ equation and applying the convolution theorem, when we find that

$$\tilde{H}(k) = \tilde{C}(k) + \tilde{H}(k).\ D.\ \tilde{C}(k) \ , \tag{30}$$

where $\tilde{H}(k)$ and $\tilde{C}(k)$ are matrices whose elements are the fourier transforms of the h and c functions, respectively, and D is the diagonal matrix whose elements are the concentrations c_i of the ions in solution. This equation can be solved for either $\tilde{C}(k)$ or $\tilde{H}(k)$ if one or the other is known, which proves our assertion.

An important characteristic of the direct correlation function c(r) is that it usually has a simpler dependence on r than h(r), and is accordingly easier to approximate. It can be written in terms of the pair potential and the distribution function as

$$c_{ij}(r) = -\beta\, u_{ij}(r) + h_{ij}(r) - \ln g_{ij}(r) + \zeta_{ij}(r) \ , \tag{31}$$

where the last term $\zeta_{ij}(r)$, called the bridge function, is usually small and has the expansion

$$\zeta_{ij}(r) = \;_i\!\diamond_j + \cdots . \tag{32}$$

Equation (31) is an exact relation. Away from the critical point, the asymptotic form of the direct correlation function is dominated by the pair potential, i.e.,

$$c_{ij}(r) = -\beta\, u_{ij}(r) \qquad (r \rightarrow \infty) . \tag{33}$$

Two integral-equation approximations, which are useful in electrolyte theory, are the following:

(a) The Mean-Spherical Approximation (MSA)

$$c_{ij}(r) = -\beta\, u_{ij}(r) \qquad (r > \sigma) \tag{34}$$

$$g_{ij}(r) = 0 \qquad (r < \sigma) . \tag{35}$$

The second of these relations is exact; the first is an approximation which assumes the asymptotic form of the direct correlation function for all separations beyond the hard core diameter. By implication the solutes molecules are assumed to have hard cores, e.g., charged hard spheres (RPM) or sticky charged hard spheres (SEM). An advantage to the MSA is that the thermodynamic properties can be determined analytically, and are quite accurate for low valence electrolytes in aqueous solution at room temperature. The thermodynamics of the MSA for simple fluids is discussed by Høye and Stell (1977).

(b) The Hypernetted Chain Approximation (HNC)

Here the bridge functions, which are assumed to be small, are neglected, i.e.,

$$\zeta_{ij}(r) = 0 \text{ for all } r , \tag{36}$$

and the resulting equation is solved simultaneously with the Ornstein-Zernike equation. It can be applied equally well to solutes with continuous potentials or hard cores, but the solutions have to be obtained numerically. The HNC approximation is one of the most accurate integral equations available for electrolytes.

Another approximation used in the theory of fluids is the Percus-Yevick (PY) approximation (1958). It assumes that the bridge functions are zero and, in addition, that $\exp(h_{ij}(r)-c_{ij}(r)) \sim 1 + h_{ij}(r) - c_{ij}(r)$, from which it follows that

$$c_{ij}(r) = [1 - \exp(\beta\, u_{ij}(r))]g_{ij}(r) \qquad \text{(PY)} . \tag{37}$$

For hard spheres, the PY and mean-spherical approximations are identical. This identity does not extend to other potentials (charged hard spheres, for instance). The PY approximation is useful for molecules with short-ranged potentials, and the analytic solutions for hard spheres have been obtained by Wertheim (1963) and Thiele (1963).

(c) The Reference-Hypernetted-Chain Approximation.

The reference-hypernetted-chain approximation (RHNC) proposed by Lado (1973) approximates the bridge function $\zeta_{ij}(r)$ by the bridge function $\zeta_{ij}^{*}(r)$ for the corresponding short-range potential:

$$c_{ij}(r) = -\beta\, u_{ij}(r) + h_{ij}(r) - \ln g_{ij}(r) + \zeta_{ij}^{*}(r)\ . \tag{38}$$

Rosenfeld and Ashcroft (1979) proposed that this is essentially the same function for all potentials, and Lado et al. (1983) suggested that the optimum short-range potential can be determined by minimizing the free energy. A variational principle for integral equations associated with the Ornstein-Zernike equation has been derived by Olivares and McQuarrie (1976).

The distribution functions for electrolytes must satisfy two important moment conditions:

(a) The zeroth-moment condition

This follows from the electroneutrality of the solution which implies that the total charge on an ion must be equal and opposite in sign to the charge in its surroundings; hence

$$-e_i = \int \rho_i(r)\,d\vec{r}\ . \tag{39}$$

(b) The second-moment condition

This was first derived by Stillinger and Lovett (1968) and states that

$$-\frac{3\varepsilon}{2\pi\beta} = \sum_{i=1}^{\sigma} c_i e_i \int \rho_i(r)\, r^2\, d\vec{r}\ . \tag{40}$$

Mitchell et al. (1977) showed that the zeroth- and second-moment conditions follow from the asymptotic form of the direct-correlation function [Eq. (33)]. The MS, HNC, RHNC, and Debye Huckel limiting law approximations satisfy both these conditions; the Percus-Yevick equation does not obey the second-moment condition and is less useful for electrolytes than it is for uncharged systems.

The solutions to the integral equation approximations for electrolytes are discussed in the section entitled SOME THEORIES OF ELECTROLYTES. The numerical solutions to the HNC and PY approximations are usually obtained with Fast Fourier transform routines; while the solutions to the MSA for the RPM electrolyte are determined analytically. In the next section we discuss an application of the Wiener-Hopf factorization to symmetrically charged electrolytes which allows the numerical solutions to the integral equations to be obtained in real space. The analytic solutions to the MSA for the RPM electrolyte can also be obtained with the same equations and is discussed as part of the section under SOME THEORIES OF ELECTROLYTES. The technique was first introduced by Baxter for molecules with short-range potentials, and was later extended to electrolytes independently by Tibavisco (1974), Blum (1975, 1980), and Thompson (1978).

SYMMETRICALLY CHARGED ELECTROLYTES: WIENER-HOPF FACTORIZATION

Baxter (1968b) showed that the Ornstein-Zernike equation could, for some simple potentials, be written as two one-dimensional integral equations coupled by a function q(r). In the PY approximation for hard spheres, for instance, the q(r) functions are easily solved, and the direct-correlation function c(r) and the other thermodynamic properties can be obtained analytically. The pair-correlation function g(r) is derived from q(r) through numerical solution of the integral equation which governs g(r) for which a method proposed by Perram (1975) is especially useful. Baxter's method can also be used in the numerical solution of more complicated integral equations such as the hypernetted-chain (HNC) approximation in real space, avoiding the need to take Fourier transforms. An equivalent set of relations to Baxter's equations was derived earlier by Wertheim (1964).

Baxter's integral equations for the correlation functions, h(r) and c(r) of a one-component fluid, were derived using the Wiener-Hopf technique to factorize the function $[1 - \rho\tilde{c}(k)]$, where $\tilde{c}(k)$ is the Fourier transform of c(r), and $[1 - \rho\tilde{c}(k)]$ is assumed not to have any zeros on the real axis. The potential energy of interaction u(r) was taken to be short ranged, which implies that the direct-correlation function c(r) is also a short-ranged function, since it is generally assumed that the asymptotic form of c(r) is $-\beta\, u(r)$ for all states that are not near the critical point. The technique was later extended by Baxter to multicomponent fluids. Since the coulomb potential is long ranged, special methods are necessary to apply this method to electrolytes. The results of this factorization for the symmetrically charged RPM electrolyte are summarized in this section with the details of the derivation provided in an Appendix following the work of Thompson (1978). Blum (1975) and Tibavisco (1974) had earlier extended the analysis to asymmetrically charged hard spheres of unequal size.

The symmetry of the system allows some simplification:

$$e_+ = e_- = e \; ; \quad c_+ = c_- = \rho/2 \; ; \tag{41}$$

$$h_{++}(r) = h_{--}(r) \; ; \quad c_{++}(r) = c_{--}(r) \; ; \tag{42}$$

where we use the symbol ρ instead of c here for the total ionic concentration to avoid confusion with the direct-correlation function. Defining the sum and difference functions,

$$F_S(r) = [F_{+-}(r) + F_{++}(r)]/2 \quad \text{and} \quad F_D(r) = [F_{+-}(r) - F_{++}(r)]/2 \; , \tag{43}$$

of the indirect and direct correlation functions $h_{ij}(r)$ and $c_{ij}(r)$, the Ornstein-Zernike equation separates into two equations

$$h_S(r) = c_S(r) + \rho\, c_S(r) * h_S(r) \quad \text{and} \tag{44}$$

$$h_D(r) = c_D(r) - \rho\, c_D(r) * h_D(r) \; , \tag{45}$$

where for the restricted primitive model (RPM)

$$h_S(r) = -1 \qquad 0 < r < \sigma \tag{46}$$

$$h_D(r) = 0 \qquad 0 < r < \sigma \; , \tag{47}$$

and * denotes a convolution integral. Note that $h_S(r)$ and $h_D(r) \rightarrow 0$ as $r \rightarrow \infty$, and $c_{ij}(r) = -\beta\, e_i e_j/\varepsilon r$ for large r provided the system is not too close to the critical point. It follows that $c_S(r)$ is a short-ranged function and $c_D(r)$ is long ranged which allows us to write,

$$c_D(r) = c_D{}^{\circ}(r) + \beta\, e^2 \exp(-zr)/\varepsilon_0 r \; , \tag{48}$$

where $c_D{}^{\circ}(r)$ is short ranged. The exponential function with $z > 0$ is a convergence factor and z is set equal to zero in the final equations for $h_D(r)$ and $c_D(r)$. Define

$$J_D(r) = \int_r^{\infty} t h_D(t) dt \; , \tag{49}$$

and J_S by a similar definition. The integral of $J_D(r)$ is determined by the electroneutrality condition which can be written as

$$-e = \frac{\rho e}{2} \int_0^{\infty} [h_{++}(r) - h_{+-}(r)] 4\pi\, r^2 dr \; , \tag{50}$$

or

$$1 = 4\pi\rho \int_0^{\infty} h_D(r)\, r^2 dr = -4\pi\rho \int_0^{\infty} r J_D{}'(r)\, dr \; . \tag{51}$$

Integrating by parts we find that

$$1 = 4\pi\rho \int_0^{\infty} J_D(r)\, dr \; . \tag{52}$$

It follows easily from the Eq. (25) that the excess energy is given by

$$E^{ex}/NkT = E^{ex,*}/NkT - \{1 + d\ln\varepsilon_0/d\ln T\}\kappa H/2 , \quad (53)$$

where $H = -\kappa J_D(\sigma)$ and $E^{ex,*}$ is the contribution of the short-range potential to the excess energy. This is zero for charged hard spheres in the RPM but is equal to the binding energy in the SEM.

The main equations which follow from the application of Baxter's method to symmetrically charged electrolytes are a pair of equations each for the sum and difference functions:

(a) For the sum equations with $r > 0$

$$rc_s(r) = -q_s'(r) + 2\pi\rho \int_r^\infty q_s'(t)q_s(t-r)dt , \quad (54)$$

$$rh_s(r) = -q_s'(r) + 2\pi\rho \int_0^\infty q_s(t)(r-t)h_s(|r-t|)dt , \quad (55)$$

where $q_s(r) = 0$, $r < 0$, and the prime in $q_s'(r)$ denotes differentiation with respect to the variable r.

(b) For the difference equations with $r > 0$

$$rc_D^\circ(r) = q_D^{\circ\prime}(r) + 2\pi\rho\, [Mq_D^\circ(r) - \int_0^\infty q_D^\circ(s)\, q_D^{\circ\prime}(r+s)ds] , \quad (56)$$

$$rh_D(r) = q_D^{\circ\prime}(r) + 2\pi\rho \int_0^\infty [q_D^\circ(s)+M](r - s)h_D(|r-s|)ds , \quad (57)$$

with $q_D^\circ(r) = 0$, and $2\pi\rho M = -\kappa$, where κ is the inverse Debye length. The integrated forms of Eqs. (55) and (57) are

$$J_s(r) = q_s(r) + 2\pi\rho \int_0^\infty q_s(t)\, J_s(|t - r|)dt , \quad (58)$$

$$J_D(r) = q_D^\circ(r) - M/2 + 2\pi\rho \int_0^\infty q_D^\circ(t)J_D(|r-t|)dt - \kappa \int_0^r J_D(t)\, dt . \quad (59)$$

These equations are derived in the Appendix. In the MSA the $q_{S,D}^\circ(r)$ functions are zero for $r > \sigma$, and the upper limits of integration equal to ∞ in Eqs. (54) to (59) are replaced by σ.

SOME THEORIES OF ELECTROLYTES

(a) The Debye-Huckel Theory

The model used is the restricted primitive model (RPM). The average electrostatic potential $\psi_i(r)$ at a distance r away from an ion i is related to the charge density $\rho_i(r)$ by Poisson's equation

$$\nabla^2 \psi_i(r) = -\frac{4\pi\rho_i(r)}{\varepsilon_o} = -\frac{4\pi}{\varepsilon_o}\sum_j c_j e_j g_{ij}(r) \, . \tag{60}$$

Debye and Huckel (1923) assumed that the ion distribution functions $g_{ij}(r)$ are given by $\exp(-\beta\, e_j\psi_i)$ which is an approximation. This leads to the Poisson-Boltzmann equation:

$$\begin{aligned}\nabla^2 \psi_i(r) &= -\frac{4\pi}{\varepsilon_o}\sum_j c_j e_j \exp[-\beta\, e_j \psi_i(r)] \qquad (r > \sigma)\\ &= 0 \qquad (r < \sigma)\, ,\end{aligned} \tag{61}$$

which on linearization gives the Debye-Huckel differential equation

$$\begin{aligned}\nabla^2\psi_i(r) &= \kappa^2\, \psi_i(r) \qquad (r > \sigma)\\ &= 0 \qquad (r < \sigma)\, ,\end{aligned} \tag{62}$$

where κ is defined by Eq. (20). The solution to this differential equation is

$$\begin{aligned}g_{ij}(r) &= 1 - \frac{\beta\, e_i e_j}{\varepsilon_o r}\,\frac{\exp[-\kappa(r-\sigma)]}{(1+\kappa\sigma)} \qquad (r > \sigma)\\ &= 0 \qquad (r < \sigma)\, ,\end{aligned} \tag{63}$$

which obeys the zeroth-moment or electroneutrality condition, but not the second-moment condition.

Debye and Huckel calculated the contribution of the electrical interactions to the mean activity coefficient γ_{+-} of a single electrolyte, and found that

$$\ln \gamma_{+-} = \frac{-A|z_+ z_-|\sqrt{I}}{1 + B\,\sigma\sqrt{I}}\, , \tag{64}$$

where A and B are constants determined by the temperature t, ε_o is the dielectric constant of the solvent, and z_+ and z_- are the valencies of the positive and negative ions, respectively. This expression is widely used to calculate the activity coefficients of simple electrolytes in the usual preparative range, although the contributions of the hard cores to non-ideal behavior are ignored. This is called the "excluded-volume" effect which is present even in the absence of electrical interactions between the solute molecules.

When $\kappa\sigma \ll 1$ (i.e., at very low concentrations), Eq. (63) simplifies to the Debye-Huckel limiting-law (DHLL) distribution function

$$\begin{aligned}g_{ij}(r) &= 1 - \beta\, e_i e_j \exp(-\kappa\, r)/\varepsilon_o r \qquad (r > \sigma)\\ &= 0 \qquad (r < \sigma)\, .\end{aligned} \tag{65}$$

Equation (65), unlike its predecessor Eq. (63), is consistent with both the zeroth- and second-moment conditions and has an interesting physical interpretation. The total charge $P_i(r)dr$ in a shell of radius r and thickness dr around an ion is

$$P_i\ (r)dr = \rho_i\ (r)\ 4\pi\ r^2 dr = -\kappa^2 e_i\ r\ \exp\ (-\kappa\ r)dr\ , \tag{66}$$

which has a maximum at $r = 1/\kappa$, called the Debye length or the radius of the ionic atmosphere. Each ion is pictured as surrounded by a cloud or "ionic atmosphere" whose net charge is opposite in sign to the central ion. The charge $P_i(r)$ has a maximum at a distance $1/\kappa$ from the central ion. For a univalent electrolyte in aqueous solution at 298 K,

$$1/\kappa = 3.043/\sqrt{C}\ \text{angstroms}\ , \tag{67}$$

where in this equation C is the total electrolyte concentration in moles per liter.

The thermodynamic properties derived from the limiting law distribution functions are:

$$\frac{E^{ex}}{NkT} = -\frac{\kappa^3}{8\pi c}\left[1 + \frac{d\ln\varepsilon_o}{d\ln T}\right] , \tag{68}$$

$$\ln\gamma_\pm = \ln\ \gamma^\circ - \frac{\kappa^3}{8\pi c} , \tag{69}$$

$$\phi = \phi^\circ - \frac{\kappa^3}{24\pi c} ,\ \text{and} \tag{70}$$

$$\frac{A^{ex}}{NkT} = \frac{A^{ex,\ o}}{NkT} - \frac{\kappa^3}{12\pi c} , \tag{71}$$

where the superscript zero refers to the properties of the corresponding uncharged hard-sphere system which Debye assumed to be ideal ($\phi^\circ = \gamma^\circ = 1$ and $A^{ex,\ o} = 0$), and $c = \Sigma c_i$ is the total ionic concentration. As pointed out earlier the contributions of the hard cores to the thermodynamic properties of the solution at high concentrations are not negligible. Using the Carnahan and Starling (1969) equation of state the osmotic coefficient of an uncharged hard sphere solute (in a continuum solvent) is given by

$$\phi^\circ = 1 + \frac{4\eta - 2\eta^2}{(1-\eta)^3} , \tag{72}$$

where $\eta = \pi c\sigma^3/6$. For a 1 molar solution this contributes ~ 0.03 to the deviation of the osmotic coefficient from ideal behavior.

The Debye-Huckel limiting law predicts a square-root dependence on the ionic strength $I = 1/2 \Sigma c_i z_i^2$ of the logarithm of the mean activity coefficient ($\log \gamma_{+-}$), the heat of dilution (E^{ex}/VI), and the excess volume (V^{ex}); it is considered to be an exact expression for the behavior of an electrolyte at infinite dilution. Some experimental results are shown in Figs. 4 and 5 for aqueous solutions of NaCl and $ZnSO_4$ at 25 C; the results are typical of the observations for 1-1 (e.g., NaCl) and 2-2 (e.g., $ZnSO_4$) aqueous electrolyte solutions at this temperature. The thermodynamic properties approach the limiting law at infinite dilution but deviate from it, at low concentrations, in different ways for the two charge types. Evidence from the ionic conductivity of 2-2 electrolyte solutions suggests that the deviations observed for these solutions are due to ion pairing or association. The opposite behavior found for aqueous 1-1 electrolytes, for which ion pairing is negligible at room temperature, is caused by the finite size of the ions and is called the excluded-volume effect. The Debye-Huckel theory ignores ion association and treats the affect of the sizes of the ions incompletely. The limiting law slopes and deviations from them depend strongly on the temperature and dielectric constant of the solvent and on the charges on the ions. An aqueous solution of sodium chloride, for instance, behaves like a weak electrolyte near the critical temperature of water because the dielectric constant of the solvent decreases dramatically with increasing temperature.

Frank and Thompson (1954) examined the concentration region over which the theory is valid and concluded that Debye and Huckel's treatment of the ionic atmosphere breaks down when $1/\kappa < 1$, the average distance between neighboring ions. For a univalent electrolyte in aqueous solution at 25 C this occurs at concentrations above 1×10^{-3} molar!

Attempts to improve the theory by solving the Poisson-Boltzmann equation present other difficulties first pointed out by Onsager (1933); one consequence of this is that the pair distribution functions $g_{+-}(r)$ and $g_{-+}(r)$ calculated for unsymmetrically charged electrolytes (e.g., $LaCl_3$ or $CaCl_2$) are not equal as they should be from their definitions. Recently Outhwaite (1975) and others have devised modifications to the Poisson-Boltzmann equation which make the equations self-consistent and more accurate, but the labor involved in solving them and their restriction to the primitive model electrolyte are drawbacks to the formulation of a comprehensive theory along these lines. The Poisson-Boltzmann equation, however, has found wide applicability in the theory of polyelectrolytes, colloids, and the electrical double-layer. Mou (1981) has derived a Debye-Huckel-like theory for a system of ions and point dipoles; the results are similar but for the presence of a

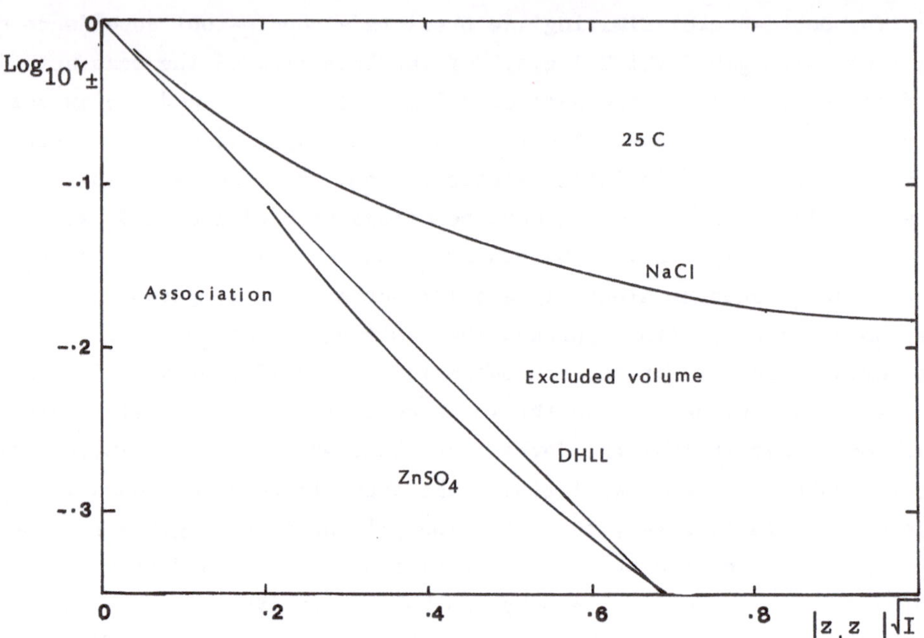

Fig. 4. The mean activity coefficients of NaCl and $ZnSO_4$ in aqueous solution at 25 C (DHLL - Debye-Huckel limiting law).

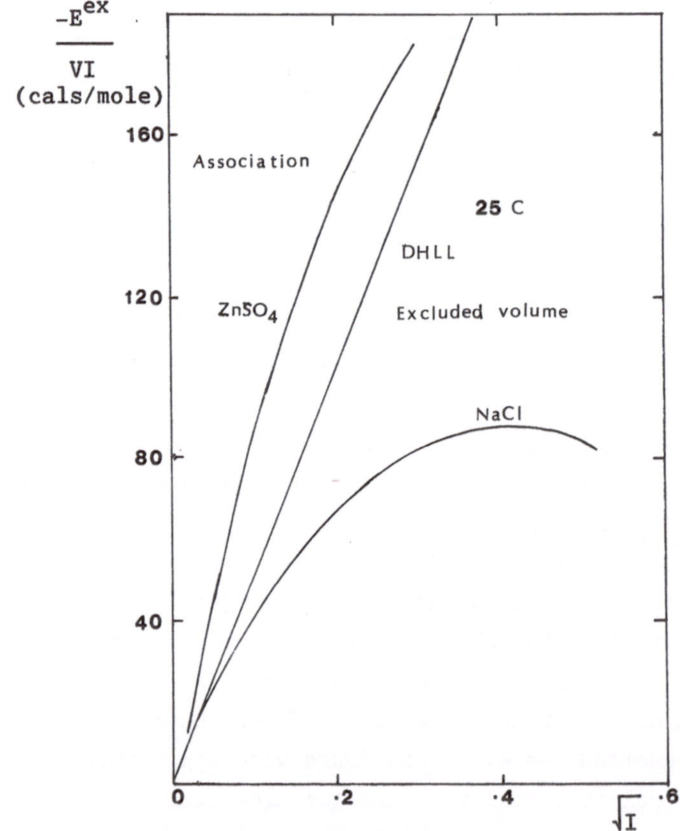

Fig. 5. The heats of dilution of NaCl and $ZnSO_4$ in aqueous solution at 25 C (DHLL - Debye-Huckel limiting law).

density-dependent dielectric constant and Debye shielding with a modified density dependence in the equations.

(b) The Mayer Theory

The main problem with the virial expansion when applied to ionic solutions is that virial coefficients diverge. This difficulty was resolved by Mayer (1950) who showed how the series could be resumed to cancel the divergences and yield a new expansion for a charged system. The terms in the new series are ordered differently from those in the original expansion and Mayer showed that the Debye-Huckel limiting law follows as the leading correction to the ideal behavior for ionic solutions. In principle, the theory enables systematic corrections to the limiting law to be obtained as the concentration of the electrolyte increases for any Hamiltonian which defines the short-range potential $u_{ij}^*(r)$, not just the one which corresponds to the RPM. A modified (or renormalized) second virial coefficient was tabulated by Porrier (1953), while Meeron (1957) and Abe (1959) derived an expression for this in closed form. Extensions of the theory to non-pairwise additive solute potentials have been discussed by Friedman (1962).

Mayer's paper was an important milestone in the development of electrolyte theory and the principle ideas behind this theory and the main results at the level of the renormalized second virial coefficient will be presented below. It follows from Eqs. (3) and (11) that the Mayer f-function for the solute pair potential can be written as the sum of terms:

$$f_{ij}(r) = f_{ij}^*(r) + [1 + f_{ij}^*(r)] \sum_{n=1}^{\infty} \frac{1}{n!} (-\beta\, e_i e_j/\varepsilon_o r)^n , \tag{73}$$

where $f_{ij}^*(r)$ is the Mayer f-function for the short-range potential $u_{ij}^*(r)$, which we represent graphically as ${}_i$o----o${}_j$, and $\beta = 1/kT$. Then the above expansion can be represented graphically as

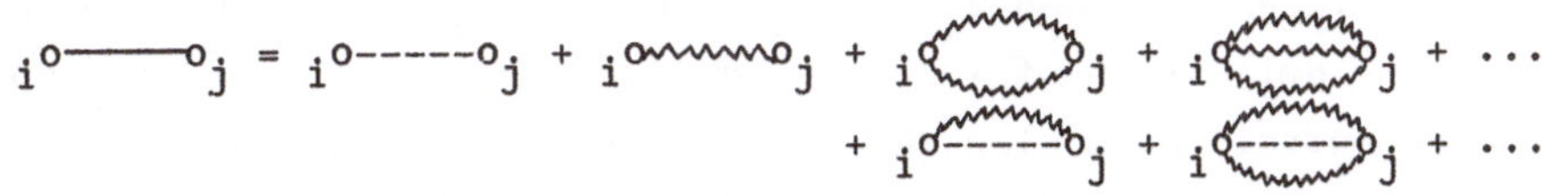

${}_i$o——o${}_j$ = ${}_i$o-----o${}_j$ + ${}_i$o∿∿∿o${}_j$ + [graph: two coulomb bonds in parallel]${}_j$ + [graph: three coulomb bonds in parallel]${}_j$ + ...
+ [graph: f* bond and one coulomb bond in parallel] + [graph: f* bond and two coulomb bonds in parallel] + ... ,

where ${}_i$o-----o${}_j$ represents $f_{ij}(r)$, the Mayer f-bond for the pair potential $u_{ij}(r)$, and ${}_i$o∿∿∿o${}_j$ $\equiv -\beta\, e_i e_j/\varepsilon_o r$ represents a coulomb potential multiplied by $-\beta$. The graphical representation of the virial coefficients in terms of Mayer f-bonds can now be replaced by an expansion in terms of f^* bonds (o-----o), and coulomb bonds (o∿∿∿o), where each f-bond is replaced by an f^* bond and the sum of one or more coulomb bonds in parallel, with or without an f^* bond also in parallel. The virial coefficients then have the following graphical representation:

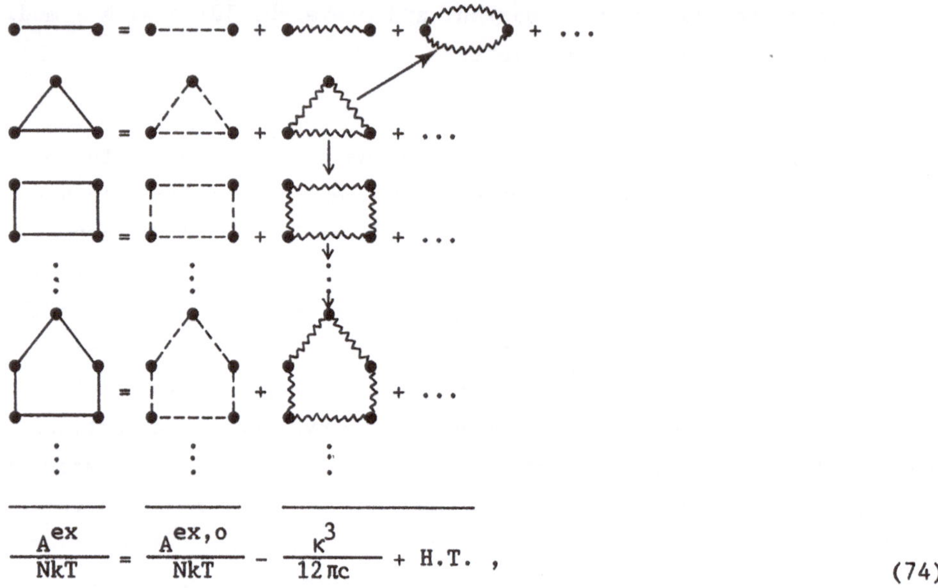

$$\frac{A^{ex}}{NkT} = \frac{A^{ex,o}}{NkT} - \frac{\kappa^3}{12\pi c} + \text{H.T.}\ , \tag{74}$$

where H.T. stands for higher-order terms. There is a symmetry number associated with each graph which we do not need to consider explicitly in this discussion. Each black circle denotes summation over the concentration c_i and integration over the coordinates of species i. The sum over all graphs in which the f-bond is replaced by an f^*-bond gives the free energy $A^{ex,*}$ of the corresponding uncharged system. The effect of the coulomb potential on the expansion is more complicated because of its long range. The second term in the expansion of the second virial coefficient is the bare coulomb bond multiplied by $-\beta$. If we multiply this by a screening function and do the integral the result is finite but contributes nothing to the overall free energy because of electroneutrality. This is because the contribution of the charge e_i from the single coulomb bond at a vertex when multiplied by c_i and summed over i is zero [see Eq. (1)]. The result for a graph with a cycle of coulomb bonds, however, is finite. Each vertex in these graphs has two coulomb bonds leading into it and instead of $\Sigma\, c_i e_i$ we have $\Sigma\, c_i e_i^{\,2}$ [which appears as a factor in the definition of κ^2 - see Eq. (20)]. This is not zero unless the ion concentration is also zero. Mayer summed all graphs with cycles of coulomb bonds and found that it leads to the Debye-Huckel limiting-law expression for the excess free energy! The essential mechanism behind this astonishing result is that the long range nature of coulomb interaction requires that the ions be considered collectively rather than in pairs, triplets, etc., which is implied by the conventional virial expansion. The same mechanism is also responsible for the modification of the interaction between two charges by the presence of others, which is called "screening". The sum of all chains of coulomb bonds between two ions represents the direct interaction, as well as the sum of indirect interactions (of the longest range) through other ions. The latter is a subset of the graphs which contribute to the correlation function $h_{ij}(r) = g_{ij}(r) - 1$ and has the graphical representation

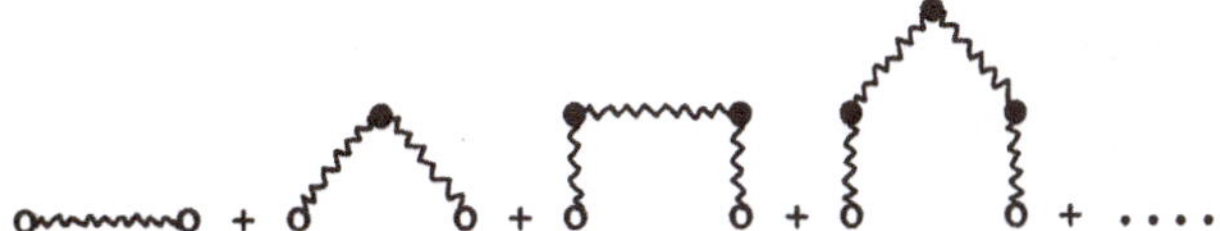

Explicit calculation of this sum (Meeron, 1958) shows that it is the Debye screened potential

$$q_{ij}(r) = -\beta\, e_i e_j \exp(-\kappa\, r)/\varepsilon_o r$$

defined earlier in Eq. (19).

Going beyond the limiting law it is found that the modified (or renormalized) virial coefficients in Mayer's theory of electrolytes are functions of the concentration through their dependence on κ. The ionic second virial coefficient $B_2(\kappa)$ is given by

$$B_2(\kappa) = -\frac{1}{2}\sum_{i=1}^{\sigma}\sum_{j=1}^{\sigma} c_i c_j \int\left[\exp[-\beta\, u_j^*(r)+q_{ij}(r)]-1 - q_{ij}(r) - \frac{q_{ij}^2}{2}\right]d\vec{r}\ . \quad (75)$$

This expression contains the contribution of the short-range potential included earlier in $A^{ex,*}$ so that the excess free energy, to this level of approximation, is

$$\left(\frac{A^{ex}}{NkT}\right)_{DHLL+B_2} = -\frac{\kappa^3}{12\pi c} + \frac{B_2(\kappa)}{c}\ . \quad (76)$$

This is called the DHLL+B_2 approximation. On doing the integrals over $q_{ij}(r)$ and $q_{ij}(r)^2/2$, and using Eqs. (1) and (20), this can be rewritten as

$$\left(\frac{A^{ex}}{NkT}\right)_{DHLL+B_2} = -\frac{5\kappa^3}{96\pi c} + \frac{S_2(\kappa)}{c}\ , \quad (77a)$$

where

$$S_2(\kappa) = -\frac{1}{2}\sum_{i=1}^{\sigma}\sum_{j=1}^{\sigma} c_i c_j \int\{\exp\,[-\beta\, u_{ij}^*(r) + q_{ij}(r)] - 1\}\, d\vec{r}\ . \quad (77b)$$

Equation (77b) has the form of a second virial coefficient in which the Debye screened potential has replaced the coulomb potential. Expressions for the other excess thermodynamic properties are easily derived (Rasaiah, 1972).

The great advantage of Mayer's theory is that it is formally exact within the radius of convergence of the virial series and it predicts the properties characteristic of all charge types without the need to introduce any additional assumptions. Unfortunately, the difficulty in calculating the higher virial coefficients limits the range of concentrations to which the theory can be applied with precision. The DHLL+B_2 approximation is qualitatively correct in reproducing the association effects observed at low concentrations for higher valence electrolytes and the excluded volume effects observed for all electrolytes at higher concentrations. Examples of this are shown in Fig. 6.

(c) The Stell-Lebowitz (SL) Theory

Equation (20) for κ^2 shows that it is essentially the product of the ionic concentration c and $\alpha = e^2/\varepsilon_0 kT$, called the Bjerrum parameter. The virial series is then seen as an expansion in the total ionic concentration at a fixed value of α. The Stell and Lebowitz (1968) theory on the other hand is an expansion of the free energy in the Bjerrum parameter α at constant c. This may also be considered to be an

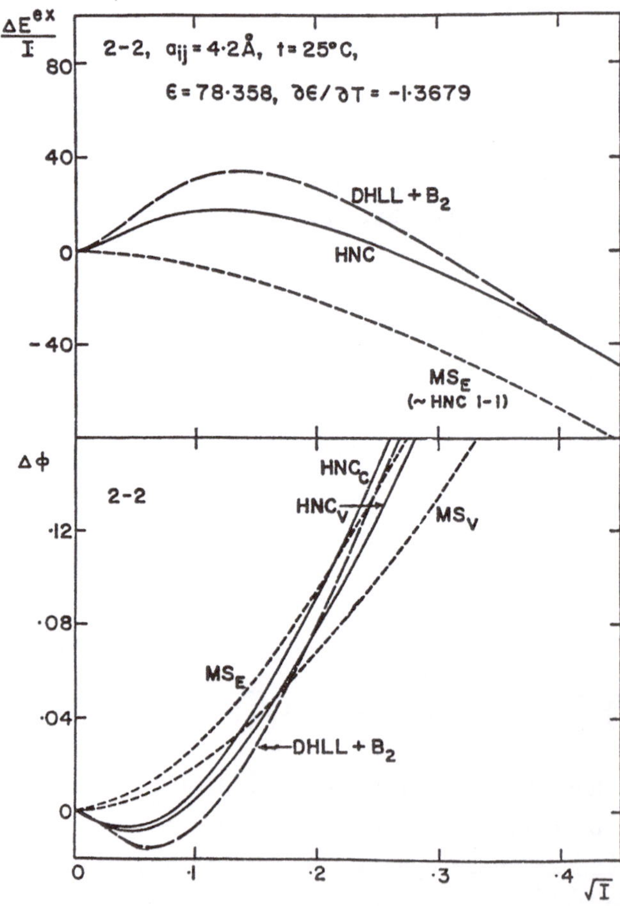

Fig. 6. Deviation of the heat of dilution E^{ex}/I and the osmotic coefficient ϕ from the Debye-Huckel limiting law for 1-1 and 2-2 RPM electrolytes in aqueous solution at 25 C according to the DHLL + B_2, HNC, and MS approximations (from Rasaiah, 1972).

expansion in the inverse temperature T^{-1}, or the product of the charges $e_i e_j$, or of e^2; in fact $(\varepsilon_o T)^{-1}$ and e^2 always occur together in pairs like two peas in a pod (see the definition of α), so that an expansion in one is equivalent to an expansion in the other. The leading terms in the excess free energy are given as

$$\frac{A^{ex}}{NkT} = \frac{A^{ex,*}}{NkT} + \frac{e^2}{2\varepsilon_o kT} \sum_{i=1}^{\sigma} \sum_{j=1}^{\sigma} c_i c_j z_i z_j \int \frac{h_{ij}(r)}{r} d\vec{r} - \frac{\kappa_1^3}{12\pi c} + \ldots , \quad (78)$$

where $h_{ij}^*(r)$ is the correlation function for the corresponding uncharged system which is determined by the short-range potential $u_{ij}^*(r)$, and κ_1^{-1} is a modified Debye length defined by

$$\kappa_1^2 = \kappa^2 + \frac{4\pi e^2}{\varepsilon_o kT} \sum_{i=1}^{\sigma} \sum_{j=1}^{\sigma} c_i c_j z_i z_j \int h_{ij}^*(r) d\vec{r} . \quad (79)$$

When the short-range potentials for all ion pairs are the same (e.g., the RPM) the integral may be taken outside the summations, and, since

$$\sum_i \sum_j c_i c_j e_i e_j = (\Sigma\, c_i e_i)^2 = 0 , \quad (80)$$

we find that $\kappa_1 = \kappa$. Rasaiah and Stell (1970) showed that the first two terms in the Stell-Lebowitz expansion form an upper bound for the free energy so that, unlike the Debye-Huckel limiting law, which can be approached from above or below as $c \rightarrow 0$ at fixed temperature T (e.g., $ZnSO_4$ and HCl in aqueous solutions), the limiting law as $T \rightarrow \infty$ at constant c must always be approached from one side. For the RPM, electrolyte electroneutrality ensures that the second term in the expansion for the free energy is zero and the inequality reduces to the statement that the excess free energy of the charged system is less than the corresponding free energy for the uncharged system. A lower bound for the free energy difference ΔA between the charged and uncharged RPM system was derived by Onsager (1939); this states that $\Delta A/N > -e^2/\varepsilon\sigma$.

Larsen et al. (1977) have examined the terms to $O(\kappa^6)$ in the SL expansion for the free energy and found that the convergence is extremely slow for a RPM 2-2 electrolyte in "aqueous solution" at room temperature (see Fig. 7). Nevertheless, the series can be summed using a Padé approximant (Stell and Wu, 1975; Larsen et al., 1977) which gives results that are comparable in accuracy to those obtained from the mean-spherical approximation (MSA). This is similar in spirit to the Padé approximant of the free energy which works so well for dipolar hard spheres (Stell, et al., 1972). However, unlike the DHLL + B_2 approximation, neither of these approximations produces the negative deviations in the osmotic and activity coefficients from the DHLL observed for higher valence electrolytes at low concentrations. This can be traced to the absence of

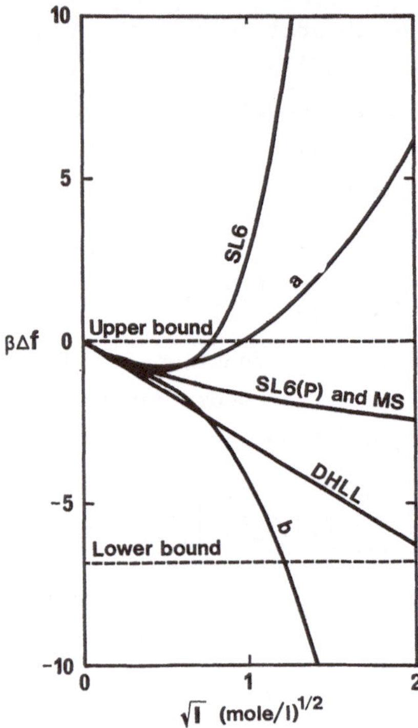

Fig. 7. The excess free energy for a 2-2 RPM electrolyte (σ = 4.2 A, t = 25 C, and ε = 78.358) versus the square root of the ionic strength according to different approximations. DHLL is the Debye-Huckel limiting law. The curves a = SL4, b = SL5, and SL6 are the free energies in the Stell-Lebowitz (SL) theory to orders κ^4, κ^5, and κ^6, respectively. SL6(P) is the free energy of the Padé approximant based on the SL6, and MS is the free energy in the mean spherical approximation; they are indistinguishable on this scale and show positive deviations from the limiting law. The two bounds shown are Rasaiah and Stell's upper bound, $\beta\Delta f \leq 0$, and Onsager's lower bound, $\beta\Delta f \geq \beta e^2/\varepsilon\sigma$ (from Larsen et al., 1977).

the complete renormalized second virial coefficient in these theories; it is present only in a linearized form. The union of the Padé approximant [SL6(P)], derived from the Stell-Lebowitz theory to $O(\kappa^6)$, and the Mayer expansion, carried as far as the DHLL + B_2, produces the right behavior at low concentrations and has an accuracy comparable to the MSA at high concentrations:

$$\text{SL6(P)}\cup B_2 = \text{SL6(P)} + B_2 - \text{SL6(P)}\cap B_2 \ . \tag{81}$$

Larsen et al. (1977) have discussed the union of these two expansions in detail; and Stell (1977) has discussed an array of other approximations for which similar unions with B_2 are considered. Figure 8 shows the coexistence curve for the RPM electrolyte (Stell, et al., 1976) predicted by SL6(P) and the SL6(P)$\cup B_2$.

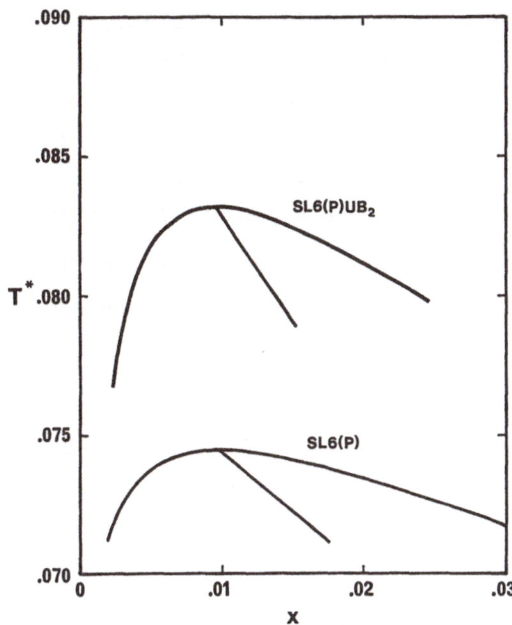

Fig. 8. Coexistence curve for the RPM predicted by the SL6(P) and SL6(P)UB_2 approximations. The reduced temperature is $T^* = \varepsilon\sigma/\beta e^2$; $x = \rho\sigma^3$ is the reduced density (from Larsen et al., 1977).

By integrating over the hard cores in Eq. (78) and collecting terms it is easily shown (Henderson and Blum, 1980) that SL expansion may be viewed as a correction to the mean spherical approximation, which still lacks the complete second virial coefficient. Since the mean spherical approximation (MSA) has a simple analytic form with an accuracy comparable to the Padé [SL6(P)] it is more convenient to consider the union of the MSA with the Mayer theory. Systematic improvements to the MSA for the free energy are discussed by Stell and Larsen (1979). The review by Hafskjold and Stell (1982) summarizes this work; the "stellar array" and several other theories, some of which are discussed below, are compared with the Monte Carlo data by Valleau et al. (1980).

(d) Integral Equation Approximations for Electrolytes

The convergence of the Mayer expansion and the Stell-Lebowitz expansions for the free energy is slow, and accurate estimates of the thermodynamic properties for a model electrolyte at concentrations near 1 M are difficult to obtain. A way out of this difficulty is to consider approximations for the radial distribution functions which correspond to the summation of a certain class of terms which contribute to all of the virial coefficients. The integral-equation approximations, such as the HNC, PY, and MS approximations, attempt to do just this. They also provide information on the structure of the solutions to varying degrees of success. We have already considered the HNC, PY, and MS approximations for simple fluids in the section on THE DIRECT CORRELATION FUNCTION AND INTEGRAL EQUATION APPROXIMATIONS. The first complete solutions to the PY and HNC approximations for the primitive model electrolyte were obtained by Carley (1967) but were not sufficiently complete to establish the superiority of one approximation or the other.

The procedure adapted by Mayer to derive the free energies of ionic solutions from the virial expansion has been applied to the pair-correlation functions by Meeron (1957). The pair-distribution function can then be written as

$$g_{ij}(r) = \exp[-\beta\, u^*_{ij}(r) + q_{ij}(r) + \tau_{ij}(r) + \zeta_{ij}(r)] \ , \tag{82}$$

where

$$\tau_{ij}(r) = h_{ij}(r) - c_{ij}(r) - q_{ij}(r) - \frac{\beta\, e_i e_j}{\varepsilon_o r} \ , \tag{83}$$

and $\zeta_{ij}(r)$ is the bridge function discussed earlier - see Eq. (31). Defining

$$X_{ij}(r) = h_{ij}(r) - q_{ij}(r) - \tau_{ij}(r) \ , \tag{84}$$

Allnatt (1964) showed that an expression equivalent to the Ornstein-Zernike relation can be written in matrix form as

$$\tau = X * h + q * X + q*X*h \ , \tag{85}$$

where the matrix convolution A*B is defined by the elements

$$(A*B)_{ij} = \sum_{k=1}^{\sigma} c_k \int A_{ik} B_{kj}\, d\vec{r}_k \ . \tag{86}$$

Equations (82) and (85) are specialized for electrolytes; the main difference, from the equations applicable to any simple fluid mixture, is that the equations are formulated in terms of the Debye screened potential rather than full coulomb pair potential. This implies that the sum over chains of coulomb bonds in the diagrams contributing to the potential of average force have already been taken [see Eq. (72)]. The analog of the HNC approximation is obtained by setting

$$\zeta_{ij}(r) = 0 \qquad \text{(HNC and PYA)} \ , \tag{87}$$

while the analog of the PY approximations uses the additional assumption that

$$\exp\,[\tau_{ij}(r)] = 1 + \tau_{ij}(r) \qquad \text{(PYA)} \ . \tag{88}$$

The solutions to the HNC approximation discussed in the section on DIRECT CORRELATION FUNCTIONS AND INTEGRAL EQUATION APPROXIMATIONS are the same as those of its analog when applied to the same ionic system, but the solutions to the PY approximation are quite distinct from the results for the analog of the PY approximation in Allnatt's formulation. To distinguish between the two we call the latter the Percus-Yevick-Allnatt (PYA) approximation after Allnatt who first derived it along with the analog of the HNC approximation. The advantage of Allnatt's formulation is that the Debye screened potential facilitates convergence of the numerical solutions to these equations. As noted earlier, the HNC approximation obeys the zeroth- and second-moment conditions; the PY and PYA approximations fail to satisfy the second-moment condition.

A systematic study of the HNC and PYA approximations for electrolytes was carried out by Rasaiah and Friedman (1968c) who solved these equations numerically in Fourier space using fast fourier transform algorithms. They showed, by means of self consistency tests, that the HNC approximation was the more accurate of the two. The effects of ion association for higher valence electrolytes, and the excluded-volume effect for all electrolytes at higher concentrations, are natural consequences of the HNC theory and need no further theoretical elaboration in the usual preparative concentration range (0.1 to 2M). Also the HNC (and PYA) approximations are not specialized to charged hard spheres and refinements to the short-range potential are readily incorporated into the computer programs for the distribution functions and other thermodynamic properties. Further studies (Rasaiah, 1972) revealed that the accuracy of the HNC approximation for the pressure or osmotic coefficient calculated from the virial and compressibility equations diminishes as the charges on the ions are increased. For highly charged ions the distribution functions for oppositely charged ions at contact become very large because of the tendency to form ion pairs (Fig. 9). In contrast to this, the small hump in the HNC distribution functions of like ions at low concentrations appears to be an artifact of this approximation (Fig. 10). These conclusions have been confirmed by the Monte Carlo simulations (Fig. 11) carried out by Card and Valleau (1970) and Larsen (1974, 1976) for 1-1 electrolytes, and by van Megan and Snook (1980), Valleau and Cohen (1980), Valleau et al. (1980), and Rogde and Hafskjold (1985) for 2-2 electrolytes. The excess energy determined from the HNC approximation is in good agreement with the MC results for all of the valence types studied. This implies that the most accurate route to the thermodynamic properties is through the energy equation. This is true not only for the HNC approximation but also for the MSA which is discussed below (Fig. 12). In Figs. 13 to 15, HNC calculations for the RPM and Gurney-overlap models are compared with experiment. Table 1 lists the thermodynamic properties at 25 C of a 2-2 RPM electrolyte in

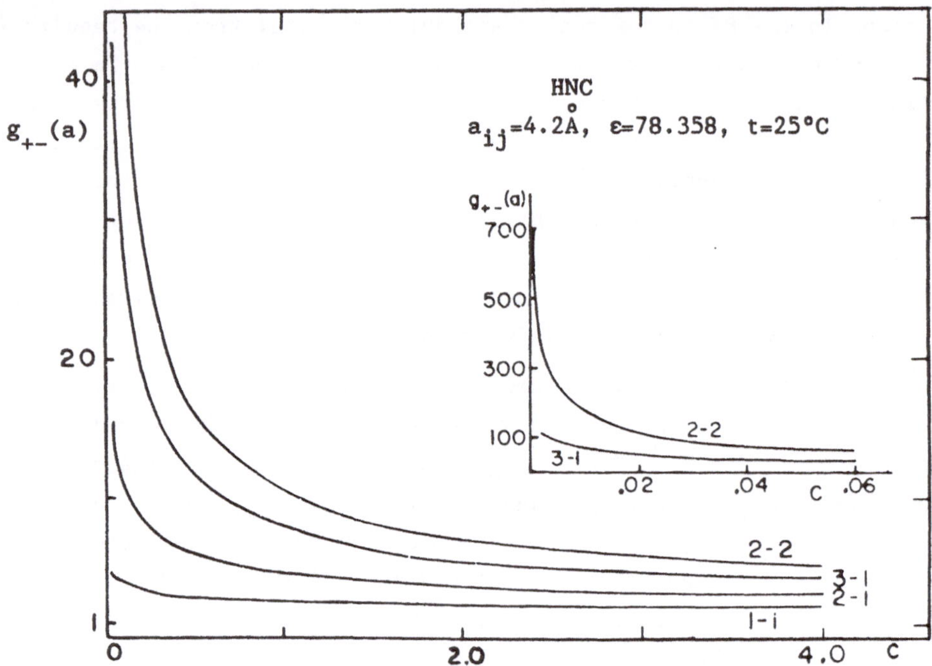

Fig. 9. The distribution functions $g_{+-}(\sigma+)$ at contact for oppositely charged ions in the HNC approximation as a function of the total ion concentration (from Rasaiah, 1972).

the HNC approximation. These calculations (Rasaiah and Larsen, 1978) are more accurate than the results reported earlier by Rasaiah (1972).

The use of the HNC approximation to study the equilibrium properties of electrolytes and polar fluids is now widespread. Recent examples are the investigations of multipolar fluids by Fries and Patey (1985), the study of the TIPS (transferable intermolecular potentials) model for water and alkali halides in water by Pettitt and Rossky (1982, 1986), and a central-force model for water by Thuraisingham and Friedman (1983). Studies of the rod-like polyelectrolytes (Bacquet and Rossky, 1984) using the HNC approximation have shown qualitative agreement with Manning's (1969, 1978) counter-ion condensation theory, but some quantitative predictions of the theory are not borne out. In the section on WEAK ELECTROLYTES AND DIPOLAR DUMBBELLS, we discuss the sticky electrolyte model for weak electrolytes and acids, which has also been solved numerically in the HNC approximation (Rasaiah and Lee, 1985a).

Improvements to the HNC approximation which incorporate the bridge diagrams in some form have also been pursued (Rossky et al., 1980; Iyotomi and Ichimaru, 1982; and Bacquet and Rossky, 1983). Rossky et al. (1980) replaced the f-bonds in the simplest bridge diagram [see Eq. (32)] by h-bonds calculated from the HNC approximation. This

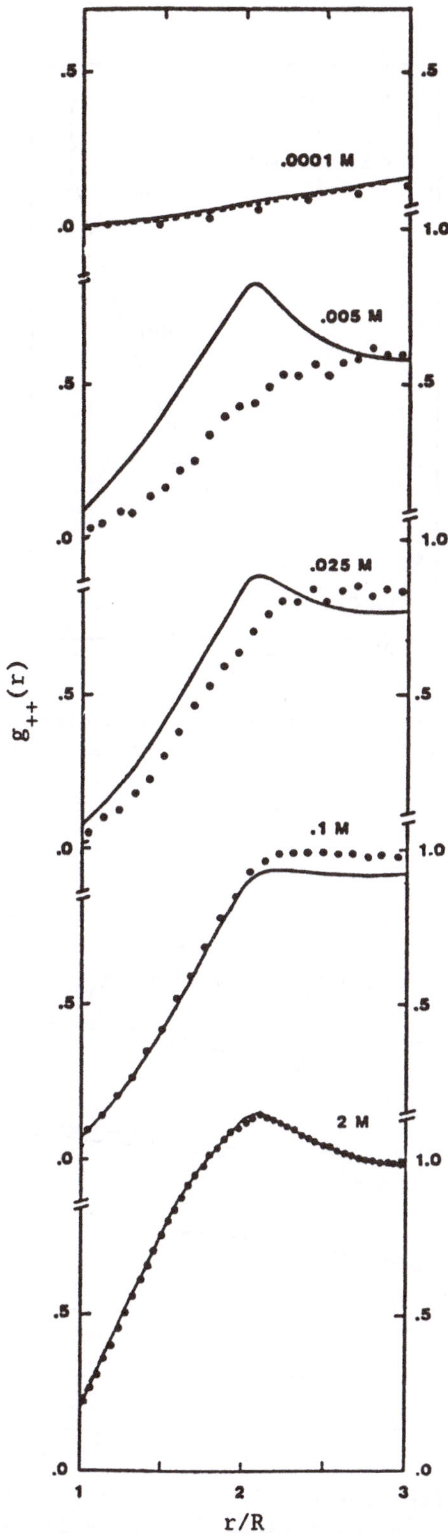

Fig. 10. Like charge correlation functions $g_{++}(\sigma+)$ for a 2-2 RPM electrolyte (σ = 4.2 A, t = 25 C, and ε = 78.358) calculated by the HNC approximation (solid line) and Monte Carlo simulations (circles) (from Rogde and Hafskjold, 1983).

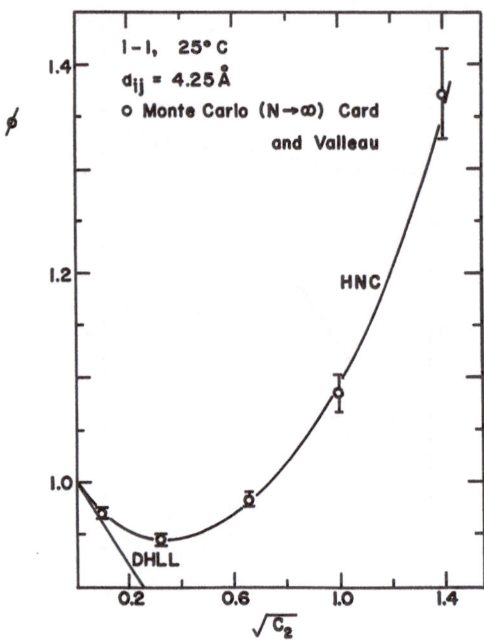

Fig. 11. The osmotic coefficient of a 1-1 RPM electrolyte in water compared with the Monte Carlo results (from Card and Valleau, 1970).

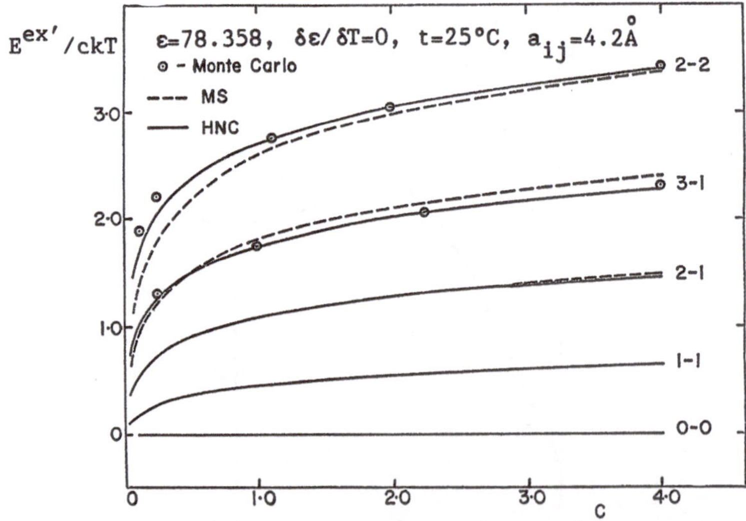

Fig. 12. The excess free energy of the 1-1, 2-1, 3-1, and 2-2 RPM electrolytes in water at 25 C. The solid and dashed curves are calculated according to the HNC and MS approximations, respectively. The Monte Carlo results of Card and Valleau (1970) are also shown (Rasaiah, 1970).

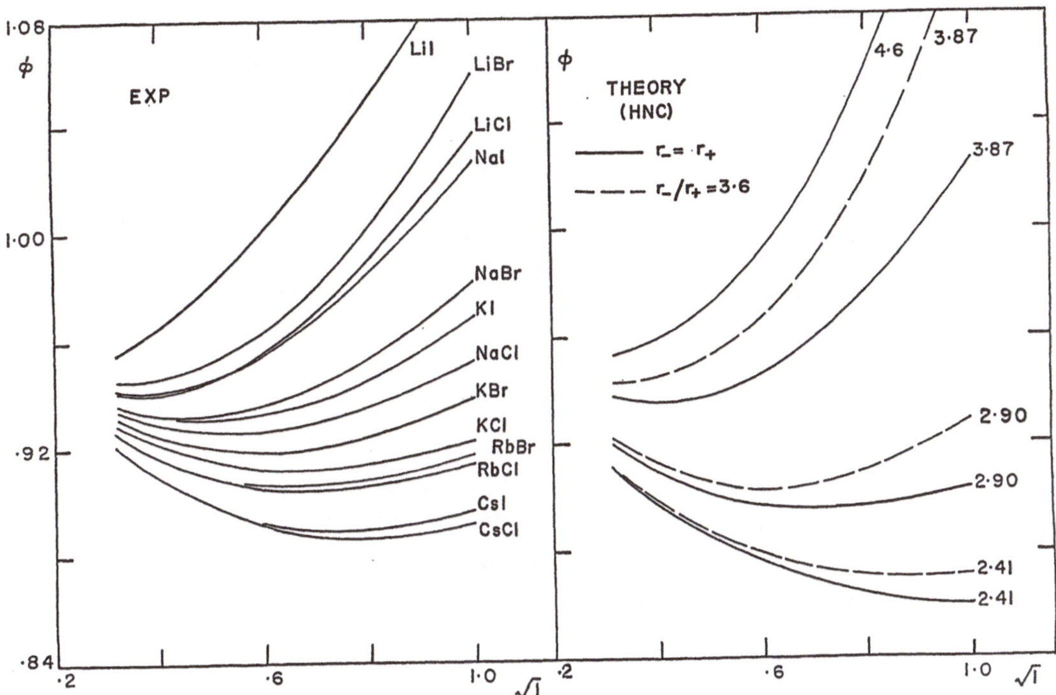

Fig. 13. Comparison of the HNC calculations of the osmotic coefficients for the primitive-model (charged hard spheres) electrolyte with experiment. All of the results are for the McMillan-Mayer system (Rasaiah, 1970).

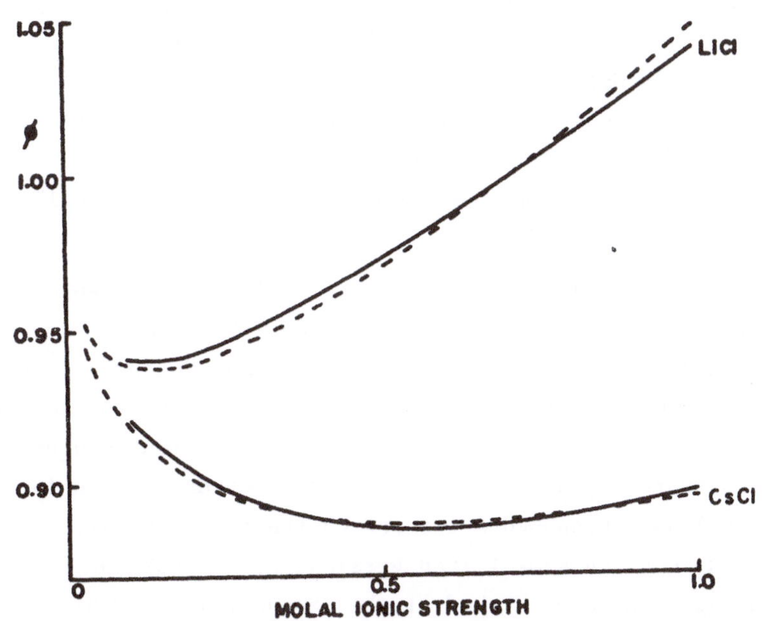

Fig. 14. Osmotic coefficients for the overlap model calculated with the HNC approximation compared with experimental results for CsCl and LiCl (from Ramanathan and Friedman, 1971).

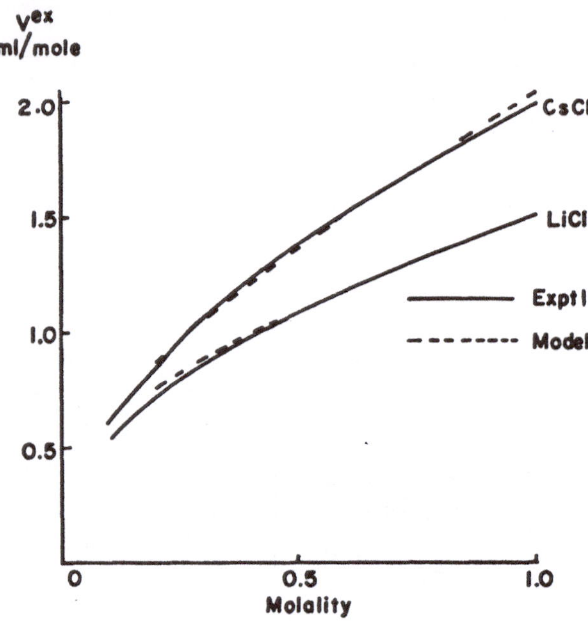

Fig. 15. Excess volumes for the overlap model calculated with the HNC approximation and compared with experiment for CsCl and LiCl (from Ramanathan and Friedman, 1971).

modification leads to a significant improvement in the correlation functions for like ions [$g_{++}(r)$] of 2-2 aqueous electrolytes where the HNC approximation predicts a spurious maximum at low concentrations (see Fig. 10). The RHNC approximation discussed earlier has been applied to electrolytes and molten salts by Larsen (1978) and used by Lado (1973) to study the one-component plasma. Here the bridge function is calculated for a short-range potential; the absence of a contribution from the coulomb interactions between the ions in the RHNC bridge function makes it unlikely that the HNC predictions for 2-2 electrolytes at low concentrations, where anomalous properties are observed, will be improved significantly.

The Born-Green-Yvon integral equation with the Kirkwood superposition approximation has been solved for 1-1 and 2-2 electrolytes by Croxton and McQuarrie (1979). Ramanathan and Woodbury (1982) have studied the BBGKY hierarchy of equations for electrolytes and shown how the Poisson Boltzmann, EXP, PY, and HNC approximations can be derived from this hierarchy providing a theory of theories. Ramanathan and Jensen (1986) have proposed a two region theory for higher valence electrolytes in which symmetrized versions of the BGY and Poisson Boltzmann equations are matched at the boundary inside which ion association occurs. The agreement with the HNC and MC results is good. Ion-pairing, clustering and the coexistence curve for the RPM electrolyte

Table 1. Equilibrium properties of a 2-2 restricted primitive model electrolyte in the HNC approximation. T=298.16K, ε=78.358, $\partial\varepsilon/\partial T$=0 and σ=4.2 A.*

C_{st}	$\partial \ln \gamma_{+-}/\partial \ln c$	ϕ_V	$-E^{ex'}/ckT$	$g_{+-}(\sigma+)$	$g_{++}(\sigma+)=g_{--}(\sigma+)$
0.0001	-.0559	.9631	.1311	726.7	.0004
0.0004	-.1079	.9266	.2817	548.1	.0193
0.001	-.1528	.8891	.4438	395.8	.0507
0.0016	-.1791	.8658	.5471	322.2	.0718
0.0025	-.2044	.8425	.6555	261.2	.0836
0.005	-.2429	.8013	.8504	181.4	.0975
0.01	-.2800	.7568	1.072	122.1	.0952
0.015	-.3005	.7304	1.213	95.93	.0884
0.020	-.3137	.7113	1.318	80.46	.0842
0.025	-.3233	.6966	1.402	70.08	.0817
0.0625	-.3560	.6396	1.772	39.29	.0729
0.1	-.3672	.6138	1.976	29.09	.0727
0.2	-.3715	.5872	2.292	18.70	.0781
0.3	-.3625	.5812	2.484	14.51	.0853
0.4	-.3473	.5836	2.624	12.17	.0927
0.5625	-.3146	.5966	2.793	9.937	.1044
0.8	-.2538	.6268	2.973	8.142	.1211
1.0	-.1930	.6586	3.089	7.230	.1352
1.4	-.0443	.7350	3.269	6.139	.1654
1.7	+.0900	.8020	3.376	5.650	.1882
2.0	+.2453	.8733	3.468	5.319	.2115
2.4	+.4874	.9913	3.574	5.028	.2472
2.7	+.6995	1.0888	3.644	4.889	.2760
3.0	+.9376	1.1963	3.709	4.801	.3073

* Recomputed by J.C. Rasaiah and B. Larsen (1978) - unpublished work. C_{st} is the electrolyte concentration in moles per litre.

have been discussed by Friedman and Larsen (1979). We next turn to a discussion of approximations for charged hard spheres which can be solved analytically and are an improvment over the Debye Huckel theory.

(e) The Mean Spherical and ORPA Approximations for the Restricted Primitive Model

Waismann and Lebowitz (1970, 1972) solved the mean spherical approximation (MSA) for the RPM electrolyte using Laplace transform techniques. We give the detailed solution to this approximation using the modification of Baxter's Wiener-Hopf factorization discussed in the section on SYMMETRICALLY CHARGED ELECTROLYTES: WIENER-HOPF FACTORIZATION and in the Appendix. This will serve as the starting point for the solutions of this approximation for the sticky electrolyte model (SEM). In the MSA

$$c_{ij}(r) = -\beta\, e_i e_j/\varepsilon_o\, r \qquad r > \sigma\,, \tag{89}$$

$$h_{ij}(r) = -1 \qquad 0 < r < \sigma\,. \tag{90}$$

The first relation is an approximation and the second is exact. They imply that

$$c_s(r) = 0 \qquad r > \sigma , \tag{92}$$

$$h_s(r) = -1 \qquad 0 < r < \sigma , \tag{93}$$

and

$$c_D(r) = \beta\, e^2/(\varepsilon_o r) \qquad r > \sigma , \tag{94}$$

$$h_D(r) = 0 \qquad 0 < r < \sigma . \tag{95}$$

The closure for the sum relations is identical to the PY approximation for hard spheres. Baxter applied his method to obtain the solutions found earlier by Wertheim (1963, 1964) and Thiele (1963). This is reviewed briefly before taking up the solution to the difference equations.

The closure $c_s(r) = 0$ for $r > 0$ implies $q_s(r) = 0$ for $r > 0$, which enables the upper limit of integration in Eqs. (54), (55), and (58) to be replaced by σ. On integrating the differential equation obtained from Eq. (54) for $r < \sigma$ one finds that

$$\begin{aligned} q_s(r) &= a(r^2 - \sigma^2)/2 + b(r - \sigma) \qquad 0 < r < \sigma \\ &= 0 \qquad r > 0 , \end{aligned} \tag{96}$$

where the continuity condition $q_s(0) = 0$ has been used to determine the integration constant, and

$$a = 1 - 2\pi\rho \int_0^\sigma q_s(t)dt , \tag{97}$$

and

$$b = 2\pi\rho \int_0^\sigma q_s(t)\, tdt . \tag{98}$$

Substituting for $q_s(t)$ and solving the simultaneous equations it is found that

$$a = (1 + 2\eta)/(1 - \eta)^2 , \tag{99}$$

$$b = -3\sigma\eta\, /[2(1 - \eta)^2] , \tag{100}$$

where $\eta = \pi\rho\sigma^3/6$. It follows from Eq. (54) for $rc_s(r)$ that $c_s(r)$ is a cubic polynomial in r for $r < \sigma$:

$$c_s(r) = -\lambda_1 - 6\eta\, \lambda_2\, (r/\sigma) - (1/2)\, \lambda_1 \eta (r/\sigma)^3 \qquad 0 < r < \sigma , \tag{101}$$

where

$$\lambda_1 = (1 + 2\eta)^2/(1 - \eta)^4 , \tag{102}$$

$$\lambda_2 = -(1 + \eta\, /2)^2/(1 - \eta)^4 . \tag{103}$$

The pressures calculated from the virial and compressibility equations are

$$(P/\rho kT)_v = (1 + 2\eta + 3\eta^2)/(1 - \eta)^2 , \tag{104}$$

and

$$(P/\rho kT)_c = (1 + \eta + \eta^2)/(1 - \eta)^3 . \tag{105}$$

These are also the pressures calculated from the PY equation since the MS and PY approximations are identical for hard spheres. The discrepancy between them is a measure of the accuracy of the approximation for this system. Interestingly, the Carnahan and Starling equation [Eq. (72)] of state is the weighted sum of these equations with a weight 2/3 for the compressibility and 1/3 for the virial equation.

To solve the difference equations we first note that since $c_D{}^\circ(r) = 0$ for $r > \sigma$, it follows from Eq. (56) that $q_D{}^\circ(r) = 0$ for $r > \sigma$, which allows the upper limit of integration in Eqs. (56), (57), and (59) to be replaced by σ. Then since $h_D(r) = 0$ for $r < \sigma$ it follows that $J_D(r)$ is a constant equal to J for $r \leq \sigma$. On subtracting from each other the two relations obtained for J from Eq. (59) for $0 < r < \sigma$ and $r = \sigma$ it is found that

$$q_D{}^\circ(r) = -\kappa J(r - \sigma) \qquad 0 < r \leq \sigma , \tag{106}$$

with

$$J = \int_0^\infty s h_D(s) ds . \tag{107}$$

Also at $r = \sigma$ it follows from Eq. (59) that

$$J = -M/2 + 2\pi\rho J \int_0^\sigma q_D{}^\circ(t)dt - \kappa J\sigma . \tag{108}$$

Substituting Eq. (106) for $q_D{}^\circ(r)$ in the above equation leads to a quadratic equation for J:

$$2(2\pi\rho\sigma)^2 \kappa J^2 - 4\pi\rho(1 + \kappa\sigma)J + \kappa = 0 , \tag{109}$$

whose solution is

$$J = \frac{(1 + \kappa\sigma) \pm (1 + 2\kappa\sigma)^{1/2}}{2\pi\rho\sigma^2\kappa} . \tag{110}$$

The excess energy $\beta E^{ex'}/N = -\kappa^2 J/2$, where we have conveniently set the temperature derivative of the dielectric constant to be zero. The correct Debye-Huckel limit as $\kappa \to 0$ is only obtained with the negative sign before the square root. Thus

$$\frac{E^{ex'}}{NkT} = \frac{-x[(1 + x) - (1 + 2x)^{1/2}]}{4\pi\rho\sigma^3}, \tag{111}$$

where $x = \kappa\sigma$. In the section on WEAK ELECTROLYTES AND DIPOLAR DUMBBELLS we show that the MSA for the sticky electrolyte model (SEM) for a weak electrolyte has a similar form. Integration with respect to β, between $\beta = 0$ and finite β, provides the excess Helmholtz free energy:

$$\beta(A^{ex} - A^{ex,o})_E/N = -[6x + 3x^2 + 2 - 2(1 + 2x)^{3/2}]/(12\pi\rho\sigma^3) , \tag{112}$$

where $A^{ex,o}$ is the Helmholtz free energy of the hard sphere system, and subscript E identifies A as being obtained from the energy equation. The osmotic coefficient derived from this is given as

$$\phi_E = \phi^o + [3x + 3x(1 + 2x)^{1/2} - 2(1 + 2x)^{3/2} + 2]/12\pi\rho\sigma^3 . \tag{113}$$

Integrating the Gibbs-Duhem equation, or from

$$A^{ex} = NkT[\ln\gamma_{+-} + (1 - \phi)] , \tag{114}$$

the logarithm of the mean activity coefficient is found to be

$$\ln \gamma_{\pm,E} = \ln\gamma^o + [x(1 + 2x)^{1/2} - x - x^2]/(4\pi\rho\sigma^3) , \tag{115}$$

where again the subscript E again indicates that ϕ and $\ln\gamma_{+-}$ are derived from the energy equation. Note that the difference in the logarithm of the activity coefficient is related to the excess energy by

$$\ln\gamma_{+-} - \ln\gamma^o = \beta E^{ex}/N . \tag{116}$$

This is true not only for the MSA but for any approximation or theory in which the free energy difference $\beta[A^{ex} - A^{ex,o}]$ is a function only of $\kappa\sigma$ (Rasaiah et al., 1972).

The thermodynamic properties calculated through the virial and compressibility equations (Waismann and Lebowitz, 1970, 1972) are:

$$\phi_v = \phi^o + x^2 B/(12\pi\rho\sigma^3) \tag{117}$$

and

$$\phi_c = \phi^o . \tag{118}$$

They are not as accurate as the properties derived from the energy equation.

To determine the direct correlation function substitute the expression $q_D{}^o(r)$ from Eq. (106) into Eq. (56) where, after some algebra, it is found that

$$rc_D{}^o(r) = -\kappa J + \kappa^2 J(r - \sigma) + \pi\rho\kappa^2 J (r^2 - \sigma^2) \qquad 0 < r < \sigma . \tag{119}$$

Using the quadratic equation for J [Eq. (109)], this can be simplified to

$$rc_D{}^o(r) = -\kappa^2/4\pi\rho + \kappa^2 Jr(1 - \pi\rho Jr) \qquad 0 < r < \sigma , \tag{120}$$

from which it follows that

$$c_D(r) = [c_D^\circ(r) + \beta e^2/\varepsilon_o r] = \kappa^2 J(1 - \pi\rho Jr) \qquad 0 < r < \sigma . \tag{121}$$

Since $B = 2\pi\rho\sigma J$, one finds that

$$c_D(r) = (\beta e^2/\varepsilon_o\sigma)\,[2B-B^2(r/\sigma)] \qquad 0 < r < \sigma ,$$
$$= \beta e^2/\varepsilon_o r \qquad r > \sigma . \tag{122}$$

All of these solutions for the MSA were first obtained by Waismann and Lebowitz (1970, 1972) who used Laplace transforms to solve this problem. Blum (1975) has extended the analysis described here to study asymmetric electrolytes in the primitive model. Further details of the thermodynamics and correlation functions for these electrolytes have been discussed by Blum and Høye (1977), Hiroike (1977), and by Høye and Blum (1978).

The equilibrium properties of several approximations (HNC, PYA, MS) for the RPM have been compared with Monte Carlo simulations by Rasaiah et al. (1972) and Hafskjold and Stell (1982). The HNC approximation is found to be the most accurate of the integral equations for electrolytes. The MSA distribution functions, especially at contact, are poor and in some instances even negative! However, the MSA predicts the thermodynamic properties of RPM electrolytes in aqueous solution quite accurately when they are calculated from the energy equation. Blum and coworkers (Grigera and Blum, 1976; Triolo et al., 1976, 1977; Triolo and Floriono, 1980) have compared their results with experiment for simple 1-1 electrolytes. The MSA for the screened coulomb potential has been discussed by Larsen and Rodge (1980).

The MSA does not show the correct behavior at low concentrations for 2-2 electrolytes and can be improved by taking the union of this approximation with the Mayer expansion. The ORPA+B_2 approximation, developed earlier by Andersen and Chandler (1972), does just this, since the optimized random phase approximation (ORPA) is identical to the MSA, when the reference uncharged system obeys the Percus-Yevick approximation. For 1-1 electrolytes the ORPA+B_2 approximations for ϕ and E^{ex} agree very well with the MC simulations up to 2M. The agreement of the osmotic coefficient ϕ is not as good for the higher 2-2 valence electrolytes but the accuracy of the excess energy remains. The distribution functions are given quite accurately by the EXP approximation (Andersen and Chandler, 1972), which combines the PY approximation for hard spheres with the MSA for the charged species in the argument of the exponential function. Ramanathan and Woodbury (1982) discuss how many of these approximations can be derived from the BBGKY hierarchy and obtained a perturbative correction to the EXP approximation. Other improvements to the MSA (Stell and Larsen, 1979; and Larsen et al., 1977), including the generalized mean-spherical approximation (GMSA), proposed by Høye et al. (1974), are reviewed by Hafskjold and Stell (1982). The MSA for a mixture of charged hard spheres and dipoles has also been solved (Blum, 1974a, 1974b, 1978; Vericat and Blum, 1980, 1982, 1985; Blum and Vericat, 1985; Høye and Lomba, 1988).

Ebeling and Grigo (1980) have treated the problem of ion association in the RPM for higher valence electrolytes by a modification of the classical Bjerrum theory (1926) in which the free ions are treated in the mean-spherical approximation, and the association constant for pairing is defined by splitting the second virial coefficient into hard-sphere, long-range and bound-state parts. The ratio of free ions to ion pairs is obtained by minimizing the free energy. An equation of state and critical point for the fluid are located. Similar theories of associating electrolytes have been discussed recently by Tani and Henderson (1983), Corti and Fernandez-Prini (1986) and Corti et al. (1987). A disadvantage of the MSA is that it applies only to charged hard-sphere models. Models with more realistic or elaborate potentials are not easily treated by this approximation except by using it as the leading term in a perturbation theory.

WEAK ELECTROLYTES AND DIPOLAR DUMBBELLS

In a weak electrolyte (e.g., an aqueous solution of acetic acid) the solute molecules AB are incompletely dissociated into ions A^+ and B^- according to the familiar chemical equation

$$AB = A^+ + B^- \ .$$

The forces binding the atoms A and B together in AB are chemical in nature and must be introduced, at least approximately, in the Hamiltonian. Then it should be possible to apply the same theoretical methods (e.g., HNC and MS approximations) used to study strong electrolytes to investigate incomplete dissociation in weak electrolytes as well. The binding between A and B is quite distinct from the ion pair formation observed for higher valence electrolytes (Fig. 9). In these cases no alterations in the Hamiltonian models already discussed were required to account qualitatively for the experimental observations.

The model for weak electrolytes that I will discuss here is the sticky electrolyte model (SEM) in which a delta-function interaction is introduced into the Mayer f-function for the oppositely charged ions at a distance $L \leq \sigma$, where σ is the hard sphere diameter. This model was first introduced by Cummings and Stell (1984, 1985) to study association in uncharged systems and is closely related to Baxter's model (1968a) for adhesive hard spheres. They solved this in the PY approximation using Baxter's (Wiener-Hopf) factorization.

In the SEM the Mayer f-function for ions of opposite sign is defined by

$$f_{+-} = -1 + L\zeta\delta(r - L)/12 \qquad r \leq \sigma , \tag{123}$$

where ζ is called the sticking coefficient; the delta function in Eq. (123) mimics bonding. The presence of a delta function in the f-function induces a delta function in the correlation function $h_{+-}(r)$ with a different coefficient λ, called the association parameter:

$$h_{+-} = -1 + L\lambda\delta(r - L)/12 \qquad r \leq \sigma . \tag{124}$$

The interaction between ions of the same sign is pure hard-sphere repulsion for $r \leq \sigma$. It follows from simple steric considerations that an exact solution will predict dimerization only if $L < \sigma/2$, but polymerization may occur for $\sigma/2 < L \leq \sigma$. However, an approximate solution may not show the full extent of polymerization that could occur when $L \geq \sigma/2$.

The association ratio K defined by $K = c_{AB}/c_{+}c_{-}$ is easily found to be (Cummings and Stell, (1984, 1985))

$$K = \frac{\pi\lambda(L/\sigma)^3}{3(1 - \langle N\rangle)^2} , \tag{125}$$

where the average number of dimers $\langle N\rangle = \eta\lambda(L/\sigma)^3$. We can now distinguish three different cases;

$\lambda = 0$	No Dimers	Strong Electrolyte (RPM)
$\lambda = (\sigma/L)^3/\eta$	All Dimers if $L < \sigma/2$	Dipolar Dumbbells
$0 < \lambda < (\sigma/L)^3/\eta$	Ions + Dimers	Weak Electrolyte (SEM)

Either the same or different approximations may be used to treat the binding at L and the interactions between the ions. The use of PY approximation for the weak electrolytes leads to negative values for λ prompting Lee et al. (1985) to employ the HNC approximation for the binding, and the MSA for the remaining interactions between the ions. Analytic solutions to this hybrid (HNC/MSA) approximation were obtained by an extension of factorization techniques discussed in a previous section (SYMMETRICALLY CHARGED ELECTROLYTES: WIENER-HOPF FACTORIZATION) and in the Appendix for $L = \sigma/n$, with the integer n varying from 1 to 5 (Lee et al., 1984; Rasaiah and Lee, 1985b; Lee and Rasaiah, 1987). Numerical solutions to the HNC approximation for both types of interactions present in this model with n = 2 and 3 were also obtained (Rasaiah and Lee, 1985a) by a simple modification of the procedure employed for strong electrolytes (Rasaiah and Friedman, 1968c).

Surprisingly, the association parameter λ obtained from the HNC/MS and HNC approximations are very nearly the same although the distribution functions (g_{++} and g_{+-}) are quite different particularly at low concentrations. It follows from the energy equation that

$$\frac{E^{ex}}{NkT} = \frac{\langle N \rangle}{2}\left(\frac{d\ln\zeta}{d\beta}\right) - \left(1 + \frac{d\ln\varepsilon_o}{d\ln T}\right)\frac{\kappa H}{2} , \tag{126}$$

where $H = \kappa J(\sigma)$. The first term in this equation is the binding energy; the second is the energy due to the interactions between the charges which can be determined analytically in the MSA. For any integer n, Rasaiah and Lee (1985a) have shown that $H' = H/\sigma$ has the form

$$H'\ (\text{SEM/MSA}) = \frac{\left(a_1 + a_2 x\right) - \left(a_1^2 + 2xa_3\right)^{1/2}}{24a_4\eta} , \tag{127}$$

where $a_i (i = 1 \text{ to } 4)$ are functions of the reduced ion concentration η, the association parameter λ, and $n = L/\sigma$. When $\lambda = 0$, $a_i = 1$, the average number of dimers $\langle N \rangle = 0$, and the energy of the RPM electrolyte in the MSA is recovered [see Eq. (111)]. The energies of a weak electrolyte calculated analytically from the HNC/MSA approximation (HNC for the binding and MSA for the electrical interactions) and numerically from the HNC approximation alone are very nearly the same. This is illustrated in Fig. 16 for n = 2 and 3 where comparison is also made with the energy of the corresponding RPM electrolyte ($\lambda = 0$). The more negative excess energy of the SEM is mainly due to the binding energy. The osmotic coefficients for the same systems calculated from the HNC approximation using the virial equation are compared in Fig. 17; as expected the weak electrolyte has a smaller osmotic coefficient than the strong electrolyte. We leave it to the reader to reason why the excess energies and osmotic coefficients are closer to the RPM results for n = 3 than for n = 2. The distribution functions for a 2-2 SEM electrolyte are shown in Fig. 18.

I have already mentioned the limit $\lambda = (\sigma/L)^3/\eta$ with $L < \sigma/2$ when the system should consist of dipolar dumbbells. In the absence of a solvent, the asymptotic form of the direct correlation function [defined through the Ornstein-Zernike (OZ) equation] for this system is given by (Rasaiah and Lee, 1985a)

$$c_{ij}(r) = -\beta A e_i e_j / r , \tag{128}$$

where $A = \varepsilon/(\varepsilon - 1)$ and ε is the dielectric constant of the system of dipolar dumbbells. It follows from Eqs. (126) and (127) that the energy of dipolar dumbbells, excluding the binding energy, is given in the MSA by (Lee and Rasaiah, 1987)

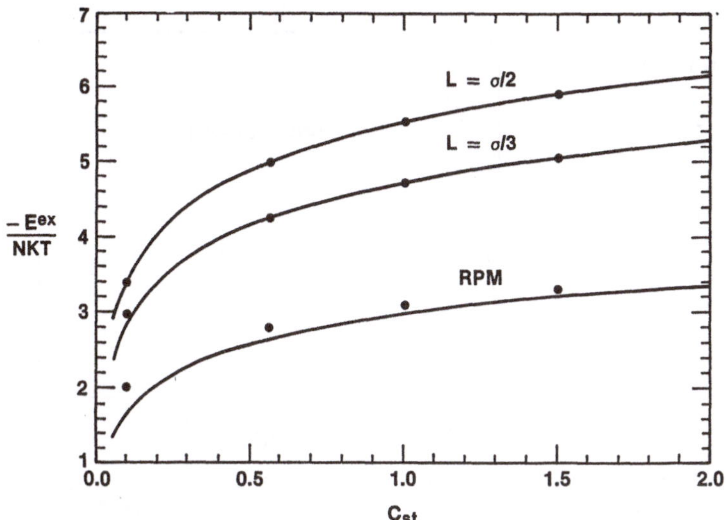

Fig. 16. The excess energy E^{ex} in units of NkT as a function of the electrolyte concentration c_{st} for RPM and SEM 2-2 electrolytes. The lines and points are the results for the HNC/MS and HNC approximations; σ = 4.2 A, t = 25 C, and ε = 78.358. The sticking coefficient ζ is 1.63 x 10^6 and 2.44 x 10^6 for L = σ/2 and σ/3, respectively (from Rasaiah and Lee, 1985a).

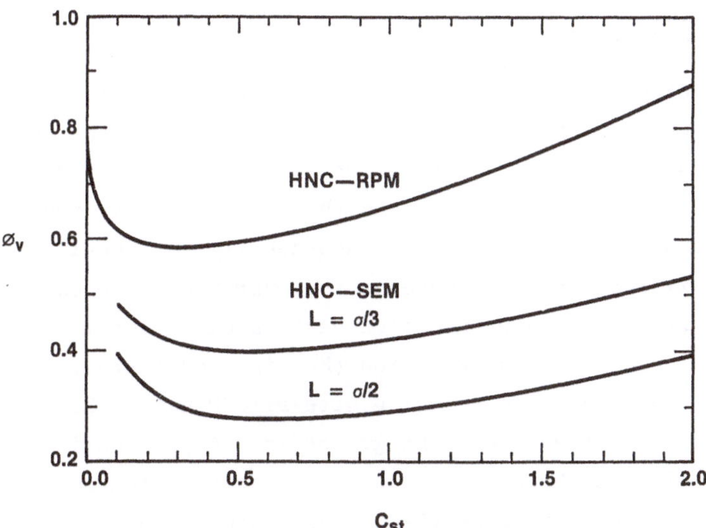

Fig. 17. The osmotic coefficient ϕ calculated from the virial equation as a function of the electrolyte concentration c_{st} for RPM and SEM 2-2 electrolytes using the HNC approximation. The parameters are the same as in Fig. 15 (from Rasaiah and Lee, 1985a).

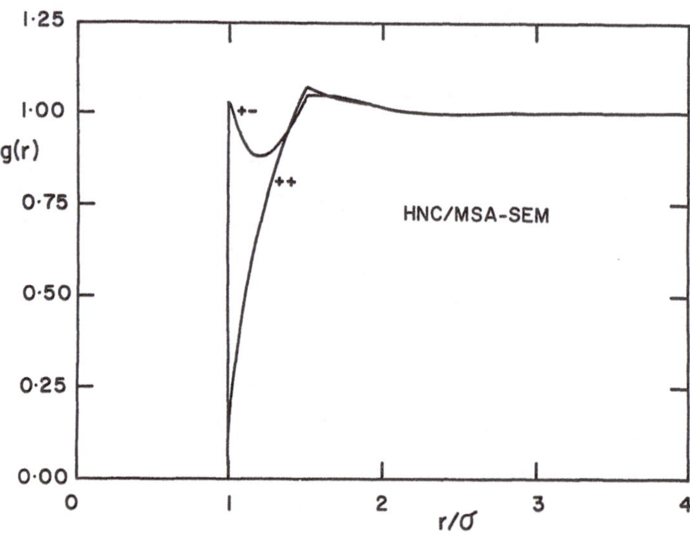

Fig. 18. The distribution functions for a 2-2 SEM electrolyte with $L = \sigma/2$ according to the HNC/MSA approximation. $\sigma = 4.2$ A, $t = 25$ C, and $\varepsilon = 78.358$. The sticking coefficient ζ is 1.63×10^6 (from Lee et al., 1985).

$$\frac{E^{ex}}{N_D kT} = \frac{-x\left[\left(c_1 + c_2 x'\right) - \left(c_1^2 + 2c_3 x'\right)^{1/2}\right]}{24\eta} , \tag{129}$$

where x' is the reduced dipole moment defined by

$$x' = \kappa\sigma = 2n\ (A\pi\rho/kT)^{1/2}\mu , \tag{130}$$

in which the dipole moment $\mu = eL = e\sigma/n$, c_i ($i = 1$ to 3) are numbers which depend on the dipole elongation, and $N_D = N/2$ is the number of dipoles. This is the analytic solution for the energy of dipolar dumbbells in the MSA; it suffers from the defect that it tends to a small but finite constant in the limit of zero density and should strictly be applicable only for $L < \sigma/2$. However the energy at zero charge is given correctly as zero. The approximation bears a close resemblance to the analog of the zero-pole approximation (Morriss and Perram, 1981; Morriss and Isbister, 1984) derived from the site-site Ornstein-Zernike (SSOZ) equation. However, the excess energy for this approximation is not zero in the limit of zero charge. In Fig. 19 we compare the energy with the Monte Carlo results of Morriss (1982) and energies calculated by Morriss and MacGowan (1986) using the MSA and HNC approximations for the direct correlation function defined through the SSOZ equation. In Fig. 20, we compare the sum and difference functions $g_S(r)$ and $g_D(r)$ with the Monte Carlo calculations for dipolar dumbbells (Morriss and Cummings, 1982; Morriss and Isbister, 1984). The agreement for this simple approximation is quite satisfactory. The methods discussed here can be

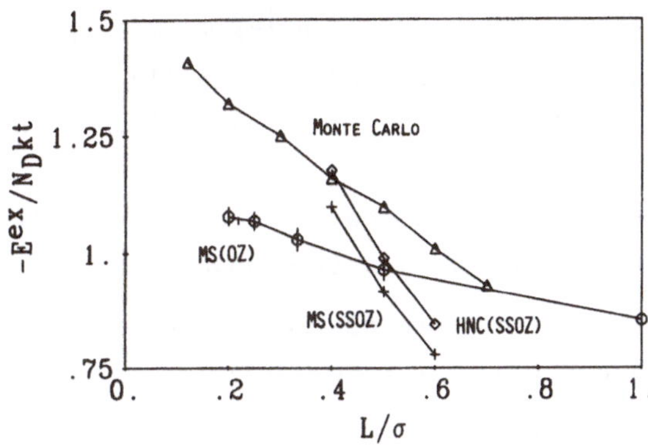

Fig. 19. The excess energy as a function of the elongation L for dipolar dumbbells (from Lee and Rasaiah, 1987).

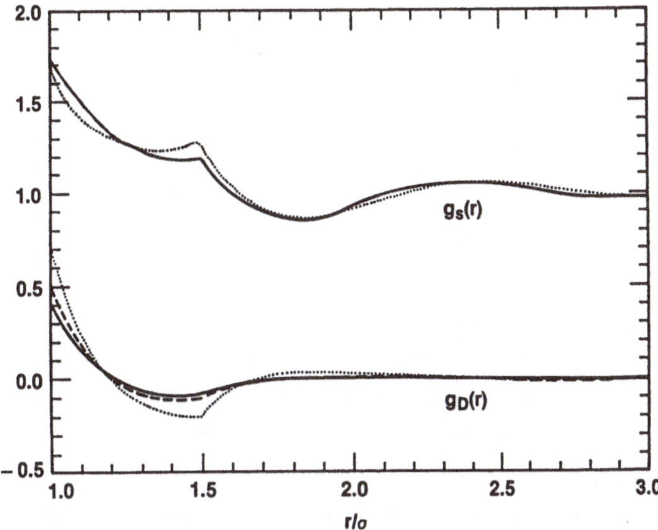

Fig. 20. The sum and difference distribution functions $g_S(r)$ and $g_D(r)$ for dipolar dumbbells with $L = \sigma/2$. The dotted line is the Monte Carlo result and the solid and broken lines are the MSA calculations with A = 1.0 and 1.25, respectively. The model parameters are T = 253 K, $p\sigma = 3.5$ A, $\rho_D \sigma^3 = 0.462$, and $\mu = 6.2 \times 10^{-30}$ Cm, where ρ_D is the density of the dipoles. $\mu^* = (\mu^2/kT\sigma^3)^{1/2} = 1.52$ (from Rasaiah and Lee, 1985a).

extended to other approximations (e.g., HNC) for polar fluids with extended dipoles and provide an alternative to the theories discussed by Fries and Patey (1985).

ACKNOWLEDGMENTS

I am grateful for the support provided by the National Science Foundation and the collaboration of many scientists whose names appear on these pages. The hospitality and computing facilities at the National Bureau of Standards, Gaithersburg, are gratefully acknowledged. I thank Robert Dunlap and Jianjun Zhu for a careful review of the manuscript.

APPENDIX: Wiener-Hopf factorization of the Ornstein-Zernike equation for a symmetrically charged electrolyte (Thompson, 1978; Lee, 1985).

Taking the Fourier transform of the decoupled Ornstein-Zernike equations, Eq. (44) and (45), we have

$$[1 + \rho\tilde{h}_S(k)]\,[1 - \rho\tilde{c}_S(k)] = 0\ , \tag{A.1}$$

$$[1 - \rho\tilde{h}_D(k)]\,[1 + \rho\tilde{c}_D(k)] = 0\ , \tag{A.2}$$

where

$$\tilde{c}_D(k) = \tilde{c}_D{}^{\circ}(k) + 4\pi\beta\, e^2/[\varepsilon_0(k^2 + z^2)]\ . \tag{A.3}$$

Since $\tilde{h}_s(k)$ is bounded on the real axis, $1 - \rho\tilde{c}_s(k)$ has no zeros on this axis and can be factorized to give

$$[1 - \rho\tilde{c}_S(k)] = \tilde{q}_S(k)\ \tilde{q}_S(-k)\ , \tag{A.4}$$

where $\tilde{q}_S(k)$ has zeros on the lower half plane, $\tilde{q}_S(-k)$ has zeros only on the upper half-plane, and $\tilde{q}_S(k)$ tends to unity as $|k| \longrightarrow \infty$. This is an example of the famous Wiener-Hopf factorization. Since $c_S(r)$ is short-ranged, Eq. (A.1) can be solved using Baxter's method. Similarly, as long as z is not equal to zero, we can argue that $\tilde{h}_D(k)$ is finite along the real axis and, hence, $[1 + \rho\tilde{c}_D(k)]$ has no zeros along the real axis and is a regular function of k, which tends to unity as $|k| \longrightarrow \infty$, so that the criteria of Wiener-Hopf factorization are realized and

$$[1 + \rho\tilde{c}_D(k)] \equiv [1 + \rho\tilde{c}_D{}^{\circ}(k) + 4\pi\beta\, e^2/[\varepsilon_0(k^2 + z^2)]$$

$$= \tilde{q}_D(k)\ \tilde{q}_D(-k)\ , \tag{A.5}$$

where $\tilde{q}_D(k)$ has properties analogous to $\tilde{q}_S(k)$. Since $\tilde{q}_{S,D}(k) \longrightarrow 1$ as Re $k \longrightarrow \infty$, $1 - \tilde{q}_{S,D}(k)$ have Fourier transforms, and we can define $q_{S,D}(r)$ through the relations

$$2\pi\rho q_{S,D}(r) = (1/2\pi) \int_{-\infty}^{\infty} [1 - \tilde{q}_{S,D}(k)]\exp(-ikr)dk\ , \tag{A.6}$$

from which it follows

$$\tilde{q}_{S,D}(k) = 1 - 2\pi\rho \int_{-\infty}^{\infty} q_{S,D}(r)\exp(ikr)dk\ . \tag{A.7}$$

By closing the contour in the upper half plane, it is seen that for $r < 0$

$$\int_{-\infty}^{\infty} [1 - \tilde{q}_{S,D}(k)]\exp(-ikr)dk = 0 , \tag{A.8}$$

which implies [see Eq. (A.6)] that

$$q_{S,D}(r) = 0 \qquad r < 0 . \tag{A.9}$$

The solution to the Ornstein-Zernike equation for the sum functions can be solved using Baxter's method, which will be discussed briefly before taking up the analysis of the difference equations. The factorizations in Eq. (A.4) can be rewritten as

$$\rho\tilde{c}_S(k) = 1 - \tilde{q}_S(k) + 1 - \tilde{q}_S(-k) - [1 - \tilde{q}_S(k)]\,[1 - \tilde{q}_S(-k)] . \tag{A.10}$$

The Fourier transform of $c_S(r)$ is

$$\begin{aligned}\tilde{c}_S(k) &= (4\pi/k)\int_0^{\infty} \sin(kr)rc_S(r)dr \\ &= 4\pi\int_0^{\infty} \cos(kr)\,S_S(r)dr \\ &= 2\pi\int_0^{\infty} \exp(ikr)S_S(|r|)dr ,\end{aligned} \tag{A.11}$$

where the second equation follows from the first on integration by parts and $S_S(r)$ is defined by

$$S_S(r) = \int_r^{\infty} tc_S(t)dt . \tag{A.12}$$

Multiplying Eq. (A.10) by $\exp(-ikr)/2\pi$, and integrating over k between $-\infty$ and ∞, one finds that

$$\begin{aligned}2\pi\rho S_S(r) &= 2\pi\rho\, q_S(r) + (1/2\pi)\int_{-\infty}^{\infty} [1 - q_S(-k)]\exp(-ikr)dk \\ &\quad - (1/2\pi)\int_{-\infty}^{\infty} [1 - \tilde{q}_S(k)]\,[1 - \tilde{q}_S(-k)]\exp(-ikr)dk .\end{aligned} \tag{A.13}$$

By closing the contour in the lower half plane we find that the first integral is zero for $r > 0$. Using Eq. (A.7) to substitute for $[1 - \tilde{q}_S(k)]$ and $[1 - \tilde{q}_S(-k)]$ one finds (using the properties of the Dirac delta function) that

$$S_S(r) = q_S(r) - 2\pi\rho\int_{-\infty}^{\infty} q_S(t)q_S(t-r)dt . \tag{A.14}$$

The lower limit of integration in the above equation has been raised to r because $q_s(r) = 0$ for $r < 0$. Differentiation with respect to r leads to an equation for $c_s(r)$:

$$rc_s(r) = -q_s'(r) + 2\pi\rho \int_r^\infty q_s'(t)\, q_s(t - r)dt \qquad 0 < r < \sigma . \tag{A.15}$$

To obtain the corresponding equation for $h_s(r)$ we return to Eq. (A.5) which, with the aid of Eq. (A.6), can be written as

$$\tilde{h}_s(k) = [1 - q_s(k)] + \tilde{h}_s(k)[1 - \tilde{q}_s(k)] + [1 - \tilde{q}_s(-k)]/\tilde{q}_s(-k) . \tag{A.16}$$

By analogy with Eq. (A.12) we define

$$J_s(r) = \int_r^\infty th_s(t)dt , \tag{A.17}$$

so that

$$\tilde{h}_s(k) = 2\pi \int_{-\infty}^\infty \exp(ikr)J_s(|r|)dr , \tag{A.18}$$

$$2\pi\rho\, J_s(r) = 2\pi\rho q_s(r) + (1/2\pi) \int_{-\infty}^\infty h_s(k)[1 - q_s(k)]\exp(-ikr)dk . \tag{A.19}$$

Substituting for $\tilde{h}_s(k)$ and $[1 - \tilde{q}_s(k)]$ from Eqs. (A.18) and (A.7) one finds

$$J_s(r) = q_s(r) + 2\pi\rho \int_0^\infty q_s(t)\, J_s(|t - r|)dt , \tag{A.20}$$

which on differentiation with respect to r gives

$$rh_s(r) = -q_s'(r) + 2\pi\rho \int_0^\infty q_s(t)(r - t)h_s(|r - t|)dt, \qquad r > 0 . \tag{A.21}$$

Equations (A.15) and (A.21) are the basic equations for the sum-correlations functions. They are identical to the equations derived by Baxter (1968) for a one-component fluid.

The Ornstein-Zernike equation for the difference functions is solved in a similar manner. We first note that, like the corresponding sum function, the fourier transform is:

$$\tilde{h}_D(k) = 2\pi \int_{-\infty}^\infty \exp(ikr)J_D(|r|)dr , \tag{A.22}$$

where

$$J_D(r) = \int_r^\infty th_D(t)dt . \tag{A.23}$$

The integral of $J_D(r)$ is determined by the electroneutrality condition as shown earlier in Eq. (52).

We also note that the fourier transform of the short-range component of the direct correlation function can be written as

$$\tilde{c}_D{}^{\circ}(k) = 2\pi \int_0^{\infty} \exp(ikr) S_D(|r|) dr , \tag{A.24}$$

where

$$S_D(r) = \int_r^{\infty} t c_D{}^{\circ}(t) dt . \tag{A.25}$$

The factorization of $1 + \rho\tilde{c}_D(k)$, given in Eq. (A.5), can be written as

$$\rho\tilde{c}_D{}^{\circ}(k) + 4\pi\rho\beta\, e^2/[\varepsilon_0(k^2 + z^2)] = -[1 - \tilde{q}_D(k)] - [1 - \tilde{q}_D(-k)]$$
$$+ [1 - \tilde{q}_D(k)][1 - \tilde{q}_D(-k)] . \tag{A.26}$$

Multiplying Eq. (A.26) by $\exp(-ikr)/2\pi$, integrating over k between $-\infty$ and ∞, and closing the contour in the lower half-plane, we find that the integral of the second term on the right-hand side of the above equation is zero for $r > 0$ since $\tilde{q}_D(-k)$ is regular in the lower half plane. Since the contribution from the semicircle arc is zero for $r > 0$ we find that

$$S_D(r) + \beta\, e^2 \exp(-zr)/(\varepsilon_0 z) = -q_D(r) + 2\pi\rho \int_r^{\infty} q_D(t) q_D(t - r) dt . \tag{A.27}$$

Differentiating with respect to r and changing variables with $s = t + r$ we obtain

$$r c_D{}^{\circ}(r) + \beta\, e^2 \exp(-zr)/\varepsilon_0 = q_D{}'(r) - 2\pi\rho \int_0^{\infty} q_D(s) q_D{}'(s + r) ds . \tag{A.28}$$

If we let $z \rightarrow 0$ it is apparent that $q_D(r)$ has a term which includes a constant. The solution for the Yukawa potential is known to be an exponential function so we write

$$q_D(r) = q_D{}^{\circ}(r) + M\exp(-zr) , \tag{A.29}$$

where M is a constant and $q_D{}^{\circ}(r) = 0$ as $r \rightarrow 0$. Differentiating and substituting in the equation for $r c_D{}^{\circ}(r)$ we get

$$r c_D{}^{\circ}(r) + \left[\beta\, e^2/\varepsilon_0 + Mz - \pi\rho M^2\right]\exp(-zr) = q_D{}^{\circ\prime}(r)$$
$$- 2\pi\rho \int_0^{\infty} q_D{}^{\circ\prime}(r + s)\, q_D{}^{\circ}(s) ds - 2\pi\rho M\left[\int_0^{\infty} \exp(-zs) q_D{}^{\circ\prime}(r + s) ds\right.$$
$$\left. - z \int_0^{\infty} \exp[-z(r + s)] q_D{}^{\circ\prime}(s) ds\right] . \tag{A.30}$$

On taking the limit $z \longrightarrow 0$ and $r \longrightarrow \infty$, we find that

$$\pi \rho M^2 = \beta e^2/\varepsilon_0 , \qquad (A.31)$$

from which it follows, on substitution in Eq. (A.31), that for finite r and $z \longrightarrow 0$

$$rc_D°(r) = q_D°'(r) + 2\pi\rho \left[\int_0^\infty q_D°'(r + s)q_D°(s)ds \right.$$

$$\left. + M \int_0^\infty q_D°'(r + s)ds \right] . \qquad (A.32)$$

The last integral is simply $q_D°(r)$; hence, for $r > 0$

$$rc_D°(r) = q_D°'(r) + 2\pi\rho \left[Mq_D°(r) - \int_0^\infty q_D°(s)\, q_D°'(r + s) \right] ds . \qquad (A.33)$$

To get the corresponding equation for $h_D(r)$ we start with

$$[1 - \rho\tilde{h}_D(k)]\, \tilde{q}_D(k) = 1/[\tilde{q}_D(-k)] , \qquad (A.34)$$

and rearrange this to

$$[\tilde{q}_D(k) - 1] - \tilde{h}_D(k) + \tilde{h}_D(k)[1 - \tilde{q}_D(k)] = [1 - \tilde{q}_D(-k)]/\tilde{q}_D(-k) . \qquad (A.35)$$

For $r > 0$, $\tilde{q}_D(-k)$ is regular in the lower half-plane, and $1 - \tilde{q}_D(-k)$ vanishes as $|k| \longrightarrow \infty$ on the real axis. Multiplying by $\exp(-ikr)/2\pi$ and integrating from $-\infty$ to ∞, we find for $r > 0$ that

$$J_D(r) = -q_D(r) + 2\pi\rho \int_r^\infty q_D(r - t)\, J_D(|t|)dt , \qquad (A.36)$$

where we have made use of the fact that $q_D(r) = 0$ for $r < 0$. Changing variables to $s = r - t$, we have

$$J_D(r) = -q_D(r) + 2\pi\rho \int_0^\infty q_D(s)\, J_D(|r - s|)ds . \qquad (A.37)$$

Differentiating this equation with respect to r we obtain

$$rh_D(r) = q_D'(r) + 2\pi\rho \int_0^\infty [q_D°(s) + M]\, (r - s)h_D(|r - s|)ds . \qquad (A.38)$$

We have yet to discuss the magnitude and sign of M. It follows from Eq. (A.31) that

$$2\pi\rho M = \pm \kappa , \qquad (A.39)$$

where κ is the inverse Debye length defined in Eq. (2). We will show that only the negative sign is consistent with the Debye-Huckel limiting law (DHLL). The limiting law follows from the assumption that $c_D(r) =$

$\beta e^2/(\varepsilon_o r)$ for all $r > 0$, i.e., the ions are assumed to be point charges. This implies that the short-ranged function $c_D°(r)$ is zero for all positive r. It follows from Eq. (A.33) that $q_D°(r) = 0$ for $r > 0$, which on substitution in the equation for $h_D(r)$ leads to

$$rh_D(r) = 2\pi\rho M \int_0^\infty (r - s)h_D(|r - s|)ds \ . \tag{A.40}$$

The solution to this is

$$h_D(r) = A\exp(2\pi\rho Mr)/r \ , \tag{A.41}$$

where A is a constant. Since $h_D(r)$ must be a decreasing function of r, the negative sign in Eq. (A.39) is the correct one. Using the definition of M we see that the equation for $h_D(r)$ agrees with the Debye-Huckel limiting law distribution functions.

The equation for $J_D(r)$ can be written in terms of $q_D°(r)$ and M. Substitution of Eq. (A.29) in Eq. (A.37), taking the limit $z \longrightarrow 0$ and making use of the electroneutrality condition, leads to

$$J_D(r) = -q_D°(r) - M/2+2\pi\rho \int_0^\infty q_D°(s)J_D(|r-s|)\, ds - \kappa \int_0^r J_D(s)ds \ . \tag{A.42}$$

REFERENCES

Abe, R., 1959, Prog. Theoret. Phys. (Kyoto), 22:213.
Adelman, S.A., 1976a, J. Chem. Phys., 64:724.
Adelman, S.A., 1976b, Chem. Phys. Lett., 38:724.
Allnatt, A.R., 1964, Mol. Phys., 8:533.
Andersen, H.C., and Chandler, D., 1972, J. Chem. Phys., 57:1918; Andersen, H.C., Chandler, D., and Weekes, J.D., 1972, J. Chem. Phys., 57:2626.
Arrhenius, S., 1887, Z. Physik. Chem., 1:631.
Bacquet, R.J., and Rossky., 1983, J. Chem. Phys., 79:1419.
Bacquet, R.J., and Rossky, 1984, J. Phys. Chem., 88:2660.
Baus, M., and Hansen, J.P., 1980, Physics Reports, 59:1.
Baxter, R.J., 1968a, Austr. J. Phys., 21:563.
Baxter, R.J., 1968b, J. Chem. Phys., 49:2770.
Berkowitz, M., Karim, O.A., McCammon, J.A., and Rossky, P.J., 1984, Chem. Phys. Lett., 105:577.
Berry, R.S., Rice, S.A., and Ross, J., 1980, "Physical Chemistry", John Wiley, NY.
Belch, A.C., Berkowitz, M. McCammon, 1986, J. Am. Chem. Soc., 109:1755.
Blum, L., 1974a, J. Chem. Phys., 61:2129.
Blum, L., 1974b, Chem. Phys. Lett., 26:200.
Blum, L., 1975, Mol. Phys., 30:1529.
Blum, L., 1978, J. Stat. Phys., 18:451.
Blum, L., 1980, Primitive electrolytes in the mean spherical approximation in: "Theoretical Chemistry Vol. 5", ed., D. Henderson, Academic Press, New York.
Blum, L., and Høye, J.S., 1977, J. Phys. Chem., 81:1311.
Blum, L., and Vericat, F., 1985, Molecular description of ionic solvation and ion-ion interactions in dipolar solvents in: "Chemical Physics of Ionic Solvation", R.R. Dogonadze, E. Kalman, A.A. Kornishev and J. Ulstrup, eds., Elsevier, Amsterdam.
Bjerrum, N., 1926, K. Dan. Vidensk. Selsk., 7:1.

Card, D.N., and Valleau, J., 1970, J. Chem. Phys., 52:6232.
Carley, D.D., 1967, J. Chem. Phys., 46:3783.
Carnahan, N.F., and Starling, K.E., 1969, J. Chem. Phys., 52:6232.
Corti, H.R., and Fernandez-Prini, R., 1986, J. Chem. Soc, Faraday Trans 2, 82:921.
Corti, H.R., Fernandez-Prini, R., and Blum, L., 1987, J. Chem. Phys., 87:3052.
Croxton, T.L., and McQuarrie, D.A., 1979, J. Chem. Phys., 83:1840.
Cummings, P.T., and Stell, G., 1984, Mol. Phys., 51:253; ibid., 1985, 55:33.
Dang, L.X., and Pettitt, B.M., 1987, J. Chem. Phys., 86:6560.
Debye, P., and Huckel, E., 1923, Z. Physik., 24:185.
Ebeling, W., and Grigo, M., 1980, Ann. Phys. (Lepzig), 37:21.
Frank, H.S., and Thompson, P., 1954, J. Chem. Phys., 31:1086.
Friedman, H.L., 1962, "Ionic Solution Theory", John Wiley, New York.
Friedman, H.L., and Ramanathan, P.S., 1970, J. Chem. Phys., 74:3756.
Friedman, H.L., 1972, J. Soln. Chem., 1:387; ibid. 1:413.
Friedman, H.L., and Larsen, B., 1979, J. Chem. Phys., 76:1092.
Friedman, H.L., and Dale, W.D.T., 1977, Electrolyte solutions at equilibrium in: "Statistical Mechanics, Part A: Equilibrium Techniques", B.J. Berne, ed., Plenum Press, New York.
Friedman, H.L., 1981, Current trends in the fundamental theory of ionic solutions in: "Thermodynamics of Aqueous Systems with Industrial Applications", S.A. Newman, ed., ACS Symposium Series 133, American Chemical Society, Washington D.C.
Friedman, H.L., 1982, J. Chem. Phys., 76:1092.
Friedman, H.L., 1985, "A Course in Statistical Mechanics" Prentice Hall, New Jersey.
Friedman, H.L., 1986, The structure of aqueous electrolyte solutions model calculations for various probes at equilibrium in: "Water and Aqueous Ionic Solutions", G.W. Neilson, and J.E. Enderby, eds., Adam Hilger, Bristol.
Friedman, H.L., 1987, Theory of ionic solutions at equilibrium in: "The Physics and Chemistry of Aqueous Ionic Solutions", M.C. Bellissent-Funel and G.W. Neilson, eds., NATO ASI Series, Reidel, Boston.
Friedman, H.L., 1988, Faraday Discuss. Chem. Soc., 85:1.
Fries, P.H., and Patey, G.N., 1985, J. Chem. Phys., 82:429.
Grigera, G., and Blum, L., 1976, J. Chem. Phys., 38:486.
Hafskjold, B., and Stell, G., 1982, The equilibrium studies of simple ionic liquids in: "The Liquid State of Matter: Fluids, Simple and Complex", E.W. Montroll and J.L. Lebowitz, eds., North Holland, NY.
Hansen, J.P., and McDonald, 1986, "Theory of Simple Liquids", Academic Press, London.
Harned, H.S., and Owen, B.B., 1950, "The Physical Chemistry of Electrolytic Solutions", American Chemical Society Monograph, Reinhold, New York.
Henderson, D., and Blum, L., 1980, Mol. Phys., 40:1509.
Hiroike, K. 1977, Mol. Phys., 33:1195.
Høye, J.S., Lebowitz, J.L., and Stell, G., 1974, J. Chem. Phys.,61:3253.
Høye, J.S., and Blum, L., 1978, Mol. Phys., 35:299.
Høye, J.S., and Stell, G., 1977, J. Chem. Phys., 67:439.
Høye, J.S., and Lomba, E., 1988, J. Chem. Phys., 88:5790.
Hovarth, A.L., 1985, "Handbook of Aqueous Electrolyte Solutions", Ellis Horwood Limited, Chichester.
Iyetomi, H., and Ichimaru, S., 1982, Phys. Rev., A25:2434.
Karim, O.A., and McCamman, J.A., 1986a, J. Am. Chem. Soc., 108:1762.
Karim, O.A., and McCamman, J.A., 1986b, Chem. Phys. Lett., 132:219.
Kirkwood, J.G., and Buff, R.P., 1951, J. Chem. Phys., 19:774.
Lado, F., 1973, Phys. Rev., A8:2548.
Lado, F., Foiles, S.M., and Ashcroft, N.W., 1983, Phys. Rev., A28:2374.
Larsen, B.J., 1974, Chem. Phys. Lett., 27, 47.
Larsen, B.J., 1976, J. Chem. Phys., 65:3431.
Larsen, B.J., Rasaiah, J.C., and Stell, G., 1977, Mol. Phys., 33:987.

Larsen, B.J., 1978, J. Chem. Phys., 78:1309; ibid., 68:4511.
Larsen, B.J., and Rogde, R.A., 1980, J. Chem. Phys., 72:2578.
Larsen, B.J., Stell, G., and Wu, K.C., 1977, J. Chem. Phys., 67:530.
Lee, S.H., 1985, "Statistical Mechanical Studies of Charged Systems and Polar Fluids," Ph.D Thesis, University of Maine.
Lee, S.H., Rasaiah, J.C., and Cummings, P.T., 1985, J. Chem. Phys., 83:317.
Lee, S.H., and Rasaiah, J.C., 1987, J. Chem. Phys., 86:1983.
Manning, G., 1969, J. Chem. Phys., 51:924; ibid., 51:934; ibid., 51:3249.
Manning, G., 1978, Q. Rev. Biophys., 11:179 and references therein.
Mayer, J., 1950, J. Chem. Phys., 18:1426.
McDonald, I., and Rasaiah, J.C., 1975, Chem. Phys. Lett., 34, 382.
McMillan, W.G., and Mayer, J., 1945, J. Chem. Phys., 13:276.
McQuarrie, D.A., 1976, "Statistical Mechanics", Harper and Row, New York.
Meeron, E., 1957, J. Chem. Phys., 26:805; 1958, ibid., 28:630.
Mitchell, D.J., McQuarrie, D.A., Szabo, A., and Groeneveld, H., 1977, J. Stat. Phys., 17:15.
Morriss, G.P., and Perram, J., 1981, Mol. Phys., 43:669.
Morriss, G.P., 1982, Mol. Phys., 47:833.
Morriss, G.P., and Cummings, P.T., 1982, Mol. Phys., 45:1099.
Morriss, G.P., and Isbister, D.J., 1984, Mol. Phys., 52:57.
Morriss, G.P., and MacGowan, D., 1986, Mol. Phys., 58:745.
Mou, C.Y., 1981, Proc. Nat. Sci. Council (Taiwan), 5:134.
Olivares, W., and McQuarrie, D.A., 1975, Biophys. Journal, 15:143.
Olivares, W., and McQuarrie, D.A., 1976, J. Chem. Phys., 65:3604.
Onsager, L., 1933, Chem Rev., 13:73.
Onsager, L., 1939, J. Phys. Chem., 43:189.
Onsager, L., 1964, J. Amer. Chem. Soc., 86:3421.
Ornstein, L.S., and Zernike, F., 1914, Proc. Acad. Sci. (Amsterdam), 17:793.
Outhwaite, C.W., 1975, Equilibrium theory of electrolyte solutions in: "Statistical Mechanics Vol. 2", Specialist Periodical Reports, Chemical Society, London.
Percus, J.K., and Yevick, G.J., 1958, Phys. Rev., 1:110.
Perram, J., 1975, Mol. Phys., 30:1505.
Perry, R.L., Cabezas, H. Jr., and O'Connell, J.P., 1988, Mol. Phys., 63:189.
Pettitt, B.M., and Rossky, P., 1982, J. Chem. Phys., 77:1451.
Pettitt, B.M., and Rossky, P., 1986, J. Chem. Phys., 84:5836.
Pitzer, K.S., 1973, J. Phys. Chem., 77:268.
Pitzer, K.S., and Mayorga, G., 1973, J. Phys. Chem., 77:2300.
Pitzer, K.S., and Kim, J.J., 1974, J. Am. Chem. Soc., 96:5701.
Porrier, J., 1953, J. Chem. Phys., 21:965.
Ramanathan, P.S., and Friedman, H.L., 1971, J. Chem. Phys., 54:1086.
Ramanathan, G.V., and Woodbury, C.P. Jr., 1982, J. Chem. Phys., 77:4120; ibid, 77:4133.
Ramanathan, G.V., and Jensen, A., 1986, J. Chem. Phys., 84:3472.
Rasaiah, J.C., and Friedman, H.L., 1968a, J. Chem. Phys., 48:2745.
Rasaiah, J.C., and Friedman, H.L., 1968b, J. Chem. Phys., 72:3352.
Rasaiah, J.C., and Friedman, H.L., 1968c, J. Chem. Phys., 50:3965.
Rasaiah, J.C., 1970a, J. Chem. Phys., 52:704.
Rasaiah, J.C., 1970b, Chem. Phys. Lett., 7:260.
Rasaiah, J.C., and Stell, G., 1970, Mol. Phys., 18:249.
Rasaiah, J.C., Card, D.N., and Valleau, J.P., 1972, J. Chem. Phys., 56:248.
Rasaiah, J.C., 1972, J. Chem. Phys., 56:3071.
Rasaiah, J.C., 1973, J. Soln. Chem., 2:301.
Rasaiah, J.C., and Larsen, B., 1978 - unpublished work.
Rasaiah, J.C., and Lee, S.H., 1985a, J. Chem. Phys., 83:5870.
Rasaiah, J.C., and Lee, S.H., 1985b, J. Chem. Phys., 83:6396.
Robinson, R.A., and Stokes, R., 1959, "Electrolyte Solutions," Butterworths, London.
Rogde, S.A., and Hafskjold, B., 1983, Mol. Phys., 48:1241.
Rosenfeld, Y., and Ashcroft, N.W., 1979, J. Chem. Phys., A20:1208.

Rossky, P., Dudowicz, J.B., Tembe, B.L., Friedman, H.L., 1980, J. Chem. Phys., 73:3372.
Stell, G., and Lebowitz, J.L., 1968, J. Chem. Phys., 48:3706.
Stell, G., Rasaiah, J., and Narang, H., 1972, Mol. Phys., 48:1241.
Stell, G., 1973, J. Chem. Phys., 59:3926.
Stell, G., and Wu, K.C., 1975, J. Chem. Phys., 63:491.
Stell, G., Wu, K.C., and Larsen, B., 1976, Phys. Rev. Lett., 37:1369.
Stell, G., 1977, Fluids with long-range forces: Towards a simple analytic theory in: "Statistical Mechanics, Part A, Equilibrium Techniques", B.J. Berne, ed., Plenum Press, New York.
Stell, G., and Larsen, B., 1979, J. Chem. Phys., 70:361.
Stillinger, F.H., and Lovett, R., 1968, J. Chem. Phys., 48:3858; ibid., 48:3869; ibid., 49:1991.
Tani, A., and Henderson, D., 1983, J. Chem. Phys., 79:2390.
Tembe, B.L., Friedman, H.L., and Newton, M.D., 1982, J. Chem. Phys., 76:1490.
Thiele, E., 1963, J. Chem. Phys., 39:474.
Thompson, N., 1978, "Wiener-Hopf Factorization of the Ornstein-Zernike Equation for the Restricted Primitive Model of Electrolytes", University of Maine, Chemistry Department, Report No. 2001.
Thuraisingham, R.A., and Friedman, H.L., 1983, J. Chem. Phys., 78:5772.
Tibavisco, H., 1974, M.Sc Thesis, Univ of Puerto Rico, Rio Piedras.
Triolo, T., Grigera, J.R., and Blum, L., 1976, J. Chem. Phys., 80:1858.
Triolo, T., Blum, L., and Floriano, M.A., 1977, J. Chem. Phys., 67:5956.
Triolo, T., and Floriano, M.A., 1980, Classical ionic fluids in the mean spherical approximation in: "Advances in Solution Chemistry", I. Bertini, L. Lunazzi, and A. Dei, eds., Plenum Press, New York.
Valleau, J.P., and Cohen, L.K., 1980, J. Chem. Phys., 72:5935.
Valleau, J.P., Cohen, L.K., and Card, D.N., 1980, J. Chem. Phys., 72:5942.
van Megan, W., and Snook, I.K., 1980, Mol. Phys., 39:1043.
van't Hoff, J.H., Z. Physik. Chem., 1:481.
Vericat, F., and Blum, L., 1980, J. Stat. Phys., 22:593.
Vericat, F., and Blum, L., 1982, Mol. Phys., 45:1067.
Vericat, F., and Blum, L., 1985, J. Chem. Phys., 82:1492.
Waismann, E., and Lebowitz, J.L., 1970, J. Chem. Phys., 52:430.
Waismann, E., and Lebowitz, J.L., 1972, J. Chem. Phys., 56:3086, 3093.
Watts, R.O., 1973, Integral equation approximations in the theory of fluids in: "Statistical Mechanics Vol. 1", Specialist Periodical Reports, Chemical Society, London.
Watts, R.O., and McGee, I.J., 1976, "Liquid State Chemical Physics", John Wiley, New York.
Wertheim, M., 1963, Phys. Rev. Lett., 10:321.
Wertheim, M., 1964, J. Math. Phys., 5:643.
Zhong, E., and Friedman, H.L., 1988, J. Phys. Chem., 92:1685.

THEORETICAL STUDIES OF ELECTRONS IN FLUIDS

Neil R. Kestner

Chemistry Department
Louisiana State University
Baton Rouge, LA 70803 USA

INTRODUCTION

While we could probably trace the appearance of excess electrons in fluids to prehistoric thunderstorms, a more likely starting point for our discussion is man's discovery of metal ammonia solutions. It had been widely believed that the German chemist Weyl (1864) was the first to study this interesting system in 1863 since he published the first accounts of his studies in the Annals of Physics in 1864. However, Peter Edwards (1982) discovered in a laboratory notebook dated November 1808, almost fifty years earlier, that Sir Humphry Davy had made very detailed studies of potassium exposed to ammonia gas (dry gas as he noted most precisely) leading to potassium-ammonia solutions. These studies were apparently never published, although he clearly described the unique features of the blue and bronze solutions.

In the 1920's these systems were subjected to a very extensive study by Kraus and others. Kraus (1908, 1914, 1921, and 1931) suggested that the low concentration solutions consisted of an electron in a cavity or disordered solvent region since the density of the solution was so low. He also interpreted much of the data at low concentrations as an electrolyte solution while at high concentrations they were discussed as liquid metal. Much of the earlier studies of trapped electrons were dominated by the study of metal ammonia solutions, in part because of their exceptional stability. These studies were first collectively presented in the proceedings of Colloque Weyl I in 1963 (Lepoutre and Sienko, 1963).

Beginning in the early fifties it was realized that one could inject electrons into a variety of fluids, gaseous and liquid and study their properties. In most cases the resulting species are short lived and thus require special techniques to study them. Nevertheless in many cases they have well defined properties, but properties are often very sensitive to temperature, pressure, and solvent composition, in contrast to normal anions, e. g. chloride ion in water. In some cases the nature of the electronic structure depends on the density in such a dramatic way that the properties differ radically from one density range to another. For example, in dilute sodium ammonia solutions, the blue color can be simply modelled in a naive fashion as did Ogg (1946) in the mid 1940's by a particle in an infinite walled box model while the electrons in liquid xenon or in concentrated metal ammonia solutions behave as almost free electrons, not spatially localized.

It was really the study of the radiation chemistry of water which lead to the general investigation of electron localization in a variety of media, especially in polar solvents. Platzman (1955) in 1953 suggested that the hydrated electron might be studied experimentally. The observation ten years later by Hart and Boag (1962, 1963) of the transient optical absorbtion spectrum of the hydrated electron confirmed that prediction and began the intense study of trapped electrons in various media. It also made the pulse radiolysis technique a very common and convenient tool for the study of optical spectroscopy of localized electrons in a vast variety of solvents. This method in modern form still provides most of the available data on the energetics of trapped electrons in fluids.

Before discussing in detail any examples it is appropriate to discuss the range of systems which can be studied as well as the issues which must be faced in treating these rather unusual species. First of all, the range of systems which could be considered in this paper could include the following:

a) Metal ammonia solutions [and the similar case of electrons in hexaphosoamine (HMPA)].

b) Excess electrons in polar solvents (e.g. water, alcohols, ethers, acetonitrile).

c) Excess electrons in simple dense nonpolar fluids (liquid rare gases, liquid H_2).

d) Excess electrons in complex nonpolar liquids (liquid hydrocarbons).

e) Excess electrons in nonpolar vapors (helium and argon).

f) Excess electrons in polar dense vapors (supercritical water and ammonia vapors).

g) Excess electrons in polar clusters.

h) Concentrated metal - ammonia solutions.

i) Metal-molten salt systems.

j) Liquid metals.

From the point of view of the experimental chemist, the following subdivision, originally due to Onsager, is relevant:

1) Chemically stable systems corresponding to classes (a), (g), (h), and (i) which can be studied by conventional physical and chemical methods.

2) Chemically metastable systems which correspond to classes (b), (c), (d), (e), and (f), where the lifetimes of excess electrons are limited by efficient (electron-cation or electron-ion) recombination reactions, or electron-electron reactions in water.

From the theoretical point of view two subdivisions are pertinent:

a) Electron localization

1) The energetically stable ground state of the excess electron is quasi-free. Such a solution prevails at low density of excess electrons in liquid Ar, Kr, and Xe and in high density systems such as concentrated metal solutions in ammonia and in molten salts.

2) Localized excess electron states are energetically stable in some nonpolar solvents such as liquid He and liquid Ne. In polar solvents and in polar dense gases electron localization occurs at low concentrations.

b) Excess electron concentrations

1) Low density of excess electrons. These systems involve dilute metal solutions, electrons in polar and nonpolar solvents. In these systems interactions between electrons are negligible; a single isolated electron localization centre can be considered.

2) High density of excess electrons. Concentrated metal ammonia, metal-molten salt solutions and liquid metals exhibit a transition from a localized to the metallic state.

One way in which to organize the types of states observed is to consider the density dependence and the effective strength of the electron-molecule interaction. Milton Cohen (1964) once suggested an idea which has been expanded upon by Kestner (1976) in several publications. If the interactions are very strong we will get a molecular anion. There are a few such cases known. In some ethers, neutral alkali metal atoms have been documented. In the case of weaker interactions which are more common, we might also see a transition from a localized to a delocalized state. If the interaction is just a little stronger we might see only localized states.

Theoretical studies of excess electron states in polar and nonpolar liquids and dense fluids are pertinent for the elucidation of the electronic states of disordered systems. Localized excess electron states in nonpolar and in polar dense fluids at low electron densities yield interesting information regarding electron-medium interaction and configurational changes in the medium. The quasi-free electron provides an excellent probe for the interaction of a single charge carrier with an unperturbed disordered system. We will ignore in this review all examples involving a high concentration of excess electrons (systems h, i, and j above).

This chapter will concentrate on theoretical studies of excess electron states in polar and non-polar fluids. In view of the complexity of this field **a-priori** theoretical schemes are not practical and we shall emphasize the interplay between theoretical models and experimental facts of life.

In addition, there are numerous very basic issues we must consider before discussing specific examples.

a) What is the definition of a localized state, with reference to a delocalized one?

b) Is the definition of localized state or the transition between regimes as a function of temperature, density, or pressure for example the same for transport i.e. mobility as for energetic properties, such as spectra?

c) What is the effect of disorder or inhomogeneity in the medium?

d) What is the role of the molecular structure and forces?

We will spend a great deal of time on item d later in this chapter. But first we consider a and c and then use these discussions to make some preliminary remarks about the role of the measurement technique on the definition of the states.

LOCALIZED VERSUS DELOCALIZED STATES

It is absolutely vital for our discussion to have a good definition of a localized electronic state as opposed to one which is delocalized. The later will also be referred to as a quasi-free electron. Our primary definition will be that the localized electron is characterized by a wave function which decays exponentially with distance; the same type of definition used for simple atomic wave functions. The quasi-free electron is delocalized, being best represented by plane waves states (often a number of them, not only one).

While this definition is theoretically sound, one usually does not always have the detailed models to use it. A more practical definition is desired. We will state some of the characteristics of localized and free states here but emphasize that this is not a nice clean distinction.

Tentative Definitions:

A localized state usually has an electronic absorption spectrum, a "size" corresponding to a molecular size, and the usually low mobility.

A Quasi-free state is best characterized by high mobility and no evidence of an electronic absorption spectrum.

The most serious problem with these definitions is the words "high" and "low" mobility. That is highly dependent on the type of medium and the temperature. Nevertheless these definitions are reasonable zero-order statements if we take mobility to imply values over 10 cm^2 V^{-1} sec^{-1}.

After a brief discussion of disordered media we will consider how to apply these definitions to the problems of excess electrons in fluids where it is not clear what state we should expect and why.

DISORDERED, AMORPHOUS MEDIA

Perfect Metals; Pseudopotentials

In a perfect metallic solid the conduction states of electrons are relatively easy to classify. The real situation is, of course, more complex but ideally the electrons are in a perfectly periodic potential of scattering sites. Their electronic distribution is characterized by Block waves and long range order. The most serious issue faced in the early years was why the electrons experimentally were so well represented by plane waves, i.e. quasi-free electrons. The electronic wave functions are highly delocalized with no distinct exponential fall off nor any appreciative variations near the ions which form the bulk of the material. In fact, the electronic structure is, for many properties, well represented by a jellium model which ignores completely any details relating to the atomic nature of the underlying ions. Phillips and Kleinman (1959) in their analysis of the OPW (orthogonalized plane wave) approach to electronic structure of metals observed that the quantum mechanical requirements of orthogonality forced upon electrons by their Fermion character could be thought of as an effective repulsive potential which in large part cancels the strong coulomb attractive of the ions. The net potential is therefore much smoother that one might have first imagined.

Actually the first attempt to derive an effective potential, in general, was by Hellman (1935, 1936) in order to simplify the orthogonality requirements in quantum calculations. The first real exact theory for a one valence electron atom was due to Szepfalusy (1955) but it is best understood in the paper of Szasz and McGinn (1966). From a very different formalism and background Phillips and Kleinman derived similar results. The entire theory is presented very nicely in the book by Szasz (1985).

There are other advantages to using a pseudopotential, namely we can approximate it as the effective "semi-classical" interaction between a point electron and the atom or ion. We all know that this is an approximation but it does enable us within certain constraints such as low energy electrons to compare how two species interact with an electron located at a point in space, or even to use it as a potential in semi-classical studies such as path integrals.

The concept of a pseudopotential or effective interaction between electrons and atoms or ions was investigated by many others following those initial works. The most thorough paper with reference to our presentation is the work of Cohen and Heine (1961). We shall use their formulation in many ways so it is worth considering immediately since knowledge of it will enable us to consider the "apparent" strength of the interaction between an electron and an atom or ion.

A proper description of the electron-molecule interaction potential, V, can be provided in terms of pseudopotential theory, which accounts for the nature of the short range repulsive interactions originating from exclusion-orthogonality effects. The excess electron wavefunction, Ψ, can be determined from some one electron equation

$$(\hat{T} + v)\ \Psi = E\Psi\ , \tag{1}$$

where $\hat{T}$ is the kinetic energy operator, v is the one electron potential and E the effective orbital energy. This energy must be determined in a manner compatible with the Pauli exclusion principle. This restriction can be formulated mathematically by imposing orthogonality relations between all the molecular core spin-orbitals, X_c, and the excess electron wave function, i.e.

$$\langle \Psi / X_c \rangle = 0\ , \tag{2}$$

for all X_c. Application of Eq. (2) raises practical difficulties, as this orthogonality condition introduces violent oscillations into the excess electron wavefunction within the core region. It will be convenient to be able to ignore these oscillators, and still satisfy the restrictions imposed by the exclusion principle. This dilemma is resolved by introducing the pseudopotential. We introduce a pseudo wavefunction for the valence electron, which is in one-to-one correspondence with the actual excess (or valence) electron wavefunction. The pseudo wavefunction is not required to satisfy the orthogonality relation which implies that the valence electron must be excluded from the region of space occupied by the core electrons. Instead we shall modify the potential. The electron described by the pseudo wavefunction ϕ is kept out of the core by adding a non-local repulsive potential, V_R, to the Hamiltonian, thereby defining the eigenvalue equation satisfied by ϕ to be

$$(\hat{T} + v + V_R)\ \phi = E\ \Phi\ , \tag{3}$$

where the valence electron energy E is the same as in Eq. (2). The potential $v + V_R$ defines the pseudopotential

$$V_{ps} = v + V_R\ . \tag{4}$$

Cohen and Heine (1961) have shown that there is no unique way to introduce ϕ and V_R by relaxing the orthogonality condition. They have taken advantage of this arbitrariness by requiring that V_R cancel the core contribution as much as possible. In particular, they have shown that the smoothest orbital (in the sense of having the lowest kinetic energy) is obtained from the following potential

$$V_R^{CH}\ \phi = -\sum_c X_c\ \langle X_c|V|\phi\rangle + \frac{\langle\phi|V_{ps}|\phi\rangle}{\langle\phi|\phi\rangle} \sum X_c\ \langle X_c|\phi\rangle\ . \tag{5}$$

The second term on the right-hand-side of Eq. (5) is small and can be neglected. This form of the pseudopotential can be derived from the general expression obtained by Austin, Heine and Sham (1962)

$$V_R^{AHS} = \sum_c X_c\ \langle X_c|F\phi\rangle\ , \tag{6}$$

where F is any arbitrary operator and we may set F=-V. We have thus managed to simplify the wavefunction by introducing the pseudopotential which depends on ϕ explicitly in two places: in the exchange potential and in V_R.

The general expression for the pseudopotential for an excess electron is conveniently displayed in the form (Kestner et al., 1965)

$$\begin{aligned} V_{ps}\phi = {} & \left(V_{coul} + V_{nuc} + V_{exc} + V_{pol}\right)\phi \\ & - \sum_c \langle X_c \left| V_{coul} + V_{nuc} + V_{exc} + V_{pol} \right| \phi\rangle\ X_c\ , \end{aligned} \tag{7}$$

where V_{coul}, V_{nuc} and V_{exc} are the coulomb, nuclear and exchange interaction potentials of the excess electron, and V_{pol} is the polarization potential which may be omitted however in the calculation of only the short range interactions.

Concepts of Localization in Disordered Media

The prediction of electronic structure is not nearly so clear if we deal with amorphous and disordered materials. Even if the disorder is only substitutional, the medium is almost homogeneous, and the electron interaction is weak, the situation immediately became more complex and long range disorder is no longer obvious. This classic problem was seriously studied in the sixties and the most important theory for such materials was derived by Anderson (1958), although it was more simply presented by Ziman (1969) and was extensively studied by Mott (1956, 1961, 1967) and Cohen (1970, 1972) and their coworkers among many, many others. Thouless (1972) has discussed in greater detail the mathematical assumptions needed. Economou and Cohen (1972) have extended the arguments especially those related to electron mobility.

An excellent brief presentation is contained in the chapter by M. H. Cohen in Electrons in Fluids (1973). This article is highly relevant to this study institute. We will now review in qualitative form the results of the more general theory. The authors consider the following simple model of a disordered system. It consists of a series of potentials everywhere positive, but of random magnitude between V_1 and V_2, i.e. no value less than V_1 nor larger than V_2 possible. The potentials about any random point are relatively short ranged and uncorrelated with other well separated points. The exact distribution of potentials is unimportant.

The most important result can be summarized as follows. If we have a pure metal with a free particle density of states we would have a density of states which had the simple $(E-V)^{1/2}$ dependence. However, in this material we have no states below V_1 and thus there is a band edge at $V=V_1$. This is followed by a tail which depends on the form of P (V) but grows relatively slowly. Finally sufficiently far from V_1 in energy we will have the free particle density of states but shifted by V_1 plus an additional contribution. There is no square root singularity at either the origin or at V_1. Above V_2 all energies are shifted by a small amount. Ziman (1969) quickly recognized that this is like a classical percolation problem and thus those ideas must explain the nature of the electronic motion in these systems as a function of energy. Basically a point or region is prohibited (P) to an electron of energy E if the energy is less than the potential, but the region is allowed (A) if the energy exceeds the potential.

We can make this somewhat more quantitative by speaking in terms of volume of energy space rather than volumes in real space. We will define C(E) as the fraction of the energy states (potential states) in the system which are above the minimum energy.

$$C(E) = \int_{V_1}^{E} P(V)\, dV \ , \tag{8}$$

whose P(V) is the probability of a potential V in the material. C(E) is zero below V_1 and then increases first slowly, then faster but always monotonically toward one which it achieves for energies of V_2 and upward.

We can imagine a typical fluid in terms of the energy of specific regions. We show various typical states of the fluid in Fig. 1. These separate pictures can differ in the density, temperature, or other physical or chemical factors. For regions below V_1 we have only prohibited states. In terms of mobility this would mean that there would be only highly localized states with zero mobility in this classical rather rigid picture. As the energy gets larger, the electrons are allowed more freedom to move but the states so formed are still localized although with large extents. Only through random hopping or thermally assisted motion or quantum tunneling do we expect any sizable mobility. As the number of allowed states grow, the allowed regions in phase space can also grow and as one might expect when there are enough allowed spaces we can form long channels, channels which can even extend across the material. These very extended states are high mobility states. This pattern can, in fact, be demonstrated in very simple examples like resistor networks or graphite films or computer models. Based on various computer simulations on lattice studies, Cohen (1973) and others suggest that the critical percolation threshold for this system is about 0.2. That is, when C exceeds 0.2 we expect open channels and high mobility to be observed even for these amorphous materials. We know that high mobility can be observed in these materials. To complete the picture at energies near V_2 there are mainly delocalized states with some prohibited regions which act more like scattering centers or generators of resistance. Above V_2, of course, the material conducts with ease as there are no prohibited regions.

At energies near E_c the electrons are confined to channels and are thus not typical three dimensional free electrons. But, we have ignored quantum mechanics and in reality the electrons can tunnel from one channel to another and between various allowed regions. Furthermore as already indicated if we allow the electrons to hop there will be an additional component due to thermally assisted processes (or phonon

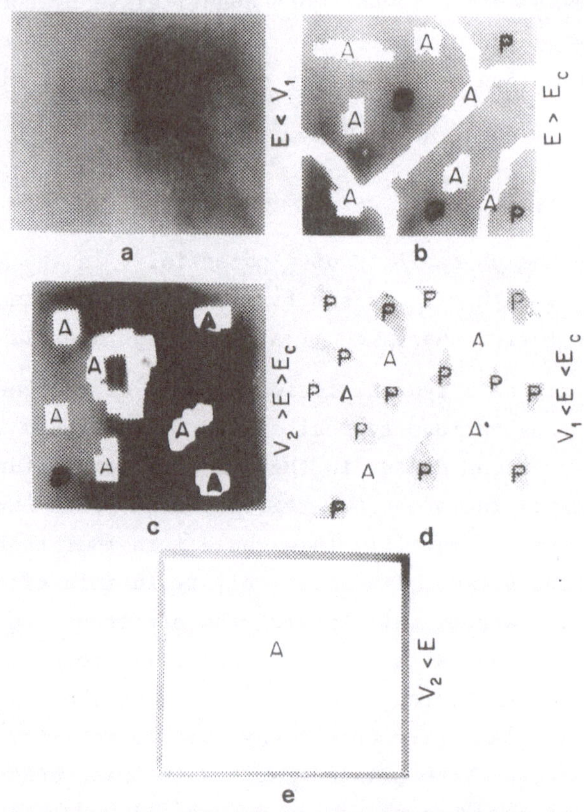

Fig. 1. Regions in a typical microscopically inhomogeneous amorphous material as a function of energy. There are random barriers of energies between V_1 and V_2 in this material, $V_2 > V_1$. E_c is the percolation threshold.

assisted tunneling). This will make some of the regions and pictures less well defined (and temperature dependent).

The conclusions we need to remember is that in disordered materials like liquids we should expect a set of localized states at energies just above the allowed energy band but for higher energies these became extended conductive states. This localization will always occur, as Anderson (1958) showed, if the disorder is great enough. The transition from the localized states to extended states was predicted by Mott (1967). In 1969 Cohen et al. (1969) pointed out that E_c should, in fact, be identified with the mobility edge found in disordered materials. If the electron-medium interaction is strong enough it can also distort the medium to create real trapping sites. We expect to see this behavior in polar fluids particularly. In a polar fluid the potential varies rapidly from position to position and it is <u>not</u> of short range nor is it weak. In hydrocarbon and rare gas fluids the situation is more closely related to amorphous materials of the type discussed above.

In the next sections we will first consider the electron-medium interaction for some disordered materials and observe how it determines the nature of the species in the material. We will then return to these percolation and localization concepts in a few typical examples.

GENERAL FEATURES OF ELECTRON-MEDIUM INTERACTIONS

Quasifree and Localized Excess Electron States

When an excess electron is introduced into a nonpolar or polar liquid, it is not immediately apparent what is the energetically stable state of the electron. The total ground state energy E_t of the system can be written always in terms of two contributions: the electronic energy, E_e, and the medium arrangement energy E_m, so that

$$E_t = E_e + E_M \ . \tag{9}$$

The electronic energy E_e has to be computed in the spirit of the Born-Oppenheimer approximation for each (fixed) nuclear configuration of the medium. The second term in Eq. (9) involves the structural modifications induced in the medium due to the presence of the excess electron, so that in general $E_M>0$. In this context two limiting extreme cases should be distinguished:

a) The quasifree electron state. The excess electron can be described by a plane wave (e.g. a wave packet) which is scattered by the atoms or molecules constituting the dense fluid. Under these circumstances the liquid structure is not perturbed by the presence of the excess electron, so that $E_M=0$. The electronic energy of the quasifree electron state, which we shall denote by V_o, is determined by a delicate balance between short range repulsions and long range polarization interactions, so that

$$E_t(\text{quasifree}) = V_o \ . \tag{10}$$

This energy corresponds to the bottom of the conduction band in the liquid relative to the vacuum level. V_o would correspond to V_2 in the models considered in the second section.

b) The localized excess electron state. The wavefunction for the excess electron tends to zero at large distances from the localization center. In this case the liquid structure has to be modified to form the localization center. Such a liquid rearrangement process requires the investment of energy. It will be convenient at this stage to specify the configuration of the solvent by a single "configurational coordinate" R; the simplest choice is, of course, the mean cavity radius. Although one really implies some multidimensional space. The total energy of the system can be written

$$E_t(R)[\text{localized}] = E_e(R) + E_M(R) \ . \quad (11)$$

The most stable configuration of the localized state is obtained by minimizing the total energy [Eq. (11)] with respect to R

$$\partial E_t{}^{(R)}/\partial R = 0$$

$$\text{at } R = R_o \ . \quad (12)$$

$$\partial^2 E_t(R)/\partial R^2 < 0 \ .$$

This condition establishes the configurational stability of the localized excess electron. The resulting energy E_t (R_o) can be, of course, either positive or negative as the liquid rearrangement process requires the investment of energy $(E_M(R_o)>0)$ which is (wholly or partially) compensated by the electronic energy term $E_e(R_o)$.

The absolute sign of the minimum electronic energy of the localized state does not determine whether this localized state will be energetically stable. Cases are encountered in nonpolar liquids when $E_t(R_o) > 0$ for the localized state and still the localized state is stable relative to the quasifree electron state. To assess the energetic stability of the localized excess electron state in a liquid one has to compare the energy of the localized state with that of the quasifree state. The general energetic stability criterion for the ground localized state implies that

$$E_t(R_o) < V_o \ . \quad (13)$$

Equation (13) disregards entropy effects. The medium rearrangement process accompanying electron localization is associated with an entropy charge ΔS_M. Thus the free energy of medium rearrangement processes is approximately

$$\Delta G_M = E_t \ (R_o) - T\Delta S_M \ . \quad (14)$$

While the free energy of the localized state is given by

$$G_t = E_e(R_o) + \Delta G_M \ . \quad (15)$$

The thermodynamic stability for the ground state is thus

$$G_t < V_o \ . \quad (16)$$

TYPICAL CALCULATIONS

Typical Pseudopotentials

The general form of a pseudopotential consists of an electrostatic, polarization, and the all important orthogonalization terms (Eq.7). Somewhere, either in these later terms or separately, one must also include the exchange interactions.

If we begin by looking only at rare gas atoms, the pseudopotentials have a simple form. There are only two major factors involved. First of all, we have the interaction of the excess electron with the electrons in the atom, compensated by the exclusion effect. This is repulsive in all cases studied by various methods. Then there is the electron polarization interaction which increases monotonically from helium to xenon. When put together the result depends dramatically on the relative strength of the pieces.

In the case of helium the potential is repulsive for all separations. In fact it is very strongly repulsive as shown in Fig. 2. So repulsive that the electron-helium atom interaction can be treated with a high degree of vigor as a hard sphere interaction. In many ways this makes its study so very simple. However, it is also very misleading as it is a very special case. The best result is in the paper by Kestner et al. (1965). The more general case is typified by electron-argon or electron-xenon interactions. In this case the interaction energy is something like an exponential repulsion and an inverse fourth power attraction, modified by the need to cut it off at some finite distance. One representation of such a potential is

$$Ae^{-ar} \frac{B}{(r^2+d^2)^2} \tag{17}$$

or

$$\frac{A}{r^4}\left[\frac{B}{(C+r^6)} -1\right] . \tag{18}$$

Coker et al. (1987) used the latter form with A=12.59, B=4920 and C=3793 (all atomic units) to model the electron-xenon pseudopotential. A typical potential is plotted in Fig. 3. Notice that it does have a substantial attractive portion. However, at very short distances the repulsive exclusion effect again takes over and effectively prevents the electron from existing close to the nucleus.

If we now consider electron-water interactions, which have been actively studied of late, the picture is a little different. We now have a major electrostatic component due to the electron charge interactions which are primarily electron-dipole and electron-quadruple. However, it

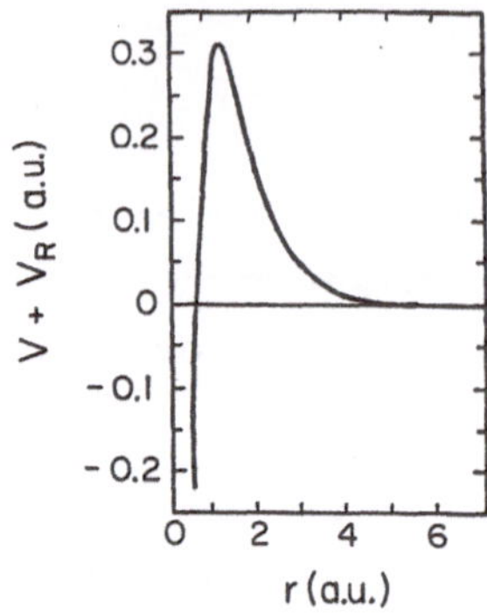

Fig. 2. Electron-helium pseudopotential. Based on the results of Kestner et al. (1965).

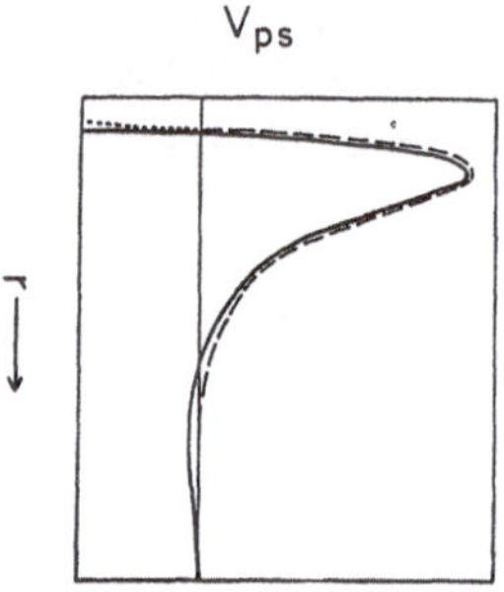

Fig. 3. Typical electron-rare gas pseudopotential. Dotted line is Hartree-Fock results only. Total including polarization is the solid line. Depth of well increases for argon to xenon. The effective depth is very small for neon.

is best to represent these as an electron interacting with a series of point charges on the water - charges which duplicate the electric moments of the water since this potential will be more accurate at smaller separations. This leads to a pseudopotential composed of electrostatic, polarization, and effective repulsion. The latter term may implicitly or explicitly include exchange terms. There are three such potentials published. The most elaborate studies of the potential have been published by Schnitker and Rossky (1987a) who include no explicit exchange terms. Landman et al. (1987) use a density functional approach which explicitly includes exchange. It is remarkable that these two results are so similar. Even more surprising is that the general results of the Wallqvist et al. (1987) paper are rather similar. In Figs. 4 and 5 we plot the potentials of those two papers along the H-O-H bisector and along the O-H bond, a plot which comes close to the absolute minimum of the potential. The potential minimum for the better potentials lies at an angle of about 45 degrees from the H-O-H bisector and is roughly 1.76 eV deep and about 2.2 Å from the oxygen (remember the bond length is about 1 Å). By symmetry there are two equivalent minima, of course. Of greater interest and importance is the fact that the electron-water

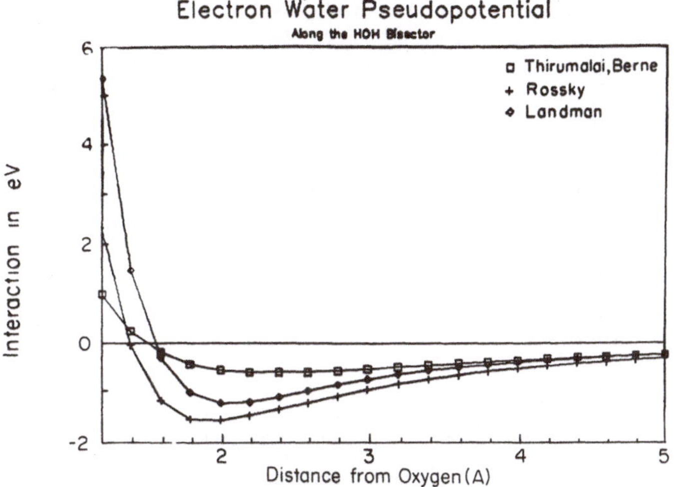

Fig. 4. Electron water pseudopotentials by three authors along the H-O-H bisector (Schnitker and Rossky, 1987a; Landman et al., 1987; Wallqvist et al., 1987a).

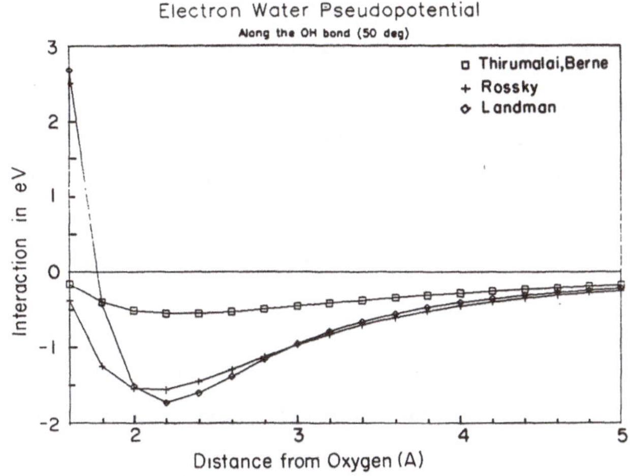

Fig. 5. Electron-water pseudopotentials by three authors along 50 degrees from the H-O-H bisector. This is almost along the O-H bond (references same as in Fig. 4).

potential is dominated by the interaction with the dipole moment, once the electron is excluded from the region of the major electron distribution of the molecule. This is shown in Fig. 6.

The fact that the interaction is dominated by the dipole interaction explains why the semi-empirical models used to explain electrons in water and ammonia have been so successful. To a first approximation one can treat the electron water interaction as a soft sphere repulsion with a dipole moment. Such an extension of the simpler models developed by Jortner (1962, 1964, 1970), Kestner (1973, 1976), Kevan et al. (Fueki,

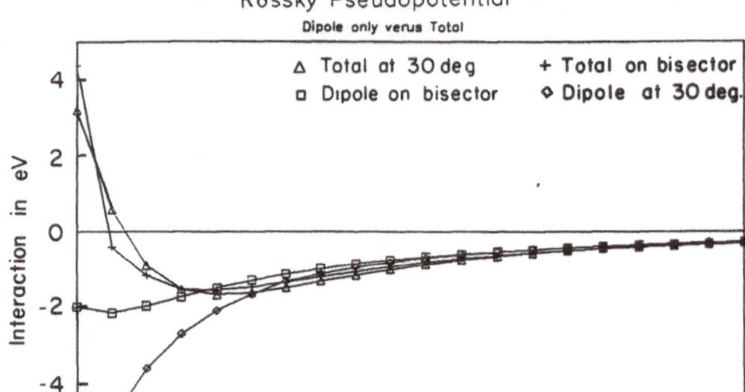

Fig. 6. Comparison of total electron-water pseudopotentials and dipole only contributions using the Schnitker, Rossky potential (Schnitker and Rossky, 1987a).

1971a, 1971b, 1972a, 1972b, 1973), and Feng and Kevan (Feng, 1980) was done by Gaathon and Jortner (1973) with good success.

In other systems we can readily see that similar considerations should apply. There will be electron-multiple interactions, dominated by dipole interactions and then attractions due to polarization of the molecule. The latter effect is taken as proportional to the dipole polarizability of the molecule. If the molecule does not have a net dipole there could be local bond moments which could interact with the electron as well as the contributions from the polarizabilities of the bonds. It is important to realize that these can be large, even larger than dipole moment contributions. Furthermore, in the case of hydrocarbons we must include the polarizability of carbon-carbon bonds which are very important. For large molecules it is important to realize that the electron might be localized in a region dominated by only part of the molecule. For highly localized electrons when we consider a long chain primary alcohol, the electron might experience the O-H dipole or the hydrocarbon, but possibly not both. When the electron is very delocalized it might experience both environments.

Let us consider now a model of a trapped electron in a fluid. We will first discuss a very simple model and then look at the more detailed calculations.

Total Energy

The following factors must be considered:

a. Electron-molecule interactions for molecules in the first coordination layer.

b. Electron-molecule interactions for electrons further removed from the trapping site. In many cases this will be treated as a continuum.

c. Molecule-molecule interactions, a complicated function of the orientation of the molecules relative to the radius vector drawn from the center of the trapping site. It can be derived using the various intermolecular potentials used to study liquid structure.

d. Any effects which relate to solvent reorganization not included in item c.

These are a listing of what one needs to consider for a model semi-classical calculation. For quantum or cluster type calculations some of these effects are combined. In any case the most serious problems relate to electron-solvent interactions, especially those at large distances. Furthermore the exact statistical mechanics of such regions and the proper inclusion of temperature is difficult to treat properly.

For simple model calculations we can write the total energy of the system as

$$E_t = E_E + E_M + E_S \; . \qquad (19)$$

We will discuss each term separately for simple polar molecules first. Later we will see that the same ideas can be applied to the situation involving an electron in a more complex fluid such as that of a large hydrocarbon or alcohol.

The first term in Eq. (19) is the electronic energy for the electron interaction with molecules.

We must also consider the long range polarization term to the total interaction potential in the fluid [Eq. (7)]. The polarization potential in the liquid can be evaluated by application of the electronic adiabatic approximation. The velocity of the excess electron considerably exceeds the atomic velocities, while on the other hand the velocity of the electrons is considerably higher than that of the nearly free excess electron. Under these circumstances, the motion of the excess electron can be ignored in calculating the induced polarization of the atoms, while the atomic motions can be disregarded in calculating the mutual screening effects of the neighboring atoms. However, the polarization field in the liquid has to be modified to include local field effects.

The polarization potential in the gas phase $V_p^g(R)$ was sometimes represented in the semi-empirical form

$$V_p^g(R) = -\alpha\, e^2/2(R^2+d^2)^2 \; , \qquad (20)$$

or alternatively

$$V_p^g(R) = -\alpha e^2/R^4 \; ; \; R > d$$

$$= 0 \; ; \qquad R < d \; , \tag{21}$$

where α is the polarizability and the cut-off parameter d (of the order of 1 a.u.) was chosen to fit the adiabatic polarization potential at the nucleus. In a fluid, the electron-atom polarization interaction is screened by the contribution of the local field induced on the neighboring atoms. Following Lekner (1969) the polarization in a liquid V_p (r) is given in the form

$$V_p(R) = V_p^g(R) \; f(R) \; , \tag{22}$$

where f(r) is the screening function which was calculated by Lekner. Springett (1968) approximated this in the form.

$$f(R) = 1 \; ; \qquad R < b$$

$$f(R) = \frac{1}{1+(8\;/3)\pi\alpha} \; ; \qquad R > b \; . \tag{23}$$

The discontinuity in f(r) at r = b is unimportant because f(r) is used only within an integral.

We have yet to define the two remaining terms in Eq. (19). The energy contribution E_M is the molecule-molecule interactions which must be included explicitly in the model but, of course, are automatically included in **a-priori** calculations which include all of the electrons of all of the molecules.

This completes the evaluation of the contribution to Eq. (19). Each V had to be replaced by a model potential while each V_p is properly given by Eqs. (22) and (23). The total electron-medium potential is determined by the liquid configuration, which is different for the quasi-free and for the localized states.

Localization Issues

What then determines what the state of the electron is in the various media? To state it simply as we said before it is the balance between the electron attractions and the "steric effects", those repulsions between the molecules which have to be modified when the electron tries to maximize its orientation. If the molecule has to be rotated in order to interact strongly and it is asymmetric then it will displace other molecules when it turns. For this reason it is not surprising that a primary long chain alcohol will have a different structure around the electron than a tertiary one. The bulkiness of the molecule when the dipole is oriented is a major factor in determining the properties of the trapped species. With very large groups one expects that the energy of the trap is shallower and thus one would find a lower energy electronic transition. This has been observed and is well documented for alcohols as well as other classes of compounds. There are, of course, the effects due to differences in the dipole moment or more precisely, the electron-molecule (or electron region) interaction.

Calculational Methods

How do we solve the quantum mechanical problem presented by the Hamiltonian presented earlier? Basically one first decides what part of the solution will be quantum mechanical and what part classical.

Entirely quantum mechanical-

This approach is of limited usefulness because the problem rapidly becomes too large to handle easily unless semi-empirical quantum chemistry models are used. Then one has the issues of the accuracy of such methods to consider. The best all electron calculations were first done by Newton (1975) for small (4 or 5) water and ammonia clusters. Later Rao and Kestner (1984) used similar procedures to increase the size of the clusters to six and eight. Even then the calculations are limited since the basis set does not yield the correct dipole moment for the isolated water molecule. Newton (1975) did add in terms to account for the rest of the medium polarization effect and thus achieve better accuracy. The method is important since no pseudopotential has to be defined and all interactions are treated rigorously within the basis set used. These calculations are therefore the standard by which other methods must be judged when they attempt to treat larger systems. However, there are other problems since geometry optimization on large systems at this level of calculation is also complicated and impossible to achieve completely. Temperature is also completely ignored, i.e. the system is at absolute zero.

Classical molecules plus a quantum mechanical excess electron-

a. Model calculations. Under this heading one could classify all of the models from the first crude particle in a box used by Ogg (1946) to the more sophisticated systems used more recently (see Kestner, Jortner, Kevan, and Fueki references). In this case one considers the electron-molecule as represented by a pseudopotential and the molecule-molecule interactions are modeled from intermolecular force considerations. Usually only the first one to two coordination layers of molecules around a trapped electron are explicitly taken into account and the rest modeled by continuum polarization expressions. The long range many body scattering effects are taken into account through the use of the energy of the bottom of the conduction band, i.e., V_0. These calculations, especially the self consistent work by Kevan, Fueki, Feng (Feng and Kevan, 1980) or the adiabatic version by Kestner, Jortner et al. (Copeland, 1970; Kestner, 1973; Kestner, 1976, 1987) seem to explain the general features and trends in the properties of trapped electrons. However, there are serious problems with the models as they are usually used: first of all, disorder or thermal effects are only approximately included and so not all possible liquid configurations are properly explored, and secondly, the calculations are not being done in a proper thermodynamic framework. However, they do readily yield excited state energies for the configurations sampled and give a way of estimating line shapes, although those are usually found to be too small.

b. Quantum Path Integral Methods. Very recently we have been able to achieve a very good solution to the problem of trapped electrons using these methods which are semi-classical molecular dynamics or Monte Carlo simulations of the liquid state including the electron by a "necklace" of semi-classical electron-molecule pseudopotential interactions in the manner proposed by Feynman and Hibbs (1965). With modern fast computers these many body systems can be studied for enough steps to achieve convergence. The method is ideally suited to obtaining the ground state properties at a finite temperature T. It yields all of the thermodynamic information and by design includes all possible configurations of the liquid, properly weighted. These methods can not include more than one electron rigorously and so are ideal for studying isolated trapped electrons. They depend on a number of extrapolations and on the accuracy of the pseudopotentials but those issues can be tested. Furthermore, recently there has been a flurry of activity to obtain excited state properties by selecting large numbers of these configurations and performing on each of them a simple quantum chemistry calculation, then averaging the results to compare them to experiment. This seems to be yielding very good results. Other workers are trying to generate mobility results from various correlation functions which can be generated. This latter work has very significant potential but needs to be explored more to see if it can be done routinely.

In the following section we will review very briefly some of the results of such calculations on a variety of systems.

SAMPLE RESULTS

Rare Gases

The simplest gaseous system to study is a nonpolar gas, and the simplest of these is gaseous helium. The electron-atom interaction is characterized by a zero energy scattering length which is large and positive (see Section TYPICAL CALCULATIONS). This means that the electron-medium interaction will be positive (repulsive) regardless of the density of helium. The height of the barrier for trapping an electron is given by the optical

$$V_o = (h^2/2m)\, 4\pi\rho a \ , \tag{24}$$

where ρ is the density of scatterers and a is the scattering length, multiple scattering being ignored. Simple physical arguments indicate that at high gaseous density the electron prefers to exist in a bubble. The slight loss in entropy and gain in medium rearrangement energy are more than compensated by the decrease in the electron medium energy from the quasi-free state. At low gaseous density the gain in electron energy is not sufficient and thus the state of lowest free energy is the quasi-free state of the electron in which the medium is not disturbed from its state with no electron present. Putting it another way, when the zero point energy, V_o, becomes greater than kT, the electron prefers to dig a hole in the gas since this lowers the free energy of the entire electron-gas system.

Sanders and Levine (Sanders, 1962, 1964, 1967; Levine, 1962) in a very important experiment determined the mobility of electrons in helium gas by a time of flight method. They found that the mobility changes by many orders of magnitude for relatively small charges in the density. Harrison and Springett (1971,1973) extended this work over larger temperature ranges finding an only slightly less dramatic behavior at higher temperatures.

As in similar studies of disordered media we can approach the problem by calculating the energy (or free energy) or attempt to calculate mobility directly. Since the former is easier we shall describe it first. At low temperatures we can consider only the enthalpy of the quasifree as compared with the bubble state. In early work (Jortner, 1965, 1966), a very approximate pseudopotential calculation was used to estimate the transition point between the two domains as about 9 x 10^{20} atoms/cm^3 at 4.2K or (P/Psat = 0.35). Using the better model of Springett et al. (1968), the electronic energy is that of a particle in a box with a well depth of V_o. This model leads to similar transition densities.

The above models are somewhat naive since there are large fluctuations in low density gases and these fluctuations in the density can lead to regions favorable or unfavorable for electron localization even though the average density would be far from the transition region. However, to determine the transition range for mobility one needs to develop a theory for mobility directly. Using a percolation model with density fluctuations Eggarter and Cohen (1971, 1972) achieved excellent agreement with the experimental data as a function of density and temperature. Hernandez (1972) has extended the Eggarter and Cohen work to include a more complete study of the temperature dependence of the mobility, low field Hall mobility, and also the field dependence of the drift velocity.

Another fluid which is a candidate for a system in which the localized-delocalized states might be observed is gaseous H_2 (or D_2). The scattering length is large and positive and thus at high densities a bubble state should exist. Harrison and Springett (1973) have given experimental data to support these qualitative arguments.

It is possible that gaseous neon also exhibits such a transition region but since the scattering length is so small it would be a very gradual transition and much harder to study. Some experimental data support this conclusion.

In the case of atoms with negative scattering lengths, i.e. attractive long range interactions with an electron, the situation is more complicated. The attractive scattering length is the result of a very large long range (asymptotically going as r^{-4}) attractive energy dominating the short range pseudopotential repulsive energy. As discussed in section GENERAL FEATURES OF ELECTRON-MEDIUM INTERACTIONS, these long range interactions are screened in a dense medium and thus the effective scattering length is density dependent. Thus one can get large variations in the scattering as a function of density. At certain densities one expects extremely high values for the mobility. As we have seen earlier, however, fluctuations in the liquid density must be considered. Thus although the average liquid is at this critical density value, most regions are either above or below this point. In many cases it is observed that if the scattering length is negative at high energies the cross section (relative to an effective scattering length squared) goes through a minimum. At higher energies the electron interacts more strongly with the short range repulsive regions and less strongly with attractive or long range interactions, thus providing a type of "screening" also. In normal electron scattering this phenomenon is referred to as the Ramsauer - Townsend effect.

The first experiments to suggest that these mobility maxima should occur were conducted by Schnyders, Rice, and Meyer (1965, 1967) especially in their results on liquid krypton. The definitive experimental work is that of Jahnke et al. (1971) on fluid argon.

The most complete study of electrons in argon and similar liquids is the work of Basak and Cohen (1979) who were able to model the mobility behavior above the critical point using a deformation potential in conjunction with percolation arguments. This has been refined by Ascarelli (1985). Since other chapters will deal with this we will not expand on it here.

It is very apparent that many nonpolar systems might show these unique density dependences. However, for many reasons experimental data is lacking. Freeman (1987) has presented a broad compilation of such data. He also presents some of his own data which shows such effects in hydrocarbon fluids.

Very recently path integrals have been applied to electrons in rare gas fluids. From these methods one obtains the electron distribution in space from a molecular dynamics or Monte Carlo program which treats the electron as many small charges distributed throughout space and treats the molecules in the medium as interacting with each other through standard intermolecular potentials. The electron-medium interaction is introduced by a pseudopotential approach.

There were two early attempts to use such methods to study electrons in helium (or an equivalent hard sphere fluid): Bartholomew et al. (1985) and Chandler et al. (1984) using integral equations. Since then Sprik et al. (1985), Nichols and Chandler (1984) and Coker et al. (1987) have done much more extensive calculations. The first two works are for a classical hard sphere fluid and also for polarizable hard sphere fluids. While Coker et al. have presented path integral Monte Carlo studies for electrons in helium and electrons in xenon using good pseudopotentials and good molecule-molecule interactions. Coker et al. considered these two qualitatively different media. We will comment on the Coker results and on unpublished comments by Laria and Chandler (1987) on the comparison of the earlier results with Coker's. Basically the path integral methods confirm the early model studies as well as the quantum statistical formulation of Hiroike et al. (1965) using boson fluid perturbation theory, a theory with some similarities to modern RISM theory. They are able to obtain additional information related to the statistical nature of the problem that the models could not obtain.

They find that the electron in helium is essentially confined to occupy a restricted region in space because of the repulsive nature of the interactions. At low densities the electron is quite delocalized while at the higher densities the electron is essentially prevented from getting any closer to the helium atoms than a distance of about the same magnitude as the scattering length (a quantity due solely to s-wave scattering). Another indication of the transition as a function of density is the evidence of ground state dominance above their reduced density of about 0.3. Ground state dominance means that the electron has a ground state energy level well separated from higher excited states at the temperature of the study. This shows up in the imaginary time correlation functions as a simple behavior (dependent on relative times for short times and then independent of it longer times). The correlation functions at low densities are similar to the free electron predictions. The resulting density profiles in helium are like the "fuzzy" bubbles in the study of Hiroike et. al. (1965). Laria and Chandler (1987) were able to use RISM theory and effective hard sphere simulations to generate results similar to those of Coker.

The results for electrons in xenon are not so dramatic. Coker considered two isoterms, one above the critical temperature (309 K) and one below (248 K). The electrons here have a kinetic energy almost independent of xenon density and the effective "size" of the electron (the second moment of the bead distribution) is likewise not density dependent. Furthermore no major distortion of the liquid structure occurs nor is there any evidence of ground state dominance. There is some evidence from some studies they made at low densities and lower temperatures that the electron via its long range polarization interaction does cause some structural changes in the fluid. This is also suggested by the deviations of the imaginary time correlation functions from their free particle values. They have discussed the mobility versus density versus temperature profiles using percolation theory and their findings of some small structural changes in the medium are compared to the experimental work of Huang and Freeman (1978). Laria and Chandler in their RISM-polaron and simulation studies conclude that the repulsive forces are the dominant factor in deciding upon electron mobility behavior and the effective size of the electron in xenon is very small leading to free electron like behavior.

Electrons in Polar Liquids

The latest works are those involving path integral methods. We will not review earlier work since it can be found in reviews by Kestner (1976) or Feng and Kevan (1980). They include the work of Klein and co-workers (Sprik, 1985; Marchi, 1988), Jonah et al. (1986), Schnitker and Rossky (1987b), and Wallqvist et al. (1986a, 1986b, 1987b). The pseudopotentials used by these groups have been discussed earlier.

The work of Schnitker and Rossky is the most thorough and they have compared their findings with the earlier models. The basic result is a very disordered structure of the water around the electron. The local order must fluctuate extensively. The water molecules are primarily ordered with their O-H bonds pointing inward, just as the pseudopotential would require. Estimates have been made of most of the thermodynamic properties and they are very good.

The previous models have been very deficient in regard to their predictions of the excitation line shape, previous models had very narrow lines since they allowed very little disorder of the solvent. Rossky and Schnitker (1988), Schnitker (1987b), and Wallqvist et al. (1988) using various techniques have studied the predicted line shapes in their calculations. Two features seem to emerge. Rossky (1988) has found that while most of the shape is dominated by several strong predominantly s to p transitions, the tail of the spectrum arises in part from a continuum contribution. Also there is some evidence that it may be necessary to consider the electron-medium polarization interaction self consistently as this can lead to slightly deeper bonding and a narrower line width. This conclusion is not proven yet but Rossky does obtain an excitation spectrum which is too broad, a remarkable feat since previous theories errored in the opposite direction. With these latest calculations it is becoming clear that our understanding of the nature of the trapped species is improving rapidly and those previous simple models did indeed contain all of the major features, except for the solvent disorder.

The study of electrons in other polar fluids is quite extensive but theoretical studies have been limited to model calculations. Extensive reviews of these are found in Kestner (1976, 1987) and Feng and Kevan (1980). The general features are remarkably explained by steric restrictions and the magnitude dipole moments, permanent and induced.

Electrons in Hydrocarbons

This system has been studied mostly for those cases where the electrons are weakly bound and when the mobility of the excess electron is high. In such cases the mobility can be rationalized by considering that the medium is microscopically inhomogeneous in the manner discussed earlier. In some regions it is localized due to the optimal fluctuations of density or molecular conformations, while in other regions it can not be localized. If those traps are very shallow and the electron density very diffuse one can use the semi-classical percolation pictures. A very naive theory was discussed by Kestner and Jortner (1973) and shows the correct trends versus V_0. Others such as Holroyd (1987) have analyzed this simple theory in detail and have pointed out its deficiencies.

Nevertheless, it is a simple model which seems to reconcile a lot of experimental data. In contrast to this is the two carrier model in which two types of carriers are present and the electron hops between trapping sites. The recent Hall measurements by Munuo and Holroyd (1988) find no evidence for two carriers, at least, in medium and fluids where the electrons have medium to high mobilities.

Using typical geometries for the CH_2 groups, Feng et al. (1974) showed that a typical arrangement of about four CH_2 groups leads to a bound state of roughly 0.5 eV. The charge distribution, however, is quite diffuse relative to that found in polar media. Typically only about 25% of the charge is in the cavity, relative to the much larger amount in polar fluids. The theory predicts optical excitation energies of only 0.3-0.4 eV versus the experimental values which are usually near 0.6 eV; the calculation is quite sensitive to the value of V_0 used and to the long range polarization energy used. Ichikawa and Yoshida (1980) have extended this work by approximating the potential well formed in 3-methylpentane glass and performing a better calculation of both the wave function and energy. The results suggest that the first excited state is unbound, explaining the broad experimental absorption spectra. Nishida (1970) and Kimura et al. (1970) evaluated the trapping well depth from exchange interactions, an approximation to the pseudopotential. Reasonable energies are obtained, despite some criticism that some integrals might be overestimated (remarks by Carmichael, 1981).

This is known experimentally also (Kenney-Wallace and Jonah, 1981). These traps, whether transitory or permanent, are important in determining the mobility of electrons in hydrocarbon liquids. This situation is quite different from that in the gaseous state since the configurations required are unlikely at low solvent density.

Electrons in Water Clusters

Experimentally we now have a great deal of information about electrons attached to water clusters. From the work of Haberland et. al. (1984) and Knapp et al. (1986) we know that an electron can be attached to a cluster of about 11 water molecules or more. It can also be attached very loosely to clusters of two or three water molecules but that special case will not be discussed here. [See Wallqvist et al. (1986) or Landman et al. (1987).] We believe it has a rather simple explanation. Bowen and colleagues (Coe, 1986) had also measured the photoionization spectra of those larger clusters. Their ionization potentials are of the order of one volt and increase monotonically with cluster size for the cases studied. These experimental data, however, yield very little information about the structure of the complexes and it is unclear if the electron is attached in some way to the surface or is deeply buried inside.

Rao and Kestner (1984) studied the small clusters using the buried electron geometry like that used by Newton for the hydrated electron. Their results suggested that the species of eight or so molecules was still unstable. Their ionization potentials though were in fair agreement with Bowen if the proper extrapolations were made.

Landman et al. (1987), as part of a continuing comprehensive study of clusters using the path integral formalism, studied water clusters of various sizes in regard to their ability to stabilize an attached electron. Some of their results are now in press. Dhar and Kestner have been collaborating with them in order to fine tune their pseudopotentials. We have been performing large scale quantum chemistry calculations on a number of their configurations in order to check out the parameterizations. Thus far those simulations are in good agreement when both are scaled to the same dipole moment. In fact, they are better than the quantum chemistry results since their dipole moments agree with experimental data, and we can readily show that the dipole moment is very important (see earlier discussion on pseudopotentials). Their findings to date are as follows. For clusters of the order of eight water molecules the electron is very weakly bound to the exterior of the water cluster with a binding energy of about 0.73 ± 0.22eV. As the cluster size grows the surface state deepens to 0.92 ± 0.19eV, and appears to distort the cluster geometry. In the case of eight water molecules, the water cluster is not markedly distorted by the presence of the electron, i.e. the water molecules are ordered or disordered as they might be in a neutral cluster. It does appear that the electrons are attached predominantly to clusters with larger dipole moments but this is still being checked out. For eighteen water molecules there also exists an interior state, slightly less stable, with an ionization energy of 1.86 ± 0.32 eV, with properties reminescent of the electron in bulk water. Work is now in progress collaboratively to investigate this system more extensively, especially as a function of number of water molecules and try to evaluate the point at which bulk properties appear to be attained. In all cases the electrons are in rather diffuse orbitals, usually with mean square radius of gyration of the electron bead chain of order of 5 Å.

CONCLUDING REMARKS

The above comments have dealt with the general features of electron-molecule interactions and the behavior of electrons in inhomogeneous media in general. It has deliberately been kept at the survey level.

This field has recently had rebirth and much exciting new work continues to appear. The new methods provide the justification for previous work but more importantly they include the important statistical mechanical features which could not be properly included previously, while there have not been many real surprises as yet the new work had provided much better understanding of the whole field.

Special thanks are due to the support of the Department of Energy and its antecedent agencies and the many hours spent with J. Jortner, M. H. Cohen, S. A. Rice, K. Hiroike, M. Newton, D. A. Copeland, U. Landman as well as the exciting research institutes at the University of Chicago, the University of Tel-Aviv and Brookhaven National Laboratories.

REFERENCES

Anderson, P.W., 1958, Phys. Rev., 109:1492.
Ascarelli, G., 1985, Comm. Sol. Stat. Phys., 11:179.
Austin, B.J., Heine, V., and Sham, L.J., 1962, Phys. Rev., 127:276.
Basak, S., and Cohen, M.N., 1979, Phys. Rev. B, 20:3404.
Bartholemew, J., Hall, R., and Berne, B.J., 1985, Phys. Rev., B32:4234.
Berne, B.J., and Thirumalai, D., 1986, On the simulation of quantum systems: path integral methods, Ann. Rev. Phys. Chem., 37:401.
Boag, J.R., and Hart, E.J., 1963, Nature, 197:45.
Carmichael, I., 1981, Abstract PH161 at 28th Congress of the IUPAC, Vancouver, B.C.
Chandler, D., and Wolynes, P.G., 1981, J. Chem. Phys., 74:4078; Chandler, D., 1982, in: "Studies in Statistical Mechanics VIII," E.W. Montrall and J.L. Lebowitz, eds., North Holland, Amsterdam, p. 275.
Chandler, D., Singh, Y., and Richardson, D.M., 1984, J. Chem. Phys., 81:1975; Nichols, A.L., III, Chandler, D., Singh, Y., and Richardson, D.M., 1984, J. Chem. Phys., 81:5109.
Coe, J.V., Worsnop, D.R., and Bowen, K.H., 1986, J. Chem. Phys., (submitted).
Cohen, M.H., and Heine, V., 1961, Phys. Rev., 12:1821, 122:1821.
Cohen, M.H., 1964, unpublished notes, University of Chicago.
Cohen, M.H., Fritzsche, H., and Ovshinsky, S.R., 1969, Phys. Rev. Let., 22:1065.
Cohen, M.H., 1970, in: "Proceedings of the Tenth International Conference on the Physics of Semiconductors", Cambridge, Mass., S.K. Keller, S.C. Hensel, and R. Stern, eds., (USAEC, Division of Technical Information, Springfield, VA) p.645.
Cohen, M.H., 1970, J. Non Crystalline Solids, 4:391.
Cohen, M.H., and Sak, J., 1972, in Amorphous and Liquid Semiconductors; M.H. Cohen and G. Lucousky, eds. (North Holland Publ Co., Amsterdam), p.696.
Cohen, M.H., 1973, The electronic structures of disordered materials in: "Electrons in Fluids," J. Jortner and N.R. Kestner, eds., Springer-Verlag, Berlin.
Coker, D.F., Berne, B.J., and Thirumalai, D., 1987, J. Chem. Phys., 86:5689.
Copeland, D.A., Kestner, N.R., and Jortner, J., 1970, J. Chem. Phys., 53:1189.
Economou, E.N., and Cohen, M.H., 1972, Phys. Rev., 5:2931.
Edwards, P.P., 1982, Advances in Inorg. and Radiochem., 25:135.
Eggarter, T.P., and Cohen, M.H., 1970, Phys. Rev. Let., 25:807; 1971, 27:129.
Eggarter, T.P., 1972, Phys. Rev., 5A:2496; ibid., A5:2496.

Feng, D.F., and Kevan, L., and Yoshida, H., 1974, J. Chem. Phys., 61:4440.
Feng, D.F., and Kevan, L., 1980, Chem. Rev., 80:1.
Feynman, R.P., and Hibbs, A.R., 1965, "Quantum Mechanics and Path Integrals", McGraw Hill, New York.
Freeman, G.R., 1987, Ionization and charge separation in irradiated materials, in: "Kinetics of Inhomogeneous Processes", G.R. Freeman, ed., J.W. Wiley and Sons, New York.
Fueki, K., Feng, D.F., Kevan, L., and Christoffersen, R., 1971a, J. Chem. Phys., 75:2291.
Fueki, K., Feng, D.F., and Kevan, L., 1971b, Chem. Phys. Let., 10:504; 1972a, J. Chem. Phys., 56:5351; 1972b, 57:1253.
Fueki, K., Feng, D.F., and Kevan, L., 1973, J. Amer. Chem. Soc., 95:1398.
Gaathon, A., and Jortner, J., 1973, in: "Electrons in Fluids", J. Jortner and N.R. Kestner, eds., Springer-Verlag, Berlin, pp.429-446.
Haberland, N., Schindler, H.G., and Worsnop, D.R., 1984, Ber. Bunsenges. Chem., 8:270; 1984, J. Chem. Phys., 88:3903; 1984, J. Chem. Phys., 81:3742.
Harrison, H.R., and Springett, B.E., 1971, Phys. Let., 35A:73.
Harrison, H.R., and Springett, B.E., 1973, Chem. Phys. Let., 10:418; ibid., 1973, 19:231.
Hart, E.J., and Boag, J.W., 1962, J. Amer. Chem. Soc., 84:4090.
Hellmann, H., 1935, J. Chem. Phys., 3:61; 1935, Acta Fizicochem. USSR, 1:913; 1962, 4:225; and with W. Kassatotschkin, 1936, 5:23.
Hernandez, J.P., 1972, Phys. Rev., 5A:635; 1972, 5A:2696.
Hiroike, K, Kestner, N.R., Rice, S.A., and Jortner J., 1965, J. Chem. Phys., 43:2625.
Holroyd, R.A., The electron: its properties and reactions in: "Radiation Chemistry: Principles and Applications," M. Rodgers and Farahatziz, eds, VCH Publishers, New York, pp. 237-262.
Huang, S.S.S., and Freeman, G.R., 1978, J. Chem. Phys., 68:1355.
Ichikawa, T., and Yoshida, H., 1980, J. Chem. Phys., 73:1540.
Jahnke, J.A., Meyer, L., and Rice, S.A., 1971, Phys. Rev., A3:734.
Jonah, C.D., Romero, C., and Rahman, A., 1986, Chem. Phys. Let., 123:209.
Jortner, J., 1962, Mol. Phys., 5:257.
Jortner, J., Rice, S.A., and Wilson, E.G., 1964, in: "Solutions Metal-Ammoniac:Proprietes Physicochimiques, Colloque Weyl," G. Lepoutre, M.H. Sienko, eds., W.A. Benjamin, pp.222-276.
Jortner, J., Kestner, N.R., Rice, S.A., and Cohen, M.H., 1965, J. Chem. Phys. 43:2614; "Modern Quantum Chemistry-Istanbul Lectures," O. Sinanoglou, ed., (Academic press, New York, 1966) p. 129.
Jortner, J., and Kestner, N.R., 1970, in: "Metal Ammonia Solutions - Colloque Weyl II", J. Lagowski, M., Sienko, eds., Butterworths:London.
Kenny-Wallace, G., and Jonah, J., 1981, Advanc. Chem. Phys., 47:535.
Kestner, N.R., Jortner, J., Rice, S.A., and Cohen, M.H., 1965, Phys. Rev., 140:A56.
Kestner, N.R., 1973, Theory of electrons in polar liquids in: "Electrons in Fluids," J. Jortner and N.R. Kestner, eds, Springer Verlag, Berlin, pp. 1-25.
Kestner, N.R., and Jortner, J., 1973, J. Chem. Phys., 77:1040.
Kestner, N.R., 1976, Theoretical Studies of Electron-Solvent Interactions:Solved and Unsolved Problems, in: "Electron-Solvent and Anion-Solvent Interactions", L. Kevan, and B.C. Webster, eds. Elsevier, Amsterdam, pp. 1-43.
Kestner, N.R., 1987, "Solvated Electrons in Radiation Chemistry: Principles and Applications, M. Rodgers and Farahatziz, eds, VCH Publishers, New York, pp. 237-262.
Kimura, T., Fueki, K., Narayana, P.A., and Kevan, L., 1970, Can. J. Chem., p. 55.
Knapp, M., Echt, O., Kreiste, E., and Recknagel, E., 1986, J. Chem. Phys., 85:636; 1986, J. Chem. Phys., (preprint).
Kraus, C.A., 1908, J. Am. Chem. Soc., 30:1323; 1914, 36:864; 1921, 43:749; 1931, J. Franklin Inst., 212:537.

Krebs, P., Bukowski, K., Giraud, V., and Heintze, M., 1982, Ber. Bunsenges. Phys. Chem., 86:879. For low density water studies, see also Christophorou, L.G., Carter, J.G., and Maxey, D.V., 1982, J. Chem. Phys., 76:2653.
Krebs, P., and Heintze, M., 1982, J. Chem. Phys., 76:5484.
Landman, U., Barnett, R.N., Cleveland, C.L., Sharf, O., and Jortner, J., 1987, J. Chem. Phys., 88:4421; 1987, Int. J. Quant. Chem. Symposium Volume, 88:4429; Phys. Rev. Let., 59:811.
Laria, D., and Chandler, D., 1987, Comparative Study of Theory and Simulation Calculations for Excess Electrons in Simple Fluids, preprint.
Lekner, J., 1969, Phys. Rev., 158:130.
Lepoutre, G., and Sienko, M.E., eds., "Solutions Metal-Ammoniac; Proprietes Physicochimiques:Colloque Weyl I, June 1963". (W.A. Benjamin, Inc., New York, 1964).
Levine, J., and Sanders, T.M., 1962, Phys. Rev. Let., 8:159.
Marchi, M., Sprik, M., and Klein, M.L., 1988, J. Chem. Phys., 92:3625.
Mott, N.F., 1967, Advances in Phys., 16:49.
Mott, N.F., 1949, Proc. Phys. Soc., A62:416; 1956, Can. J. Phys., 34:1356; 1961, Phil. Mag., 6:287.
Munuo, R., and Holroyd, R.A., 1988, these proceedings.
Newton, M., 1975, J. Phys. Chem., 79:2795.
Nichols, A.L., III, and Chandler, D., 1984, J. Chem. Phys., 84:398.
Nichols, A.L., III, Chandler, D., Singh, Y., and Richardson, P.M., 1984, J. Chem. Phys., 81:5109.
Nishida, M., 1970, J. Chem. Phys., 67:2760; 1970, 67:2760; 1977, 67:4786.
Ogg, R.A., 1946, J. Amer. Chem. Soc., 68:155; 1946, J. Chem. Phys., 14:114,295; 1946, Phys. Rev., 69:243, 668.
Parrinello, M., and Rahman, A., 1984, J. Chem. Phys., 80:860 and comments at Colloque Weyl VI, 1983, by A. Rahman.
Phillips, J.C., and Kleinman, L., 1959, Phys. Rev., 116:187.
Platzman, R.L., 1955, "Physical and Chemical Aspects of Basic Mechanisms in Radiobiology", Natl. Res. Council Publ., 305:34; 1953, Radiation Research, 2:1.
Rao, B.K., and Kestner, N.R., 1984, J. Chem. Phys., 80:1587.
Rossky, P.J., and Schnitker, J., 1988, J. Chem. Phys., 92:4277.
Sanders, T.M., 1962, Bull. Amer. Phys. Soc. Ser. II, 7:606; 1964, J. Levine, Ph.D. Thesis, University of Minn. (unpublished); Levine, J.L., and Sanders, 1967, Phys. Rev., 154:138.
Schnitker, J., and Rossky, P.J., 1987a, J. Chem. Phys., 86:3462.
Schnitker, J., and Rossky, P.J., 1987b, J. Chem. Phys., 86:3471.
Schnyders, H., Rice, S.A., and Meyer, L., 1967, Phys. Rev. Let., 15:187.
Schnyders, H., Rice, S.A., and Meyer, L., 1965, Phys. Rev., 150:127.
Sprik, M., Impey, R.W., and Klein, M.L., 1985, J. Chem. Phys., 83:5802.
Springett, B.E., Cohen, M.H., and Jortner, J., 1968, J. Chem. Phys., 48:2720.
Springett, B.E., 1968, Phys. Rev., 155:138.
Szasz, L., and McGinn, G., 1966, J. Chem. Phys., 45:2898.
Szasz, L., 1985, "Pseudopotential Theory of Atoms and Molecules", John Wiley and Sons, New York.
Szepfalusky, P., 1955, Acta Phys. Hung., 5:325.
Thirumalai, D., Wallqvist, A., and Berne, B.J., 1986, J. Stat. Phys., 43:973.
Thouless, D.J., 1972, J. Non-Crystalline Solids, 8-10:461.
Wallqvist, A., Thirumalai, D., and Berne, B., 1986a, J. Stat. Phys., 43:1986.
Wallqvist, A., Thirumalai, D., and Berne, B.J., 1986b, J. Chem. Phys., 85:1583; ibid., 1987b, 86:6404; ibid., 1987a, 86:5689.
Wallqvist, A., Martyna, G., and Berne, B.J., 1988, J. Phys. Chem., 92:1721.
Weyl, W., 1864, Pogg. Ann., 121:601.
Ziman, J.M., 1969, J. Phys. C, 1:1532; 1969, 2:1230.

GEOMETRICAL PERSPECTIVES OF A SOLVATED ELECTRON

David Chandler

University of California
Department of Chemistry
Berkeley, CA 94720 USA

The path integral formulation of quantum theory provides a framework to describe the behavior of solvated electrons. Feynman used the approach to treat the slow moving electron in ionic crystals -- the prototypical polaron problem. We have extended this theory, drawing on theories of the liquid state, to analyze the localization transition and related phenomena found with excess electrons in fluids.

Excess electrons in liquids exhibit a strikingly wide variety of behaviors. For example, the mobilities of e^- in two very similar liquids, neo- and normal pentane, differ by a factor of 500. To understand this variability -- akin to phase transition behavior -- it is helpful to transcribe the physical issues into those of a mathematically isomorphic problem, that of a classical Gaussian ring polymer dissolved in the liquid as illustrated in Fig. 1. This isomorphism is derived from the Feynman path integral formulation of quantum mechanics (Feynman and Hibbs, 1965; Feynman, 1972). Wolynes and I have written at length about this isomorphism (Chandler and Wolynes, 1981). A tutorial is given in my 1984 article (Chandler, 1984). This paper outlines my lecture reviewing the principal ideas and some of the literature pertaining to this approach. It is a recent and very rapidly growing literature.

The units (or beads) of the isomorphic ring polymer (or necklace) correspond to the possible positions of the single quantal electron. That is, the configuration of the isomorphic necklace corresponds to a quantum path of the electron. Parrinello and Rahman (1984) used this perspective in their study of the solvated electron in molten potassium chloride. Their simulation calculations illustrate the power of this methodology in studying the transition between extended and self-trapped states. Many groups have followed. For example, Klein and coworkers

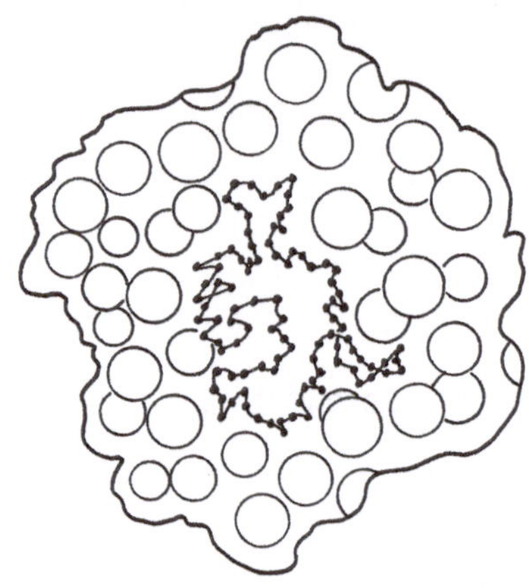

Fig. 1. Electron "polymer" in a fluid of classical atoms.

have examined e^- in liquid ammonia (Sprik et al., 1985), Schnitker and Rossky (1987) have studied e^- in water, Berne and coworkers have studied e^- in water clusters (Wallqvist et al., 1986), and also in the nonpolar liquids helium and xenon (Coker et al., 1987). In the latter cases, the results are well understood in terms of the hard-sphere models studied earlier by Klein and coworkers (Sprik et al., 1985; Laria and Chandler, 1987). These examples are not all inclusive, but serve to illustrate the recent activity in this area.

When studying this literature, the reader should be aware of a significant detail that must be addressed before one can be confident with the accuracy of the simulation results. In particular, the isomorphic polymer becomes an exact representation only in the limit of a continuum of points on the ring. A discrete number of beads is an approximation to path integrals analogous to numerical quadrature of ordinary integration. As with ordinary integration, one must devise checks of the accuracy of discretization; further, to accurately approach the continuum limit, efficient numerical procedures are often crucial in the implementation. The so-called "staging" algorithm introduced by Sprik et al., (1985) was historically the first application of a new algorithm to the path integral study of the solvated electron. A modification of that procedure was adopted with great success by Coker et al., (1987), and in their application to helium, they describe inaccuracies of earlier work which did not correctly approach the continuum limit.

By examining through simulation the structures of the isomorphic polymers, researchers gain a geometrical perspective of electronic states in a liquid environment. It is a perspective that Feynman usefully exploited long ago in his treatment of the slow moving electron in an ionic crystal -- the original polaron problem (Feynman, 1955; Feynman and Hibbs, 1965; Feynman, 1972). It is a perspective very different than the earlier theories created by Jortner, Rice, Cohen and others in their treatments of electrons in liquids (Davis and Brown, 1975; Jortner and Kestner, 1973; Kevan and Webster, 1976).

The interpretation of the new simulations have been guided by analytical theory and its focus on a particular class of correlation functions. In particular, the analysis of the electronic structure is conveniently described in terms of the mean square displacement

$$R^2(t-t') = \langle |\underset{\sim}{r}(t) - \underset{\sim}{r}(t')|^2 \rangle$$

$$= \sum_n A(\Omega_n)\left\{1 - \cos\,[\Omega_n(t-t')]\right\}, \quad \Omega_n = 2\pi\, n/\beta\hbar\,. \tag{1}$$

Here, $\underset{\sim}{r}(t)$ is the position on the path of the electron at the imaginary time t. (Imaginary time in quantum theory corresponds to real time for the classical isomorphic system.) The periodicity of the polymer implies the quantized frequencies, Ω_n. On analytical continuation, $A(i\omega) = \Phi(\omega)$, where $\omega\Phi''(\omega)$ is the absorption spectrum for the electron (Baym and Mermin, 1961).

Approximate theoretical calculations based on the RISM-polaron theory (Chandler et al., 1984) have shown how the behavior of $R^2(t-t')$ provides the signature of self trapped localization and the dominance of a ground state (Nichols et al., 1984; Nichols and Chandler, 1986; Malescio and Parrinello, 1987). These behaviors as predicted by the theory have been observed in simulation work as well (Sprik et al., 1985; Schnitker and Rossky, 1987; Coker et al., 1987; Sprik et al., 1985). The connections between ground state dominance and low mobility and other spectral features have been analyzed with theory (Nichols and Chandler, 1987), but here simulation work has not yet proved entirely successful. In the real time domain, the sampling of quantum paths involves oscillating positive and negative weights which are problematical for any currently known Monte Carlo procedure. Thus we must await new developments in time-dependent path integral quantum Monte Carlo.

The mechanism for self trapped localization is treated in the RISM-polaron theory in terms of the influence functional contribution to the action:

$$S_{inf} = \frac{1}{2}\int_0^{\beta\hbar} dt \int_0^{\beta\hbar} dt'\; v[\underset{\sim}{r}(t) - \underset{\sim}{r}(t')]\,, \tag{2}$$

where $v[\underset{\sim}{r}(t) - \underset{\sim}{r}(t')]$ has the diagrammatic structure of a reaction field or self energy,

$$v(\underset{\sim}{r}-\underset{\sim}{r}') = \text{[diagram]} \quad (3)$$

r r'

(The reader may consult my first lecture in this volume for the meaning of this picture.) With the convolution theorem, we can consider separately the different wave vector contributions to $v(\underset{\sim}{r}-\underset{\sim}{r}')$. Each is proportional to the response function or density-density correlation function, χ , at that particular wave vector. Thus, we arrive at the familiar physical picture that confinement or localization is a measure of the disorder in the material. The RISM-polaron theory adds to this picture, in that it provides the quantitative means for estimating the importance of different types of randomness in a topologically disordered system within the context of a single theoretical approach.

There is still much to be done before this theory, and also the recent simulation work, will provide fully satisfactory descriptions of the experimental phenomena. To date, few realistic systems and few experimentally observed properties have been analyzed. There are unresolved and important issues concerning the correct single-electron or pseudo-potential models for electron-solvent particle interactions. Perhaps an even more difficult and significant issue is that of analytic continuation. Without its resolution, computer simulation studies will be limited to the analysis of equilibrium structural properties, and only the more approximate theoretical approaches will provide information about dynamics.

REFERENCES

Baym, G., and Mermin, N.D., 1961, Determination of thermodynamic Green's functions, J. Math. Phys., 2:232.

Chandler, D., 1984, Quantum theory of solvation, J. Phys. Chem., 88:3400.

Chandler, D., Singh, Y., and Richardson, D.M., 1984, Excess electrons in simple fluids. I. General equilibrium theory for classical hard sphere solvents. J. Chem. Phys.,, 81:1975.

Chandler, D., and Wolynes, P.G., 1981, Exploiting the isomorphism between quantum theory and classical statistical mechanics of polyatomic fluids, J. Chem. Phys., 74:4078.

Coker, D.F., Berne, B.J., and Thirumalai, D., 1987, Path integral Monte Carlo studies of the behavior of excess electrons in simple fluids, J. Chem. Phys., 86:5689.

Davis, H.T., and Brown, R.B., 1975, Low-energy electrons in nonpolar fluids, Adv. Chem. Phys., 31:329.

Feynman, R.P., 1955, Slow electrons in a polar crystal, Phys. Rev., 97:660.

Feynman, R.P., 1972, "Statistical Mechanics," W.A. Benjamin, Reading, MA.

Feynman, R.P., and Hibbs, A.R., 1965, "Quantum Mechanics and Path Integrals," McGraw-Hill, New York.

Jortner, J., and Kestner, N.R., eds., 1973, "Electrons in Fluids," Springer, Berlin.

Kevan, L., and Webster, B.C., eds., 1976, "Electron-Solvent and Anion-Solvent Interactions," Elsevier, Amsterdam.

Laria, D., and Chandler, D., 1987, Comparative study of theory and simulation calculations for excess electrons in simple fluids, J. Chem. Phys., 87:4088.
Malescio, G., and Parrinello, M., 1987, Polaron theory of electrons solvated in molten salts, Phys. Rev. A., 35:897.
Nichols, III, A.L., Chandler, D., Singh, Y., and Richardson, D.M., 1984, Excess electrons in simple fluids. II. Numerical results for hard sphere solvents, J. Chem. Phys., 81:5109.
Nichols, III, A.L., and Chandler, D., 1986, Excess electrons in simple fluids. III. Role of solvent polarization, J. Chem. Phys., 84:398.
Nichols, III, A.L., and Chandler, D., 1987, Excess electrons in simple fluids. IV. Real time behavior, J. Chem. Phys., 87:6671.
Parrinello, M., and Rahman, A., 1984, Study of an F center in molten KCl, J. Chem. Phys., 80:860.
Schnitker, J., and Rossky, P.J., 1987, Quantum simulation study of the hydrated electron, J. Chem. Phys., 86:3471.
Sprik, M., Klein, M.L., and Chandler, D., 1985, Simulation of an excess electron in a hard sphere fluid, J. Chem. Phys., 83:3042.
Sprik, M., Impey, R.W., and Klein, M.L., 1985, Study of electron solvation in liquid ammonia using quantum path integral Monte Carlo calculations, J. Chem. Phys., 83:5802.
Wallqvist, A., Thirumalai, D., and Berne, B.J., 1986, Localization of an excess electron in water clusters, J. Chem. Phys., 85:1583.

ELECTRON LOCALIZATION AND FEMTOSECOND NONLINEAR OPTICAL RESPONSES IN LIQUIDS

Geraldine A. Kenney-Wallace

Lash Miller Laboratories
University of Toronto
Canada M5S 1A1

INTRODUCTION AND ISSUES

The liquid and its electrical properties embrace concepts and theories from many scientific and engineering disciplines, whose different perspectives do not always appear compatible in interpreting microscopic behavior. In this respect, given my title, while the electrical properties of the liquid state are clearly related to electron transport phenomena, and photoionization, electron scattering, localization and solvation are important mechanistic steps linked to photoconductivity, conduction, electron mobility, space-charge effects and dielectric breakdown, precisely what are the roles of nonlinear optical responses? Indeed, do we fully understand nonlinear optical processes? For details on nonlinear responses, the reader is referred to a text on nonlinear optics and applications, which also summarizes the previous literature (Reintjes, 1984). Second harmonic generation, although usually not observed in isotropic systems through symmetry restrictions, can be a tool to observe surface structures via symmetry-breaking processes on surface films, the dynamic melting and recrystallization. The induced-birefringence associated with strong space-charge effects around electrodes can be optically monitored through the optical Kerr effect. However, the goal of this chapter goes one step further: we propose to link the response of a disordered medium to a rapidly scattering excess electron to the intrinsic nonlinear polarizability response of the atoms or molecules along its trajectory. An excess electron undergoing multiple scattering interactions at times 10^{-15} to 10^{-11} seconds prior to its trapping and possible solvation in a liquid will experience a sequence of events governed by the nature of the dynamical structure of the host liquid (Kenney-Wallace, 1981). Or to put the case in a macroscopic way: a transient current in xenon is quite distinct in temporal profile and amplitude from a current in water. However, by examining these disparate systems from the microscopic α_0 polarizability perspective, and by understanding the nature of the optical response of the liquid to its sudden field-induced $[\vec{E}(\omega,t)]$ perturbation, perhaps much insight can be shed on how to link the myriad of specific electron mobility observations into a more general, microscopic and unified theory of electrical properties of the liquid state. The ultimate goal is a model which has microscopic predictive capability for atomic and molecular fluids.

In recent papers the unresolved general issues were outlined as follows (Kenney-Wallace et al., 1988a, 1987).

Electrons in disordered media, traveling freely as waves and subsequently localizing as particles, have raised problems which are at the core of contemporary discussions of the nature of quantum behavior and dynamical structure in dense and disordered media. Interaction-induced effects, density fluctuations, and collisional perturbations are similar concepts with time and frequency-dependences that sometimes distinguish them in scattering models, if not in nature. Considerable experimental effort has been focused on the picosecond (ps) and now femtosecond (fs) spectroscopic attempts to follow the electron localization and solvation steps (Kenney-Wallace, 1981; Christophorou, 1981; Kenney-Wallace et al., 1988a, 1987; Wiesenfeld and Ippen, 1980; Gaudel et al., 1984, 1987; Kenney-Wallace and Jonah, 1982; Wang et al., 1980). Most recently, electrons have been observed trapped in as phase beams of molecular clusters and observed as $(M)_n^-$ via mass spectroscopic techniques (Arbruster et al., 1984; Coe et al., 1987, 1988; Alexander et al., 1986), where in the case of water n appears to be 11. Other systems reveal comparably small numbers for NH_3 and CO_2 electron clusters (Coe et al., 1987, 1988; Alexander et al., 1986). The equilibrium properties and formation of a wide range of clusters, from hydrogen-bonded to metallic clusters, has been recently reviewed in detail (Castleman and Keese, 1986).

Once again, in raising the profile of the simplest anion in nature in principle, these data focus our attention on one of the most intriguing quantum species in practice, for which a fully quantitative and predictive quantum theory has yet to be developed. Recent molecular dynamics simulations (Nichols and Chandler, 1986; Wallquist et al., 1986; Schnitker et al., 1986, 1987, 1988; Sprik et al., 1985, 1988; Jonah et al., 1986, Parrinello and Rahman, 1984) are now exploring various aspects of electron-cluster interactions to determine the cluster size, stability and structure of hydrated electrons in particular. In the chapter following this, we outline some of our initial attempts to introduce simulations as a tool for studying nonlinear responses and dynamical structures in liquids (Golombok et al., 1985).

In an earlier, comprehensive study of electron solvation via ps spectroscopy and ^{13}C NMR, we investigated how the localization and solvation of electrons in polar fluids and binary systems was influenced by a range of molecular properties (Kenney-Wallace, 1981; Christophorou, 1981; Kenney-Wallace and Jonah, 1982). Data from alcohols and alcohol/alkane systems in particular revealed unexpected hydrodynamic-like correlations of the solvation time and the viscosity with the unperturbed fluid dynamics. Yet the macroscopic masks the microscopic; and it was concluded that solvation clusters, not long-range dielectric continuum properties alone, could explain the ps data. But as microscopic probes of the molecular dynamical structure of the host fluid, trapped and solvated electrons also highlighted the imperative to understand the dynamics of the liquid <u>prior</u> to perturbation. Thus the central issue became the motion of molecules in the condensed phase.

Within that issue the priority must be given to the key questions which we have posed (Kenney-Wallace and Jonah, 1982). First, how does the electronic motion couple into the fluctuating dynamical structure of the fluid, and second, how do the intermolecular forces and intrinsic motion of the atomic or molecular liquid respond to the sudden perturbation represented by an intense, transient field of a bare electron? It is the second question which was the driving force in focusing our attention on nonlinear optical responses of liquids, via nonlinear laser spectroscopy at femtosecond timescales.

If, as we have postulated elsewhere (Kenney-Wallace, 1981; Christophorou, 1981, Kenney-Wallace and Jonah, 1982), above a certain critical number density there occur density and configurational fluctuations, which at times $<< 10^{-12}$s are frozen into the liquid structure, then these will present an array of potential minima along the path of the quasifree electron during its multiple scattering interactions as Fig. 1(a) illustrates. The optical absorption spectra of classes of electrons in various media are also shown in Fig. 1(b). Prior to localization, these transitory structures reflect a subpicosecond snapshot of the unperturbed statistical and dynamical structure of the medium. The molecules themselves will also experience fluctuating intermolecular potentials and the molecular motion will also reflect this statistical structure as the system spontaneously moves about the equilibrium positions.

The only local response to the moving field of the electron is the instantaneous electronic polarization. Field-induced librational and

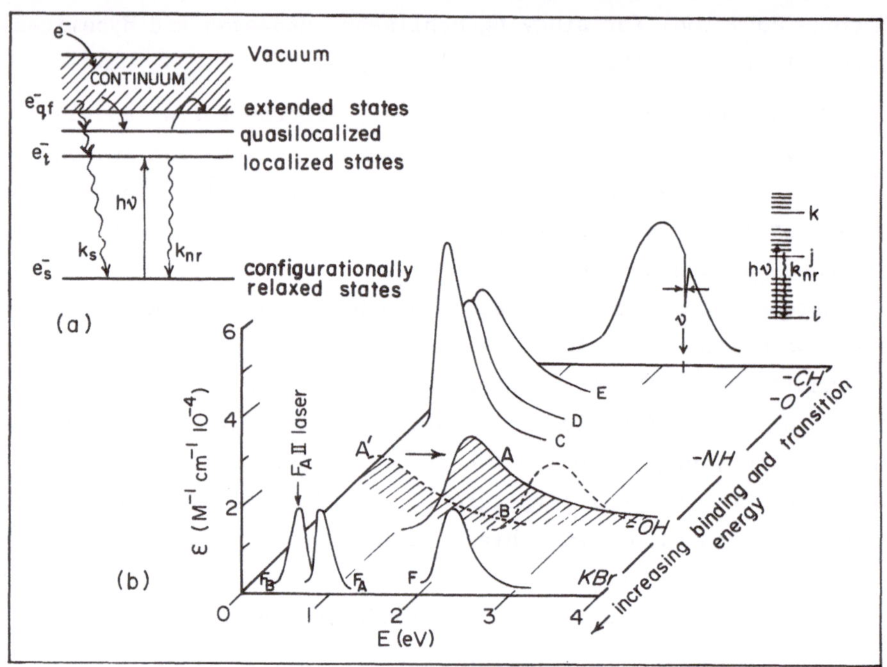

Fig. 1. Excess electronic states in liquids: (a) A schematic of electronic energy levels in disordered media illustrating transitions between continuum (quasifree states), localized, and fully solvated or configurationally relaxed e_s^-. (b) Optical spectra recorded for electrons in n-alcohols at picosecond (A′) and nanosecond (A) times, diols (B), amines (C), ethers (D), alkanes (E), and color centers (F, F_A). The color center laser emission in KBr is shown as F_B (from Phil. Trans. Roy. Soc., London, A299:309, 1980).

reorientational responses of the molecules must occur at longer times. In alcohols, these reorientations drive the final solvation process over 10^{-12}s to 10^{-10}s (Kenney-Wallace and Jonah, 1982). In small molecule liquids such as CS_2 or substituted methanes, the collisions, librational motions and a range of interaction-induced effects and reorientational times begin to overlap within the narrow window of 100fs to 5ps (Kenney-Wallace, 1984; Kalpouzos et al., 1987; Greene and Farrow, 1982; Halbout and Tang, 1982; Ruhman et al., 1987a,b; McMorrow et al., 1988a,b). In studying ultrafast, laser-driven nonlinear optical responses via $\chi^{(3)}_{ijkl}$, the nonlinear optical susceptibility third-order in the applied laser field, we have recently observed molecular systems initially respond with electronic hyperpolarization, generating induced dipoles. Subsequently molecular librational, translational and reorientational motions can occur as a consequence of coherently-driven processes and torques, imposed by these transient induced-dipoles interacting back with the applied, pulsed laser field. Can these responses be linked to the a priori liquid dynamical structure (Kenney-Wallace et al., 1988a; Kenney-Wallace et al., 1987; McMorrow et al., 1988a)? What are the non-diffusive motions we observed (Kalpouzos et al., 1987) at $< 10^{-13}$ seconds? Can the polarization accompanying and in the wake of the electron be predicted or interpreted in terms of the nonlinear optical response of the system? Some of these questions are addressed in this chapter, others await more experimental data and theoretical development.

The remainder of this chapter is organized in four parts. ELECTRON SOLVATION TIMES IN POLAR LIQUIDS briefly summarizes the electron solvation data available from various laboratories by the NATO ASI 1987 conference date; FEMTOSECOND LASER SPECTROSCOPY outlines the novel femtosecond laser spectroscopy techniques for studying ultrafast molecular motion; EXPERIMENTAL RESULTS ON FEMTOSECOND KERR RESPONSES presents the femtosecond nonlinear optical data recently obtained by us for several simple organic liquids and, in concluding, examines how these ultrafast responses could be linked to future experiment and theory of electron localization and solvation.

ELECTRON SOLVATION TIMES IN POLAR LIQUIDS

By late 1987, the picosecond (ps) and femtosecond (fs) data available on electron solvation in water, amines, and alcohols at room temperature, obtained from ps pulsed electron beam spectroscopy (PEBS), from 2 photon photoionization (PI) of guest impurity molecules or ions in liquids, and from 2 photon direct photoionization (DPI) of pure liquids with fs and ps resolution, were all consistent with the general physical features of the cluster model of solvation which we had first proposed in 1976 (Kenney-Wallace and Jonah, 1976, 1982; Kenney-Wallace, 1976). The main conclusions from the present database, some of which is shown in Table 1, and Fig. 2 are briefly summarized in six points.

a. At the earliest times observed in all liquids, the initial IR spectrum ($\lambda \geq 1200$ nm) has been present instantaneously that is $< 10^{-12}$ s, excitation pulse limited. At room temperature, spectroscopic evidence indicates that the IR spectrum gradually disappears. The final visible spectrum, whose peak absorption position more reflects the short-range e^- - dipole interactions (see Fig. 1) than the long range dielectric properties of the fluid, evolves over a timescale that correlates with the reorientation time (rot) assigned to the picosecond monomer rotations in alcohols (Kenney-Wallace and Jonah, 1982; Kenney-Wallace, 1976). In water the DPI times indicate that the electron thermalizes and reaches a localized trap within 110fs

Table 1. Electron solvation data* in polar liquids at 300K

	PEBS IR; VIS	PI IR	PI VIS	DPI IR	DPI VIS
Water			300 fs	240 fs	240 fs ± 40 fs
	a; b	c	d		e
Methanol	10.7;10 ps	-	17 ± 3 ps	-	10 ps
Ethanol	23;18 ps	20 ps	26 ± 5 ps	-	18 ± 2 ps
1-propanol	34;24 ps	20 ps			22 ± 2 ps
2-propanol	25 ps	30 ± 3 ps			23 ± 2 ps
1-butanol	39;30 ps				21 ± 4 ps
2-butanol	40 ps				– –
t-butanol	54 ps				45 ± 5 ps
pentanol	34 ps				22 ± 6 ps
c-hexanol	36 ps				
1-octanol	45 ps				
2-octanol	55 ps				
1-decanol	51 ps				
2-decanol	54 ps				
$(CH_2OH)_2$	5 ps				

* IR decay times, and visible (VIS) spectrum rise times recorded experimentally.
Water results from Wiesenfeld and Ippen, 1980; and Gaudel et al., 1987. (Column a - Chase and Hunt, 1975; b - Kenney-Wallace and Jonah, 1982, quoted to ± 10%; c - Huppert et al., 1981; d - Wang et al., 1980; e - Matsuoka et al., 1987.)

(Gaudel et al., 1984, 1987) or faster. The decay of e^-_t is 240 fs, close to librational motions. In ethylene glycol, τ_s << 5 ps, too fast and in contradiction to any expectations of viscosity (η) correlations. This time is in agreement with the predictions of preexisting localization sites in a clustering model, in which prior two or three dimensional network structures, comprising at least a dimer, could act as the initial localization site (Kenney-Wallace and Jonah, 1982). Other earlier research results (Chase and Hunt, 1975; Huppert et al., 1981) are also consistent with these ideas of the dominance of *a priori* microscopic liquid structure as the physical root of the observed spectral responses.

$$e_{qf} \rightarrow e_{t_{IR}} \rightarrow e_{s_{VISIBLE}}$$

The IR spectrum diminishes in intensity as the visible spectrum grows in with the similar rate (Gaudel et al., 1984, 1987; Kenney-Wallace and Jonah, 1982), and to date no experimental evidence for a continuous shift from IR to the visible has yet

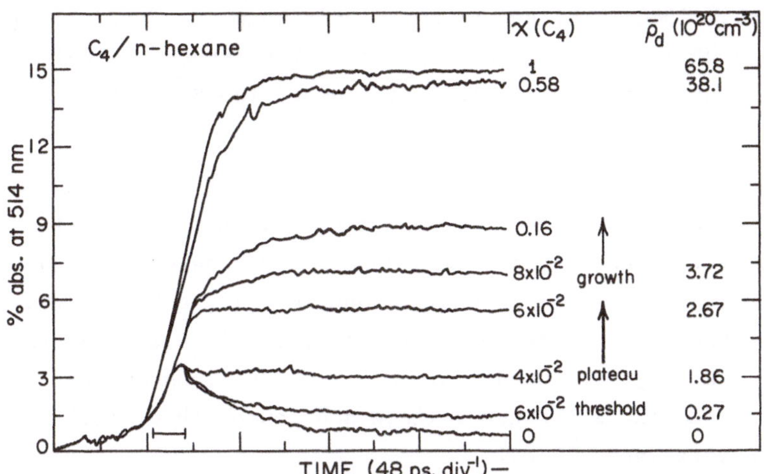

Fig. 2. Profile of electron solvation as a function of alcohol density in 1-butanol/n-hexane systems. The alcohol average dipole density (<ρ>d) and mole fraction alcohol $X(C_4)$ are also shown. The three distinct kinetic regions (absorption and decay, plateau, and growth) and the threshold effects for electron localization in alcohol dimers are discussed extensively in Kenney-Wallace and Jonah (1982).

been observed in polar liquids at 300K. This is in contrast to the time-dependent stokes shifts experimentally seen and interpreted as solvent relaxation around dipolar solute molecules, electronically excited to give a change in the magnitude or direction of the dipole (Simon, 1988).

b. In solvation clusters, if e^-_s is $(ROH)^-_n$, where $6 < n < 20$ typically, then the structure of the fully configurationally relaxed, final quantum state of e^-_s is primarily governed by the shorter range attractive interactions and the a priori packing structure of the liquid, reorganized slightly in the strong (but possibly screened) field or in the presence of the excess charge of the electron.

c. The solvation process, which occurs on a 10^{-11} to 10^{-10}s timescale primarily reflects the reorientational (or more generally dynamical) diffusive motions normally associated with the unperturbed liquid.

d. The method of generating "free" electrons to study localization and solvation may well influence the results in the sense of "fine tuning", but not the coarse grained results. Here we refer to finite lifetimes of precursor states, autoionizing states, Franck-Condon factors and so forth (Kenney-Wallace, 1981; Christophorou, 1981; Kenney-Wallace and Jonah, 1982; Wang et al., 1980; Robinson et al., 1986; Hameka et al., 1987). More research is required on this area. The issues are only answered on a subpicosecond timescale if indeed the initial localization site is as postulated under point a.

e. In alcohol/alkane (ROH/RH) experiments, the solvation times were intimately linked to τ_{rot}, and a good linear correlation was observed (Kenney-Wallace and Jonah, 1982) between τ_s and $\eta V(kT\alpha_o)^{-1}$ for pure alcohols, thus implying a "hydrodynamic" response to the electron. If longitudinal relaxation times τ^1 = τ_{rot} (E_{op}/E_s) are used, only small corrections appear for most of the alcohols (Kenney-Wallace and Jonah, 1976, 1982; Kenney-Wallace, 1976).

f. Despite such hydrodynamic correlations, ps spectroscopic and NMR studies (Holubov and Kenney-Wallace, 1986-88; Holubov, 1988; Dais et al., 1980) in dilute systems (ROH/RH) indicate clear evidence of molecular clustering and preexisting trapping sites as illustrated in Fig. 2. We have interpreted this as evidence that the microscopic details of e_s^- in pure liquids are masked by a continuum viewpoint. The apparent paradox presented by the hydrodynamic correlation is removed if one realizes that both the dynamics and the viscosity are linked to the <u>structure</u>, and thus microscopically it is the intermolecular forces within the liquid on a length scale over a few molecular diameters that determine the temporal events. In the case of strong hydrogen-bonding, this length scale extends into the "bulk". The issue of whether or not to use longitudinal versus transverse dielectric relaxation times has been long recognized and is being currently re-examined in the context of dielectric friction theories and computer simulations of dielectric relaxation responses (Kenney-Wallace and Jonah, 1982, 1976; Kenney-Wallace, 1976; Chase and Hunt, 1975; Huppert et al., 1981; Simon, 1988).

Thus the issue of whether or not preexisting trapping sites in liquids exist is more accurately addressed if separated into two questions. First, are there preexisting localization sites? Second, are there preexisting <u>deep</u> traps for which no further and substantial medium reorganization is required? The first answer is then "yes", preexisting clusters in the liquid can act as initial shallow trapping sites (Kenney-Wallace and Jonah, 1982; Schnitker et al., 1986; Schnitker and Rossky, 1987; Schnitker et al., 1988). The second answer is "possibly", but most evidence supports significant rotational configurational relaxation during the solvation process. Thus a dominance of preexisting <u>deep</u> traps is really unlikely for room temperature liquids, in which many degrees of motional freedom are continually reflected in the complex, interdependent evolution of the dynamical structures of the liquid. In expansions from supersonic jets, or in glasses at low temperature, the electron can only become localized and trapped (one presumes) in a structure that is energetically already favorable, since the usual motional degrees of freedom, which are necessary to adjust to the presence and field of the electron, are "frozen". Solvated anions rather than e_s^- are often the result; or alternatively fragmentation occurs to give new ionic species. This is another issue on which further research is required to assist theoretical developments on solvation phenomena in jets (Hager and Wallace, 1985; Hager et al., 1987; Matsuoka et al., 1987; Jortner and Even, 1983).

Given this brief summary of the electron solvation data, the reader is urged to consult the references cited for the more extensive arguments that lie behind the points made above. In order to examine the microscopic response of the molecules in the liquid to the electron, we designed a series of ultrafast nonlinear optical experiments (in the absence of e_s^-) in which we would specifically investigate the rotation, libration, vibration, and other (as then unknown) dynamical responses of a molecule suddenly polarized in a strong, transient laser field. This naturally led to the field of nonlinear optical phenomena and quantum electronics, which we will briefly discuss in the following section.

FEMTOSECOND LASER SPECTROSCOPY

While still a frontier art, the generation and application of femtosecond laser pulses to explore fundamental dynamical responses, particularly in dense media as diverse as biological cells to semiconductors, has emerged successfully as a strong spectroscopic, real-time tool for the 1990s. We briefly outline the femtosecond laser system constructed for the nonlinear optical Kerr experiments outlined in EXPERIMENTAL RESULTS ON FEMTOSECOND KERR RESPONSES and refer the reader to more detailed references for further discussion on the background theory and applications of quantum electronics (Reintjes, 1984; Kenney-Wallace, 1984; Kalpouzos et al, 1987; McMorrow et al., 1988a).

In all these time-resolved experiments the principles of <u>pump-probe laser spectroscopy</u> are the key element in the experimental design. A laser pulse is optically split into two components of unequal amplitude. The intense fraction, acting as a <u>pump</u> pulse, is directed towards the target or sample cell to trigger the molecular event under study. The much attenuated <u>probe</u> pulse monitors the absorption, raman scattering, polarization, coherence, or phase shift, which is linked explicitly to the dynamical observable under investigation. Extremely precise time resolution comes from the fact that both pump and probe pulse originate from the same parent laser pulse. Optical delay lines utilizing mirrors, prisms, and beam splitters of precisely known optical thickness and dispersion can direct the two replicas of the ultrashort pulse to the target sample with a precision of femtoseconds. The time resolution in ps or fs is a function of equivalent optical distance traveled and is measured to a micron, where $1\mu = 3.5$ fs (in vacuo). Data are recorded usually as the intensity of the probe pulse as a function of the delay time (τ) between pump and probe, namely the difference in the relative, optically equivalent distances traveled by the pump and probe pulses before crossing in the sample.

The femtosecond anti-resonant-ring dye laser, designed and constructed in this laboratory and pumped synchronously at 82 MHz with an optoacoustic mode locked Nd:YAG laser, which is frequency-doubled to give pulses of 532 nm radiation to pump the dye laser. The interpulse spacing on the subsequent 60fs dye laser pulses matches the mode locking frequency of the Nd:YAG and is thus 12 ns. The average power in the dye laser output is typically 25 mW for 615 nm $< \lambda <$ 638 nm, or kW peak power per pulse. In the optical Kerr cell, the typical power density was 150 MW cm^{-2} at the overlap of the pump and the probe beams, which were focused as shown in Fig. 3. The configuration shown also illustrates the modulation on the pump pulse with a kHz chopper. The transmission signal $T(\tau)$ is recorded on a fast photomultiplier coupled to a lock-in amplifier (LIA), which is an integral part of the heterodyne optical technique described in detail elsewhere. For a discussion on the role of each element in the dye laser and its specifications, the reader is referred to earlier publications (Kalpouzos et al., 1987; McMorrow et al., 1988a).

The nonlinear optical responses induced in these Kerr experiments are a manifestation of the polarization generated by the third-order nonlinear susceptibility, a response cubic in the applied field $\vec{E}(\omega,t)$. In a lossless medium,

$$P_{NL}^{(3)} = \chi_{ijkl}^{(3)} \vec{E}(\omega,t)\, \vec{E}(\omega,t)\, \vec{E}(\omega,t) \ , \tag{1}$$

(e.g., no absorption) only the imaginary part of $\chi_{ijkl}^{(3)}$ is utilized. A consequence of the passage of the intense, propagating pump pulse is to generate a macroscopic nonlinear polarization $P_{NL}^{(3)}$, which arises from microscopic realignment of a significant number of molecules. The torques originate from the interaction of the transient, induced dipole (initially created) back with the applied laser field $\vec{E}(\omega,t)$ as the next chapter shows. The larger the polarizability anisotropy of the molecule, the greater the magnitude of $\chi_{ijkl}^{(3)}$. We have selected CS_2 as a reference

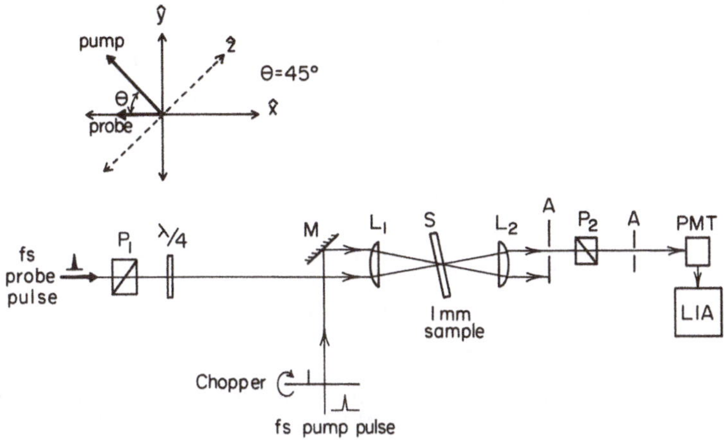

Fig. 3. Schematic of experimental optical Kerr configuration. P_1, P_2 crossed polarizers; λ/4 mica quarter wave plate; L_1, L_2 lenses; A aperture, M beam steering mirrors; PMT, fast photomultiplier coupled to kHz lock-in amplifier (LIA). Insert shows 45° positioning of relative pump and probe beam polarization axes (from McMorrow et al., 1988a).

molecule, which has a high degree of spectroscopic symmetry ($D_{\infty h}$) and also a high nonlinear polarizability. The femtosecond responses are shown in Fig. 4, in which a delayed response to the laser pump pulse is clearly evident, and an ultrafast relaxation precedes a longer decay going on into picoseconds (Kenney-Wallace et al., 1987).

From our experimental femtosecond laser studies on molecular liquids, the torques experienced in the electric field can be modeled in terms of a second-order differential equation, linking the medium response to the intrinsic properties of the molecules. Equation (2) shows the expression originally employed to model the elastic vibrations of linear molecules subjected to intense fields and leading to the formation and self-trapping of intense light filaments in dense media. The polarizability anisotropy is $\alpha = (\alpha_{\parallel} - \alpha_{\perp})$, the moment of inertia (I), the coefficient of internal friction (ξ), the elastic restoring force for librators (μ), and the angular displacement (δθ) is included with respect to the molecular axis initially, positioned at angel θ to the applied, polarized laser field, $\vec{E}(\omega,t)$. The field amplitude is A and L is the local field correction for which we employ the Lorentz 1/3 (n_o^2 + 2) prescription.

$$I \frac{d^2\delta\theta}{dt^2} + \xi \frac{d\delta\theta}{dt} + \mu\delta\theta = \frac{\alpha}{3} L^2 A^2 \sin\theta \cos\theta \, . \tag{2}$$

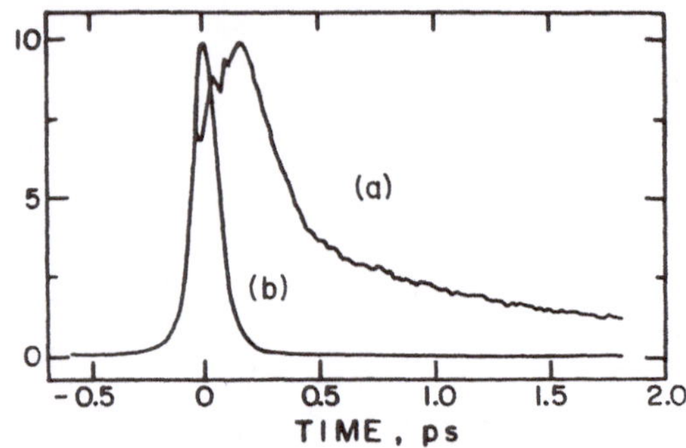

Fig. 4. Transmission of Kerr cell containing CS_2 liquid at 300K, following 65 femtosecond pump pulse (a) Optical Kerr signal showing significant intertial delay, after (b) peak of laser pulse intensity, represented by second-order auto-correlation of laser pulse in KDP crystal. The two signals are normalized at peak intensity (from Kenney-Wallace et al., 1987).

The following discussion is taken from recent literature papers (Kenney-Wallace et al., 1988a, 1987; McMorrow et al., 1988a) in which the conceptual and mathematical development are fully analyzed in the context of femtosecond nonlinear optical responses via $\chi^{(3)}_{ijkl}$. Equation (2) can be evaluated by the Green's function method with appropriate choice of the laser forcing function and varying the magnitudes of the coefficients.

$$\Delta n_i(t) = \frac{n_{2i}\beta_i}{(I/\mu)^{1/2}} \int_{-\infty}^{t} \sinh\left[(t-t')/\beta_i\right] \exp\left[-(t-t')/\tau_i\right] I_{pump}(t')dt' , \quad (2a)$$

where $\tau_i = 2I/\beta_i$; $\beta_i = 2[(\xi/I)^2 - 4(\mu/I)]^{-1/2}$

$$\Delta n_i(t) = \frac{n_{2i}\beta_i}{(I/\mu)^{1/2}} \int_{-\infty}^{t} \sin\left[(t-t')/\beta_i\right] \exp\left[-(t-t')/\tau_i\right] I_{pump}(t')dt' . \quad (2b)$$

The solution to Eq. (2) in the overdamped limited [Eq. (2a)] naturally accounts for the observed Kerr profiles by including the possibility of delays in the peak Kerr response due to inertial lag while maintaining the exponential character of the long-time diffusive relaxation behavior. The microscopic angular displacement $\delta\theta$ is manifested as a change in the macroscopic nonlinear refractive index and a transitory $\delta n(t)$ birefringence. The solution in the underdamped limit [Eq. (2b)] describes oscillatory motion of a molecule in a harmonic potential.

Any transient birefringence $\delta n(t)$ that arises in the liquid as a result of the field-induced molecular alignment can be sensitively probed in an optical Kerr pump-probe laser spectroscopy experiment (Kenney-Wallace et al., 1988a; Kalpouzos et al., 1987; McMorrow et al., 1988a,b; Golombok et al., 1985) or a four-wave mixing, phase conjugate experiment (4 WM) which also operates via $\chi^{(3)}$. We can treat the observable $\Delta n(t)$ as the linear superposition of the medium polarization responses $r_m(t)$ with respect to the intense femtosecond, linearly polarized laser pump pulse, $I_p(t) = \langle\vec{E}(\omega,t)\rangle^2$. A separation of timescales argument underlies our approach,

$$\Delta n(t) = \sum_m \Delta n_m(t) = \sum_m \int^{\infty} I_p(t')r_m(t-t')dt' , \tag{3a}$$

$$\Delta n(t) = c\phi(t)\ (\omega l)^{-1} , \tag{3b}$$

which presumes that electronic, vibrational, and reorientational polarization responses occur on separable timescales spanning three orders of magnitude, from 10^{-15} to 10^{-12}s. For optically anisotropic molecules such as the linear molecules CS_2 and CO_2, computer simulations have shown a clear time separation between collision-induced and reorientational dynamics (Madden and Tildesley, 1985; Ladanyi, 1983; Frenkel and McTague, 1980).

The transmission $T(\tau)$ of the optical Kerr cell is given by Eq. (4), in which the phase shift $\phi(t)$ experienced by the probe over pathlength l leads to a rotation of the plane of polarization of the probe pulse; this in turn is linked to $r_m(t)$ in Eq. (4). In our optical heterodyne Kerr scheme, the measured $T(\tau)$ is given by

$$T(\tau) = \int_{-\infty}^{+\infty} I_{probe}\ (\tau-t)\ \{1+ \sin\ [\phi(t)]\}\ dt , \tag{4}$$

where τ is the delay time between the pump and probe laser pulses.

Finally, this transmission is given by Eq. (4), where $G_o^{(2)}(t)$ is the zero background, second-order autocorrelation functions of the pump laser pulse

$$T(\tau) = \sum_m \int_{-\infty}^{\infty} G_o^{(2)}\ (t)\ r_m(\tau-t)dt . \tag{5}$$

The simple relationship between the linear (n_o) and nonlinear (n_2) refractive indices is given by Eq. (5) for the total index n:

$$n = n_o + n_2 \langle \vec{E}(\omega,t)^2 \rangle . \tag{6}$$

EXPERIMENTAL RESULTS ON FEMTOSECOND KERR RESPONSES

Figure 4 shows (a) the Kerr profile recorded in pure CS_2 at 298K, with (b) 100fs autocorrelation of the laser pulse $G_o^{(2)}(t)$ from second harmonic radiation (SHG) generated from a 100 micron KDP crystal, superimposed on the Kerr profile to establish $t = 0$. The SHG is normalized in amplitude to the Kerr response. The data were recorded at $150MWcm^{-2}$. Figure 5 compares CS_2 in 5(a) to chlorobenzene and nitrobenzene in (b) and (c) respectively, under similar experimental conditions.

Three features clearly stand out in the temporal Kerr profile in Fig. 4. First, we observe a clear time delay in the appearance of the peak of the CS_2 response with respect to the calibrated $t = 0$, established by the peak of the laser pulse. The growth-in of the Kerr signal is over about 70fs after the laser pulse. Secondly, the overall relaxation time of the induced anisotropy is clearly nonexponential at femtosecond times but appears to decay exponentially at picosecond, longer timescales. Thirdly, the ultrafast fs relaxation is substantially

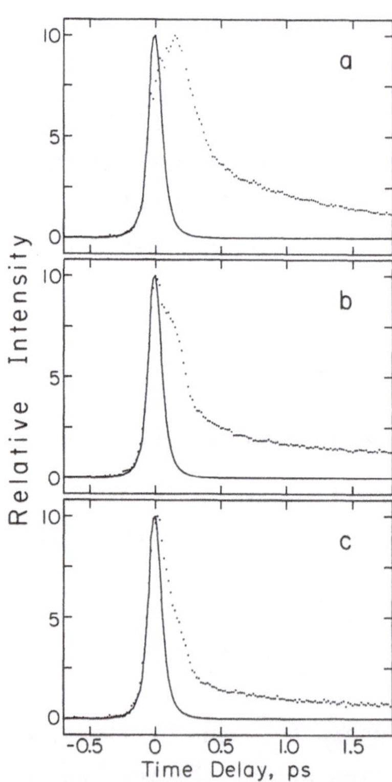

Fig. 5. Optical Kerr signals (dots) induced by 65 fs laser pulses in (a) CS_2, (b) chlorobenzene and (c) nitrobenzene. The laser pulse autocorrelations are the solid curves representing $G_o^{(2)}(\tau)$ at the sample position.

longer than the 60fs laser driving pulse, and the empirical relaxation time of ~140fs appears to be independent of the laser pump intensity. In earlier experiments (Kalpouzos et al., 1987; McMorrow et al., 1988a) we showed that birefringence followed the expected quadratic field dependence. When CS_2 is diluted in alkanes, which have negligible $\chi^{(3)}$ values in comparison, the same qualitative features are observed as Fig. 6 shows. In the series of liquids comprising benzene and substituted benzenes with F, Cl, I, and NO_2 as substituents, and also in halomethanes and alkylnitriles, again the same general features are observed although the superficial appearance of the profiles does vary from liquid to liquid (Kalpouzos et al., 1987; McMorrow et al., 1988a). In summary, we have concluded that all liquids we have studied possessing anisotropic molecules reveal empirically similar profiles to date. They can be modeled successfully using the general formalism outlined in the previous section. Figure 7(a) illustrates the CS_2 Kerr profile analyzed into four components with this phenomological theory, and 7(b) the profile observed in pure $CHCl_3$ at 300K, analyzed in a similar manner. The experimental results in pure $CHCl_3$ illustrate a distinct oscillatory component to the long time tail of the relaxing birefringence, an effect observed even more clearly in pure CCl_4, despite the 100 fold weaker signals from that liquid. This component proved to be an additional vibration modulation coherently excited as a Raman transition within our broad band (~300

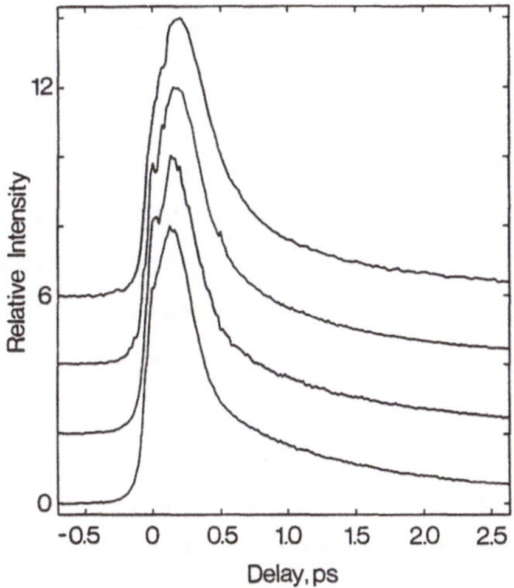

Fig. 6. Femtosecond optical Kerr profiles of CS_2 in isoheptane, an isoviscous system at 298K. Transmission is recorded as a function of time for CS_2 successively diluted in isoheptane, mole fraction CS_2 = 1.0 (lowest-trace), 0.75, 0.50, 0.20 (upper-trace) respectively. The traces are displaced for clarity (from Kenney-Wallace et al., 1988b).

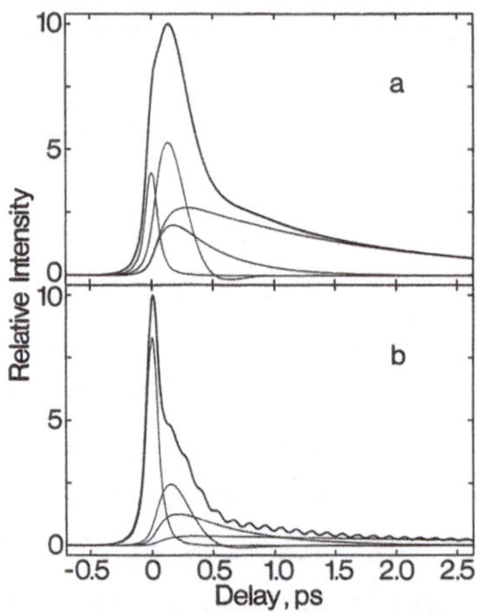

Fig. 7. Theoretical curves necessary to successfully match full fs and ps experimental profile of Kerr responses. Curves follow sequence (left to right) = electronic, librational, translational, and orientational contributions (from McMorrow et al., 1988a). See text for discussion on (a) CS_2 and (b) $CHCl_3$.

cm^{-1}) 60fs pulse width and superimposed on the optical nonlinearity. Potential applications of such a vibrational nonlinearity are currently being explored (Kenney-Wallace et al., 1988a; Ruhman et al., 1987; McMorrow et al., 1988a,b).

While the detailed analysis of the Kerr profiles in pure liquids leads to the conclusion that four distinct responses underlie the overall rise and decay of the field-induced polarization anisotropy $\Delta n(t)$, naturally in a real liquid these responses are coupled together. For clarity of discussion we will continue to describe them separately, since the timescale separation assumption in Eq. (3) proved quite valid. Summarized as follows in terms of $r_m(t)$, which are associated with the instantaneous electronic and noninstantaneous, field-driven nuclear responses linked to intra- and intermolecular motions, the responses are labeled r_1 (electronic) and r_2, r_3, r_4 (nuclear). The underlying assumption of a harmonic potential v_{ij}, and a continuum fluid $[g(r) = 1]$ might clearly lead to some over simplifications but if molecules reside at or near the bottom of the potential well, the situation is not unreasonable. The results are very instructive for modeling short-time behavior for molecular solvents at room temperature, which is the next, vastly more complex step (see references on simulations and the next chapter).

The electronic polarization, $r_1(t)$, is clearly separable from the overall molecular response and for the data analysis presented here we have assumed it is a scalar multiple of the $G_o^{(2)}(t)$ autocorrelation of the laser pulse. In all of our data we assume 100fs (fwhm) contribution from our 60 to 65fs laser pulse. Note we are far from one or two photon resonances, thus no electronic state population lifetime arguments need be considered (Kenney-Wallace, 1981; Christophorou, 1981).

The three remaining, nuclear terms have relaxation times as an ultrafast component of 80fs < $r_2(t)$ < 160fs; and intermediate component of 400fs < $r_3(t)$ < 700fs; and a slow component of $r_4(t) \geq$ 1ps. This latter component is most clearly observed in Fig. 6 and is identified with the diffusional orientational relaxation time or Debye time. This decay rate usually scales with viscosity but not necessarily under hydrodynamic stick boundary conditions (Kenney-Wallace, 1984; McMorrow et al., 1988a; Kenney-Wallace and Wallace, 1983). The diffusional relaxation can persist for up to tens of picoseconds depending on the size and interactions of the molecules.

When CS_2 is diluted in isoheptane, a liquid which at 295K is isoviscous with CS_2, η = 0.366 cp, while the amplitude and the temporal profile change, the long time tail remains almost invariant, τ_4 = 1.62 ± 0.01ps. In contrast, dilution over a similar range, down to mole fraction $m(CS_2)$ = 0.2 in tetradecane for which η = 1.154 cp, the diffusional relaxation slowed notably to 2.71ps. In n-pentane, dilution to m = 0.2 led to a solution of η = 0.259 but an increase in the rate of relaxation to 1.3ps. This is consistent with the earlier 4 WM phase conjugation results (Kenney-Wallace and Wallace, 1983) and the fast subpicosecond orientation rate extrapolated at infinite dilution to 0.9ps, compatible with the 0.85ps time calculated for the mean free rotor time in the gas phase. Clearly in a weakly interacting molecular milieu, the CS_2 molecules can reorient in the bath with little barrier to rotation at room temperature. In a strongly interacting case, the times are slowed down. The interactions are both dipolar and quadrupolar, or simply steric hindrance from structural considerations (Holubov and Kenney-Wallace, 1986-88; Holubov, 1988; Dais et al., 1980).

The behavior of the diffusional component of CS_2 at temperatures from 160K to 330K and over these dilution ranges in alkanes, substituted alkanes, benzenes, methanes, substituted methanes, and acetone has been completed using ^{13}C and ^{33}S Fourier transform NMR to measure the T_1, spin-lattice relaxation times (Holubov and Kenney-Wallace, 1986-88; Holubov, 1988; Dais et al., 1980). From all of these data the packing structures arising from the short range potential, and their concomitant influence on the stick/slip boundary conditions of the CS_2 motion, has been examined and the results discussed elsewhere (Holubov and Kenney-Wallace, 1986-88; Holubov, 1988; Dais et al., 1980). Such diffusional τ_4 motions are too slow for the initial localization of the quasifree electron, and so we return to the faster femtosecond responses observed in the optical Kerr signals.

The measurements of τ_4 from $r_4(t)$ measured over several lifetimes and τ_1 from $r_1(t)$ provide the two fixed reference points for the analysis which are model free. Thus these contributions can be removed mathematically from the overall Kerr profile (for details see Kenney-Wallace et al., 1988a) and the remaining biexponential subpicoseconds response was analyzed as follows in r_2 and r_3.

The ultrafast component r_2 is modeled in our analysis as librational response of the molecules to the external laser field. We envisage that the molecules reside at 10^{-14}s in a frozen but transitory solvent structure or "cage", whose short range intermolecular potential between any pair of molecules is defined by their local solvent packing structure. The bottom of this librational potential well is reasonably described by a harmonic potential, and the population of the librational levels is thermalized. The internal friction and elastic restoring force variables in Eq. (2) determine the period of librational oscillation. Frequently such motion is also described as a hindered or frustrated rotation. the need to include inhomogeneous broadening (Ruhman et al., 1987; McMorrow et al., 1988b) in this librational frequency is a natural consequence of the site-to-site configurational variations in the energy minimum of this potential. The inhomogeneous broadening can be quite large in these liquids. Coherently driven, these librational responses are rapidly dephased and strongly dampened by high frequency density fluctuations, which should correlate with the collision frequency or density of the fluid. The density of CS_2 at 295K is 1.262 gcm^{-3} and decreased to 0.852 in m = 0.2 in tetradecane, and 0.776 in the n-pentane analogue.

In CS_2, this librational response should occur from our analysis at $\bar{\nu} = 35.4$ cm^{-1} with $\Delta\bar{\nu} = 37.9$ cm^{-1}, and the libration exhibits a damping rate of 140fs. A very weak band has been observed through stimulated Raman gain frequency domain data (Scarparo et al., 1981) of the low frequency spectrum of CS_2; it is centered at 35 cm^{-1} and has $\Delta\bar{\nu} = 40$ cm^{-1}, in close agreement with our observations. These conclusions on the ultrafast librational component in CS_2 are consistent with other recent femtosecond observations (Ruhman et al., 1987a), including a very recent report following our own experimental results in which the temperature dependence and the inhomogeneity in CS_2 was discussed (Ruhman et al., 1987b), and agree with earlier conjectures on the 300fs ultrafast response reported in 1982-84 from experiments using much longer (150-300fs) laser pulses (Green and Farrow, 1982; Halbout and Tang, 1982; Etchepare et al., 1982).

In nitrobenzene we determine the libration to be at 55.8 cm^{-1}, (Kalpouzos et al., 1987; McMorrow et al., 1988a) and it is of interest to see if this band can be observed in far infra-red light scattering spectroscopy, such as that reported for CH_2Cl_2 in CCl_4 solutions ranging from 50 to 100 cm^{-1} (Yarwood, 1982).

The third intermediate time component r_3 varies from liquid to liquid and is present in all atomic and molecular liquids, whether or not they possess an intrinsic molecular anisotropy. We believe, following a detailed analysis of femtosecond Kerr responses in liquid CCl_4, $CHCl_3$, and CH_2Cl_2 (McMorrow et al., 1988a,b), that r_3 reflects the dynamical nature of the solvation "cage". In other words, the transitory nature of the solvent structure is now manifested as small displacements of center-of-mass due to collision-induced perturbations of the molecules. These perturbations readily influence the effective molecular polarizability, which leads to high-frequency fluctuations in the linear and nonlinear polarizability. In turn, these motional effects should correlate with the timescales seen for collisions or high frequency density fluctuation. Collision-induced relaxations observed on computer simulations of small molecule systems do indeed occur in simulations ca 500fs (Nichols and Chandler, 1986; Wallquist et al., 1986;Schnitker et al., 1986; Schnitker and Rossky, 1987; Schnitker et al., 1988; Sprik et al., 1985; Marchi et al., 1988; Golombok et al., 1985; Madden and Tildesley, 1985; Ladanyi, 1983; Frenkel and McTague, 1980).

CCl_4 is isoelectronic with Ar, about 3.4 Å in diameter, and from the neutron scattering studies determining S(k) and g(r) the molecules appear to interlock in the liquid at 298K. In liquid CS_2 the molecules (3.17Å in length) appear to assume parallel axes preferred configurations at room temperature. In comparing the Kerr profiles of the rare gas atomic liquids and these two molecular liquids, the experimental data and subsequent theoretical analyses indicate that whereas the librational component r_2 is absent in both the spherically symmetric Ar and CCl_4, all liquids exhibit component r_3. If such "translational anisotropy" is also field-driven, then its relaxation must also be associated with the relaxation time of the density fluctuations once the field is off, and this implies a limiting rate of both $r_2(t)$ and $r_3(t)$ linked to the density, and thus temperature too, of the material. Further experiments are in progress to investigate these effects in liquids with different molecular symmetry and packing structures. Clearly as we noted earlier in this chapter, the components r_2 and r_3 are not uncoupled in reality. But for the present analysis, we assume that all components to a first approximation can be treated separately, as light scattering analysis have always presumed (Madden and Tildesley, 1985; Ladanyi, 1983; Frenkel and McTague, 1980; Berne and Decora, 1976). Thus the collisional damping (r_3) occurs either more slowly or at the same rate as the dephasing of the libration (r_2), although density fluctuations are involved in both mechanisms. For this reason, the role of coherence must be especially critical since the librational motions are driven coherently with the applied laser field (McMorrow et al., 1988a). Again, this is a critical issue for which more research is required to determine the necessary and sufficient conditions for the coherence in the relaxation phenomena.

At this point, having outlined the nature of the femtosecond molecular motions which underlie the optical responses to a strong, sudden electric field it is appropriate to review what motions may be linked to the localization and trapping of the quasifree electron. These motions appear to be generic to a wide range of organic solvents we have studied, and presently we are attempting to similarly analyze ultrafast Kerr responses from more polar liquids such as water and alcohols, which are well-known electron traps. The implications for electron scattering, localization and solvation in disordered media are clear, if one now conceptually substitutes the electron for the laser pulse as the origin of the intense, applied field.

The recent femtosecond experimental data on electron solvation in water (Gaudel et al., 1984, 1987) and observations in a mass spectroscopy on the formation of electrons stabilized in molecular beam clusters (Arbruster et al., 1984) has rekindled extensive interest on the microscopic details of the dynamics and structure of e^-_{aq} in particular. Since the appearance of the visible spectrum of e^-_{aq} has now been observed from the two photon photoionization of pure water (e.g., no dopant molecules or ions were present) we must focus on what responses can be induced from the medium on this timescale. From our present databank, it is evident that following the instantaneous electronic polarizability (linear and nonlinear) response to the moving charge and/or field, it is the librational responses that must be the key motion. We assume, for the moment, that the lifetime of the autoionizing level in H_2O is not a significant factor. As we have discussed elsewhere (Kenney-Wallace, 1981; Christophorou, 1981) the lifetime of the precursor electronic state may of course be very significant, but in the absence of evidence either way, we will leave this point aside. Also, we will assume no ordered liquid structure from solvation of a parent negative ion, such as $Fe(CN)_6^{-3}$, prior to photodetachment, or coulombic attraction in electron-hole pairs, such as could have played a role in earlier experiments (Wiesenfeld and Ippen, 1980; Wang et al., 1980). We return to our earlier postulate (Kenney-Wallace, 1981; Christophorou, 1981; Kenney-Wallace et al., 1987; Kenney-Wallace and Jonah, 1982) that the density and configurational fluctuations frozen in to the liquid structure at 10^{-14}s provide an array of potential minima of varying depths, into which a scattered quasifree electron can be temporarily localized. The distribution of these wells will depend on the number density of the medium. They well could be identified with the frozen disorder pictures arising from plots of free energy versus density $\rho(r)$ in glasses. The spontaneous density fluctuations above a certain critical amplitude create microscopic order and disorder in liquids over timescales which encompass both the r_2 and r_3 (Kerr) responses we have observed. Spectral evidence of e_s^- localization and solvation suggests the dominance of librational (r_2) motions in localization and trapping. Reorientational diffusion (r_4) dominates the solvation step, which leads to fully configurational relaxed e_s^- as we have earlier shown (Kenney-Wallace and Jonah, 1982). Of course, all motions are contiguous; it is the matching of time scales that we are addressing here. Indeed, the r_3 or solvent "caging' dynamics may be a measure of the ability of the preexisting well (Schnitker et al., 1986; Schnitker and Rossky, 1987; Schnitker et al., 1988) or mini-cluster to persist for periods long enough to permit small local rearrangements which deepen the well. In other words dipolar reorganization of adjacent molecules about the trapped excess electron, can build up the cluster, optimize the packing and deepen the potential well which is finally completed over several picoseconds, a reflection of the intrinsic dynamics of the liquid host system. The longer range continuum effects now come into play. Dielectric friction considerations must also be examined (Simon, 1988; Templeton and Kenney-Wallace, 1986; Kivelson and Spears, 1985; Kivelson and Madden, 1982; Hubbard and Wolynes, 1979; Fleming, 1988), and the role of longitudinal relaxation times, and degree of dielectric saturation close to the electron or ion, fully quantified. Indeed, while clusters of molecules are host to the electron, is it confined to the interior or are there surface states? Much remains to explore. Kestner (see this NATO volume) has fully reviewed the theoretical body of electron studies from V_o and continuum levels (see Fig. 1) to quasi localized and fully localized states. Thus we have made no attempt here to introduce more detailed theoretical concepts.

Feasibility experiments are now in progress in our laboratory to determine the Kerr responses of pure water, which has a rather low $\chi^{(3)}$ value, and the influence of the presence of electrons on the fs librational and vibrational motions of molecules known to solvate electrons, such as acetonitriles, DMSO and HMPA. The wide spectral band width of ultrashort laser pulses (typically 300 cm^{-1} to 800 cm^{-1} in visible laser pulses of 70 fs to 20 fs in duration) naturally couples in, via stimulated Raman interactions, the lower frequency skeletal modes of the molecule. Such evidence is clearly seen in Fig. 7(b) for $CHCl_3$. The Kerr profile includes $\bar{\nu}_2$ normal mode and $\bar{\nu}_4$ at 218 cm^{-1} and 314 cm^{-1} respectively in CCl_4 as well as the r_1 and r_3 responses (McMorrow et al., 1988b). Presumably the perturbation on the vibrational and librational motions of molecules in the stabilized clusters will lead to a shift in these frequencies and amplitudes and thus a change in the Kerr profiles in the presence of ions or e_s^-. We hope to observe such shifts given the satisfactory Kerr response in electrolyte solutions and in DMSO and HMPA in the absence of electrons.

In conclusion, we propose that the molecular motions most responsive to the excess scattering electron during the initial localization phase are the fs librational motions of molecules which are themselves trapped in librational wells. At 10^{-14}s, these become a measure of the configurational disorder of the "frozen" liquid. It is the coupling of the electronic response to these ultrafast nuclear responses that should be the challenging first step in building a quantitative model of electron localization and solvation, one which predicts or affirms as well the richness of the laser photophysics, spectroscopy, and electron transfer reactions of this species, the ubiquitous e_s^- and the simplest anion.

ACKNOWLEDGMENTS

The author acknowledges the valuable discussions and contributions from Dr. Bill Lotshaw, Dr. Dale McMorrow, and Dr. Constantinos Kalpouzos during the nonlinear optical experiments summarized here.

Financial support for this research comes from the U.S. Office of Naval Research, the Natural Sciences and Engineering Research Council of Canada (NSERC) and the Connaught Foundation of the University of Toronto.

REFERENCES

Alexander, M., Johnson, M.A., Levinger, N., and Lineberger, W.C., 1986, Phys. Rev. Lett., 57:976.

Arbruster, M., Haberland, H., and Schindler, H.G., 1984, J. Phys. Chem., 88:3903.

Berne, B.J., and Decora, R., 1976, "Dynamic Light Scattering", Wiley, NY.

Castleman, Jr., A.W., and Keese, R.G., 1986, Ann. Rev. Phys. Chem., 37:525.

Chase, W.J., and Hunt, J.W., 1975, J. Phys. Chem., 79:2835.

Christophorou, L.G. (ed.), 1981, "Electron and Ion Swarms", Pergamon Press, NY (contains chapters by this author and many others on pertinent electron scattering problems with respect to electrical properties of liquids).

Coe, J.V., Snodgrass, J., Freidhoff, C., McHugh, K., and Bower, K.H., 1987, J. Chem. Phys., 87:4302; J.V. Coe, D.R. Worsnop, and K.H. Bowen, to be published (1988).

Dais, P., Gibb, V., Kenney-Wallace, G.A., and Reynolds, W.F., 1980, Chem. Phys., 47:407 and references therein to earlier NMR work.

Etchepare, J., Kenney-Wallace, G.A., Grillon, G., Migus, A., and Chambaret, J.P., 1982, IEE J. Quant. Elec. QE, 18:1826.

Fleming, G., 1987, personal communication; 1988, Faraday Discussions, 85:(in press).

Frenkel, D., and McTague, J.P., 1980, J. Chem. Phys., 72:2801.

Gaudel, Y., Migus, A., Martin, J.L., and Antonetti, A., 1984, Chem. Phys. Lett., 108:319; 1987, Phys. Rev. Lett., 58:1559.

Green, B., and Farrow, R.C., 1982, J. Chem. Phys., 77:4779; J.M. Halbout and C.L. Tang, 1982, App. Phys. Lett., 40:765.

Golombok, M., Kenney-Wallace, G.A., and Wallace S.C., 1985, J. Phys. Chem., 89:5160.

Hager, J., and Wallace, S.C., 1985, J. Phys. Chem., 89:3833; J. Hager, G. Leach, D. Demmer, and S.C. Wallace, 1987, J. Phys. Chem., 91:3750.

Holubov, C., and Kenney-Wallace, G.A., 1986-88, unpublished data; C. Holubov, 1988, Ph.D. Thesis, University of Toronto.

Hubbard, J., and Wolynes, P., 1979, J. Chem. Phys., 69:998.

Huppert, D., Kenney-Wallace, G.A., and Rentzepis, P.M., 1981, J. Chem. Phys., 75:2265.

Jonah, C.D., Romero, C., and Rahman, A., 1986, Chem. Phys. Lett., 123:209; M. Parrinello and A. Rahman, 1984, J. Chem. Phys., 80:860.

Jortner, J., and Even, U., 1983, J. Chem. Phys., 78:3445 and references therein.

Kalpouzos, C., Lotshaw, W., McMorrow, D., and Kenney-Wallace, G.A., 1987, J. Phys. Chem., 91:2028; 1987, Chem. Phys. Lett., 136:323.

Kenney-Wallace, G.A., 1976, Chem. Phys. Lett., 43:529.

Kenney-Wallace, G.A., and Jonah, C.D., 1976, Chem. Phys. Lett., 39:596.

Kenney-Wallace, G.A., 1981, Adv. Chem. Phys., 47:585 (for a review of various issues, including spectroscopic techniques and earlier solvation work).

Kenney-Wallace, G.A., and Jonah, C.D., 1982, J. Chem. Phys., 86:2572.

Kenney-Wallace, G.A., and Wallace, S.C., 1983, IEEE J. Quant. Elec. QE., 19:719.

Kenney-Wallace, G.A., 1984, in "Applications of Picosecond Spectroscopy in Chemistry," K.B. Eisenthal, ed., Reidel, NY, pp. 139-162.

Kenney-Wallace, G.A., Dickson, T., and Golombok, M., 1987, Farad. Trans. II, part 10, p. 1825.

Kenney-Wallace, G.A., Kalpouzos, C., Lotshaw, W., and McMorrow, D., 1988a, Int. J. Rad. Phys. and Chem., (in press).

Kenney-Wallace, G.A., Paone, S., and Kalpouzos, C., 1988b, Faraday Discussions, 85:(in press).

Kivelson, D., and Madden, P.A., 1982, J. Phys. Chem., 86:4244.

Kivelson, D., and Spears, K.G., 1985, J. Phys. Chem, 89:1899.

Ladanyi, B.M., 1983, J. Chem. Phys., 78:2189.

Madden, P.A., and Tildesley, D.J., 1985, Mol. Phys., 55:969 and references therein.

McMorrow, D., Lotshaw, W., Kenney-Wallace, G.A., 1987, Proceedings of International Laser Science Conference ILS-III, (in press, 1988) and 1988, Chem. Phys. Lett., 150:138.

McMorrow, D., Lotshaw, W., and Kenney-Wallace, G.A., 1988a, IEEE J. Quant.-Elect. QE - 24, pp. 443-454.

McMorrow, D., Lotshaw, W., and Kenney-Wallace, G.A., 1988b, Chem. Phys. Lett., 145:309.

Miyasaka, H., Masuhura, H., and Mataga, N., 1987, Laser Chem., 7:119.

Nichols, A., and Chandler, D., 1986, J. Chem. Phys., 84:398 (D. Chandler in this NATO ASI volume).

Reintjes, J., 1984, "Nonlinear Optical Parametric Processes in Liquids," Academic Press, New York.

Robinson, G.W., Thistlewaite, P.J., and Lee, J., 1986, J. Phys. Chem., 90:4244; H.F. Hameka, G.W. Robinson, and C.J. Marsden, 1987, J. Phys. Chem., 91:3150.

Ruhman, S., Kohler, B., Joly, A., and Nelson, K., 1987b, Chem. Phys. Lett., 141:16.

Ruhman, S., Williams, L.R., Joly, A., Kohler, B., and Nelson, K.A., 1987a, J. Phys. Chem., 91:2237.
Scarparo, M.A.F., Lee, J.H., Song, J.J., 1981, Opt. Lett. 6:193.
Schnitker, J., Rossky, P., and Kenney-Wallace, G.A., 1986, J. Chem. Phys., 85:2986; J. Schnitker and P. Rossky, 1987, J. Chem. Phys., 86:3462; J. Schnitker, P. Rossky, and K. Mota Kabbiri, 1988 (in press).
Simon, J.D., 1988, Acc. Chem. Res., 21:128-134 (on time-resolved solvation studies of large polar molecules and comments on dielectric friction approaches).
Sprik, M., Impey, R., and Klein, M.L., 1985, J. Chem. Phys., 83:5802; M. Marchi, M. Sprik, and M.L. Klein, 1988, Faraday Discussion 85 no. 15 (in press), and references to earlier theoretical work.
Templeton, E., and Kenney-Wallace, G.A., 1986, J. Phys. Chem., 90:5441; ibid., 1986, 90:2896.
Wallquist, A., Thirumalai, D., and Berne, B.J., 1986, J. Chem. Phys., 85:1583.
Wang, Y., Crawford, M.H., McAuliffe, M.J., and Eisenthal, K.B., 1980, Chem. Phys. Lett., 74:160.
Wiesenfeld, J., and Ippen, E., 1980, Chem. Phys. Lett., 73:47.
Yarwood, J., 1982, Annual Reports C, The Royal Society of Chemistry, London, pp. 157-197.

STRONG FIELD EFFECTS AND MOLECULAR DYNAMICS SIMULATIONS

Michael Golombok and Geraldine A. Kenney-Wallace

Lash Miller Laboratories
University of Toronto
Canada M5S 1A1

INTRODUCTION

As the preceding chapter has shown, ultrafast laser spectroscopy and the study of molecular relaxation phenomena in the 10^{-13}s - 10^{-10}s time regime have become an important route to identify and quantify those earliest molecular events, whose intrinsic interactions and dynamics comprise the microscopic character of condensed media. Computer simulations have played an increasingly significant role in recent years as an "experimental" testing ground for equilibrium and transport properties of liquids as a function of the choice and shape of the intermolecular potential and other system variables. The links between spectroscopy, electromagnetic field-induced interactions and the theory of molecular motion, are most generally sought through the formalism of time-correlated functions (Steele, 1984). Thus, the simulation of correlation functions could offer several test points of comparison for the response of a specific atomic or molecular liquid to a given external perturbation, such as a pulsed laser field. Several groups are already examining various aspects of electron-cluster interactions to determine the cluster size, stability, structure and ultimately, one hopes, the absorption spectrum of electrons solvated in the simpler liquids, such as water, ammonia and methanol. Chandler reviews some of this work earlier in this volume.

But what of the field effects? We describe now a simulation of the nonlinear optical responses of a real liquid, CS_2, subjected to an intense laser field. We examined the specific contributions of the microscopic polarizability to the time evolution of $\chi^{(3)}$, the third-order nonlinear susceptibility, and compare the results to data from recent four-wave mixing and femtosecond optical Kerr experiments (Greene and

Farrow, 1982; Etchepare et al., 1982; Halbout and Tang, 1982; Ho et al., 1976). Motivated by our previous experimental work on probing interactions and dynamics via nonlinear laser spectroscopy and four-wave mixing (Kenney-Wallace and Wallace, 1983; Golombok et al., 1985; Dickson et al., 1987; Kenney-Wallace et al., 1987), we have particularly focused on the intense laser interaction with the molecules and subsequent field-driven changes in their orientational distribution. The application of this kind of result is not only to advance the interpretation of intramolecular and intermolecular dynamical events revealed through $\chi^{(3)}$ and even field-dependent $\chi^{(3)}$ or $\chi^{(5)}$ interaction, but also to electron trapping and solvation events (Kenney-Wallace et al., 1987) and to chemical reactions driven by strong laser fields. In the latter case, the induced-dipole moments and transient orientational ordering could well influence the height and the pathway over the reaction barrier, as noted in a preliminary account of this work (Golombok and Kenney-Wallace, 1984; Golombok, 1984). Field-induced alignment effects in gas phase collision dynamics have also been observed (Vasudev et al., 1984).

Following a brief synopsis of nonlinear optical interactions in which we set the context of the simulation (Golombok, 1984), the theoretical framework for the simulation is outlined in THEORY. The details of the actual calculations are given in RESULTS AND INTERPRETATION OF SIMULATION DATA, and the results and interpretation in CONCLUSION examine the energy and fields, the torques and orientations, and finally the molecular correlations. We emphasize at this point that, as this was the prototype study, we were interested in establishing the key physical concepts and linking the results to experimental observables first, before optimizing the full dynamics calculation through the more sophisticated potentials in use for equilibrium problems.

Four calculations have been reported in related areas of molecule-field interactions. Evans (1982) has simulated the bulk anisotropy for a collection of molecules, which while dipolar were not polarizable nor based on the properties of a real molecule. The field-induced force was not explicitly defined in terms of a dipole-field interaction, rather a full Lennard-Jones potential was used. Coffey has examined the inertial relaxation of dipolar molecules in intense fields (Coffey et al., 1983). Madden has focused on interaction-induced effects in dielectric absorption (Tildesley and Madden, 1983) and transient nonlinear optical measurements in CS_2 (Madden, 1987). Samios and Dorfmuller used a similar local field formalism to ours when they examined far IR absorption and utilized equilibrium fluctations to obtain correlation functions for liquid CS_2 (Gburski et al., 1987). More details on these calculations can be found in the NATO ASI volume of Molecular Liquids, Reidel and Dordrecht (1984).

In studying the intense laser interactions with the CS_2 molecules, and observing the behaviour of the orientational distribution and pair correlation functions [$g(\Omega)$] through the time-evolution of the optical field-induced anisotropy, we are emphasizing external field-induced phenomena rather than interaction-induced or collision-induced events. The latter we model stochastically. Clearly they are an intrinsic dynamical property of the system, even in the absence of a field, and can and do play a mechanistic role in relaxation of the anisotropy once the field is switched off. However, it was our primary objective to link these simulations to $\chi^{(3)}$, and to the orientational component in particular, $\chi^{(3)}_{or}$, since (in the absence of interaction-induced effects) it can be shown for symmetric molecules (Golombok et al., 1985):

$$\chi^{(3)}_{or} \sim \kappa^2 < P_2 (\cos \Theta) > , \tag{1}$$

where κ is the anisotropy probed in the experiment, Θ is the angle between two molecular axes and P_2 is the second order Legendre polynomial. From four-wave mixing (4WM) nonlinear optical experiments, we can deduce the orientational pair correlations giving rise to the actual value of $\chi^{(3)}$ given by $g(\Omega)$, in weak and strong laser field conditions (Golombok et al., 1985).

We conclude this section with a succinct reminder of the quantum electronic origins of $\chi^{(3)}$ and its link to experimental observables.

In the optical Kerr experiment, the transient birefringence $\delta n(t)$ imposed by the polarized pump pulse (ω_1) can be monitored by a much attenuated and polarized probe pulse (ω_2), which experiences a phase shift $\delta\phi(t)$,

$$\delta\phi(t) = \frac{2\pi l}{\lambda_2} \delta n(t) , \qquad n = n_0 + n_2 < E (\omega,t)^2 > . \tag{2}$$

The refractive index change δn is given in terms of a nonlinear index n_2

$$\delta n = \frac{1}{2} n_2 < E^2 > \simeq \frac{1}{n} \chi^{(3)} < E^2 > . \tag{3}$$

As seen in the previous chapter on femtosecond laser spectroscopy, the change n_2 or $\chi^{(3)}$ response of the molecules contains both an electronic and molecular (orientational, vibrational) polarization term. These can be separated where $\chi^{(3)}_{el}$ is purely the electronic susceptibility.

$$\chi^{(3)} = \chi^{(3)}_{el} + \chi^{(3)}_{nuc} , \tag{4}$$

$$\chi_{el}^{(3)} \simeq N < \gamma > . \tag{5}$$

The second-order molecular hyperpolarizability γ is related to the macroscopic $\chi_{el}^{(3)}$ (which is an average measure of the ease with which electronic distortion gives rise to third-order dipoles) but not in a simple manner because of the local field problem existing for N molecules, when $N > 2$.

The nuclear contribution $\chi_{nuc}^{(3)}$ represents a linear field response to field-induced modifications in the molecular orientation. An expansion of the classical expression for the bulk averaged differential polarizability (Bottcher, 1973), or alternatively a Born-Oppenheimer expansion over quantum states in the optical regime (Hellwarth, 1977), shows that in an isotropic medium, the principal contribution to $\chi^{(3)}$ is a polarizability correlation given by Eq. (1b) where α_{ij} is the molecular polarizability tensor:

$$\chi_{nuc(ijkl)}^{(3)} \sim < \alpha_{ij} \ \alpha_{Kl} > . \tag{1b}$$

What are the experimental observables to be included in a calculation? We performed a number of 4WM experimental studies to examine molecular orientation and dynamics in liquids under weak and strong laser fields, and under different environmental conditions, by diluting the reference molecules in a range of solvents (Golombok et al., 1985; Dickson et al., 1987). We wished to examine the relative importance of molecule-molecule interactions, interaction-induced polarizability changes, and field-molecule interactions on the dynamics of given reference molecule such as CS_2. We also compared the effects of atomic and bond substitutions within molecules of a given symmetry interacting with CS_2. We also derived angular distributions in the laser field and quantified the intermolecular correlations. One of the interesting effects we observed was that CS_2 whether diluted in pentane or dodecane behaved very similarly on an orientationally averaged time scale, while notable differences were seen for CS_2 in CH_2Cl_2 or in aromatic systems, even through all were at the same CS_2 number density.

Because of the resulting underlying relationship in dense media between molecular dynamics and the onset of nonlinear optical effects, it is highly desirable to develop a model of field-molecule interactions which demonstrates the evolution of $\chi_{nuc}^{(3)}$ as a collective effect as opposed to $\chi_{el}^{(3)}$ which is a single molecule effect modulated by local fields. $\chi_{el}^{(3)}$ in liquids constitutes the vast bulk of the total susceptibility (Reidel and Dordrecht, 1984; Golombok et al., 1985), whereas the isolated molecule only has an <u>electronic</u> response cubic in the applied field. In the condensed state, this electronic factor generally accounts for only perhaps ten percent of the total $\chi^{(3)}$ at nanosecond times but of course will be relatively larger at femtosecond times, since little orientational anisotropy will have been established.

The observed 4WM signal intensity is cubic in the applied laser field (Golombok et al., 1985), and by judicious choice of laser pulse duration and wavelength, can be interpreted directly in terms of orientational correlations between molecules in the liquid state (Buckingham, 1967).

We now develop two previous hard-core models (Street and Tildesley, 1978; Viellard-Baron, 1974) and show that experimental results can be simulated without a large amount of computation and using a simple but physically realistic model of the liquid. It is becoming increasingly recognized that nonlinear optical effects provide an as yet unmined source of significant detail on molecular interactions and liquid structure. Moreover, this detail is related to the pair orientation distribution function (Buckingham, 1967) and indeed it is thought that ultrashort nonlinear optical measurements may access higher order particle correlations which have not yet been experimentally observed (Kenney-Wallace, 1983; Buckingham, 1967).

THEORY

Nonlinear optical effects are modeled in this simulation using molecular linear dipoles which yield a collective optical nonlinearity. No explicit evaluation of $\chi^{(3)}$ phenomena in terms of real molecular systems, to our knowledge, has been obtained previously using this technique. Our objectives were (i) to use only a few significant molecular variables, (ii) to work with real conditions and systems, (iii) to obtain physically meaningful parameters, and (iv) to study the effect of the laser field on liquid dynamical structure as manifested through nonlinear optical phenomena.

(a) Spherocylinder parameters and positioning

We simulate hard core spherocylinders based on a realistic model of carbon disulphide. These spherocylinders have an anisotropic molecular polarizability, and we use anisotropic weighting functions to determine the interactions between molecules which are based on dipole-induced-dipole forces. By evaluating field-induced forces and their subsequent evolution, we probe $\chi^{(3)}$ effects as well as the validity of local field models.

Our cylinder is of length 3.55 Å with hemispherical endings of radius 1.55 Å, as illustrated in Fig. 1. Some 51 percent of the bulk

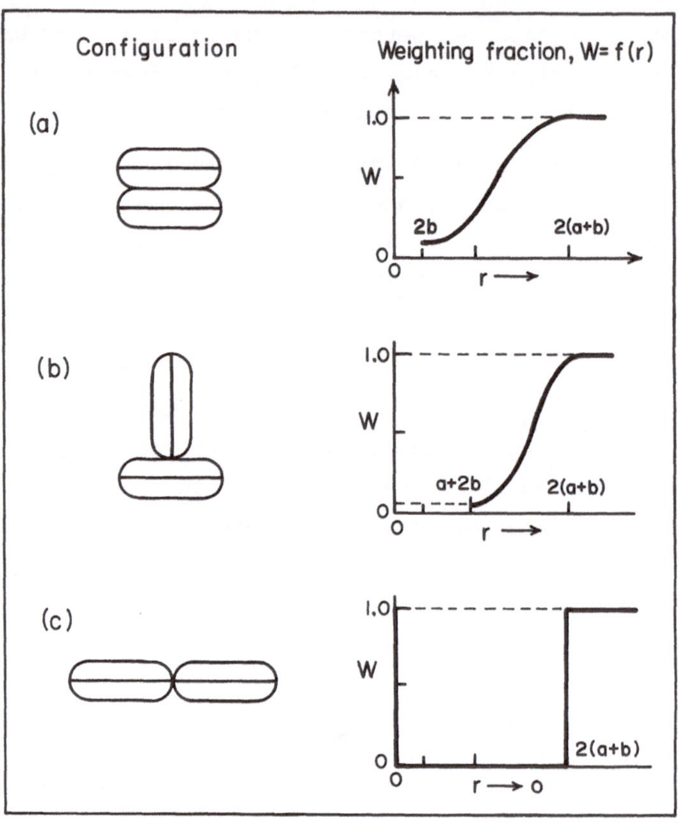

Fig. 1. Spheroidal cylinder used for hard core model of CS_2 simulation under strong field. a = 1.56 Å, b = 1.77 Å.

volume in our model is free space, but the availability of this space is highly orientation-dependent. This model was very close to the geometrical model used in a theoretical study of intermolecular potentials, which best reproduced liquid CS_2, equilibrium properties (Street and Tildesley, 1978).

The centers of mass of the molecules are randomly assigned and remain fixed. This is justified by the frozen nature of the liquid lattice over the picosecond time scale of our calculation. Molecular orientation and angular velocity (ω) are initially randomly assigned, with magnitude of the latter given by the Maxwell-Boltzmann average:

$$\omega = \left(\frac{8kT}{\pi I} \right)^{1/2} , \tag{6}$$

where $I = 2.6 \times 10^{-45}$ kgm^2 is the moment of inertia. This gives an angular velocity of 2×10^{12} s^{-1} at room temperature.

The density of the cylinder packing is set at the experimentally observed value of 1.26 gcm^{-3}. Because of the random position and orientation of these hard-core spheridal cylinders, the problems of overlapping molecules in the initial and subsequent configurations are dealt with by a weighting function, which relates shape and orientation-dependence through a "contact" function, as first outlined by Viellard-Baron (1974). The effect is essentially to add an orientation-dependent, soft outer layer to the anisotropic, hard core of each CS_2 particle. The weighted configurations are shown in Fig. 2.

The pairwise interaction between neighboring molecules is determined by their separation (r) and relative orientation. For separation $r \geq 2(a+b)$ as given in Fig. 1, no weighting function is used, shown in Fig. 2c. If the center of one molecule with respect to another lies in region 1 or 3 and the separation between orientational line segments is less than 45° or greater than 135° (it must be less than 180°), then we may model approximately a weighting function around configuration C of Fig. 2, which does not include considerations of molecules closer than 2(a+b) in order to exclude overlap.

Molecules in regions 1 and 3 with mutual orientations between 45° and 135° are given a weighting function based on configuration B. For separations lying in regions 2 and 4, configuration B also applies to mutual orientations between 45° and 135°. Configuration A is relevant for molecules in what we consider a near parallel condition, i.e., in region 2 or 4 with separation angles less than 45° or greater than 135°. The corresponding Gaussian weighting functions are also shown in Fig. 2.

Thus we have not only considered separation-dependent orientational factors but also, for constant separation, by including the randomness of the initial distribution, we have modeled the local orientational anisotropy of the closest neighbor environment of a molecule. This

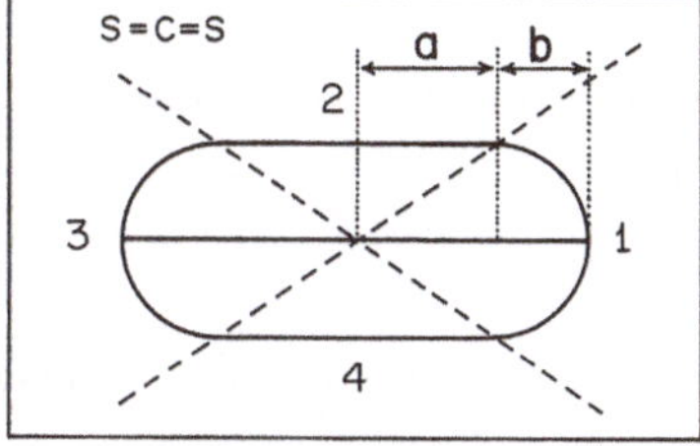

Fig. 2. Three closest approaching configurations of two CS_2 spheroidal cylinders, separation r, with associated Gaussian weighting functions w for dipole field contributions.

approach overcomes the problems associated with melting an initially ordered array of molecules.

(b) Dipoles and Torques

When an externally applied, strong electric field induces dipoles in a system, then the induced intermolecular dipolar forces dominate at distances beyond the molecular van der Waals radii. On a molecular level, different orientations of the molecules result in different local dipoles so that the orienting force experienced by each molecule is not the same. The bulk average reorientation changes the orientation distribution function $g_2(\Omega)$ (Buckingham, 1967).

In our random distribution of molecules, the instant that an electric field is turned on, the field experienced by the molecules is the same as the applied field. Thereafter, however, dipoles have been induced and the molecules now exert forces on one another through electrostatic interactions (Hellwarth, 1970). At all succeeding times following the instant the field is turned on, each molecule experiences a different local field arising from the vector sum of the local dipolar fields. Because the CS_2 molecules have no permanent dipoles, the induced dipole at every molecular center is given to first order, by

$$\mu_i = \alpha_{ij} E_j \; . \qquad (7)$$

The induced dipole gives rise to a field in its neighbourhood given by (in cgs units)

$$E_i = T_{i,j} \mu_j \; , \qquad (8)$$

where T_{ij} is the dipole tensor describing the field E of the dipole μ_i

$$T_{ij} = \frac{3r_i r_j - r^2 \delta_{ij}}{r^5} \quad , \qquad (9)$$

where r is the separation of the molecular centers. The tensor components r_i, r_j, etc., determine the orientational dependence of the interaction. Reorientational motion leads to changes in the individual interaction-induced dipoles and associate fields (Berne and Pecora, 1976).

In this model, such reorientation is due to a dipole field torque (Γ) given by (Bottcher, 1973)

$$\Gamma = \mu \times E \; . \qquad (10a)$$

We consider a molecular rod of symmetry $D_{\infty h}$, oriented in direction u. The component of force acting on the molecule, perpendicular to u leads to reorientation governed by the moment of inertia I, about the center of mass. The change in orientation $\Delta\Omega$ for a time interval Δt and at 300K is given by:

$$\Delta\Omega = \frac{1}{2}\ \frac{\Gamma}{I}\ (\Delta t)^2 + \omega\ \Delta t\ . \tag{10b}$$

As a simplification, applicable to the cylindrical axially symmetric molecule CS_2 of interest here, we apply Coffey's (Coffey et al., 1983) procedure whereby we only consider rotation arising from those components of force perpendicular to the molecular axis.

We evaluate the dipole components parallel and perpendicular to the molecular symmetry axis u for a linear molecule. We sum the dipolar fields caused at any one center by all the other induced-dipoles. Adding this to the external field now yields the local field, which will vary from molecule to molecule. This cycle of calculations is repeated at effectively "frozen" time until a satisfactory convergence of the local electric fields is obtained. The dipole field torque component which is perpendicular to the symmetry axis is calculated, and reorientation is then determined from Eq. (10b).

The cycle is repeated in time steps of h = 25fs to resolve the local fields with respect to the new orientation of the molecular axes. The laser pump field is 3×10^7 Vm^{-1} and is on for 200 fs. Typical probe intensities are $\lesssim 1\%$ of pump intensities, and this corresponds to laser light intensities of about 10 kW cm^{-2}. The Verlet algorithm for reorientation (Verlet, 1967; Ryckaert et al., 1977) for a time step h is given by

$$\Omega(t+h) = -\ \Omega(t-h) + 2\Omega(t) + \Sigma\ \frac{\Gamma}{I}\ h^2\ . \tag{11}$$

On the femtosecond time-scales of interest the molecules remain essentially undisplaced, i.e., the radial distribution is unchanged. As a consequence, we may write for the total molecular distribution function $g(r,\Omega)$

$$g(r,\Omega) = \text{constant} \times g(\Omega)\ , \tag{12}$$

over the short time scale of our experiment. This separation of time scales argument is backed up by our experiments in phase conjugate holography, where it is evident that there is insufficient time for molecular translation to destroy the contrast of the Kerr index grating (Golombok et al., 1985) is 3×10^7 Vm^{-1}.

Collisional relaxation as a dissipation channel was introduced as follows. In light of our hard-cord CS_2 model, we have utilized a phenomenological approach toward orientational relaxation based upon thermal buffeting of the molecules. We calculate the mean free time between collisions based upon thermal buffeting of the molecules and using a system of molecular dimensions, temperature and density as specified earlier. This results in a collision frequency of ~ps per molecule. Each collisional relaxation is modeled by randomizing the orientation of the axes and the angular velocity of two molecules.

The anisotropic polarizabilities of CS_2 were taken from known experimental values. The calculations were carried out on 64 molecules using a VAX computer. In 1984 this was the first time that a simulation of nonlinear optical experiments on nonpolar molecules in a strong electric field had been carried out. Because of the long time required, and the desirability of introducing many changes during the evolution of the program, we have restricted our preliminary study to a system of 64 molecules. Aware of the limitations imposed by calculations involving small numbers of particles, but motivated to get the essential dynamical physics of the system correct as a primary objective, we next compared this to a run with only 40 molecules and found only small differences, e.g., a 5 percent smaller resultant field and 5 percent shorter correlation times. We thus believe that the results that follow are a reasonably accurate physical presentation of the field-induced response, and that the introduction of the full CS_2-CS_2 intermolecular potential should be our next goal, not a larger calculation. Since this work was completed (Golombok, 1984) tremendous advances in simulation have occurred, as seen in this volume, and we hope that the key concepts here can now be treated at a higher level.

RESULTS AND INTERPRETATION OF SIMULATION DATA

(a) Energy and Fields

Within the restrictions of the calculation, the total system kinetic energy remained constant to within 1.5 percent, i.e., the temperature did not change.

Fig. 3a shows the local electric fields which are given as root-mean-square equilibrated values, and arise from recalculating dipole-induced dipole fields from a fixed spatial configuration of molecules. This typically required ten iterations to get within 0.1 percent convergence. During the 200 fs pulse, the local field is seen to increase as orientation polarization sets in, but does not equilibrate as there is insufficient time for reorientation. Figure 3b shows that if a dc external field is maintained for a protracted length of time, the equilibrium field will converge to a constant value. For our applied field of 1.05×10^7 V/m, then in the Lorentz local field model (Bottcher, 1973) we calculate 1.62×10^7 V/m for a mean local field for carbon disulphide, where we used the value of refractive index n = 1.618. Considering the spherical assumption involved in the derivation of the Lorentz model, the agreement is satisfactory. In addition, we have

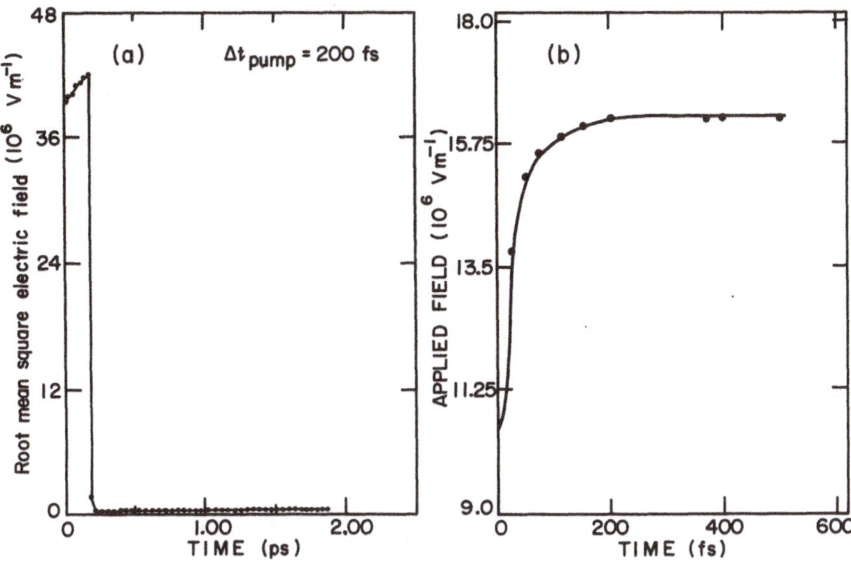

Fig. 3(a). Local electric field (root-mean-square value) at a molecule during and following a 200 fs pump pulse.
(b). Growth of local electric field to new mean internal equilibrium as a function of time for applied field of 350 SV cm^{-1}, or 10.05 Vm^{-1} (Line drawn through data points).

explicitly calculated the field at every molecular center and it is the bulk average (which is equivalent to averaging over the field modified distribution function of all the molecules) which has yielded a result similar to a more primitive (i.e., Lorentz) model. Within the confines of our model employed, this general result justifies the use of the Lorentz local field correction in the calculation of distribution functions from spectroscopic experiments (Golombok et al., 1985).

(b) Torques and Orientation

Because we have assumed that the applied torque depends only on electrostatic forces, a similar behavior to the electric field is seen in Fig. 4. The torque on a molecule increases by 23 percent compared to the 7 percent increase in local field. The induced torque arising from the field-induced-dipole is smaller in magnitude to the collision-induced torque arising from thermal buffeting (Tildesley and Madden, 1983). However, the former is consistent in direction whereas the latter is continuously changing. The coherent induced torque is sufficient to overcome the buffeting time-fluctuating forces and leads to partial intermolecular ordering. We now recall the argument of Evans (Evans, 1982) who, while dealing with polar molecules, uses the concept of a separate internal torque and an externally applied torque, i.e., forces arising from thermal buffeting and FIDID interactions, respectively. Unlike Evans calculation, our applied pump field serves two purposes.

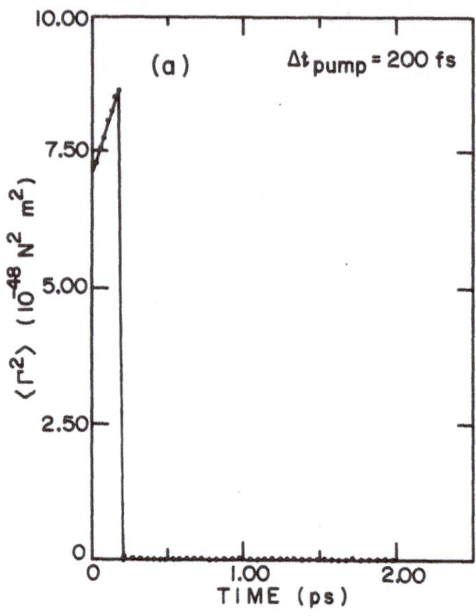

Fig. 4. Mean square torque $< \Gamma^2 >$ on a CS_2 molecule as function of time.

Not only does the dipole-field interaction provide the same molecular realignment, but the dipole is induced by the field in the first place. Such a treatment could not apply to isotropic molecules where the induced dipole would be parallel to the field and the resulting torque would vanish, as assumed by Samios and Dorfmuller (Gburski et al., 1987). Field-induced dipole-induced dipole forces (Tildesley and Madden, 1983; Nezbeda and Leland, 1979; Few and Rilgby, 1973) thus lead to the realignment, to anisotropy induction and to the third-order $\chi^{(3)}$ optical effects which we wish to model (see below).

We thus conclude that it is the duration of directional uniformity and coherence of the externally induced torques which govern orientational developments as opposed to the mere size of the internal torque at any instant.

(c) Molecular Correlations

Fig. 5a shows the induced orientational static dipole correlations. This represents the static correlations averaged between pairs of dipoles in each spatial configuration, following a 25 fs evolution period. The dipole moment density is represented by $< P_1 (\mu \cdot \mu) >$, the medium polarization (Ramshaw et al., 1971). After the 200 fs pump field is switched off, leaving only the probe field on, this polarization decreases in magnitude. The alignment $< P_2 (\mu \cdot \mu) >$, however, increases in magnitude; that is to say it becomes more negative after the field is switched off. This would indicate that either the probe field is sufficiently large to increase the dipole orientational correlation, or

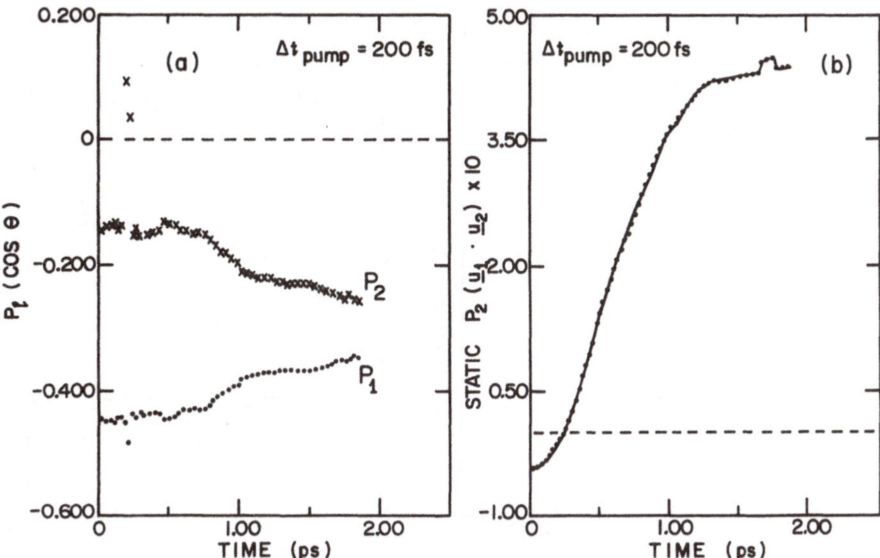

Fig. 5(a). Polarization (·) and alignment (x) in an applied, pulsed electric field.
(b). Growth of static orientation pair correlation function in time, $\langle P_2(\underline{u}_1 \cdot \underline{u}_2)\rangle$ where $\underline{u}_1$ and $\underline{u}_2$ are axes direction vectors.

that the effect of the torques generated by the strong pump laser fields persists in time to reinforce the pair alignment observed for these static configurations at different points in time. The renormalization spikes at 200 fs correspond to switching off the pump field leaving only the probe.

A related static pair correlation is that defined also by the second order Legendre polynomial calculated for the angle between pairs of molecular axes and averaged over the configuration,

$$C_s = \langle P_2 (u_i \cdot u_j) \rangle . \tag{13a}$$

In our initial configuration of 64 molecules there was at $t < 0$ a coincidental and small amount of net orientational correlation, which increased as the 200 fs field was switched on and persisted even after the pump pulse was switched off. It flattened out at around 1.30 ps, as illustrated in Fig. 5b.

The two sets of time correlation functions (in Fig. 6a) which we calculated from these simulations are based exclusively on the axis direction vector and examine autocorrelations between an axis at time zero and at successive times, i.e.,

$$C_d^1 = \langle P_1 [u(0).u(t)] \rangle , \qquad C_d^2 = \langle P_2 [u(0).u(t)] \rangle . \tag{13b}$$

This quantity is averaged over all the molecules in the simulation for each time. The time correlation of physical interest is the second order

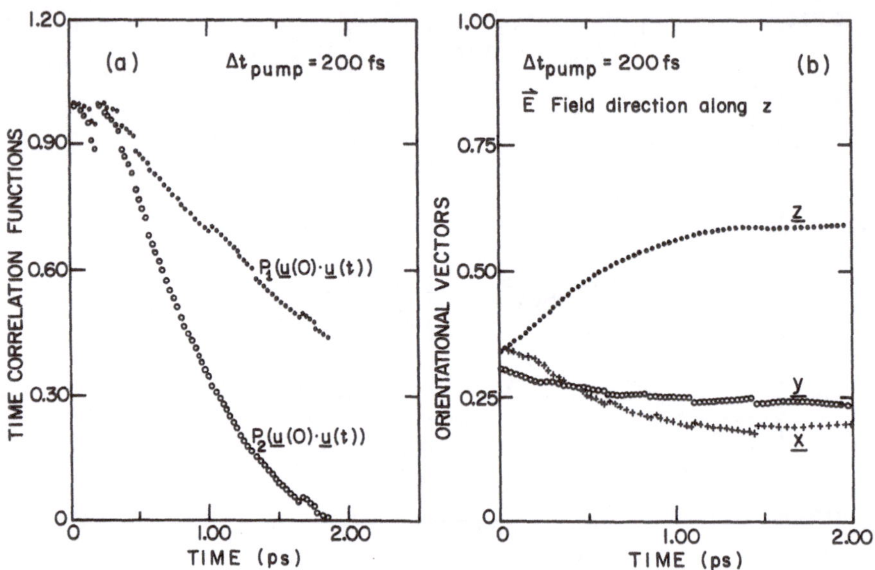

Fig. 6(a). Time correlation functions for reorientation after application of pump field. See text for discussion of P_1 and P_2 terms.

(b). Induced-anisotropy in a fluid resulting from an electric field applied along z.

one, $\langle P_2 [u(0) \cdot u(t)] \rangle$. Naturally, the P_1 term only has significance if the molecule has a permanent dipole moment. For example, in OCS, where the permanent dipole is directed along the molecular axis, the P_1 term would describe dielectric relaxation. However, our interest centers on the P_2 term, which describes the decay of the polarizability correlation function, following the termination of the pump pulse. This term has a relaxation time of 280 fs in our simulation. This number is significant because a number of recent measurements of picosecond birefringent responses in CS_2 (Greene and Farrow, 1982; Etchepare et al., 1982; Halbout and Tang, 1982; Ho et al., 1976) report 300 fs initial decays in response to pulses of comparable duration. However, while the τ responses were derived from fitting the data to a double exponential decay in these experiments, our exploratory simulation did not proceed beyond 2 picoseconds, so we were not able to check for the presence of the slower response. More recent femtosecond data clearly indicate more than 2 ultrafast components (Kalpuzos et al., 1987; Lotshaw et al., 1987; Ruhman et al., 1987). It is difficult to determine how much of this ultrafast effect is due to collision-induced anisotropy, because of the large permanent anisotropy of the CS_2 molecule. Research in progress (McMorrow et al., 1988) on the ultrafast relaxations will explore various hypotheses focusing on librational motions, vibrational hyperpolarizability, and specific interaction-induced effects in liquids of different dipole character, and other related theoretical predictions (Coffey et al., 1983; Ladanyi et al., 1983), to provide the necessary database against which to test future simulations.

The inducement of birefringence can be examined by evaluating preferred directions in the bulk liquid resulting from preferential orientation of the molecules. The initial isotropy is not quite perfect, as seen in Fig. 6b. Ideally, each direction vector autocorrelation should contribute 1/3 to guarantee isotropy in the model. Nevertheless, the appearance of anisotropy as the molecules align with respect to the externally applied z-directed field is well shown in the bulk in Fig. 6b. The growing-in of an orientation anisotropy along z, despite the more disorganized local fields, is in sharp contrast to the x, y directions. This bulk effect is seen to have a finite response time of around 0.6 ps. There is a corresponding decrease in the orientation with respect to directions perpendicular to the induced optical axis. We prefer not to make a statement regarding the possibility of trirefringence as was made on an analogous study of polar molecules (Evans, 1982); the possible errors in the starting correlations are too large for this. However, we can say that our model has indicated the induction of sufficient anisotropy to show the presence of a corresponding induced-optical axis, as is experimentally observed for carbon disulphide at times 10^{-13} - 10^{-12}s, and that the inertial effects which have now been seen in CS_2 (Kalpuzos et al., 1987) are clearly predicted. Matching of time-scales, however, must first be accompanied by a matching of conditions for both experiment and simulation in a future calculation.

CONCLUSION

We have introduced the ideas of nonlinear optical studies into molecular dynamics simulations in a prototype study. The link arises through nuclear susceptibility $\chi^{(3)}_{nuc}$. We have demonstrated reorientational motion under the effect of an applied field, and correctly simulated the fast component of $\chi^{(3)}_{nuc}$ relaxation which has been observed experimentally. We achieved this purely by considering the forces between molecules that arise from an impressed field interacting to produce linear dipoles.

We have considered explicitly the additional dipoles arising from the applied laser field. We used a shape dependent, orientational hard core model to describe the induced dipole interactions in a physically visualizable way, thus allowing the use of a physical picture to determine the significant variables. We confirmed the validity of local field approximations. An examination of the field-induced torques showed that it was not necessary that these be large in magnitude to produce an observable effect in bulk. We observed a relationship between the induced dipoles and molecular orientations. The orientational correlations agree well with experiment for the time window of transient realignment which we modeled. We also observed the onset of induced macroscopic anisotropy.

In particular, we based our field on that typically available from femtosecond and picosecond lasers and modeled the dynamics analogously to an archetypal pump probe experiment, viz. a short, intense lase field followed by a weaker probe beam. However, in light of the symmetry relations for nondispersive, nonabsorbing materials, these results can be related to the other nonlinear effects which do not deplete pump energy. The dispersion correction is estimated to be no more than 10 percent.

To gain a more precise understanding of this system, we would need to use a full potential for the intermolecular interaction. Strictly speaking we must also consider translational motion, not least because of its relevance to coherent processes in the transient phase holograms which are set up in picosecond and nanosecond 4WM experiments. Even on femtosecond time scales, translational anisotropy is a significant part of the macroscopic polarizability changes in response to sudden imposition of a 65 fs laser field (Lotshaw et al., 1987). This initial study models the laser-induced phenomena qualitatively and provides a physically insightful base from which to move to more complex calculations on CS_2 and other systems. The results will also be pertinent as a guide for designing simulations of strong-field effects and induced-polarizability changes associated with the nonlinear response of a medium to the scattering of a free electron through a dense medium at femtosecond times (Dickson et al., 1987; Kenney-Wallace et al., 1987).

ACKNOWLEDGMENTS

We gratefully acknowledge the financial support of the office of Naval Research (U.S.A.) and the Natural Sciences and Engineering Council of Canada. Geraldine A. Kenney-Wallace acknowledges the E.W.R. Steacie Fellowship and Michael Golombok the Connaught Fellowship during this work.

REFERENCES

Berne, B.J., and Pecora, R., 1976, "Dynamic Light Scattering" (Wiley).
Bottcher, C.J.F., 1973, "Theory of Electric Polarization" (Elsevier).
Buckingham, A.D., 1967, Dis. Farad., Soc., 43:205.
Buckingham, A.D., and Orr, B.J., 1967, Quat. Rev. Chem. Soc., 21:195.
Butcher, P.N., 1966, "Nonlinear Optical Phenomena", (Ohio State University Bulletin 200).
Coffey, W.T., Rybarsch, C., and Schroer, W., 1983, Chem. Phys. Lett., 99:31.

Coffey, W.T., Rybarsch, C., and Schroer, W., 1983, Intern. J. Compt. and Math. in Electrical and Electronic Eng., 2:9.

Dickson, T., Golombok, M., and Kenney-Wallace, G.A., 1987, Farad. Trans. II, 10:1825.

Etchepare, J., Kenney-Wallace, G.A., Grillon, G., Mizus, A., and Chamberet, J-P., 1982, IEEE J. Quant. Electr., 18:1876.

Evans, M.W., 1982, J. Chem. Phys., 76:5473, 5480; ibid., 77:4632.

Few, G.A., and Rilgby, M., 1973, Chem. Phys. Lett., 20:433.

Gburski, Z., Samios, J., and Dorfmuller, Th., 1987, J. Chem. Phys., 86:383 and references to earlier work therein.

Gibson, I. and Dore, J.C., 1981, Molec. Phys., 42:83.

Golombok, M., 1984, Ph.D. Thesis, University of Toronto.

Golombok, M., and Kenney-Wallace, G.A., 1984, "Ultrafast Phenomena", ed. D.H. Auston and K.B. Eisenthal (Springer-Verlag, Heidelberg), p. 331.

Golombok, M., Kenney-Wallace, G.A., and Wallace, S.C., 1985, J. Phys. Chem., 89:5160.

Greene, B.J., and Farrow, R.C., 1982, J. Chem. Phys., 77:4770.

Halbout, J.M., and Tang, C.L., 1982, Appl. Phys. Lett. 40:765.

Hellwarth, R.W., 1970, J. Chem. Phys., 52:2128.

Hellwarth, R.W., 1977, Prog. Quant. Electr., 5:1.

Ho, P.P., Yu, W., and Alfano, R.R., 1976, Chem. Phys. Lett., 37:91.

Impey, R.W., Madden, P.A., and Tildesley, D.J., 1981, Molec. Phys., 44:1219.

Kalpuzos, C., Lotshaw, W.T., McMorrow, D., and Kenney-Wallace, G.A., 1987, J. Phys. Chem., 91:323.

Kenney-Wallace, G.A., 1983, "Picosecond Laser Spectroscopy", ed. K.B. Eisenthal (Plenum).

Kenney-Wallace, G.A., Kalponzos, C., and Lotshaw, W., 1987, Int. J. Rad. Phys. and Chem., 32:573.

Kenney-Wallace, G.A., and Wallace, S.C., 1983, IEEE J. Quant. Electr., 19:719.

Ladanyi, B.M., 1983, J. Chem. Phys., 78:2189.

Lotshaw, W., McMorrow, D., Kalpouzos, C., and Kenney-Wallace, G.A., 1987, Chem. Phys. Lett., 136:323; ibid., 150:138.

Madden, P.A., 1987, in Kenney-Wallace et al. (1987); 1984, in "Ultrafast Phenomena" ed. D.H. Auston and K.B. Eisenthal (Springer-Verlag, Heidelberg) p. 244.

Marsaglia, G., 1972, Ann. Math. Stat., 43:645.

McMorrow, D., Lotshaw, W., and Kenney-Wallace, G.A., 1988, IEEE J. Quant. Electr., 24:443.

Nezbeda, I., and Leland, T.W., 1979, J. Chem. Soc. Farad. Transac. 2, 75:193.

Patterson, G.D., and Carrol, J., 1982, J. Chem. Phys., 76:4316.

Ramshaw, J.D., Schoefer, D.W., Waugh, J.S., and Deutsch, J.M., 1971, J. Chem. Phys., 54:1239.

Ruhman, S., Kohler, B., Joly, A., and Nelson, K., 1987, Chem. Phys. Lett., 141:16 and references therein.

Ryckaert, J., Ciccotti, G. and Berendsen, H.J.C., 1977, J. Comp. Phys., 23:327.

Steele, W., 1984, "Molecular Liquids", ed. A.J. Barnes, W.J. Orville-Thomas, and J. Yarwood (Reidel, Dordrecht), p. 111.

Street, W.B., and Tildesley, D.J., 1978, Disc. Farad. Soc., 66:27

Tildesley, D.J., and Madden, P.A., 1981, Molec. Phys., 42:1137; ibid., 1983, 48:129.

Vasudev, R., Zare, R.N., and Dixon, R.N., 1984, J. Chem. Phys., 80:4863.

Verlet, L., 1967, Phys. Rev., 159:98.

Viellard-Baron, J., 1974, Molec. Phys., 28:809.

ELECTRON KINETICS IN NONPOLAR LIQUIDS -- ENERGY AND PRESSURE EFFECTS

Richard Holroyd*

Brookhaven National Laboratory
Department of Chemistry
Upton, NY 11973 USA

Just as electron mobilities differ for liquids that are otherwise similar in bulk properties, the rates of electron reactions often change by orders of magnitude from one nonpolar liquid to another. Both positive and negative activation energies are observed and, as shown recently, application of high pressure has large effects on electron kinetics (Munoz et al., 1987; Nishikawa et al., 1988). Clearly it is a challenge to fully understand this diversity of behavior, and a comprehensive theory to account for electron reactions is needed.

Many electron reactions are diffusion controlled, but here we are concerned with those reactions which do not fall into this category. Considerable progress has been made on the experimental side in identifying factors which influence electron kinetics. One key factor is the energy level of the electron at the bottom of the conduction band of the liquid, denoted V_o. Other important factors are entropy and volume changes. How V_o influences electron reactions is reviewed, and the significance of volume changes as revealed by recent pressure studies is discussed.

CONDUCTION BAND EFFECTS

In order to understand how the energy of the conduction band influences electron reactions we need to examine first how this band energy changes with conditions. The energy (relative to vacuum) of the bottom of the conduction band in which quasi-free electrons move is

---

* This research was carried out at Brookhaven National Laboratory under contract DE-AC02-76CH00016 with the U.S. Department of Energy and supported by its Division of Chemical Sciences, Office of Basic Energy Sciences.

denoted by V_o. Measurements of V_o have been described elsewhere (Holroyd, 1987). As shown in Table 1 there are considerable differences in V_o among nonpolar liquids. Liquids with more symmetrical molecules typically have the lowest energy conduction bands and unbranched alkanes have higher band energies. V_o also changes with temperature, decreasing as the temperature increases (Holroyd et al., 1975b). These temperature effects are associated with the decrease in density which accompanies the temperature increase. Eventually a minimum value of V_o is reached at an intermediate density and V_o then increases to zero at zero density (see e.g., Holroyd, 1987; Nakagawa et al., 1982; Reininger et al., 1982).

Electron Attachment Reactions

In nonpolar liquids electrons react with substances with either positive or slightly negative electron affinity. Alkanes, alkenes, amines, alcohols, and substances of very negative electron affinity are generally unreactive. Substances with slightly negative or near-zero electron affinity are special cases in that reactions occur but they are often reversible.

Rate constants for attachment of electrons to certain solutes exhibit a dependence on V_o. When the rate constants are plotted vs V_o for

Table 1. Band energies at 23 °C.

Liquids	V_o^a eV
Tetramethylsilane	-0.55
Neopentane	-0.43
2,2,4,4-Tetramethylpentane	-0.33
2,2-Dimethylbutane	-0.20
2,2,4-Trimethylpentane	-0.24
2,3-Dimethylbutene-2	-0.25
Tetrakis(dimethylamino)ethylene	-0.10
Cyclopentane	-0.19
Cyclohexane	+0.01
n-Pentane	0
Benzene	-0.14
n-Hexane	+0.10
n-Decane	+0.18

[a] Data is from Allen, 1976; Holroyd, 1987; and Holroyd et al., 1985; V_o is defined relative to the energy of the electron in vacuum.

the reactions of the electron which N_2O, ethyl bromide, perfluoroalkanes, and O_2, maxima are observed at values of V_o which are characteristic for each solute (Allen et al., 1975; Holroyd and Gangwer, 1980). For example, the rate of attachment to ethyl bromide is a maximum value in 2,2,4-trimethylpentane at V_o = -0.2 eV. To demonstrate that this maximum is truly a function of the conduction band energy and not some other property of the solvents, a study was done in mixtures of n-hexane and neopentane. In such mixtures the conduction band energy is a linear function of the mol fraction, f (Holroyd & Tauchert, 1974):

$$V_o(\text{mixt}) = V_o(\text{neoC}_5)\ f_{\text{neoC}_5} + V_o(\text{hex})\ f_{\text{hex}}\ . \quad (1)$$

The rate constant for reaction with ethyl bromide varies in these mixtures and is at a maximum when f_{hex} = 0.48 (Wada et al., 1977), corresponding to V_o = -0.17 eV, in excellent agreement with studies done in the pure solvents.

Such "resonances" are attributed, as in the gas phase, to the fact that the rate maximizes when the kinetic energy of the electron matches the energy level difference required to form the negative ion. In solution the energy at which the resonance occurs is different because of the polarization energy of the product negative ion, and a match is observed at negative values of V_o.

The maximum in the rate of attachment to perfluorocyclobutane occurs at V_o = -0.32 eV (Fig. 1). In the gas phase the attachment cross section for this reaction goes through a maximum at +0.34 eV (Spyrou et al., 1985). Thus, the maximum is shifted by 0.65 eV due to the polarization energy of the $C_4F_8^-$ anion. Preliminary results for other fluorocarbons are also shown in Fig. 1. The peaks for n-C_6F_{14} and n-C_5F_{12} occur around -0.20 eV. In the gas phase the maxima are at +0.6 eV (Spyrou et al., 1983) for parent anion formation indicating a shift of 0.8 eV.

This resonance-like dependence of the attachment rate on V_o also accounts for the unusual temperature dependence of these reactions. The rates decrease with increasing temperature (corresponding to a negative activation energy) in tetramethylsilane where V_o is below the maximum; the rates increase with temperature (positive activation energy) for solvents where V_o is above the maximum. Both effects are attributed to the fact that V_o decreases with increasing temperature.

Electron Detachment

Detachment of an electron from a neutral molecule (ionization) requires less energy in a nonpolar solvent than in the gas phase. Clearly, a major reason for this shift is the polarization energy of the product ion, denoted P^+. Another factor, however, is the conduction band

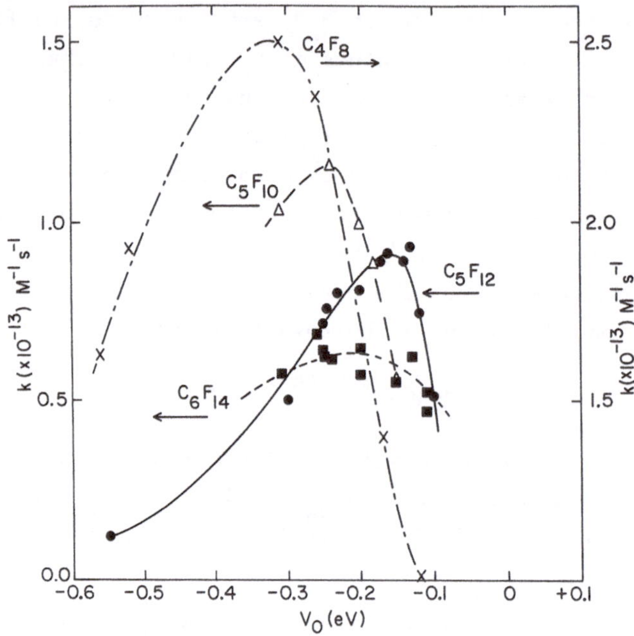

Fig. 1. Rate of attachment to fluorocarbons vs. V_o. Legend: (-·-), c-C_4F_8; (— —), C-C_5F_{10}; (—), n-C_5F_{12}; (---), n-C_6F_{14}.

energy. For TMPD (Holroyd & Russell, 1974) and other solutes the ionization threshold, E_{th}, in solution depends on V_o according to:

$$E_{th} \simeq I.P. + V_o + P^+ \; . \tag{2}$$

Since P^+ is nearly constant for liquids of comparable dielectric constant, a linear dependence of E_{th} on V_o is observed as shown in Fig. 2.

For reasons similar to the above, detachment of an electron from an anion (either thermally or by photodetachment) requires more energy in a

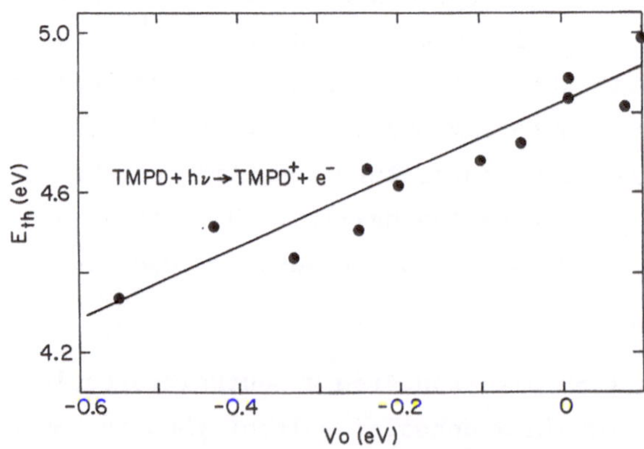

Fig. 2. Dependence of ionization thresholds on V_o. Results for TMPD (Holroyd and Russell, 1974).

nonpolar solvent than in the gas phase. Also, the threshold energy is dependent on V_0 in an analogous way; that is, for an anion:

$$E_{th}' \simeq E.A. - P^- + V_0 \ . \tag{3}$$

An interesting example of the effect of V_0 on a detachment reaction is found in a study of the reaction of electrons with p-benzoquinone (Holroyd, 1982). This reaction is fast and has a positive activation energy in n-pentane but an apparent <u>negative</u> activation energy in tetramethylsilane and neopentane (Fig. 3). The results are attributed to an equilibrium

$$e^- + BQ \longleftrightarrow BQ^{-*} \ , \tag{4}$$

$$BQ^{-*} \longrightarrow BQ^- \ , \tag{5}$$

with a short-lived excited state of the anion. Whereas the forward rate is fast in all solvents, the reverse rate, detachment from BQ^{-*}, is characterized by an activation energy which depends on the value of V_0. For tetramethylsilane for which V_0 is -.55 eV the activation energy is 0.25 eV and detachment occurs readily. For solvents where V_0 is much higher, the activation energy to reach either the conduction band or the trapped state is too large for detachment to compete with deactivation to the stable anion (reaction 5).

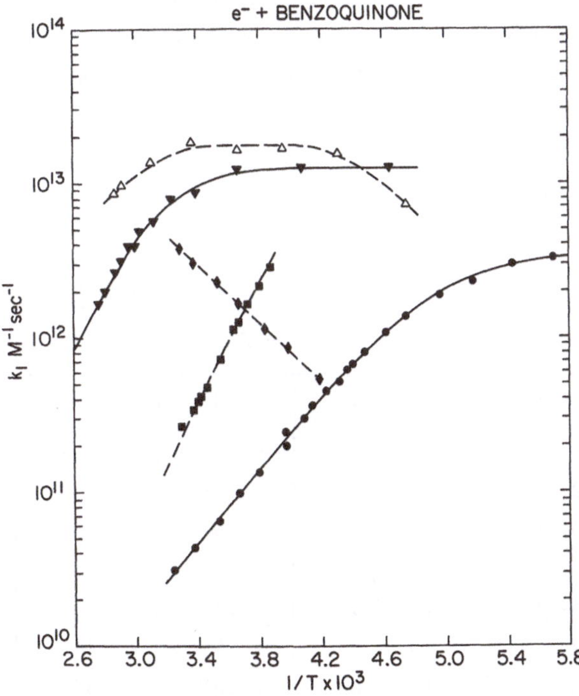

Fig. 3. Second-order rate constants, k, observed for reaction of the electron with p-benzoquinone in various solvents vs. 1/T: (●) $Si(CH_3)_4$, (■) neopentane, (◆) n-pentane, (▼) 2244-TMP, (▲) 224-TMP (Holroyd, 1982).

Detachment from ground-state anions has also been studied in equilibria reactions, Eq. (6) (Warman et al., 1975; Holroyd, 1977; Holroyd et al., 1975a):

$$X^- \longleftrightarrow X + e^- \quad . \tag{6}$$

Here X is a compound of low or slightly negative electron affinity. Certain anions autodetach readily in tetramethylsilane as solvent. As shown in Table 2 for CO_2 as solute the values of $\Delta G°$ and $\Delta H°$ increase as V_o increases. For styrene as solute the thermodynamic quantities increase with V_o between tetramethylsilane and 2,2,4-trimethylpentane but the increase in $\Delta G°$ and $\Delta H°$ is only about 0.1 eV between 2,2,4-trimethylpentane and n-hexane, whereas V_o differs by 0.34 eV between those two solvents. Here it must be noted that these equilibria data reflect the ground state energy and in n-hexane the trapped state is the ground state which lies about 0.2 eV below the V_o level for hexane.

ENTROPY EFFECTS

Studies of equilibria also show that entropy is an important factor in electron reactions. A large increase in entropy accompanies the detachment of an electron from an anion in solution. The $T\Delta S°$ term (see Table 2) is approximately 50% of the trap energy ($\Delta H°$) and thus constitutes a large driving force for the detachment reaction. The ordering of solvent molecules around the ion leads to a lower entropy state (Holroyd et al., 1975). Values of ΔS calculated from the relation

$$\Delta S = - \frac{\partial \Delta G}{\partial T} = \frac{e^2}{2R\varepsilon^2} \frac{\partial \varepsilon}{\partial T} \quad , \tag{7}$$

Table 2. Thermodynamic data for the reaction.

$$X^- \longleftrightarrow X + e^-$$

X	Solvent	$\Delta G°$	$\Delta H°$	V_o	298 $\Delta S°$	
					osb.	calc.
CO_2	2,2,4-TMP	+.55	+1.08	-.24	+.53	+.36
	neopentane	+.37	+ .97	-.43	+.60	+.56
	TMS	+.32	+ .92	-.55	+.60	+.47
Styrene	n-hexane	+.56	+1.13	+.10	+.58	+.24
	2,2,4-TMP	+.48	+ .99	-.24	+.50	+.23
	TMS	+.23	+ .67	-.55	+.43	+.31

Energies in eV; R = 2.3 Å for CO_2, 3.57 Å for styrene.
(From Holroyd et al., 1975a and Holroyd, 1977)

where R is the radius of the ion, roughly agree with the experimental entropy changes (final column, Table 2).

VOLUME EFFECTS

Studying the effect of pressure is another way of probing the mechanism of electron reactions in liquids. Recently, electron mobilities have been measured in several non-polar liquids up to pressures of 2.5 kbar (Munoz et al., 1985, 1986, 1987). The mobilities change with pressure in all liquids studied, but in different ways depending on the liquid.

For TMS, in which the mobility is high and the electron is quasi-free, a 15% decrease in mobility is observed as the pressure is increased to 2.5 kbar at 25°. At higher temperatures a somewhat greater decrease is observed. For 2,2-dimethylbutane and 2,2,4-trimethylpentane, pressure increases the mobility by 30-40% at room temperature, affects little change at temperatures near 60°, and decreases the mobility at higher temperatures.

In contrast, larger pressure effects are observed for low mobility liquids. Figure 4 shows the effect of pressure and temperature on the mobility in n-pentane. At 120° the mobility decreases about 4-fold as the pressure increases to 2.5 kbar. The results for n-hexane and 3-methylpentane are similar (Munoz et al., 1986 and 1987).

Electron trapping

In these low-mobility liquids electrons are trapped part of the time and transport occurs when the electron is activated to the quasi-free state. This two-state model assumes that an equilibrium reaction occurs between quasi-free and trapped electrons:

$$e_{qf} \longleftrightarrow e_{tr} \quad . \qquad (10)$$

If a volume change is involved in reaction 10, then according to le Chatelier's principle, pressure will shift the reaction to minimize the volume. The volume change can be deduced from the mobility in the following way. Transport occurs only when electrons are activated from traps to the quasi-free state where the mobility is μ_f. The observed mobility is therefore:

$$\mu = \mu_f/(1 + K) \quad , \qquad (11)$$

where K is the equilibrium constant for reaction 10. For low-mobility liquids it follows that:

$$d\ln K/dP = d\ln (1/\mu)/dP \qquad (12)$$

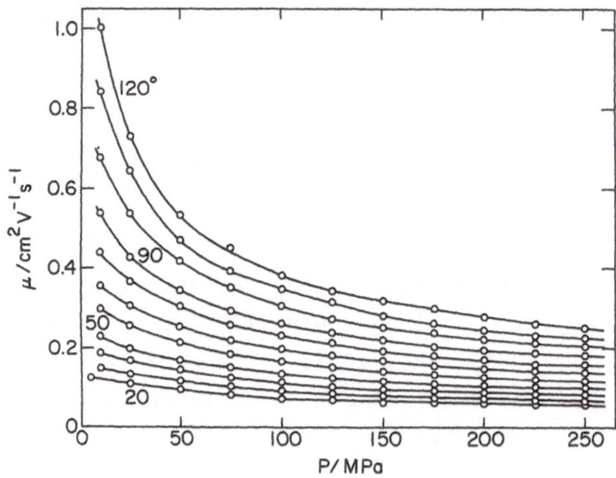

Fig. 4. Electron mobility isotherms for n-pentane as a function of pressure at the temperatures indicated (Munoz et al., 1987).

is approximately true providing μ_f does not vary with pressure. Since $\Delta V = -RT\ d\ln K/dP$, the volume change associated with electron trapping is:

$$\Delta V = RT\ d\ln \mu/dP \quad . \tag{13}$$

Volume changes for trapping calculated in this way are negative, increase in magnitude with increasing temperature, but decrease in magnitude with increasing pressure. At 25° and 1 atm the experimental volume changes are -22, -28, and -19 cc/mole for n-hexane, n-pentane and 3-methylpentane, respectively (Munoz et al., 1987).

Since electrostriction is so important in ionic reactions in nonpolar liquids (Le Noble and Klem, 1980), it is also important in the localization of an electron in nonpolar solvents. A classical model is assumed to estimate the magnitude of electrostriction. The volume change accompanying the charging of an ion, as given by Drude and Nernst (1894), is:

$$\Delta V_e = -\frac{z^2e^2}{2R}(1/\varepsilon^2)\,\partial\varepsilon/\partial P \quad , \tag{14}$$

where z is the number of charges, ε the dielectric constant of the liquid, and R the cavity radius. It must be kept in mind that the electron wavefunction extends beyond the cavity radius, R. Equation (14) can alternately be obtained from the Born energy of charging a sphere ($\Delta G = z^2e^2/2R\varepsilon$) and the relation $\Delta V = \partial\Delta G/\partial P$.

The magnitude of electrostriction around the trapped electron depends therefore on the term $(1/\varepsilon^2)\partial\varepsilon/\partial P$. This term was evaluated using the Clausius-Mosotti equation which is approximately valid over small pressure ranges (Brazier and Freeman, 1969). Since the isothermal compressibility (χ_T) is given by $-(1/V)\partial V/\partial P$, Eq. (14) can be rewritten as:

$$\Delta V_e = -\frac{z^2 e^2}{2R} \chi_T(\varepsilon - 1)(\varepsilon + 2)/3\varepsilon^2 \tag{15}$$

The values of $\chi_T(\varepsilon - 1)(\varepsilon + 2)/3\varepsilon^2$ were evaluated from density data. Figure 5 is a plot of the experimental volume changes for reaction 10 in n-hexane plotted vs. this quantity.

As predicted by Eq. (15) the derived volume changes depend linearly on $(1/\varepsilon^2)\partial\varepsilon/\partial P$ as given by $\chi_T(\varepsilon - 1)(\varepsilon + 2)/3\varepsilon^2$. The observed slope is about 60% of that predicted by Eq. (14); that is, $z^2e^2/2R$ (where values of R = 4.1 Å and z = 0.7 are assumed). This is reasonable considering the classical model assumed. Better agreement may be obtained from quantum considerations. On the other hand, this result may indicate either that electrons do not stay trapped long enough for the full electrostrictive effect to occur or that the fraction of the electron within the cavity is less than assumed.

One significant conclusion of this interpretation is that the trap changes when occupied by an electron. The cavity itself may not change size but the solvent around the trapped electron definitely constricts more than the solvent elsewhere.

Volume Changes in Electron Reactions

Previous studies of the effect of pressure on reactions of electrons have been done mainly in polar solvents. In water, electron reaction rates typically change at most by 30% for a 6-kbar pressure change (Hentz et al., 1972). However, the reaction of electrons with benzene in liquid ammonia is accelerated considerably by pressure; the volume change for this reaction is -71 cc/mole (Böddeker et al., 1969). Studies of this type have been used to provide information on the partial molar volume of the electron in polar solvents.

The reaction of the electron with CO_2 [Eq. (16)] is an equilibrium reaction in nonpolar solvents at ordinary pressure (Holroyd et al., 1975a)

$$e^- + CO_2 \underset{k_d}{\overset{k_a}{\longleftrightarrow}} CO_2^- \quad . \tag{16}$$

This reaction has recently been studied as a function of pressure and temperature in several solvents (Nishikawa et al., 1988). A pulse conductivity technique was used which allowed determination of both the attachment rate, k_a, and the detachment rate, k_d, and therefore the equilibrium constant $K = k_a/k_d$. The equilibrium shifts to the left with

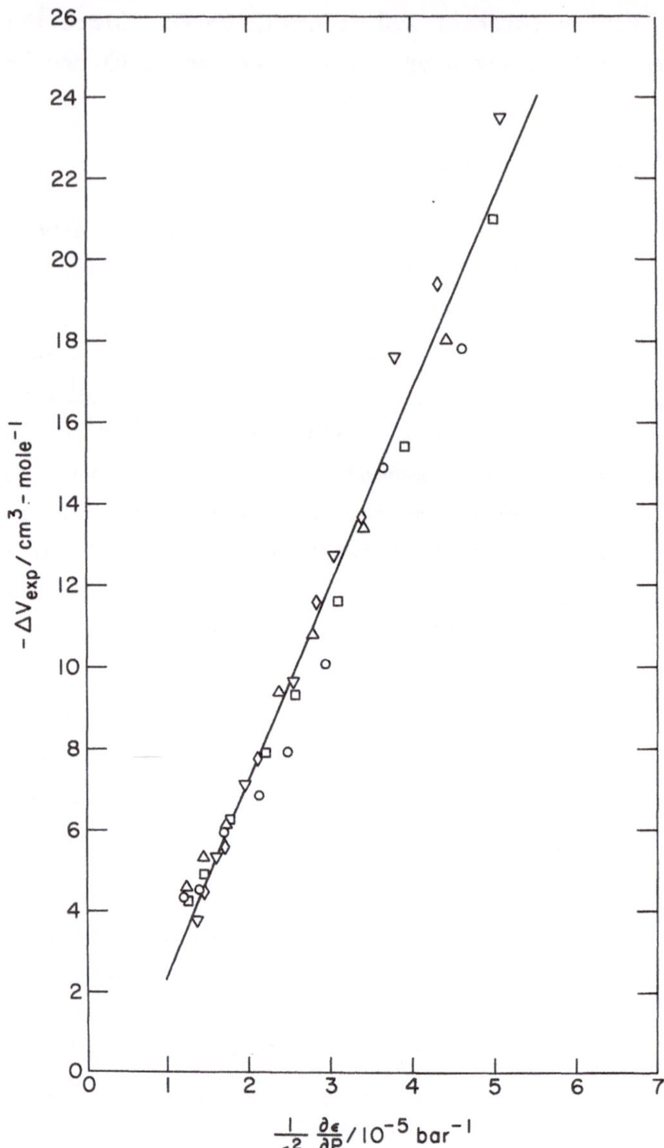

Fig. 5. Plot of ΔV_{exp} (from dlnμ/dP) vs $(\chi/3)[(\varepsilon-1)(\varepsilon+2)/\varepsilon^2]$ for n-hexane. Calculation of abscissas explained in text. Mobility data from Muñoz et al. (1986); ○ 18°, □ 30°, △ 50°, ▽ 70°, ◊ 90°.

increasing temperature, consistent with an exothermic reaction ($\Delta H_r \simeq -1$ eV in nonpolar liquids). The equilibrium shifts to the right with increasing pressure which indicates a negative volume change. The effect of pressure is dramatic; K increases approximately 10-fold for each 300-bar increase in pressure. From the slopes of ln K vs. pressure plots, ΔV_r was evaluated and found to be between -133 and -290 cc/mole, depending on solvent, pressure, and temperature.

These large volume changes are attributed mainly to electrostriction. It is customary to divide volume effects into those associated with changes in volume of the molecules reacting, or intrinsic volume changes, ΔV_i; and those associated with changes in volume of the solvent, or electrostriction effects, ΔV_e. Thus the overall change for a reaction is:

$$\Delta V_r = \Delta V_i + \Delta V_e \quad . \tag{17}$$

In solvents where the electron is quasi-free like tetramethylsilane, $V_i(e)$ is zero. The difference in volumes of CO_2 and CO_2^- is expected to be less than 1 cc/mole so that essentially $\Delta V_i \approx 0$ for tetramethylsilane as solvent. The volume change for reaction 16 is therefore due almost entirely to electrostriction by the charged species CO_2^- and the electron if trapped:

$$\Delta V_e = \Delta V_e(CO_2^-) - \Delta V_e(e_{tr}^-) \quad . \tag{18}$$

The largest term in Eq. (18) is $\Delta V_e(CO_2^-)$. The magnitude of the volume change will depend on $(1/\varepsilon^2)\partial\varepsilon/\partial P$ as shown by Eq. (14). This term is small in a solvent like water but increases in less polar solvents and is 100-fold larger in alkanes (see Table 3). Whereas the extent of electrostriction by the ions formed in the dissociation of water ($H_2O \longleftrightarrow H^+ + OH^-$) is only -1 to -2 cc/mole (Le Noble et al., 1980), in non-polar solvents ΔV_e may be 100-times larger.

SUMMARY

The energy of the bottom of the conduction band is one factor influencing electron kinetics. Anions formed by attachment are stabilized in solution, and volume changes are manifestations of the free energy

Table 3. Electrostrictive effects.

Solvent	ε	$(1/\varepsilon^2)\partial\varepsilon/\partial P$* bar^{-1}
Water	80.0	0.6×10^{-6}
Methanol	33.0	3.2
Benzene	2.3	25.0
Ethyl ether	4.3	59.0
3-Methylpentane	1.85	59.0
2,2,4-TMP	1.94	48.0
TMS	1.84	80.0

* From Hamaan (1974) for polar solvents; calculated from Eq. (15) for nonpolar solvents.

associated with polarization by the anions. This volume constriction around anions explains why equilibria shift toward the ions with increasing pressure. At the same time electrostriction around the ion corresponds to a state of low entropy, and is the source of the large increase in entropy on electron detachment -- an effect which helps shift equilibria reactions back toward the electron.

REFERENCES

Allen, A.O., Gangwer, T.E., and Holroyd, R.A., 1975, Chemical reaction rates of quasifree electrons in nonpolar liquids. II, J. Phys. Chem., 79:29.

Allen, A.O., 1976, Drift mobilities and conduction band energies in dielectric liquids, NSRDS-NBS-58.

Bőddeker, K.W., Lang, G., and Schindewolf, U., 1969, ESR measurements under pressure: effect of pressure on chemical equilibria involving solvated electrons, Angew. Chem., internatl. ed., 8:138.

Brazier, D.W., and Freeman, G.R., 1969, The effects of pressure on the chemistry, dielectric constant, and viscosity of several hydrocarbons and other organic liquids, Can. J. Chem., 47:893.

Drude, P., and Nerst, W., 1894, Uber elektrosticktion durch freie ionen, Z. Phys. Chem., 15:79.

Hamaan, S.D., 1974, Electrolyte solutions at high pressure, in: "Modern Aspects of Electrochemistry", V9:47, B.E. Conway and J.O. Bockris, ed., Plenum Press, New York.

Hentz, R.R., Farhataziz, and Hansen, E.M., 1972, Pulse radiolysis of liquids at high pressure II. Diffusion-controlled reactions of the hydrated electron. J. Chem. Phys., 56:4485.

Holroyd, R.A., and Russell, R.L., 1974, Solvent and temperature effects in the photoionization of tetramethyl-p-phenylenediamine, J. Phys. Chem., 78:2128.

Holroyd, R.A., and Tauchert, W., 1974, On the relation of electron mobility to the conducting state energy in nonpolar liquids, J. Chem. Phys., 60:3715.

Holroyd, R.A., Gangwer, T.E., and Allen, A.O., 1975a, Chemical reaction rates of quasi-free electrons in non-polar liquids. The equilibrium $CO_2 + e^- \longleftrightarrow CO_2^-$, Chem. Phys. Lett., 31:520.

Holroyd, R.A., Tames, S., and Kennedy, A., 1975b, Effect of temperature on conduction band energies of electrons in nonpolar liquids, J. Phys. Chem., 79:2857.

Holroyd, R.A., 1977, Equilibrium reactions of excess electrons with aromatics in nonpolar solvents, Berichte der Bunsen-Gesellschaft fur Phys. Chemie, 81:298.

Holroyd, R.A., and Gangwer, T.E., 1980, Electron attachment to oxygen and other solutes in non-polar liquids, Radiat. Phys. & Chem., 15:283.

Holroyd, R.A., 1982, Electron attachment to p-benzoquinone and photodetachment from benzoquinone anion in nonpolar solvents, J. Phys. Chem., 86:3541.

Holroyd, R.A., Ehrenson, S., and Preses, J.M., 1985, Electron mobility, ion yields and photoconductivity in liquid tetrakis(dimethylamino)-ethylene, J. Phys. Chem., 89:4244.

Holroyd, R.A., 1987, The electron, its properties and reactions in: "Radiation Chemistry Principles and Applications", Farhataziz and Rodgers, M.H.J., ed., V.C.H. Publishers, New York, New York, p.201ff.

Le Noble, W.J., and Klem H., 1980, Chemistry in compressed solutions, Angewandte Chemie, 19:841.

Munoz, R.C., Holroyd, R.A., and Nishikawa, M., 1985, Effect of high pressure on the electron mobility in liquid n-hexane, 2,2-dimethylbutane and tetramethylsilane, J. Phys. Chem., 89:2969.

Munoz, R.C., and Holroyd, R.A., 1986, The effect of temperature and pressure on excess electron mobility in n-hexane, 2,2,4-trimethylpentane and tetramethylsilane, J. Chem. Phys., 84:5810.

Munoz, R.C., Holroyd, R.A., Itoh, K., Nakagawa, K., Nishikawa, M., and Fueki, K., 1987, Excess electron mobility in hydrocarbon liquids at high pressure, J. Phys. Chem., 91:4639.

Nakagawa, K., Ohtake, K., and Nishikawa, M., 1982, Conduction band energy in dense propane fluid, J. Electrostatics, 12:157.

Nishikawa, M., Itoh, K., and Holroyd, R.A., 1988, Effect of pressure on the reaction of electrons with CO_2 in nonpolar solvents, J. Phys. Chem., 92:xxxx (1988).

Reininger, R., Asaf, U., and Steinberger, I.T., 1982, The density dependence of the quasi-free electron state in fluid xenon and krypton, Chem. Phys. Lett., 90:287.

Spyrou, S.M., Sauers, I., and Christophorou, L.G., 1983, Electron attachment to the perfluoroalkanes n-C_NF_{2N+2} (N=1--6) and i-C_4F_{10}, J. Chem. Phys., 78:7200.

Spyrou, S.M., Hunter, S.R., and Christophorou, L.G., 1985, A study of the isomeric ependence of low-energy (< 10 eV) electron attachment: perfluoroalkanes, J. Chem. Phys., 83:641.

Wada, T., Shinsake, K., Namba, H., and Hatano, Y., 1977, Electron reactivity in liquid hydrocarbons mixtures, Can. J. Chem., 55:2144.

Warman, J.M., deHaas, M.P., and Hummel, A., 1975, Concerning the equilibrium $e^- + \text{biphenyl} \rightleftarrows \text{biphenyl}^-$ in liquid tetramethylsilane, Chem. Phys. Lett., 35:383.

PHOTOCONDUCTIVITY, CONDUCTION ELECTRON ENERGIES, AND EXCITONS IN SIMPLE FLUIDS

I. T. Steinberger

The Hebrew University
Racah Institute of Physics
Jerusalem 91904, Israel

INTRODUCTION

The objective of this paper is to present nonpolar fluids having very simple electronic properties. It will be shown that in liquids of the heavier rare gases electron energies and electron transport are very similar to those in the corresponding crystalline solids, so that for these liquids even the nomenclature characterizing electronic states in crystals has a distinct and well-defined meaning. This will be shown with respect to the band gap, conduction band minimum, and excitons. Remarkably, for the dense liquids of argon, krypton, and xenon there is no need to invoke concepts typical to amorphous semiconductors, like different optical and mobility gaps. Thus, these liquids may serve as reference models for more involved nonpolar liquids, e.g., of hydrocarbons. Moreover, since the density of a fluid can be easily varied, studying the evolution of the electronic properties of fluid argon, krypton, and xenon with the increase of the density from that of a dilute gas up to the triple-point liquid can, in fact, serve to define the conditions for the coming into existence of "crystal-like" electronic behavior.

The pivot of this presentation will be the interrelation of the electronic properties and their evolution with increasing density. Of central importance for this subject is the variation of the electron mobility with changing pressure and temperature. This is extensively dealt with in G.R. Freeman's lecture in this Proceedings and is therefore only briefly referred to here. Moreover, because of the limited scope, a wide and extensively studied field involving doped rare gas liquids and solids will also be only briefly mentioned; a review of this field is included in the book by Schwentner et al., 1985. For the same reason, the literature quoted will be less comprehensive than in a usual review paper. The reader should also consult the very detailed article by Davis and Brown (1975) reviewing in depth earlier work on electrons in nonpolar fluids.

SOLID AND LIQUID RARE GASES

The first indications for the strong similarity between the electronic properties of rare gas liquids and solids came from measurements of the time-of-flight (TOF) electron mobility (Miller et al., 1968). These authors found that the zero-field mobility μ_o (i.e., extrapolated to zero applied field) in the triple-point liquids Ar, Kr, and Xe is very high, of the order of $10^3 cm^2$/volt-sec. In fact, the value of μ_o in these liquids is only by a factor of about two smaller than in the respective solids, as expected from theory (Schnyders et al., 1966; Cohen and Lekner, 1967) on the basis of the known long-wavelength structure factors S(0). Shinsaka and Freeman (1974) showed that neopentane behaves similarly. Thus, the zero-field mobility values in these liquids are roughly predictable on the basis of theory and known parameters.

The phase transition solid/liquid has no dramatic effect on the electronic states of Ar, Kr, and Xe. In fact, the changes observed can be essentially acounted for by the density change involved. For example, Fig. 1a shows the density dependence of electronic transitions in solid Xe (Steinberger and Asaf, 1973) and their extrapolation to liquid densities, while Fig. 1b shows density dependence of electronic transitions in fluid Kr (Laporte et al., 1987) and their extrapolation to the solid. In both figures the observed values fit well the smooth extrapolated curves. The nature of the transitions observed will be discussed below. The continuity of both the electronic excitation energies and electron transport properties across the phase transition is somewhat similar to the behavior of solid and liquid metals (Faber, 1972), but it is in sharp contrast with the very marked differences normally encountered between crystalline and amorphous semiconductors (Mott and Davis, 1979; Zallen, 1983). Other experimental data on electronic states in rare gas solids and liquids (e.g., Reininger et al., 1984b; Laporte et al., 1985) also show this continuity.

PHOTOCONDUCTIVITY

In principle, the experiment consists of applying a potential difference across the normally insulating sample, irradiating the sample with monochromatic light, and varying the wavelength. The aim is to find the photon energy I_{th} of the photoconduction threshold. It is assumed that in a pure sample $I_{th} = E_g$, the minimum energy to "free" an electron

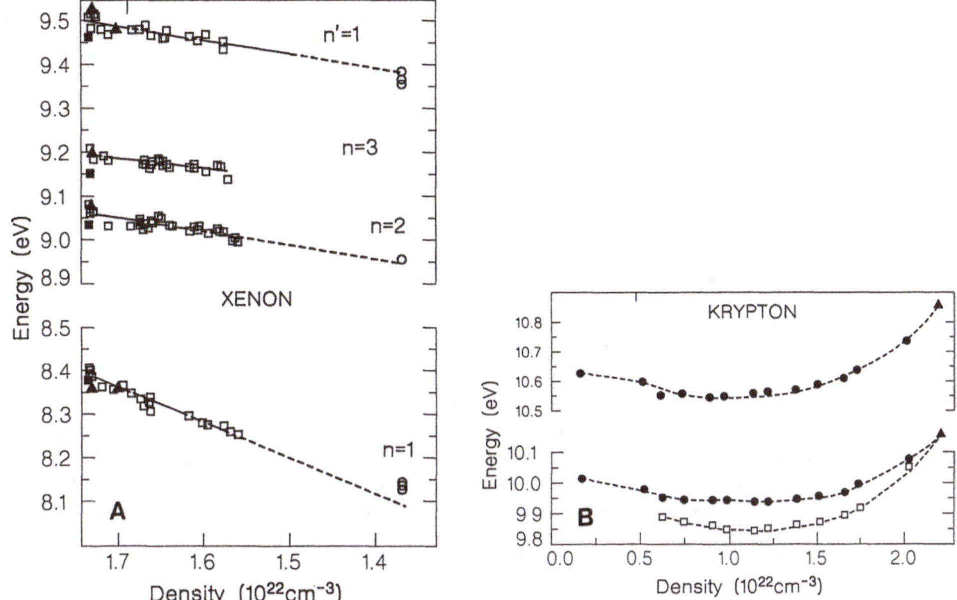

Fig. 1(a). The density dependence of various electronic transitions in solid xenon, extrapolated to the triple-point liquid. □ - solid, Steinberger and Asaf, 1973; ○ - liquid, ditto; ■ - Steinberger et al., 1970; ▲ - Baldini, 1962.

1(b). The density dependence of various electronic transitions in fluid krypton, extrapolated to the solid. Upper curve -- $4p^5 4s\ ^1P_1$ and associated n' = 1 exciton (not resolved); center -- $4p^5 4s\ ^3P_1$; lowest -- n = 1 exciton.

(= band gap in solid-state parlance). In many of the experiments described here the incident photons were supplied by synchrotron radiation. Photoconductivity of rare gas liquids has to be studied using a closed cell because of the high vapor pressure of these liquids. The need of a window sets an upper photon energy limit to the measurements with the LiF absorption edge (about 11.8 eV at room temperature). As a consequence, among the pure rare gases only with xenon is it possible to follow the variations of I_{th} in the whole density range from the dilute gas up to the triple-point liquid (Reininger et al., 1983a). However, in rare gas fluids doped with atoms or molecules having an ionization potential lower than the host, I_{th} refers to the minimum energy E_g^i needed to free an electron from the impurity level. This lies appreciably lower than the LiF cut-off for many rare gas fluid-dopant combinations making possible performing several sets of measurements (Reininger et al., 1984a, 1984b, and 1985a). Figure 2 shows one of the cells that has been used for photoconductivity. It should be mentioned that for photoconductivity and spectral transmission studies of gasses, at pressures up to 400 millibar and wavelengths well below the LiF cut-off, cells equipped with very thin

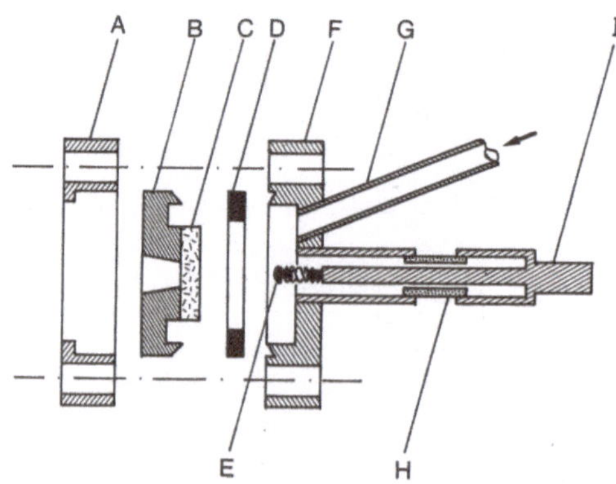

Fig. 2. Cross section of a sample cell. A - retaining ring; B and F - parts of "Conflat" seal; C - LiF window, glued to B, with sputtered gold electrodes on its inner surface; D - copper gasket; E - spring and contact head set (only one shown; the cell contains two such sets); G - filling tube; H - ceramic insulator feedthrough; I - central stalk, welded to the feedthrough.

indium windows were constructed and successfully used (Reininger, 1985; Köhler, 1987).

Figure 3a shows several representative photoconductivity excitation spectra in pure xenon; Fig. 3b for argon doped by ethane. In pure liquid and solid xenon and krypton, the value of E_g as determined by means of the photoconductivity threshold is in full accord with its value as found from the limit of Wannier-Mott excitonic series (Asaf and Steinberger, 1974; Laporte et al., 1985; see also below). This fact furnishes independent support for the applicability of solid-state concepts like energy bands and excitons for the liquids discussed. Results on I_{th} as a function of density in pure xenon (Reininger et al., 1983a) from about 10^{21} cm^{-3} to the triple point showed a monotonuous decrease of I_{th} with increasing density, in contrast with the pronounced maxima and minima exhibited by the electron mobilities of simple fluids (e.g., Holroyd and Cippolini, 1978; Huang and Freeman, 1978 and 1981; Jacobsen et al., 1986). At densities below about 10^{21} cm^{-3} photoionization processes due to the formation of Xe_2^+ molecular ions and free electrons dominate the photoresponse (Laporte et al., 1983; Reininger et al., 1985b). Recently the photoconductivity thresholds were compared (Steinberger and Baer 1987) with a fundamental theory based on the adaptation of the mean spherical approximation to electronic states (Chandler et al., 1982). It was noted that the final states of the electron and hole created involve polarization of the medium and influence the value of E_g, while the theory does not deal with such effects. In fact (see e.g., Raz and Jortner, 1971),

$$E_g = I_c + V_o + P_+ + E_v , \qquad (1)$$

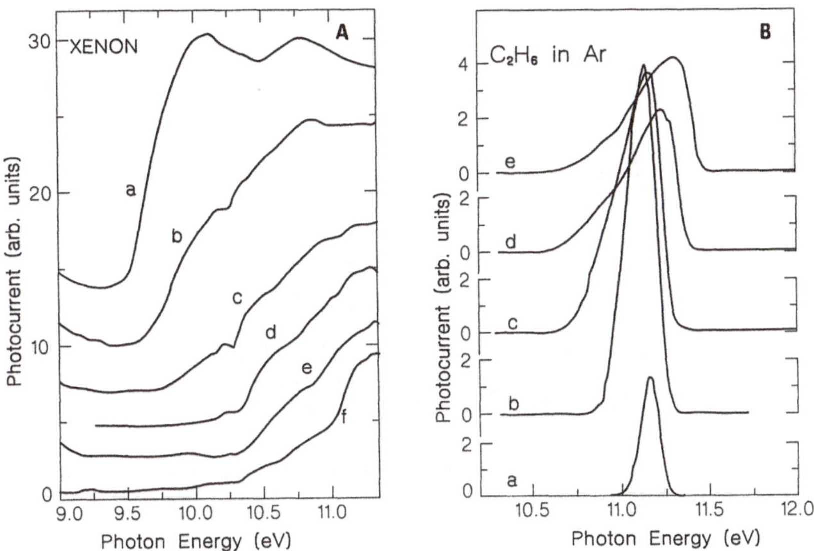

Fig. 3(a). Photoconductivity excitation spectra of fluid xenon at several number densities (in $10^{22} cm^{-3}$) and temperatures (K); a, 1.09, 242.3; b, 0.83, 284.3; c, 0.62, 296.3; d, 0.41, 296.5; e, 0.33, 295.0; f, 0.23, 296.5. Adapted from Reininger et al., 1983a. The curves are staggered vertically for clarity.

3(b). Photoconductivity excitation spectra of ethane-doped fluid argon at different number densities (in $10^{22} cm^{-3}$) of the host: a, 0.52; b, 0.75; c, 1.38; d, 1.73; e, 2.0. Adapted from Reininger et al., 1984a.

I_c being the first ionization potential of the free atom, V_o the energy of the conduction electron with respect to the vacuum level, P_+ the polarization energy of the hole (the method for the determination of these two quantities will be given below), and E_v an unknown correction term dealing with the fact that the top of the uppermost filled band is shifted with respect to the ground state of the free atom. Figure 4 shows, as a function of density, the low-energy threshold of the calculated absorption band developing from the first ionization limit of the isolated xenon atom (12.127 eV), as well as $E_g - P_+$ and $I_c + V_o$. It is seen that the theoretical curve is sandwiched between the two experimental ones; one should also note that $E_g - P_+$ extrapolates at low densities to 11.10 eV, i.e., the ionization energy of the xenon dimer, this being the major source of discrepancy between the two experimental curves. In view of the fundamental and simple nature of the theory not employing adjustable parameters, the existence of a rough agreement between theory and experiment is remarkable. However, the agreement exists only for excitations leading to a delocalized electronic state. Excitations which leave the electron on the parent atom exhibit in pure and doped rare gas fluids very often a blue shift with increasing density (e.g., Fig. 1b in

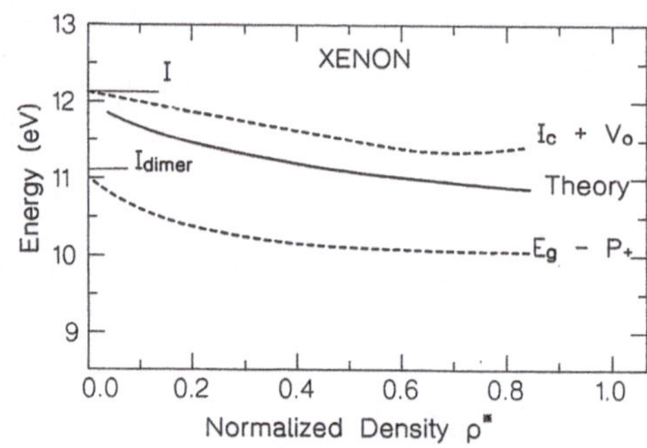

Fig. 4. Calculated values (Chandler et al., 1982) of the absorption edge evolving from the first ionization limit of fluid xenon compared with experimental data (Reininger et al., 1983a), as a function of the normalized density, $\rho^* = \rho\sigma^3$, ρ being the number density. The value of the Lennard-Jones parameter $\sigma = 0.398$nm. From Steinberger and Baer, 1987.

this paper; Messing et al., 1977), while the theory always predicts a red shift. The reason for this discrepancy seems to be that the theory in its present form does not deal in full with correlation effects (Steinberger and Baer, 1987); such effects can be expected to be much more important for the case when the excited state is a localized one rather than for an essentially free electron.

For impurity photoconductivity a relation similar to Eq. (1) holds (Raz and Jortner, 1969):

$$E_g^i = I_c^i + V_o + P_+^i \; ; \qquad (2)$$

here E_g^i is the minimum energy needed to liberate an electron from the impurity (atom or molecule) into the lowest conduction level ("conduction band") of the host, I_c^i is the first ionization potential of the free impurity atom or molecule, V_o has the same meaning as above, and P_+^i is the polarization energy due to the ionized impurity atom or molecule. Since I_c^i is known from the literature, E_g^i is determined from the threshold of impurity photoconduction, and V_o from photo injection measurements (see below), Eq. (3) furnishes an experimental procedure for finding P_+^i (Reininger et al., 1984a, 1984b, and 1985a). In fact, impurity photoconductivity excitation spectra were measured as a function of the host density for several combinations of Ar, Kr, or Xe host, and small organic molecule or xenon impurities (Reininger et al., 1984a, 1984b, and 1985a). The measurements furnished the experimental values for P_+^i and made possible comparisons between hosts differing in their dielectric properties, but all doped with the same impurity, as well as between impurity molecules of various molecular diameters in the same host. Figures 5a and 5b

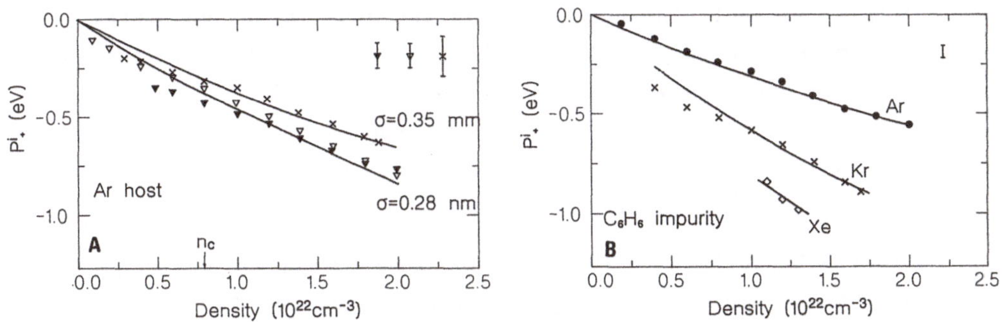

Fig. 5(a). Hole polarization energy P^i_+ for fluid argon doped with: ▼ - ethane, ∇ - propane, and X - butane, determined from Eq. (2) in the text, as a function of the hose density. The lines represent Eq. (3), with values of σ as indicated. After Reininger et al., 1984a.

5(b). Hole polarization energy P^i_+ for fluid argon, krypton, and xenon doped by benzene as a function of the host density. The lines represent Eq. (3). After Reininger et al., 1984a.

show some examples. It was found that the density dependence of P^i_+ can be represented in most cases with a simple electrostatic formula, taking into account a continuum host with a dielectric constant ε and a positive point charge, with a cavity of radius σ ("Born charging energy"):

$$P^i_+ = \frac{e^2}{2\sigma}\left(1 - \frac{1}{\varepsilon}\right) . \tag{3}$$

In the graphs σ was regarded as an adjustable parameter; its value increases with the increase of the molecular size of the impurity.

ENERGY OF THE CONDUCTION ELECTRONS

The energy V_0 of a conduction electron is equal to the electron affinity, but has the opposite algebraic sign. For a crystalline semiconductor or insulator it is the electron energy at the bottom of the conduction band, referred to the vacuum level. V_0 has been traditionally determined (e.g., Woolf and Rayfield 1965; Holroyd and Allen, 1971; Tauchert et al., 1977; for a different method see Broomall et al., 1976) from the apparent change of the work function of a metal when in contact with the substance investigated. The work functions, in turn, with and without the substance are usually determined photoelectrically. The cell is forward biased and the photoelectric current from the cathode (passing through the vacuum or injected into the sample) is recorded as a function of the energy of the incident photons; the onset is determined by means of the Fowler plot (Holroyd and Allen, 1971; Tauchert et al., 1977), taking into account the energy and momentum distribution of the electrons in the metal. In fact, the current collected and its photon energy dependence depends on the details of the electron transport in the inhomogeneous field distribution within the sample as well, (e.g., Kuntz and Schmidt, 1982); however, combination of the two distributions to calculate the actually collected current seems to be a rather difficult task that has not yet been attempted.

Another difficulty with this method of determination of V_0 is the fact that the measured work function of a metal is very sensitive to adsorbed atoms and molecules. In fact, even rare gases change the result of such measurements when adsorbed on atomically clean single-crystal metal surfaces (e.g., Chen et al., 1984). It should be mentioned in this context that the work function of the metal measured photoelectrically on the polycrystalline electrodes used is markedly lower than the values quoted in the literature for clean single-crystal surfaces. The adsorbtion and consequent change of the "reference" work function would not invalidate the correctness of the V_0 measurement, provided the degree of coverage of the electrode surface is roughly the same both in the empty cell and in the filled one, due to stabilization of the adsorbate on the electrode. However, this is very difficult to obtain at low pressures of the investigated fluid.

Figure 6a shows values of V_0 measured by the photo-injection method as a function of fluid density for fluid Kr; Fig. 6b is the same type of

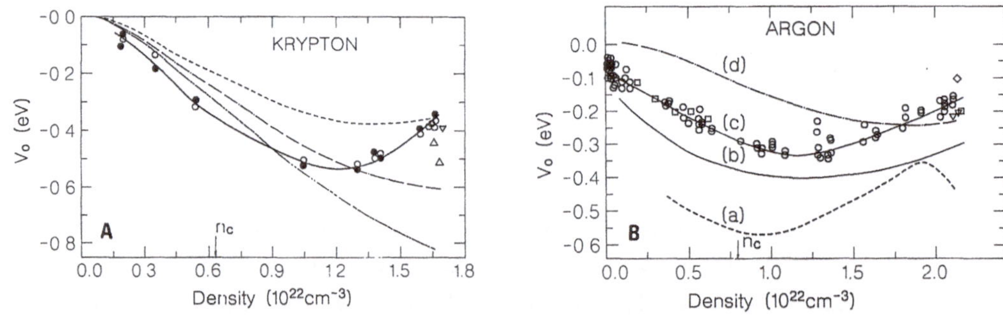

Fig. 6(a). V_0 as a function of density in fluid krypton: o and • - Reininger et al., 1982; Δ - von Zdrojewski et al., 1980; ∇ - Tauchert et al., 1977. Solid curve - guide for the eye along the experimental points; the other curves are calculated from the model by Springett et al., 1968, with the following values of the hardcore radius $\tilde{a}$: ---- 0,1073nm; -·-· 0.1030nm; -··-··- 0.099nm.

6(b). V_0 as a function of density in fluid argon. Experimental points: o - Reininger et al., 1983b; ∇ - Tauchert et al., 1977; □ - Messing and Jortner, 1977; ◇ - von Zdorjewski et al., 1980. Curves: (a) Basak and Cohen, 1979; (b) Plenkiewicz et al., 1986; (c) Eq. 5 in Reininger et al., 1983b; (d) Springett et al., 1968.

graph for fluid Ar. It is seen that the graphs vary smoothly, exhibiting a minimum. This behavior was qualitatively predicted by Springett et al. (1968). They took into account in a self-consistent manner (Lekner 1967) the long-range screened attraction between an electron and an atom as well as the short-range repulsion due to orthogonality and correlation effects. For the repulsion these authors assumed, for simplicity, a hard-core pseudopotential. As seen in Fig. 6a, with appropriate choice of the hard-core diameter one obtains a minimum, but the agreement between theory and experiment is very crude. A considerable improvement in the theory has been achieved only recently by Plenkiewicz et al. (1986). These authors used the same type of attractive term in the expression for V_o as Springett et al. (1968), but with a modern, first principles pseudopotential calculated by the local density functional approximation (Kohn and Sham, 1965). Clearly, this theory reproduces the experimental results quite satisfactorily (Fig. 6b), though the values theoretically predicted are somewhat too low. It would be very interesting to perform such calculations for other simple fluids as well.

Measurements of V_o are of relevance for testing theories on the electron mobility in fluid rare gases, since recent theories (Basak and Cohen, 1979; Ascarelli, 1986) on the electron mobility μ_o use the values of the derivatives of the V_o-vs.-density (ρ) curves as parameters for the calculations (Reininger et al., 1983b). As an illustration we present the results on fluid argon. In Fig. 7, experimental $\mu_o(\rho)$ curves (by Huang and by Freeman, 1981) are compared with several theoretical calculations based on measured values of V_o (Fig. 6b). It is seen that the Basak-Cohen theory renders properly the overall shape of the curve, but it fails quantitatively at high densities, and even more so in the

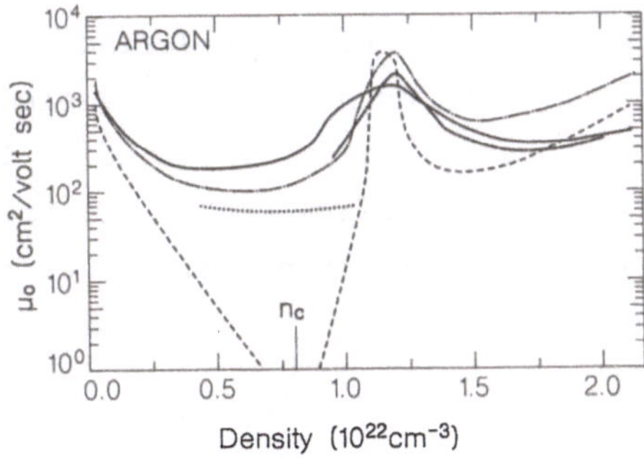

Fig. 7. Zero-field electron mobility in fluid argon: ——— experimental results by Huang and Freeman, 1981. Theoretical curves, based on parameters from Fig. 6(b): ----- Basak and Cohen, 1979; ····· Lekner and Bishop, 1973; ·-·-·- Nishikawa, 1985; --·--·--· Ascarelli, 1986 (with the assumption that the value of the effective mass, m^*/m_e, is 0.9 at $\rho = 2.1 \times 10^{22}\text{cm}^{-3}$ and 0.38 at $0.86 \times 10^{22}\text{cm}^{-3}$, varying linearly between these two limits).

critical region, while the latest theory by Ascarelli (1986) gives a satisfactory fit in the whole region where it has been applied. Figure 7 shows two more attempts to improve theory so as to obtain improved correspondence with experiment (Steinberger and Zeitak, 1985). In the approach suggested by Nishikawa (1985), the adiabatic instead of the isothermal compressibilities are used, bringing an improvement, especially in the critical region, but the substitution seems to be unjustifiable theoretically. The other approach is essentially that by Lekner and Bishop (1973), based on an expansion of the structure factor S(q) near the critical point. We refer the reader again to the paper by G.R. Freeman in this volume dealing with mobility problems.

EXCITONS

It will be recalled that an exciton is "a quantum of electronic excitation traveling in a periodic structure whose motion is characterized by a wave vector" (Dexter and Knox, 1965). The motion is usually attributed to resonant energy transfer processes. Resonant energy transfer is also known to be prominent among processes leading to fluorescence of organic solutions (Birks, 1970). On the other hand, molecular excitations broadened and shifted in the liquid, as compared with the solid, have been attributed to excitons (e.g., Inagaki, 1972; Eckhardt and Nichols, 1972; Le Sar and Kopelman, 1977). Though these assignments seem to be reasonable, broadening and shifting may have diverse causes. Thus, it is of interest to strive to a correct identification of excitons in liquids, specifying characteristic features. This has been indeed done in the liquids of the heavier rare gases, involving widening the scope of the exciton concept beyond the original definition.

An early discussion of the theoretical problem of Mott-Wannier excitons in liquids was given by Rice and Jortner (1967). We recall that these excitons can be regarded as non-localized electron-hole pairs held together by their Coulomb interactions weakened by the polarization of the medium. The Mott-Wannier exciton peaks, E_n, are usually not related to any atomic excitations, and their positions are given by the hydrogen-like formula

$$E_n = I_g - R^*/n^2 \qquad n = 1, 2, \ldots, \text{ where} \tag{4a}$$

$$R^* = 13.6 \frac{m^*}{\varepsilon^2}(\text{eV}) , \tag{4b}$$

m^* being the reduced effective exciton mass (in terms of the free electron mass), and ε the (low-frequency) dielectric constant. In accord with the tentative prediction of Rice and Jortner (1967), the n=2 level is indeed observed in dense liquid xenon and krypton (Beaglehole, 1965; Asaf and Steinberger, 1971; Laporte et al., 1985), and its energy (Fig. 1a) is in excellent accord with the extrapolation of the n=2 peak in the solid to the density of the liquid. In fact, this peak (at about 9 eV) can be observed not only in the vicinity of the triple point, but at somewhat lower densities as well; see Fig. 8d. The appearance of the exciton above a certain miminum density is explained considering that the exciton would need a sufficiently dense local environment (fluctuation) of extensions larger than its Bohr radius (10-12 nm, depending on the density; see Laporte et al, 1980). Thus the thereoretical concept has a sound experimental basis with regard to this Wannier-Mott-type exciton.

Turning to the n=1 member of the series [Eq. (4a)], a peculiarity appearing in rare-gas solids should be pointed out. The calculated Bohr radius of this exciton is small, of the order of the nearest-neighbor distance. Therefore, central-cell corrections modify its position as compared with the rest of the series (for a review, see Sonntag, 1977). Moreover, the energies of these peaks are rather close to those of atomic resonance lines $^3P_1$ and $^1P_1$. These properties are similar to those of Frenkel excitons and, hence, the designation "intermediate exciton" is often used.

Figure 1a shows that for the n=1 and n'=1 intermediate excitons too, extrapolation to liquid densities predicts correctly the position of an electronic excitation of the liquid. However, since even the transitions in the solid seem to be related to atomic transitions, the question arises whether their counterpart in the liquids can be distinguished at

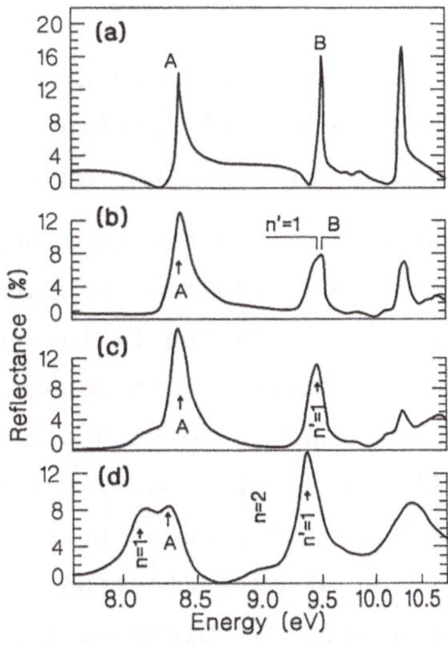

Fig. 8. Reflectance of xenon-MgF_2 interface at four number densities (in $10^{22} cm^{-3}$): (a) 0.217; (b) 0.47; (c) 0.677; (d) 1.24. From Laporte et al., 1980.

all from atomic lines broadened by molecular interactions. If the "exciton bands" were simply strongly perturbed atomic lines it would be difficult to associate with them resonant energy transfer, since, at least at low pressures, the probability for such transfer would be extremely small.

The answer to this question is found from the density dependence of excitonic bands. We refer to Fig. 8, showing the reflection spectrum of MgF_2/Xe interfaces for several xenon densities (Laporte et al., 1980). At the lowest density (a) (= $0.22 \times 10^{22} cm^{-3}$) only broadened atomic lines are seen; the shape observed is a consequence of the variation of the MgF_2/Xe reflectivity with photon energy in regions where the optical constants of Xe vary rapidly. In (b) (= $0.46 \times 10^{22} cm^{-3}$) apart from an appreciable broadening a second peak appears very close to the atomic peak marked B. In (c) (= $0.68 \times 10^{22} cm^{-3}$), peak "B" is not discernible any longer since the "extra" peak covers it and a further extra band is evident on the low-energy side of A. Comparison with (d) (= $1.24 \times 10^{22} cm^{-3}$) shows that the extra bands grow much faster with increasing density than their corresponding neighbors of atomic origin. The further peak in (d), somewhat above 9 eV, corresponding to their n=2 Wannier excitons in the solid, has been discussed above. Figure 8 and more detailed studies (Laporte and Steinberger, 1977; Laporte et al 1987) show that the extra bands seen in the immediate vicinity of atomic lines each appear at a characteristic minimum density and grow much faster with density than the respective atomic line. These facts strongly support the assumption that the extra bands are, indeed, entities different from atomic lines strongly perturbed by molecular interactions; because of the successful extrapolation from solid to liquid densities, they are identified as the n=1 and n'=1 intermediate excitons in the liquid. It seems that this is the first instance in the literature for excitons in liquids or amorphous materials that a clear distinction could be made between perturbed atomic or (molecular) transitions and excitons proper.

The coexistence of both intermediate excitons and broadened and shifted atomic lines in a rather wide density rangy is readily explained. Consider the density fluctuations in the fluid and assume that for the existence of an intermediate exciton it is essential to have a certain minimum number of atoms, near enough to each other in order to make excitation transfer between them possible (Laporte and Steinberger, 1977). This is closely connected with the work of Logan and Wolynes (1984), dealing with the localization of an excitation in a dense fluid; the exciton is the de-localized entity. Clear correlation could be found between the probability w(N) of finding in a volume V_o at least N atoms, and the ratio of the oscillator strength fraction $f_{exc}/(f_{exc}+f_{at})$, in a wide density range. This type of analysis has been performed both in fluid xenon (Laporte and Steinberger, 1977), and krypton (Laporte et al., 1987). Figure 9 shows the correlation obtained in Kr. As a result of such comparisons, the minimum conditions for the existence of the n=1 exciton in fluid xenon and krypton could be stated: in xenon, at least N=10 atoms have to occupy an elementary volume V_o=1.5nm^3; for Kr, N=12 and V_o=1.2nm^3. We note that this is an empirical criterion for the occurrence of the excitons discussed. There exists a theory about the occurrence of Wannier excitons in microcrystals (Kayanuma, 1986).

CONCLUSIONS

The electron energy levels and the main electronic transport properties of the liquids of argon, krypton, and xenon can be described in a framework similar to that used for crystalline solids. The consistency of the description is evident, e.g., when comparing band gaps obtained from exciton series limits and photoconductivity thresholds, and when considering the interrelation of the V_o and mobility data. This consistence gave confidence in the results of studies of the evolution of various entities with increasing density. However, only some aspects of this model could undergo quantitative tests using basic theories, since many relevant theoretical ideas have not been carried through to a point where comparison with experiment was possible.

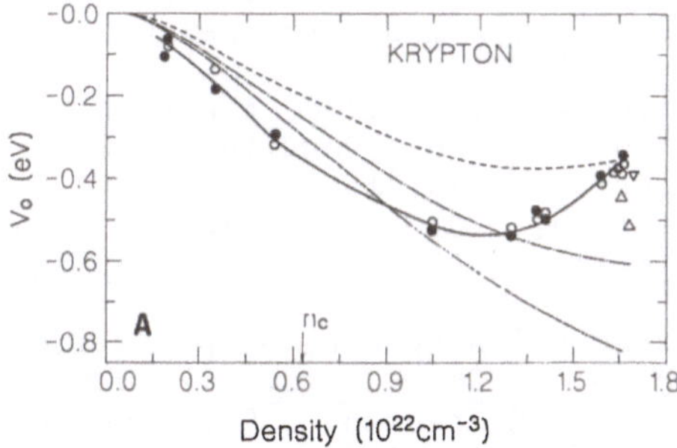

Fig. 9. The probability w(N) of finding N atoms in a local fluctuation of volume V_o for Kr as a function of the number density: □ - V_o = 12 x 10^{-22}cm^3; ◇ - V_o = 11.5 x 10^{22}cm^3; o - the oscillator strength fraction, $\alpha = f_{exc}/(f_{exc} + f_{at})$. From Laporte et al., 1987.

REFERENCES

Asaf, U. and Steinberger, I.T., 1971, Wannier excitons in liquid xenon, Phys. Lett., 34A:207.

Asaf, U. and Steinberger, I.T., 1974, Photoconductivity and electron transport parameters in liquid and solid xenon, Phys. Rev. B, 10:4464.

Ascarelli, G., 1986, Calculation of the mobility of electron injected in liquid argon, Phys. Rev. B, 33:5825.

Baldini, G., 1962, Ultraviolet absorption of solid argon, krypton and xenon, Phys. Rev., 128:1562.

Basak, S. and Cohen, M.H., 1979, Deformation potential theory for the excess electrons in liquid argon, Phys. Rev. B, 20:3404.

Beaglehole, D., 1965, Reflection studies of excitons in liquid xenon, Phys. Rev. Lett., 15:207.

Birks, J.B., 1970, "Photophysics of Aromatic Molecules", Wiley - Interscience, London, New York, Sidney, Toronto.

Broomall, J.R., Johnson, W.D. and Onn, D.G., 1976, Density dependence of the electron surface barrier for fluid ^{3}He and ^{4}He, Phys. Rev. B., 14:2819.

Chandler, D., Schweizer, K.S. and Wolynes, P.G., 1982, Electronic states of typologically disordered systems: Exact solution of the mean spherical model for liquids, Phys. Rev. Lett., 49:110.

Chen, Y.C., Cunningham, J.E. and Flynn, C.P., 1984, Dependence of rare-gas-adsorbate dipole moment on substrate work function, Phys. Rev. B, 30:7317.

Cohen, M.H. and Lekner, J., 1967. Theory of hot electrons in gases, liquids and solids, Phys. Rev., 158:305.

Davis, H.T. and Brown, R.G., 1965, Low-energy electrons in nonpolar fluids, in "Advances in Chemical Physics XXXI", I. Prigogine and S.A. Rice, eds. Wiley, New York, London, Sidney.

Dexter, D.L. and Knox, R.S., 1965, "Excitons", Interscience, New York, London, Sidney.

Eckhardt, C.J. and Nichols, L.F., 1972, Observation of excitons in a molecular liquid: specular reflection spectrum of α-methylnaphtalene, Phys. Rev. Lett., 29:1221.

Faber, T.E., 1972, "An Introduction to the Theory of Liquid Metals", Cambridge University Press.

Holroyd, R.A. and Allen, M., 1971, Energy of excess electrons in nonpolar liquids by photoelectric work function measurements, J. Chem. Phys., 54:5014.

Holroyd, R.A. and Cippolini, N.E., 1978, Correspondence of conduction band minima and electron mobility maxima in dielectric liquids, J. Chem. Phys., 69:50.

Huang, S.S.S. and Freeman, G.R., 1978, Electron mobilities in gaseous, critical, and liquid xenon: Density, electric field, and temperature effects: Quasi localization, J. Chem Phys., 68:1355.

Huang, S.S.S. and Freeman, G.R., 1981, Electron transport in gaseous and liquid argon: Effects of density and temperature, Phys. Rev. A, 24:714.

Inagaki, T., 1972, Absorption spectra of pure liquid benzene in the ultraviolet region, J. Chem. Phys., 57:2526.

Jacobsen, F.M., Gee, N. and Freeman, G.R., 1986, Electron mobility in liquid krypton as functin of density, temperature and electric field strength, Phys. Rev. A, 34:2329.

Kayanuma, Y., 1986, Wannier exciton in microcrystals, Solid State Comm., 59:405.

Knox, R.S., 1963, "Theory of Excitons", Academic Press, New York and London.

Kohler, A.M., 1987, "Density effects on Rydberg states and ionization energies of molecules", Ph.D. Thesis, Hamburg University.

Kohn, W. and Sham, L.J., 1965, Self-consistent equations including exchange and correlation effects, Phys. Rev., 140:A1133.

Kuntz, P.J. and Schmidt, W.F., 1982, A classical Monte Carlo model for the injection of electrons into gaseous argon, J. Chem. Phys., 76:1136.

Laporte, P., Saile, V., Reininger, R., Asaf, U. and Steinberger, I.T., 1983, Photoionization of xenon below the atomic ionization potential, Phys. Rev. A, 28:3613.

Laporte, P. and Steinberger, I.T., 1977, Evolution of excitonic bands in fluid xenon, Phys. Rev. A, 15:2538.

Laporte, P., Subtil, Y.L., Asaf, U., Steinberger, I.T. and Wind, S., 1980, Intermediate and Wannier excitons in fluid xenon, Phys. Rev. Lett., 45:2138.

Laporte, P., Subtil, J.L., Reininger, R., Saile, V., Bernstorff, S. and Steinberger, I.T., 1980, Evolution of intermediate excitons in fluid argon and krypton, Phys. Rev. B, 35:6270.

Laporte, P., Subtil, J.L., Reininger, R., Saile, V. and Steinberger, I.T., 1985, Wannier excitons in liquid and solid krypton, Chem. Phys. Lett., 122:525.

Lekner, J., 1967, Motion of electrons in liquid argon, Phys. Rev., 158:130.

Lekner, J. and Bishop, A.R., 1973, Electron mobility in simple fluids near the critical point, Philos. Mag., 127:297.

Le Sar, R. and Kopelman, R., 1977, Vibrational excitons, resonant energy transfer and local structure in liquid benzene, J. Chem. Phys., 66:5035.

Logan, D.E. and Wolynes, P.G., 1984, Self-consistent theory of localization in topologically disordered systems, Phys. Rev. B, 29:6560.

Messing, I. and Jortner, J., 1977, Adiabatic polarization energy in a simple dense fluid, Chemical Physics, 24:189.

Messing, I., Raz, B. and Jortner, J., 1977, Experimental evidence for Wannier impurity states in doped rare-gas fluids, Chemical Physics, 23:23.

Miller, L.S., Howe, S. and Spear, W.E., 1968, Charge transport in solid and liquid Ar, Kr and Xe, Phys. Rev., 166:871.

Mott, N.F. and Davis, E.A., 1979, "Electronic Processes in Non-crystalline Materials", Oxford University Press, 2nd ed.

Nishikawa, M., 1985, Electron mobility in fluid argon: Application of a deformation potential theory, Chem. Phys. Lett., 114:271.

Plenkiewicz, B., Jay-Gerin, J.P., Plenkiewicz, P. and Bachelet, G.B. 1986, Conduction band energy of excess electrons in liquid argon, Eurohpysics Letters, 1:455.

Raz, B. and Jortner, J., 1969, Energy of the quasi-free electron state in liquid and solid rare gases, Chem. Phys. Lett., 4:155.

Raz, B. and Jortner, J., 1971, Energy of the quasi-free electron state in dense neon, Chem. Phys. Lett., 9:224.

Reininger, R., Asaf, U. and Steinberger, I.T., 1982, The density dependence of the quasi-free electron state in fluid xenon and krypton, Chem. Phys. Lett., 90:287.

Reininger, R., Asaf, U. and Steinberger, I.T., 1983a, Photoconductivity and the evolution of energy bands in fluid xenon, Phys. Rev. B., 28:3193.

Reininger, R., Asaf, U. and Steinberger, I.T., 1983b, Relationship between the energy V_0 of the quasi free electron and its mobility in fluid argon, krypton and xenon, Phys. Rev. B, 28:4426.

Reininger, R., Saile, V., Laporte, P. and Steinberger, I.T., 1984a, Photo-conduction in rare gas fluids doped by small organic molecules, Chemical Physics, 89:473.

Reininger, R., Steinberger, I.T., Bernstorff, S., Saile, V. and Laporte, P., 1984b, Extrinsic photoconductivity in xenon doped fluid argon and krypton, Chemical Physics, 86:189

Reininger, R., 1985, Private communication.

Reininger, R., Saile, V., Findley, G., Laporte, P. and Steinberger, I.T., 1985a, Photoconduction in fluid rare gases doped with molecular impurities, p. 253 in "Photophysics and Photochemistry Above 6eV". F. Lahmani, ed., Elsevier, Amsterdam.
Reininger, R., Saile, V. and Laporte P., 1985b, Photoionization yield spectra below the atomic ionization limit in xenon, Phys. Rev. Lett., 54:1146.
Rice, S.A. and Jortner, J., 1966, Do excitons states exist in the liquid phase?, J. Chem. Phys., 44:4470.
Schnyders, H., Rice, S.A. and Meyer, L., 1966, Electron drift velocities in liquied argon and krypton at low electric field strengths, Phys. Rev., 150:127.
Schwentner, N., Koch, E.G. and Jortner, J., 1985, "Electronic Excitations in Condensed Rare Gases", Springer Tracts in Modern Physics, G. Hohler, ed., Springer, Berlin-Heidelberg-New York-Tokyo.
Shinsaka, K. and Freeman, G.R., 1974, Electron mobilities and ranges in solid neopentane: Effect of the liquid-solid phase change, Can. J. Chem., 52:3556.
Sonntag, B., 1977, Dielectric and optical properties, in: "Rare Gas Solids", M.L. Klein and J.A. Venables, eds., Academic Press, London, New York, San Francisco.
Springett, B.E., Jortner, J. and Cohen, M.H., 1968, Stability criterion for the localization of an excess electron in a nonpolar fluid, J. Chem Phys., 48:2720.
Steinberger, I.T., Atluri, C. and Schnepp, O., 1970, Optical constants of solid xenon in the VUV region, J. Chem. Phys., 52:2723.
Steinberger, I.T. and Asaf, U., 1973, Band structure parameters of solid and liquid xenon, Phys. Rev. B, 8:914.
Steinberger, I.T. and Baer, S., 1987, Electronic excitations of pure and doped rare gas fluids, Phys. Rev. B, 36:1358.
Steinberger, I.T. and Zeitak, R., 1986, Estimation of electron mobilities in simple nonpolar fluids, Phys Rev. B, 34:3471.
Tauchert, W., Jungblut, H. and Schmidt, W.F., 1977, Photoelectric determination of V_o values and electron ranges in some cryogenic liquids, Canad. J. Chem., 55:1860.
Von Zdrojewski, W., Rabe, J.G. and Schmidt, W.F., 1980, Photoelectric determination of V_o-values in solid rare gases, Z.Natureforsch., 35A:672.
Woolf, M.A. and Rayfield, G.W., 1965, Energy of negative ions in liquid helium by photoelectric injection, Phys. Rev. Lett., 15:235.
Zallen, R., 1983, "The Physics of Amorphous Solids", Wiley, New York.

ELECTRON SCATTERING AND MOBILITY IN DIELECTRIC LIQUIDS

Gordon R. Freeman

Chemistry Department, University of Alberta
Edmonton, AB, T6G 2G2 Canada

INTRODUCTION

When a pulse of X-rays hits a dielectric liquid a cascade of electrons is produced in the liquid. The absorbed X-rays set a few high energy electrons in motion. The high energy electrons ionize molecules and set many lower-energy electrons in motion, which in turn produce many more electrons of still lower energy, and so on, until there is a large number of low energy electrons in the liquid.

A single photon with 300 fJ (2 MeV) of energy interacts with a molecule and knocks out an electron with about 150 fJ (1 MeV) excess energy. This high energy electron then produces a cascade of ionization. The cascade generates about 3×10^4 electrons, with an average excess energy of about 3aJ (20 eV) each.

So we have a convenient method of producing, inside a liquid, electrons of a few attojoules of energy in numbers large enough to allow measurement of their behavior. The studies are providing a wealth of information about the interactions of electrons with liquids. They are also a driving force for studies of liquids for their own sake, and of electrons for their own sake. For example, the studies of electrons in liquids have highlighted the need for much more detailed information about the dynamics of molecules in the liquid state; information can be obtained by rotational relaxation time measurements and by computer modeling. The studies of electron behavior in fluids have also led to reinterpretation of experimental verifications of the Aharonov-Bohm effect (interaction of an electron beam with a "distant" magnetic field, Freeman, 1987a), and to reinterpretation of a single-photon interference effect (Freeman, 1987b; Grangier et al., 1985). These reinterpretations are basic to the visualization of electromagnetic effects in quantum mechanics.

Electrons produced in liquids by the radiolysis method have initial energies mainly in the region of 10^{-17}J (10^2 eV). This energy propels the electrons through the liquid, but the electrons collide with molecules and lose energy to them. The electrons ultimately reach thermal equilibrium with the liquid and have energies in the vicinity of 10^{-21}J (10^{-2} eV). Measurements allow us to estimate the distance the electrons travel through the liquid while being de-energized from 10^{-17} to 10^{-21} J; it is called the thermalization distance. The thermalization distance is sensitive to the nature of the molecules of the liquid and to how they are packed together. For example, it is sensitive to whether the molecules are polar or nonpolar, and to whether they are spherelike or nonspherelike, and to how densely they are packed together (Freeman, 1987c).

After the electrons reach thermal equilibrium with the liquid, they diffuse at random through it until they find something to react with. We can apply an electric field across the liquid and measure how fast the electrons drift in the direction of the field. From that we can deduce the diffusion coefficient of the electrons. The drift velocity is proportional to the applied field strength, as long as the field strength is not too high. The drift velocity v_d per unit field strength E is called the mobility μ:

$$\mu(m^2/V.s) = v_d(m/s)/E(V/m) \quad . \qquad (1)$$

Ordinary diffusion is driven by the thermal agitation energy $k_BT(J)$, where k_B is Boltzmann's constant and T is the temperature. Drift in an electric field is driven by the electrostatic force ξE, where $-\xi(C)$ is the charge on the electron. The relationship between the diffusion coefficient D and the mobility of thermal electrons is therefore

$$D(m^2/s) = \mu k_B T/\xi \quad . \qquad (2)$$

The mobility of thermal electrons is even more sensitive than is the thermalization distance to the nature of the molecules of the liquid and to how they are packed together.

This article describes electron thermalization distances and then mobilities under diverse conditions.

ELECTRON THERMALIZATION DISTANCES

Experimental method

A highly purified liquid is put into a cell that contains parallel plate electrodes (Dodelet and Freeman, 1977a). The cell had previously been baked out under high vacuum, and all material transfers are done in a vacuum apparatus.

A voltage is applied to one electrode and the other is grounded through a measuring circuit. A pulse of X-rays, typically 100 ns long, ionizes a tiny fraction of the molecules, about one in 10^{14}, thereby producing about 10^8 electrons and ions in the liquid between the electrodes. The drift of ions and electrons in the field between the electrodes produces a current that is integrated to obtain a measure of the amount of charge collected.

The amount of X-rays per pulse is measured with a calibrated device, and the amount of energy absorbed by the liquid is calculated by a standard procedure (Gee and Freeman, 1987).

Reactions and Dynamics

The ionization reaction induced by radiation is written

$$M \rightsquigarrow [M^+ + e^-] \ , \tag{3}$$

where the wiggly arrow signifies absorption of energy from radiation. The electron initially has excess kinetic energy, which it loses in collisions with molecules. The electron usually reaches thermal energy before it escapes the coulombic field of its sibling ion, at some distance y from the ion. The square brackets in reaction (3) indicate that the coulombic attraction $(-\xi^2/4\pi\varepsilon_o \varepsilon y)$ between the thermalized electron and ion is not negligible compared to the thermal energy k_BT; ε_o is the permittivity of vacuum and ε is the relative permittivity of the fluid between the ion and electron.

The coulombic attraction between the ion and electron tends to draw them back together. When it succeeds the reaction is called geminate neutralization.

$$[M^+ + e^-] \rightarrow M \text{ (geminate neutralization)} \ . \tag{4}$$

Normal diffusion of the ion and electron tends to make them wander away from each other. The diffusion is driven by thermal energy and it is in competition with the coulombic attraction. If the coulombic energy is not too great some of the pairs manage to diffuse further apart and their coulombic attraction becomes negligible. The ion and electron then diffuse independently of each other and they are called free ions.

$$[M^+ + e^-] \rightarrow M^+ + e^- \text{ (free ions)} \ . \tag{5}$$

In the absence of other reactants, the free ions eventually undergo neutralization at random in the bulk fluid:

$$M^+ + e^- \rightarrow M \text{ (random neutralization)} \ . \tag{6}$$

By applying an electric field across the system, free ions can be collected at electrodes.

$$\left.\begin{array}{l} M^+ \rightarrow \text{electrode} \\ e^- \rightarrow \text{electrode} \end{array}\right\} \qquad (7)$$

It is, therefore, possible to measure the extent of reaction (5).

The extent of the initial reaction (3) can be measured chemically, by adding a solute that reacts with either M^+ or e^- and measuring the product yield (Freeman, 1987d). The fraction ϕ_{fi} of ions and electrons produced in (3) that become free ions can therefore by determined. The fraction ϕ_{fi} is a function of the thermalization distance y, so the value of y can be estimated.

The separation between the ion and electron is attained in three stages. Most of the electrons initially have energies near 10^{-17}J, which is much in excess of the ~ 10^{-21}J thermal energy. The rate of energy loss by the electron is strongly dependent upon the type of excitation that the electron can produce in the molecules with which it collides. The change of excitation process as the electron energy decreases gives rise to the three stages of energy loss (Mozunder and Magee, 1967; Freeman, 1972). Energy transfer becomes much less efficient as the energy of the electron degrades from that which permits electronic excitation of the molecules to that which permits only vibro-rotational excitation, and still less efficient when only rotational and intermolecular oscillational excitation are possible:

(i) 10^{-15}s after generation in reaction (3), the electron's energy has been reduced to $\lesssim 10^{-18}$J, mainly by transfer to electronic states of molecules;

(ii) 10^{-13}s after reaction (3) the electron energy has been reduced to $\lesssim 10^{-19}$J, mainly by transfer to vibrational states of molecules; in liquids of compounds that contain hydroxy (-OH) groups, electrons form localized states (e^-_{loc}) in this time period;

(iii) 10^{-12} - 10^{-9}s later, depending on the nature of the liquid, the energy of the electron has been reduced to thermal (< 10^{-20}J), mainly by transfer to librational and intermolecular modes; electrons form localized states in liquids of polar or non-polar, nonspherelike molecules, but remain in the delocalized state if the molecules are spherelike; localized states relax to thermal equilibrium in liquids whose molecular orientational relaxation times are in this region.

Thermalization distances are related to the thermalization times. As a crude approximation one can take the scattered path to be a random walk, and the distance between the electron and its sibling ion at the instant of localization or thermalization is roughly proportional to the square root of the time it takes the electron to become localized or thermalized. As the speed of the electron decreases its mean free path tends to increase, so the effects of these changes on the thermalization distance tend to cancel each other.

Thermalization Distances

The fraction ϕ_{fi} of ions produced in reaction (3) that become free ions is a function of the electron thermalization distance y, the relative permittivity ε of the fluid, the temperature T, and the applied electric field strength E. We, therefore, write (for details see Freeman, 1987c):

$$\phi_{fi}^{E}(y) = \phi_{fi}^{o}(y)[1 + f(E,\varepsilon,T,y)] \ , \tag{8}$$

where

$$\begin{aligned}\phi_{fi}^{o}(y) &= \exp(-\xi^2/4\pi\varepsilon_o\varepsilon k_B Ty) \\ &= e^{-r_c/y} \ ,\end{aligned} \tag{9}$$

and

$$f(E,\varepsilon,T,y) = e^{-2\beta y} \sum_{m=1}^{\infty} \frac{(2\beta y)^m}{(m+1)!} \sum_{j=0}^{m-1} (m-j) \frac{(r_c/y)^{j+1}}{(j+1)!} \ , \tag{10}$$

where $\beta = \xi E/2k_B T$, and $r_c = \xi^2/4\pi\varepsilon_o\varepsilon k_B T$ is the distance at which the coulombic attraction between the ion and electron equals the thermal agitation energy $k_B T$.

In any given system the electrons do not all have the same thermalization distance, but they have a distribution F(y). The average fraction ϕ_{fi}^{E} of ion pairs that becomes free ions is obtained by averaging Eq. (8) over the distribution:

$$\phi_{fi}^{E} = \int_0^{\infty} \phi_{fi}^{E}(y) \ F(y)dy \ . \tag{11}$$

A form of F(y) that fits data for many liquids has a Gaussian body and a power tail, and is designated YGP:

$$\begin{aligned}F(y) &= 0.96 \ YG \ , && y < 2.4 \ b_{GP} \ , \\ F(y) &= 0.96 \ YG + 0.48(b_{GP}^2/y^3) \ , && y > 2.4 \ b_{GP} \ ,\end{aligned} \tag{12}$$

where 0.96 is a normalization factor, $YG = (4y^2/\pi^{1/2}b_{GP}^3)\exp(-y^2/b_{GP}^2)$, and b_{GP} is the dispersion parameter of the Gaussian and the most probable value of y.

Yields in radiation chemistry are recorded as G values, the number of species formed per 100 eV (16aJ) of energy absorbed by the system. The free ion yields are measured as a function of applied field strength and ϕ_{fi}^E is calculated from the ratio

$$\phi_{fi}^E = G_{fi}^E/G_{tot} , \tag{13}$$

where G_{tot} is the yield of reaction (3).

The value of the most probable thermalization distance b_{GP} is determined by fitting a set of G_{fi}^E, E data to Eqs. (8) - (13), with known values of ε and T and an estimated value of G_{tot} (Freeman, 1987c).

If the scattering properties of a series of liquids are the same, the thermalization distances are inversely proportional to the liquid density d. To compare scattering properties we therefore compare values of the density normalized distance $b_{GP}d$.

Molecular Structure Effects in Liquid Hydrocarbons

Scattering of electrons at energies $<10^{-19}$J (< a few tenths of an eV) is strongly dependent on the molecular structure. Most of the thermalization distance of 10^{-17}J electrons in liquid hydrocarbons is attained in the energy region below about 1×10^{-19}J. The behavior illustrated in Fig. 1 is dominated by two empirical characteristics:

(i) scattering is weaker ($b_{GP}d$ is larger) when the molecules are more spherelike;

(ii) alkyl C-H groups are much weaker scatterers than are C-C bonds (compare methane and ethane, Fig. 1), while alkenyl C-H and alkynyl C-H groups scatter progressively more strongly.

The sphericity effect (i) applies only to dense phases, where the electron interacts with several molecules at a time. In low density gases, where the electron interacts with only one molecule at a time, scattering at these low energies is stronger when the molecule is more spherelike. Thus the molecular sphericity effect reverses on going from single-body to multi-body scattering. Attempts by many people over the past 15 years to provide a theoretical interpretation of the sphericity effect in either a gas or a liquid have failed.

The increasing scattering power of C-H groups on going from alkyl to alkenyl to alkynyl is due to their increasing polarity. The proton in the H atom becomes less shielded as the electron density is attracted

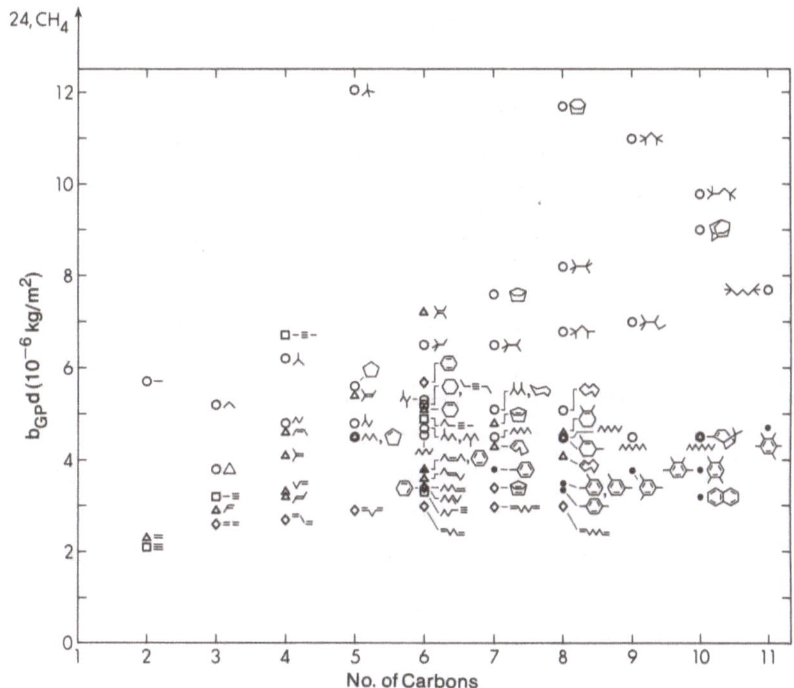

Fig. 1. Molecular structure dependence of density-normalized thermalization distances $b_{GP}d$ of ~ 10^{-17}J (~ 60 eV) electrons in liquid hydrocarbons. The temperatures were mainly near the normal boiling point or near 296 K, whichever is lower. For more information see Freeman, 1987c.

away from it by the adjacent progressively more electronegative C-C, C=C and C ≡ C groups.

The interaction of a low energy electron with an alkane molecule can penetrate the molecule to a depth of two C-C bonds in series. Groups up to 2,2-dimethylpropyl (neopentyl) in size are sensed as a unit. This includes the entire neopentane molecule because the alkyl C-H groups are nearly transparent to the electron. The large molecule 2,2,4,4-tetramethylpentane interacts only slightly more strongly than does neopentane (Fig. 1), because from most directions of approach 2244TMP feels like neopentane as far as the electron can tell with its two C-C bond reach.

Figure 1 contains a wealth of information that is awaiting formal theoretical study. The literature contains much more data (Freeman, 1987c), with crude beginnings of interpretations.

Effects of Liquid Density, the Critical Fluid, and Crystallization

If a liquid is heated under its vapor pressure the density of the liquid decreases and that of the vapor increases, until at the critical temperature T_c they become the same, d_c. The liquid phase then ceases to exist and the meniscus that characterizes the liquid surface disappears. The fluid at T_c and d_c is called the critical fluid.

Changing the density of a liquid alters the average distance between its molecules and the way they pack together. This changes the electron scattering properties of the liquid, which are reflected in $b_{GP}d$. The type of change depends on the sphericity of the molecules.

Data for a spherelike and a chainlike isomer of pentane are shown in Fig. 2. Beginning at the critical fluid and increasing the density of the spherelike isomer NP, $b_{GP}d$ increases. This is attributed to a decrease of scattering by the destructive interference of long range attractions as the electron interacts with progressively more molecules at once. The value of $b_{GP}d$ then passes through a maximum and decreases with increasing density. At these densities electron-molecule repulsive interactions dominate the scattering. However, at densities approaching the crystallization point of the liquid the intermolecular arrangement becomes more regular and scattering becomes less strong; $b_{GP}d$ increases.

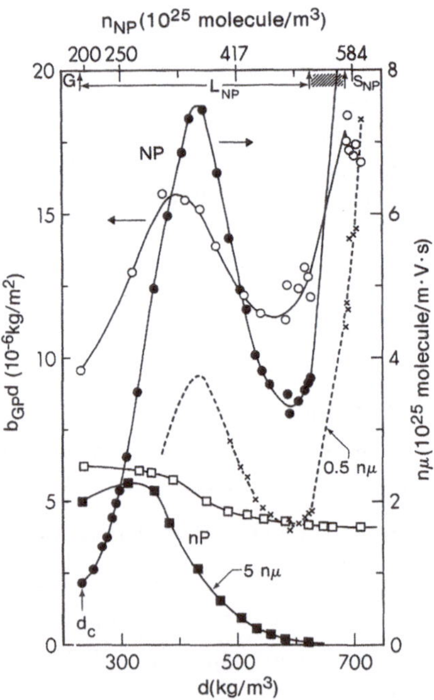

Fig. 2. Density-normalized thermalization distances $b_{GP}d$ of ~ 10^{-17}J (~ 60 eV) electrons in isomeric pentanes as a function of density: o , spherelike $(CH_3)_4C$, NP; □ , chainlike $CH_3CH_2CH_2CH_2CH_3$, nP. G = gas, L = liquid, S = solid, d_c = critical density. //// indicates abrupt change of density during liquid-solid phase change. The number-density-normalized mobilities nμ of thermal electrons are plotted for comparison: NP, ● = nμ, - - - and x = 0.5 nμ; nP, ■ = 5 nμ. For more information see Freeman, 1987c.

The value of $b_{GP}d$ in the crystalline phase is double that in the liquid near the freezing point (Fig. 2).

In the chainlike isomer nP, $b_{GP}d$ decreases with increasing density throughout the liquid region. (nP freezes at a lower temperature and higher density than those represented in Fig. 2.) In the dense liquid, where $d/d_c > 2.0$, $b_{GP}d$ is nearly independent of density. The chains do not easily pack together to form a highly regular structure, and that is why the freezing point of nP is so low, 143 K, compared to that of the spherelike NP, 257 K.

The density normalized mobilities $n\mu$ of thermal electrons display variations similar to those of $b_{GP}d$ in the respective liquids (Fig. 2). The variations in $n\mu$ are much greater than those in $b_{GP}d$, because μ involves much lower energy electrons than does b_{GP}. Lower energy electrons are more sensitive to these structure effects.

THERMAL ELECTRON MOBILITIES

Experimental method

The method described in the previous section is also used to measure mobilities, except that lower field strengths are used and the electron drift current is recorded as a function of time. The breadth of the current signal is the flight time t_d of the electrons over the distance l between the electrodes. The mobility is given by

$$\mu = l/E \cdot t_d \ . \tag{14}$$

Molecular structure effects

The ease with which an electron can move through a liquid is strongly dependent on the nature of the liquid (Fig. 3). The electron mobilities can differ by up to seven orders of magnitude in different liquids. In nonpolar liquids electron mobilities can differ by six orders of magnitude (xenon and ethylene). In the strongly polar liquids ethanol and water electrons have mobilities similar to those of ethoxide and hydroxide ions, respectively.

The correlation between values of thermal electron mobilities and secondary electron thermalization distances mentioned earlier is also observable in Fig. 3. Thus, the same two empirical factors that affect thermalization distances also affect mobilities:

(i) mobilities are greater when the molecules are more spherelike;

(ii) alkyl C-H groups are much weaker scatterers than are C-C bonds, while alkenyl C-H groups scatter more strongly.

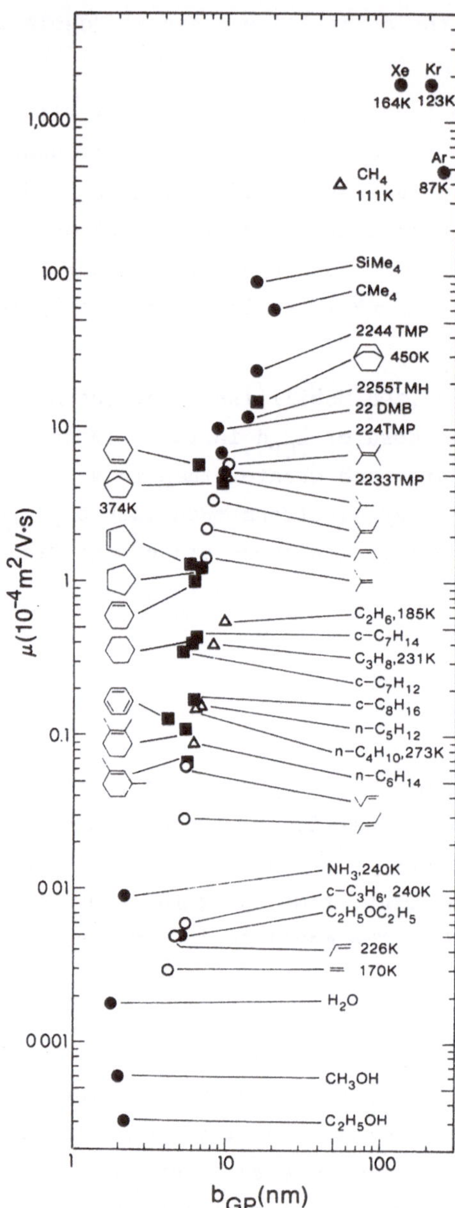

Fig. 3. Mobilities of thermal electrons and thermalization distances b_{GP} of ~ 10^{-17}J electrons in dielectric liquids. T = 296±3 K unless otherwise indicated. For much more data see Freeman, 1987c.

Alkynes (acetylenes) attract electrons so strongly that they form stable negative ions (Dodelet et al., 1973).

The molecular sphericity effect on electron mobility in alkanes reverses on going from the dense liquid to the low density gas (Fig. 4). Attempts to theoretically interpret the molecular sphericity effect on electron mobility in either the gas or liquid phase have failed. The low

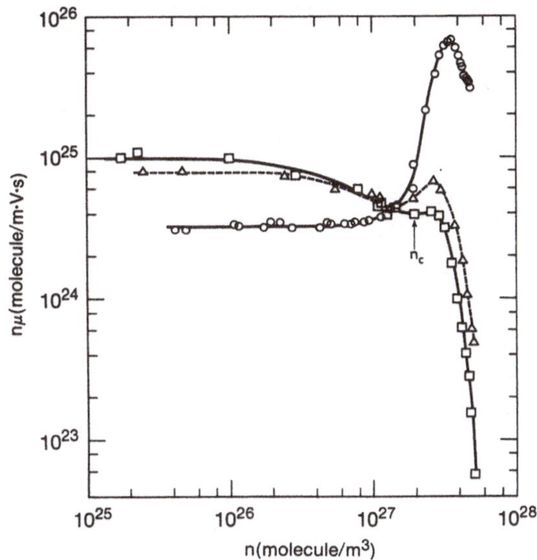

Fig. 4. Density normalized mobility of thermal electrons as functions of fluid density in the coexistence vapor and liquid of the isomeric pentanes. □, n-pentane; Δ , iso-pentane; o , neo-pentane. Data from György and Freeman, 1979.

mobilities in liquids of nonspherelike molecules are qualitatively understood in terms of localized states of electrons. The localized state is more stable when the groups interacting with the electron have greater anisotropy of polarizability and more hindered rotation (Fig. 5).

The correlation between mobility and thermalization distance has structure (Fig. 6). For a given value of b_{GP}, say 6 nm, the value of μ increases on going from n-alkanes to cycloalkanes to cycloalkenes (unbranched). Figure 5 and a recent compilation (Freeman, 1987c) contain a large amount of experimental data that awaits more appropriate theoretical interpretation.

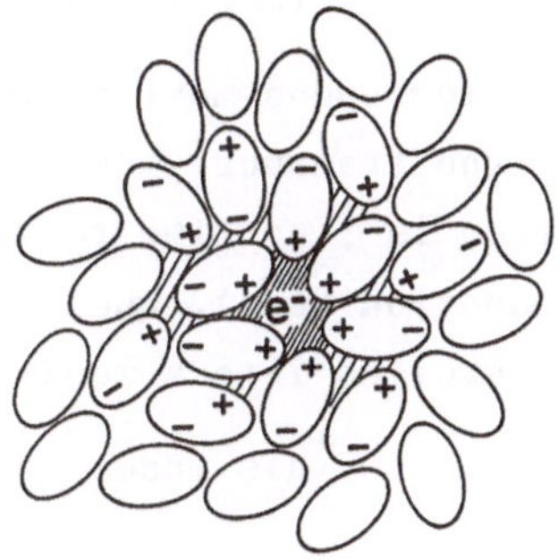

Fig. 5. Schematic of localized state of electron in liquid of nonspherical molecules.

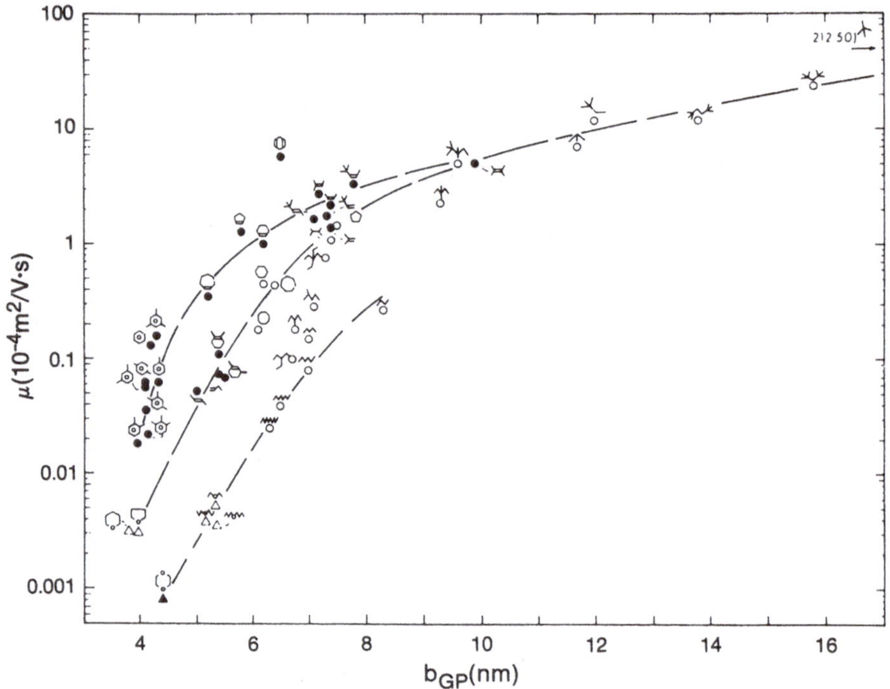

Fig. 6. Correlations between thermal electron mobilities μ and secondary electron thermalization distances b_{GP} in organic liquids at 294±2 K. ● , unsaturated hydrocarbon; o , saturated hydrocarbon; Δ, ether. Data from Dodelet et al., 1976.

Two-state Model of Electron Mobility (Liquids of Nonspherelike Molecules)

The mobility of electrons in liquid hydrocarbons that have non-spherelike molecules is illustrated by the behavior in the four isomeric butenes (Fig. 7). On increasing the temperature from near the freezing point to the critical point, with the liquid under its vapor pressure, the electron mobility increases by four orders of magnitude (Dodelet and Freeman, 1977b). The mobilities are similar in all the liquids near the freezing points, and again at the critical points, but at intermediate temperatures they differ by up to two orders of magnitude. The mobility is higher in the disc-like isomers than in the stick-like isomers.

Electrons in these liquids spend most of their time in localized states. One model of electron transport, derived from semiconductor theory, is that each electron is from time to time thermally excited into the delocalized state (conduction band), where it migrates relatively freely until it becomes de-excited into a localized state again.

Migration of the electron in this model involves successive detrapping and retrapping:

$$e^-_{loc} \rightleftarrows e^-_{deloc} \ . \tag{15}$$

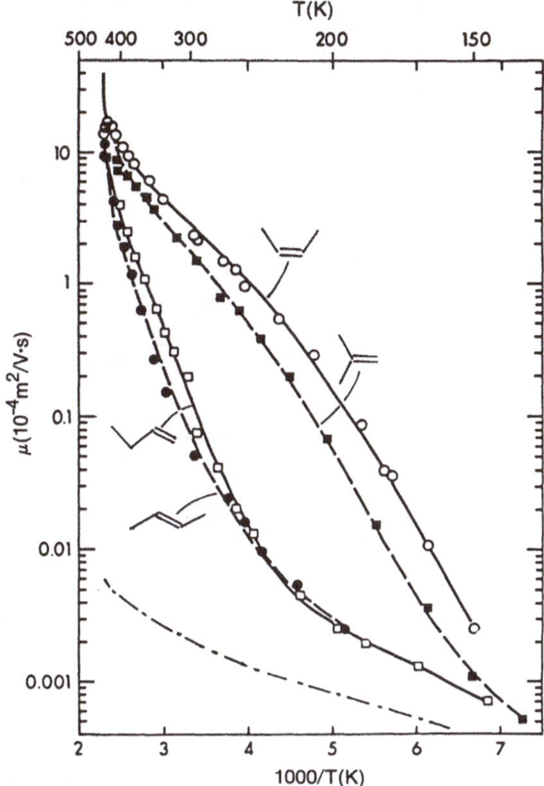

Fig. 7. Mobilities of thermal electrons in liquid butenes under their vapor pressure, at temperatures up to the critical, T_c. o , cis-butene-2; ■, iso-butene; □, butene-1; ● , trans-butene-2. The lines were calculated from Eqs. (16)-(22), using the parameter values listed in Table 1. -·-·, approximate mobilities of anions in each of the liquids, taken as 0.7 $(\mu_- + \mu_+)$ averaged for the four liquids. Data from Dodelet and Freeman, 1977b.

The detrapping and retrapping of the electron are inelastic processes that involve the dynamics of the molecules of the liquid.

The two-state model that is presented here is very crude, but it could be used to begin one that properly accounts for the molecular dynamics and electron-multimolecular interactions.

There are many electrons in the sample, each in its own trap or in the conduction band (delocalized state), and there is a distribution of trap depths. The trap depths are not all the same because orientational disorder of the molecules provides a variety of polarization potential wells. Furthermore, an electron in a trap seems to couple with vibrational and librational modes of the trapping molecules. The optical absorption band of solvated electrons is very broad (Fig. 8). Part of the broadening might be caused by the distribution of trap depths, and part by the coupling with molecular modes. These parts are sometimes

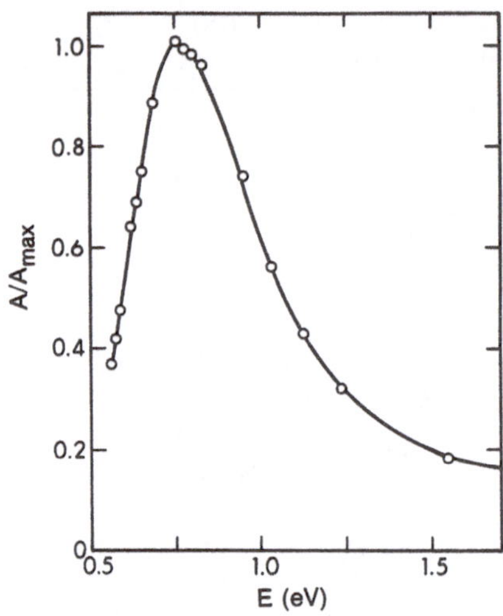

Fig. 8. Optical absorption spectrum of electrons solvated in di-*n*-propyl ether. 123 K, supercooled liquid. Jou and Freeman, 1976.

called heterogeneous and homogeneous broadening, respectively. The tail at high energies is attributed to excitation into the continuum.

The electrons with the smallest values of E are the easiest to excite thermally into the conduction band, so for simplicity we neglect the high energy tail and take a Gaussian distribution of thermal excitation energies about a most probable value E_o:

$$N(E) = (\pi^{1/2}\sigma)^{-1} \exp[-(E-E_o)^2/\sigma^2] . \tag{16}$$

By further analogy with the optical band we assume that E_o and σ are temperature dependent in the liquid under its vapor pressure:

$$E_o = E(0) - \underline{a}T ; \tag{17}$$

$$\sigma = \sigma(0) + \underline{b}T . \tag{18}$$

Typical values of E_o are near 5×10^{-20}J (0.3 eV) and of σ are near 2×10^{-20}J (0.1 eV) in hydrocarbons.

The mobility of the localized state is ionlike and is several orders of magnitude lower than that in the delocalized state. In alcohols the traps are so deep that the electrons are not thermally excited out of them to an appreciable extent; the measured electron mobilities are ionlike. However, in liquid alkanes or alkenes the traps are shallow enough that most of the transport occurs by excitation to the delocalized state.

The measured mobility is a time average of the values in the delocalized and localized states:

$$\mu = X \mu_{deloc} + (1-X)\mu_{loc} , \tag{19}$$

where χ is the fraction in the delocalized state at a given time. The expression for χ depends on the model used. To accommodate conditions where the traps are very shallow we have assumed that density fluctuations in the liquid could cause some localized states to be at energies above the delocalized state. Most of the calculations have therefore been done by integrating E from $-\infty$ to $+\infty$:

$$\chi = \int_{-\infty}^{\infty} N(E)\,[1 + \exp(E/k_BT)]^{-1}\, dE \;. \tag{20}$$

When a liquid is heated under its vapor pressure from its freezing point to the liquid-gas critical point, its density changes by a factor of about 2.5. The mobility of electrons in the delocalized state should be a function of both temperature and density. Although the temperature and density dependence of μ_{deloc} are probably not simple over this wide range of conditions (Cohen and Lekner, 1967; Lekner, 1967, 1968; Jahnke et al., 1971; Schwarz, 1972, 1975; Ostermeier and Schwarz, 1972; Lekner and Bishop, 1973; Basak and Cohen, 1979; Huang and Freeman, 1981) we assume $\mu_{deloc} \propto 1/n^2T$:

$$\mu_{deloc} = \mu_{deloc}^{ref} \left(\frac{n_{ref}}{n_T}\right)^2 \left(\frac{T_{ref}}{T}\right) \;. \tag{21}$$

The value of μ_{deloc} at some reference temperature and density is used as an adjustable parameter, but it is held constant for a series of similar compounds.

The mobility of the localized state is approximated as

$$\mu_{loc} = \mu_{loc}^{o} \exp(-E_{loc}/k_BT) \;, \tag{22}$$

where μ_{loc}^{o} and E_{loc} are respectively the pre-exponential factor and Arrhenius temperature coefficient for ion-like mobility.

These equations have been fitted to μ, T, n data for many hydrocarbons (Dodelet and Freeman, 1977a, 1977b; Ryan and Freeman, 1978; György and Freeman, 1979; Huang and Freeman, 1980; Gee and Freeman, 1983). The fit to electron mobilities in the isomeric butenes is shown in Fig. 7. To restrict the flexibility of the model, certain parameter values were kept constant and all the variation was forced into the others (Table 1). The parameters kept constant were μ_{deloc}^{ref}, E(0), μ_{loc}^{o}, and E_{loc}. Then the fitted temperature coefficient of the electron trap depth E_o was near 16 x 10^{-23}J/K (1 meV/K) in all four isomers.

The curvatures of the Arrhenius plots of μ in Fig. 7 are governed mainly by the behavior of σ, the dispersion parameter of trap depths. In this model the disc-like molecules of <u>cis</u>-butene-2 produce a σ that is directly proportional to T (Table 1). The stick-like molecules of *trans*-butene-2 produce a σ that is independent of T and is relatively small. At 300 K the energy of the most probable trap depth E_o is 5×10^{-20}J (0.3 eV) in both *cis*- and *trans*-butene-2, while the distribution width parameter σ is 5×10^{-20}J in *cis*- but only 2×10^{-20}J in *trans*-butene-2.

The constant, small value of σ in *trans*-butene-2 might indicate that the localized state in this liquid is a molecular anion, although the electron binding energy is small. The large temperature dependence of σ in *cis*- butene-2 indicates that there is a strong coupling between the electron and the thermal motions of the molecules of the liquid immediately around it. Thermal energy fluctuations of the molecules have a dispersion parameter (Dodelet et al., 1976)

$$\sigma_{th} \simeq (2k_B C_p)^{1/2} T \ , \qquad (23)$$

where C_p is the heat capacity of the liquid at constant pressure. This has the form of Eq. (18) with $\sigma(0) = 0$. The appropriate value of C_p is proportional to the number of molecules that couple with one electron, and to the magnitude of the electron-molecule couple. A formal connection between Eqs. (23) and (18) has not yet been developed. We also lack a formal understanding of the value of σ when it is constant, with $\underline{b} = 0$ as in *trans*-butene-2.

Electrons in Liquids of Spherelike Molecules

For spherical molecules the electron-molecule and molecule-molecule interactions are isotropic; density and energy fluctuations make contributions to the electron behavior. For nonspherelike molecules the electron-molecule and molecule-molecule interactions are anisotropic; the orientational fluctuations have far greater effect on electron behavior

Table 1. Parameter values in Eqs. (16)-(22) for curves in Fig. 7[a].

Solvent	a (10^{-23}J/K)	$\sigma(0)$ (10^{-22}J)	b (10^{-23}J/K)
cis-butene-2	16	0	17
iso-butene	14	13	16
butene-1	15	78	5.6
trans-butene-2	17	187	0

[a] Assume $\mu^o_{deloc} = 3 \times 10^{-3} m^2/V.s$ at $T_{ref} = 295$ K, $E(0) = 9.6 \times 10^{-20}$J (0.60 eV), $\mu^o_{loc} = 1.15 \times 10^{-5} m^2/V.s$ and $E_{loc} = 1.03 \times 10^{-20}$J (0.065 eV) for all butene isomers. Data from Dodelet and Freeman, 1977b.

than do those of density. Therefore, from the point of view of development of theory of liquids, electron behavior in liquids of spherelike molecules contains less useful information than does that described in the preceding section. This section is limited to a few vectors into the literature.

In liquid helium a thermal electron is localized in a bubble of roughly 2nm radius that is sensitive to temperature and pressure (Ostermeier and Schwarz, 1972). In superfluid helium the electron mobility can become very large (Schwarz, 1972, 1975).

In liquid argon (Jahnke et al., 1971; Huang and Freeman, 1981; Gushchin et al., 1982), krypton (Jacobsen et al., 1986), and xenon (Huang and Freeman, 1978a; Gushchin et al., 1982) a thermal electron is in a delocalized state (conduction band). However, in the low-density liquid and coexistence dense gas at $2.0 > n/n_c > 0.2$, electrons can become quasilocalized in density fluctuations of appropriate size (Huang and Freeman, 1978, 1981). A plot of $n\mu$ against n contains a high maximum.

In liquid neon the bubble state of a thermal electron might be marginally stable, but the situation is still somewhat in doubt (Freeman, 1966; Loveland et al., 1972; Kuan and Ebner, 1981; Sakai et al., 1982).

In liquids of spherelike hydrocarbons the behavior is similar to that in argon. Examples are methane (Floriano and Freeman, 1986), neopentane (Dodelet and Freeman, 1977a; Huang and Freeman, 1978b; György and Freeman, 1979; Muñoz and Ascarelli, 1983), 2,2,4,4-tetramethylpentane (Ryan and Freeman, 1978), and _iso_-butane (Gee and Freeman, 1983). Tetramethylsilane is also spherelike; electron behavior in it (Cipollini and Allen, 1977) is somewhat similar to that in neopentane (tetramethylmethane).

FUTURE DEVELOPMENTS

In liquids where the trap depths are $\lesssim k_BT$, the temperature and density dependences of electron mobilities are dominated by changes in the conduction band, not by the distribution of trap depths. In further development of the model we would ignore the possibility that localized states might lie above the delocalized level, and integrate Eq. (20) only over energies from zero to infinity. We would then need a formal couple between the values of E_0 and σ, which the present model does not have. The couple depends on the electron-molecule interactions and the liquid properties.

We assumed $\mu_{deloc} \propto 1/n^2T$, which does not apply in the low density liquids of solvents that have spherelike molecules. At densities below that of the mobility maximum the conduction band becomes progressively less smooth, scattering increases and μ_{deloc} decreases. Equation (21) also seems inadequate for many other liquids near the liquid-gas critical region. A better description of μ_{deloc} is needed for liquids of non-spherelike molecules. A place to begin would be the work of Lekner and Cohen and their coworkers, for μ_{deloc} in liquids of spherical molecules (for example see Lekner and Bishop, 1973; Basak and Cohen, 1979; Huang and Freeman, 1981; Freeman, 1987e).

The molecular property that gives the most prominent correlation with electron mobilities in liquids is the degree of sphericity. We think we understand the correlation qualitatively, but all attempts to treat it theoretically have failed. Spherelike molecules packed close together scatter low energy electrons ($<1 \times 10^{-20}$J) less strongly that do stick-like molecules. In the low density gas phase the reverse is true (György and Freeman, 1979; Gee and Freeman, 1983) and no theoretical treatment of that has succeeded either.

Finally, we need a model for the diffusion of the localized state itself, and a more accurate model for the energies and shapes of optical absorption spectra of solvated electrons.

SI UNITS FACILITATE VISUALIZATION OF MODELS

People who now study the liquid state are working in many different fields: solution chemistry, solvated electrons, condensed matter physics, statistical mechanics, spectroscopy, computer simulation, electrical engineering, and so on. Unfortunately, communication between people in different fields suffers from many jargon barriers. The International System of Units (SIU) is being developed to accommodate all fields of science and engineering so that all workers will have common ground in the units and quantities of their languages. This makes communication much easier.

A difficulty is that when a given name has been used in different fields to represent different things, and SIU selects one of them, some people continue to use the name in the non SI manner. An example is the word frequency, which in spectroscopy means ν(cycles/s), but in solid state physics is commonly used to represent ω(radians/s). In SI ν is called frequency and ω is called angular speed, but unfortunately both are presently assigned the unit s^{-1}. This makes it easy to continue calling ω frequency and sometimes causes error of 2π, since $\omega = 2\pi\nu$.

We visualize frequency as a counting, digital concept (rate of producing completed cycles). Angular speed is visualized as a continuous, analog concept (rate of change of rotation angle or phase angle). To make these visualizations inherent in the SI units we can complete the units by including cycle (cy) and radian (rad). Then perhaps everyone would call ω (rad/s) angular speed and the word frequency would be confined to ν(cy/s).

The cycle has not yet been accepted as a unit in SI, but the proposal is now being considered. I hope you will support it. Radian and steradian (sr) are already part of SI, and sr is equivalent to rad^2.

Table 2 contains a number of additions to SIU that would facilitate the visualization of processes represented by equations (Freeman, 1987f). An example of how the units proposed in Table 2 can correct common errors is the following. Plasma "frequency" is commonly reported to be (Feynman et al., 1964)

$$\omega_p = (n_e \xi^2 / m_e \varepsilon_o)^{1/2} , \tag{24}$$

Table 2. Proposed completed SI units.

new:	cycle	cy
Already SI:	radian	rad
	steradian	sr(=rad^2)

Quantity	Symbol	Units	
		Present	Proposed
angular speed	ω	rad/s	rad/s
frequency,	ν, Hz	s^{-1}	cy/s
Hooke's constant	k_H		
linear	k_{Hx}	kg/s^2	kg/s^2
sinusoidal	$k_{H\phi}$	kg/s^2	$kg\cdot rad^2/s^2$
moment arm	r	m	m/rad
moment of inertia	I	$kg\cdot m^2$	$kg\cdot m^2/rad^2$
momentum, linear	p_x	kg·m/s	kg·m/s
angular	p_ϕ	$kg\cdot m^2/s$	$kg\cdot m^2/s\cdot rad$
period	τ	s	s/cy
permittivity of vacuum	ε_o	C/V·m	C/V·m·sr
	$4\pi\varepsilon_o$	C/V·m	C/V·m
Planck's constant	h	J·s	J·s/cy
h/2π (h-bar)	$\hbar$	J·s	J·s/rad
radianlength	$\lambda\!\!\!^- = \lambda/2\pi$	m	m/rad
radiannumber	$k_r = \lambda\!\!\!^{-\,-1}$	m^{-1}	rad/m
torque	$T = Id\omega/dt$	$kg\cdot m^2/s^2$	$kg\cdot m^2/rad\cdot s^2$
wavelength	λ	m	m/cy
wavenumber	$k_\lambda = \lambda^{-1}$	m^{-1}	cy/m

where n_e is the number of electrons per unit volume of plasma, and m_e is the mass of an electron. The units of the right side of Eq. (24) are

$$\left[\frac{\text{electron}}{\text{m}^3} \cdot \left(\frac{\text{C}}{\text{electron}}\right)^2 \cdot \frac{\text{electron}}{\text{kg}} \cdot \frac{\text{V.m.sr}}{\text{C}}\right]^{1/2} = \left[\frac{\text{C.V.rad}^2}{\text{m}^2\text{kg}}\right]^{1/2}$$

$$= \text{rad/s} \ . \qquad (24')$$

Thus, ω_p is the rate of change of phase, or the angular speed, of electron density oscillations in the plasma. The angular speed of the electron oscillations is proportional to the square root of the restoring force in a nonuniform electron distribution; the restoring force $n_e \xi^2/\varepsilon_o$ is due to the electrostatic repulsion between the electrons.

The SI expression for plasma frequency, cy/s, is (Reitz et al., 1979)

$$\nu_p = \omega_p/2\pi \ . \qquad (25)$$

It was recently demonstrated that errors of 2π in the literature have caused havoc in some areas of electron physics (Gee and Freeman, 1986).

REFERENCES

Basak, S., and Cohen, M.H., 1979, Deformation potential theory for the mobility of excess electrons in liquid argon, Phys. Rev. B., 20:3404.

Cipollini, N.E., and Allen, A.O., 1977, Electron mobilities in liquid tetramethylsilane at temperatures up to the critical point, J. Chem. Phys., 67:131.

Cohen, M.H., and Lekner, J., 1967, Theory of hot electrons in gases, liquids, and solids, Phys. Rev., 158:305.

Dodelet, J.-P., and Freeman, G.R., 1977a, Electron mobilities in alkanes through the liquid and critical regions, Can. J. Chem., 55:2264.

Dodelet, J.-P., and Freeman, G.R., 1977b, Electron mobilities in fluids through the liquid and critical regions: Isomeric butenes, Can. J. Chem., 55:2893.

Dodelet, J.-P., Shinsaka, K., Kortsch, U., and Freeman, G.R., 1973, Electron ranges in liquid alkanes, dienes, and alkynes: Range distribution function in hydrocarbons, J. Chem. Phys., 59:2376.

Dodelet, J.-P., Shinsaka, K., and Freeman, G.R., 1976, Molecular structure effects on electron ranges and mobilities in liquid hydrocarbons: Chain branching and olefin conjugation: Mobility model, Can J. Chem., 54:744.

Feynman, R.P., Leighton, R.B., and Sands, S., 1964, "The Feynman Lectures on Physics," Vol. 2, Addison-Wesley, Reading, MA, p. 7-7.

Floriano, M.A., and Freeman, G.R., 1986, Electron transport in liquids: Effect of unbalancing the sphere-like methane molecules by deuteration, and comparison with argon, krypton, and xenon, J. Chem. Phys., 85:1603.

Freeman, G.R., 1966, Electrons and ions in the radiolysis of liquids, lecture notes for Gordon Conference on Radiation Chemistry, New Hampton, N.H., p. 3.

Freeman, G.R., 1972, Energy decay of energetic electrons in liquids, Quaderni dell' Area di Ricerca dell' Emilia-Romagna, 2:55.

Freeman, G.R., 1987a, Aharonov-Bohm effect observed by electron holography and by electron transmission through split conductors, Phys. Rev. Lett., (submitted).

Freeman, G.R., 1987b, Quantum interference effect for two atoms radiating a single photon, Phys. Rev. Lett., (submitted).

Freeman, G.R., 1987c, Ionization and charge separation in irradiated materials, in: "Kinetics of Nonhomogeneous Processes," G.R. Freeman, ed., Wiley, New York, Chapter 2.

Freeman, G.R., 1987d, Stochastic model of charge scavenging in liquids under irradiation by electrons or photons, in: Kinetics of Nonhomogeneous Processes," G.R. Freeman, ed., Wiley, New York, Chapter 6.

Freeman, G.R., 1987e, Estimation of electron mobilities in simple fluids near the critical point, Phil. Mag. Lett., 56:47.

Freeman, G.R., 1987f, SI units of frequency, angular velocity, Planck's constant and $\hbar$, Metrologia, 23:221.

Gee, N., and Freeman, G.R., 1983, Effects of molecular properties on electron transport in hydrocarbon fluids, J. Chem. Phys., 78:1951.

Gee, N., and Freeman, G.R., 1986, Electron transport in dense gases: Limitations on the Ioffe-Regel and Mott criteria, Can. J. Chem., 64:1810.

Gee, N., and Freeman, G.R., 1987, Electron mobilities, free ion yields, and electron thermalization distances in liquid, long-chain hydrocarbons, J. Chem. Phys., 86:5716.

Grangier, P., Aspect, A., and Vique, J., 1985, Quantum interference effect for two atoms radiating a single photon, Phys. Rev. Lett., 54:418.

Gushchin, E,M., Kruglov, A.A., and Obodovskii, I.M., 1982, Electron dynamics in condensed argon and xenon, Sov. Phys. JETP,, 55:650.

György, I., and Freeman, G.R., 1979, Effects of density and temperature on electron transport in hydrocarbon fluids, J. Electrostatics, 7:239.

Huang, S.S.-S., and Freeman, G.R., 1978a, Electron mobilities in gaseous, critical, and liquid xenon: Density, electric field, and temperature effects: Quasilocalization, J. Chem. Phys., 68:1355.

Huang, S.S.-S., and Freeman, G.R., 1978b, The gravity effect and the mobility of electrons in critical neopentane, J. Chem. Phys. 69:1585.

Huang, S.S.-S., and Freeman, G.R., 1980, Electron transport in gaseous, critical and liquid benzene and toluene, J. Chem. Phys., 72:2849.

Huang, S.S.-S., and Freeman, G.R., 1981, Electron transport in gaseous and liquid argon: Effects of density and temperature, Phys. Rev. A, 24:714.

Jacobsen, F.M., Gee, N., and Freeman, G.R., 1986, Electron mobility in liquid krypton as functions of density, temperature, and electric field strength, Phys. Rev. A, 34:2329.

Jahnke, J.A., Meyer, L., and Rice, S.A., 1971, Zero-field mobility of excess electrons in fluid argon, Phys. Rev. A, 3:734.

Jou, F.-Y., and Freeman, G.R., 1976, Optical spectra of electrons solvated in liquid ethers: Temperature effects, Can J. Chem., 54:3693.

Kuan, D.-Y., and Ebner, C., 1981, Theory of excess-electron states in classical rare-gas fluids, Phys. Rev. A, 23:285.

Lekner, J., 1967, Motion of electrons in liquid argon, Phys. Rev., 158:130.

Lekner, J., 1968, Mobility maxima in the rare-gas liquids, Phys. Lett., 27A:341.

Lekner, J., and Bishop, A.R., 1973, Electron mobility in simple fluids near the critical point, Phil. Mag., 27:297.

Loveland, R.J., Le Comber, P.G., and Spear, W.E., 1972, Experimental evidence for electronic bubble states in liquid neon, Phys. Lett., 39A:225.

Mozunder, A., and Magee, J.L., 1967, Theory of radiation chemistry. VIII. Ionization of nonpolar liquids by radiation in the absence of an external electric field, J. Chem. Phys., 47:939.

Muñoz, R.C., and Ascarelli, G., 1983, Hall mobility of electrons injected into fluid neopentane (dimethylpropane) along the liquid-vapor coexistence line between the triple and critical points, Phys. Rev. Lett., 51:215.

Ostermeier, R.M., and Schwarz, K.W., 1972, Motion of charge carriers in normal helium-four, Phys. Rev. A, 5:2510.

Reitz, J.R., Milford, F.J., and Christy, R.W., 1979, "Foundations of Electromagnetic Theory," 3rd edn., Addison-Wesley, Reading, MA, p. 308.

Ryan, T.G., and Freeman, G.R., 1978, Electron mobilities and ranges in methyl substituted pentanes through the liquid and critical regions, J. Chem. Phys., 68:5144.

Sakai, Y., Böttcher, H., and Schmidt, W.F., 1982, Excess electrons in liquid hydrogen, liquid neon, and liquid helium, J. Electrostatics, 12:89.

Schwarz, K.W., 1972, Charge-carrier mobilities in liquid helium at the vapor pressure, Phys. Rev. A, 6:837.

Schwarz, K.W., 1975, Mobilities of charge carriers in superfluid helium, Adv. Chem. Phys., 33:1.

HOT ELECTRON MOBILITY AND ELECTRON ATTACHMENT IN NON-POLAR LIQUIDS

Werner F. Schmidt

Hahn-Meitner-Institut Berlin, Bereich Strahlenchemie
D1000 Berlin 39, FRG

INTRODUCTION

Electron transport in electric fields of high strength is of fundamental and practical interest. The study of the electron mobility and of electron attachment as a function of the electric field strength can give important information on the physics of the transport process. A detailed knowledge of the elementary processes which govern electron mobility and attachment is the presupposition for the development of an electrotechnology based on liquid dielectrics. In this lecture we give a short resume of the electron transport in high mobility liquids at high electric field strength, and we review some facts on the attachment of electrons to electronegative molecules. The examples and references have been chosen in such a way that the interested reader will have no difficulty to locate most of the important publications by means of an information retrieval system.

ELECTRON MOBILITY

Electrons in a medium move under the influence of an electric field E with a drift velocity v_d given by

$$v_d = \mu_{el} E \text{ , where} \tag{1}$$

μ_{el} is the electron mobility; at low electric field strength, μ_{el} is constant. It depends on the temperature T of the medium and on the molecular structure of the liquid (Schmidt, 1984). At the same time the electrons exchange energy with the atoms or molecules of the medium and they have a random thermal velocity given by

$$v_{th} = \sqrt{\frac{3\ k_B T}{m}} \text{ , where} \tag{2}$$

k_B is the Boltzmann constant, T the absolute temperature, and m the electron mass. The path of an electron under these conditions is depicted in Fig. 1. The drift velocity v_d can be calculated from the equation of motion

$$m \frac{d\vec{v}}{dt} = - e \vec{E} \text{ , where} \tag{3}$$

m is the electron mass, and e the electronic charge. Integration of Eq. (2), taking into account a Maxwellian distribution of electron velocities, yields

$$\vec{v}_d = - \frac{e}{m} \tau_{el} \vec{E} \text{ ;} \tag{4}$$

τ_{el} is the mean free time between collisions. Comparison of Eqs. (1) and (4) gives for the mobility

$$\mu_{el} = \frac{e}{m} \tau_{el} \text{ .} \tag{5}$$

The mean free time is related to the mean free path Λ by the thermal velocity

$$\Lambda = v_{th} \tau_{el} \text{.} \tag{6}$$

At low electric field strength we have

$$v_d << v_{th} \text{ ,} \tag{7}$$

and μ_{el} is constant. At higher electric field strength the condition of Eq. (7) is no longer fulfilled. Electrons gain more energy from the field than they can dissipate in collisions. The result is an increase of the mean electron energy above $k_B T$.

The energy gain between collisions is given by

$$\Delta T_f = (e E) (\mu_{el} E) \tau_e$$

$$= \frac{e^2 E^2}{m} \frac{\Lambda^2}{v_{el}^2} \text{ , where} \tag{8}$$

Fig. 1. Schematic path of an electron in an electric field.

v_{el} is the mean electron velocity. The energy loss in a collision is given by

$$\Delta T_c = \Delta T_{kin} f \text{ , where} \tag{9}$$

T_{kin} is the kinetic energy of the electron, and f is the fractional energy loss given as

$$f = \frac{2m}{M} , \tag{10}$$

in the elastic limit; M is the mass of an atom or molecule comprising the liquid. Equilibrium is reached when

$$\Delta T_f = \Delta T_c , \tag{11}$$

or

$$\frac{e^2 E^2}{m} \frac{\Lambda^2}{v_{el}^2} = \frac{m}{2} v_{el}^2 f , \tag{12}$$

or

$$v_{el} = \sqrt{\frac{e E}{m} \Lambda \sqrt{2/f}} . \tag{13}$$

Under these conditions v_{el} increases with $\sqrt{E}$ and the mean electron energy increases with E. The drift velocity is obtained by replacing v_{th} in Eqs. (5) and (6) by v_{el}:

$$v_d = \sqrt{\frac{e E}{m} \Lambda \sqrt{\frac{f}{2}}} . \tag{14}$$

The field strength E_c at which the transition from $v_d \sim E$ to $v_d \sim \sqrt{E}$ occurs is given by the condition

$$v_d/v_{th} = \sqrt{f/2} , \tag{15}$$

or

$$E_c = \sqrt{f/2} \frac{v_{th}}{\mu_{el}} . \tag{16}$$

In considering the motion of electrons in non-polar liquids one has to take into account the influence of the liquid structure on the electron/atom or electron/molecule scattering process. For liquid argon, Cohen and Lekner (1967) and Lekner (1967) related the cross section in the liquid to the gas-phase cross section for scattering via the structure factor S(0) of the liquid. The introduced two mean free paths, Λ_o and Λ_1, for energy and momentum transfer, respectively. Both mean free paths are related by the structure factor

$$\frac{\Lambda_0}{\Lambda_1} = S(0) \ . \tag{17}$$

The limit of the low-field region where μ_{el} is independent of E, is then given by the condition

$$\frac{1}{3} (e E_c \Lambda_1)(e E_c \Lambda_o) = f (k_b T)^2 \ . \tag{18}$$

For E_c equation (16) with different numerical factor follows. The critical field strength E_c is inversely proportional to the electron mobility. The lower the electron mobility, the higher the critical electrical field strength at which the deviation from proportionality between v_d and E occurs. Examples of the electron drift velocity as a function of the electric field strength in liquefied rare gases and in some non-polar molecular liquids are shown in Fig. 2. A compilation of μ_{el} and E_c values is given in Table 1. Further increase of the electric field strength leads to a saturation of the drift velocity with respect

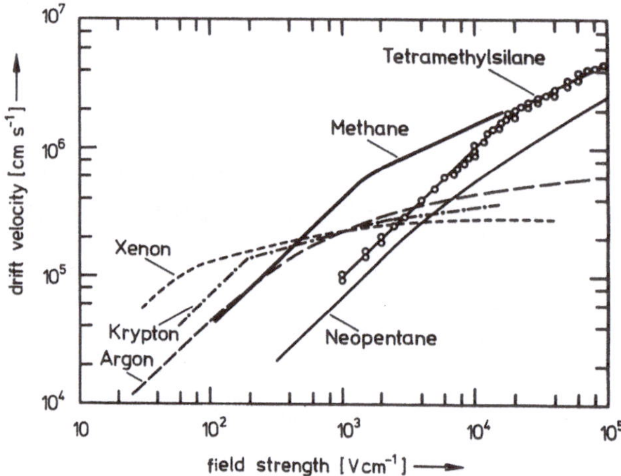

Fig. 2. Electron drift velocity as a function of the electric field strength: argon, 85 K; krypton, 121 K; xenon, 163 K; methane, 111 K; neopentane, 196 K; tetramethylsilane, 296 K.

Table 1. Zero-field mobilities and critical field strengths.

Liquid	T (K)	$\mu_{el}(cm^2V^{-1}s^{-1})$	E_c (V/cm)
Argon	85	400	300
Krypton	120	1200	180
Xenon	165	2000	50
Methane	111	400	1500
Neopentane	296	70	20000
Tetramethylsilane	296	100	20000

to the electric field. This effect has been observed in the heavier liquefied rare gases (LAr, LKr, LXe) and in liquid methane and liquid tetramethylsilane (see Figs. 3 and 4). The saturation velocities and the low field mobilities are compiled in Table 2. Different models have been developed to explain the saturation velocity.

In the molecular liquids it can be assumed that excitation of vibrations stabilizes the drift velocity. In semiconductor theory (Kireev, 1978) a relationship has been derived between the saturation drift velocity v_s and the energy quantum (phonon) W_p emitted by the electron

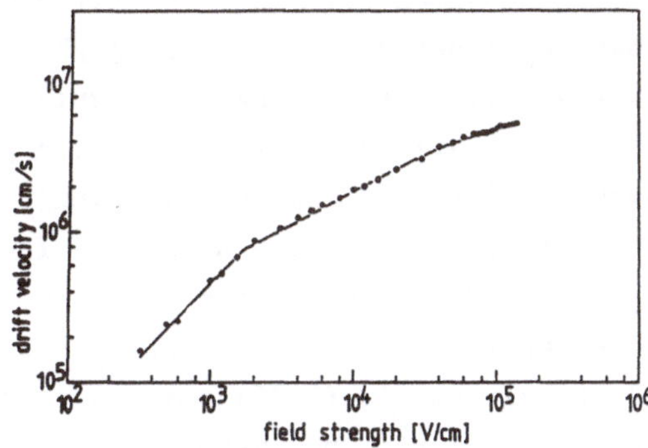

Fig. 3. Electron drift velocity in liquid methane, 111 K.

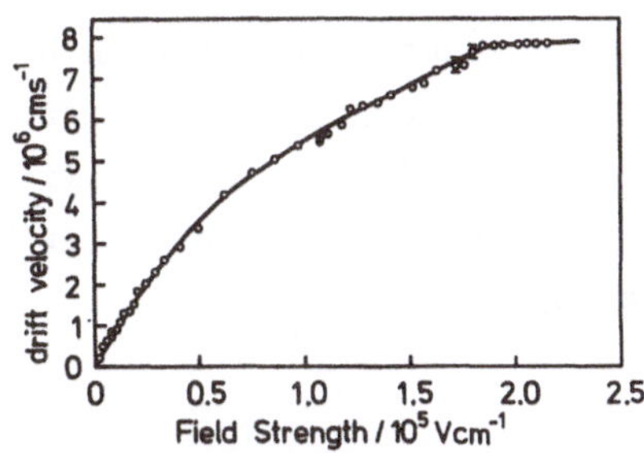

Fig. 4. Electron drift velocity in tetramethylsilane, 296 K.

Table 2. Saturation drift velocities

Liquid	T (K)	v_s (cm/s)
Argon	87	6.4×10^5
Krypton	120	4.8×10^5
Xenon	165	2.6×10^5
Methane	111	6×10^6
Tetramethylsilane	296	8×10^6

$$v_s \sim \sqrt{\frac{8\ W_p}{3\ m}}\ . \tag{19}$$

Evaluation of the saturation velocities in TMSi and methane (of Table 2) yield W_p(TMSi) = 0.04 eV and W_p(Methane) = 0.03 eV, respectively. These energies correspond to the excitation of molecular vibrations of the molecules. They are in the energy range in which electrons produced by high energy radiation achieve most of their thermalization distance.

In the heavier liquefied rare gases, different explanations for the saturation velocity have been given. Spear and LeComber (1969) assumed that the effective mass of the electron increases with electron energy. Nakamura et al. (1986) assumed that at higher electron energies inelastic collision processes become important. Inelastic light scattering modes have been observed in these liquids and they are explained with collision-induced changes of translational states of atomic pairs, triplets, etc. Within the framework of the Boltzmann collision equation Cohen and Lekner (1967) used measured drift velocity data as a function of the electric field strength in order to calculate the mean electron energy $\bar{\xi}$ at a particular field strength. Using their approach with the more detailed data of Fig. 2, Bakale et al. (1976) obtained the electron energy distribution functions shown in Fig. 5. Gushchin et al. (1982), taking the same approach, calculated $\bar{\xi}$ as a function of E for different rare gases. Their results are shown in Fig. 6. Increase of the electric field above 10^5 V/cm finally leads to excitation of electronic levels and the liquid emits light. This effect is used in scintillation counters (Doke, 1982). Further increase of the electric field strength leads to collisional ionization. This process is characterized by an ionization coefficient $\alpha(cm^{-1})$ given by

$$n(x) = n_o \exp(\alpha\ x)\ , \tag{20}$$

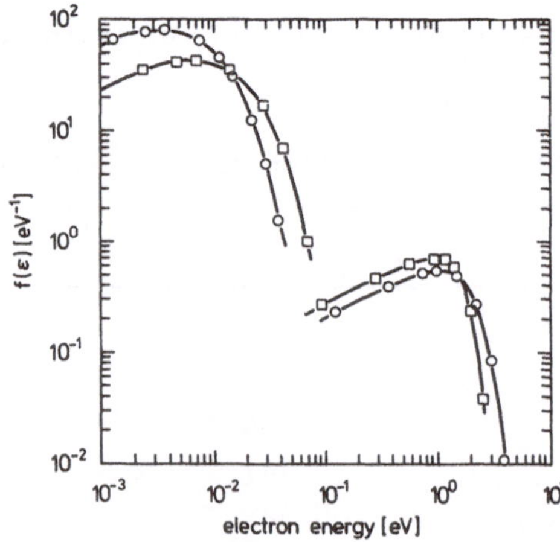

Fig. 5. Electron energy distribution in liquid xenon (□) and liquid argon (o); left 10 V/cm, right 100 kV/cm.

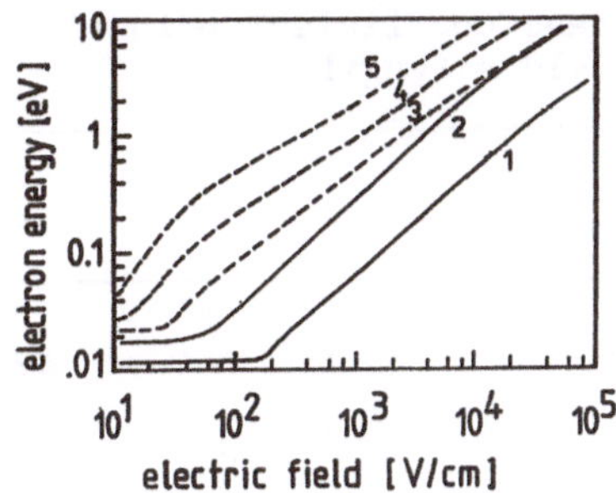

Fig. 6. Electron mean energy as a function of the electric field strength; 1-LAr, 85 K; 2-LAr, 140 K; 3-LXe, 165 K; 4-LXe, 200 K; 5-LXe, 230 K.

where n_o is the number of electrons starting at $x = 0$ in a homogeneous electric field and $n(x)$ is the augmented number at distance x due to collisions (see Fig. 7). Derenzo et al. (1974) determined α as a function of the electric field strength and obtained the values compiled in Table 3. Liquid xenon is the only liquid where such a Townsend type electron multiplication process has been observed so far. In LAr, evidence for collisional ionization seems to exist but no α -values have been reported yet.

An in-depth treatment of the problem of hot electrons in liquefied rare gases has been given by Iakubov and his co-workers (Atrazhev and Iakubov, 1981; Atrazhev and Dmitriev, 1985). The electric field dependence of the mean electron energy (cf, Fig. 6) has been verified experimentally by the measurement of the ratio of the diffusion coefficient to mobility in liquid argon and liquid xenon by Shibamura et al. (1979) and Kubota et al. (1982).

ELECTRON ATTACHMENT

If an electro-negative solute S is present in the liquid, electron attachment occurs:

$$e + S \xrightarrow[k_s]{} S^- . \qquad (21)$$

A highly mobile electron is converted into a slow ion. This process can be observed in the measurement of the radiation-induced conductivity

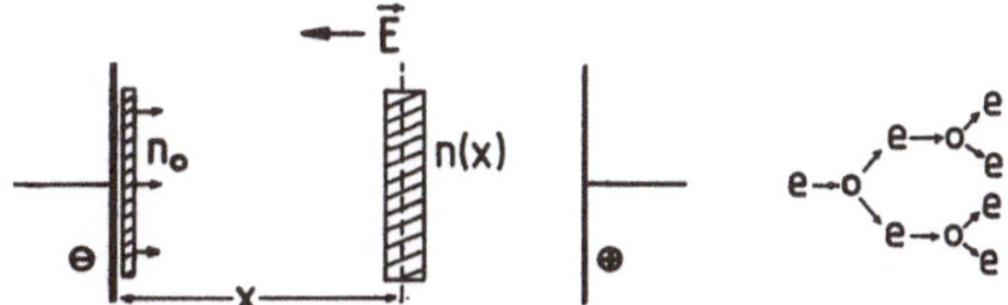

Fig. 7. Schematic representation of collisional electron multiplication.

Table 3. First Townsend coefficient α in LXe as a function of the electric field strength.

$E(kV\ cm^{-1})$	$\alpha\ (cm^{-1})$
400	$470\ ^{+\ 600}_{-\ 470}$
600	8050 ± 3500
800	$2550\ ^{+\ 7000}_{-\ 2550}$
1000	$8300\ ^{+\ 3400}_{-\ 4700}$
1200	15000 ± 6000
1400	27500 ± 6000
1600	33500 ± 4500
1800	38400 ± 4500
2000	44700 ± 2600

after a short burst of x-rays (Bakale et al., 1973). An example is given in Fig. 8. The electron concentration n_{el} decays due to attachment as

$$\frac{dn_{el}}{dt} = - k_s\ [S]\ n_{el}\ , \tag{22}$$

where [S] is the concentration of the electron-attaching compound. Integration of Eq. (22) yields

$$n_{el}(t) = n_{el}(0)\ \exp(-t/\tau)\ , \tag{23}$$

where τ represents the lifetime of the electrons given as

$$\tau = \frac{1}{k_s\ [S]}\ . \tag{24}$$

Generally, k_x is a function of the mean electron energy and in this way it depends on the electric field strength in liquids where non-thermal electrons may exist. In liquid argon and liquid xenon attachment of hot electrons to SF_6, N_2O, and O_2 has been observed (Fig. 9). The field-strength-dependent rate constant $k_s(E)$ is given by

$$k_s(E) = \int_0^\infty v\ \sigma(v)\ F(v,E)\ dv\ . \tag{25}$$

$\sigma(v)$ is the cross section for electron capture, v denotes the electron velocity, and F(v,E) is the electron velocity distribution as a function

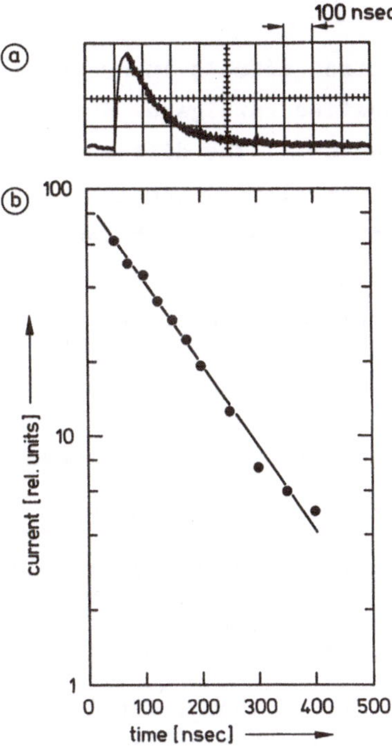

Fig. 8. Temporal variation of the radiation-induced electron current in liquid methane (111 K) due to electron attachment to oxygen.

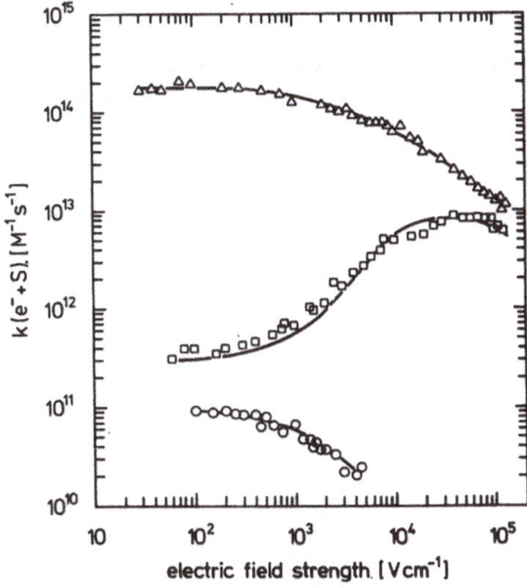

Fig. 9. Rate constant for the attachment of electrons to several solutes in liquid argon (87 K): Δ SF_6; □ N_2O; o O_2.

of the electric field strength. With the velocity or energy distribution known, σ(v) or σ(ε), respectively, can be obtained by deconvolution of the integral of Eq. (25). This procedure is, however, not so sensitive with respect to the exact variation of σ with v (cf, Bakale et al., 1976).

CONCLUSIONS

The study of hot-electron effects in non-polar liquids has been limited so far to the liquefied rare gases Ar, Kr, and Xe and to two molecular liquids, TMSi and CH_4. The difficulties lie in the application of high electric fields to the liquid samples which often leads to electric breakdown. New experimental techniques involving pulsed electric fields must be developed in order to obtain more information on this fundamentally and practically important topic.

REFERENCES

Atrazhev, V.M., and Dmitriev, E.G., 1985, J. Phys. C: Solid State Phys., 18:1205.

Atrazhev, V.M., and Iakubov, I.T., 1981, J. Phys. C: Solid State Phys., 14:5139.

Bakale, G., Schmidt, W.F., and Naturforsch, Z., 1973, 28a:511.

Bakale, G., Sowada, U., and Schmidt, W.F., 1976, J. Chem. Phys., 80:2556.

Cohen, M.H., and Lekner, J., 1967, Phys. Rev., 158:305.

Derenzo, S.E., Mast, T.S., Zaklad, H., and Miller, R.A., 1974, Phys. Rev. A9, 2582.

Doke, T., 1982, Nucl. Instrum. Methods, 196:87.

Gushchin, E.M., Kruglov, A.A., and Obodovskii, I.M., Sov. Phys. JETP, 55:650.

Kireev, P.S., 1978, "Semiconductor Physics", Mir Publishers, Moscow.

Kubota, S., Takahashi, T., and Ruan, J., 1982, J. Phys. Soc., Japan, 51:3275.

Lekner, J., 1967, Phys. Rev., 158:130.

Nakamura, S., Sakai, Y., and Tagashira, H., 1986, Chem. Phys. Letters, 130:551.

Schmidt, W.F., 1984, IEEE Trans. El. Insul. EI-19, 389.

Shibamura, E., Takahashi, T., Kubota, S., and Doke, T., 1979, Phys. Rev. A20, 2547.

Spear, W.E., and LeComber, P.G., 1969, Phys. Rev., 178:1454.

GAS/LIQUID TRANSITION: INTERPHASE PHYSICS*

L. G. Christophorou

Atomic, Molecular, and High Voltage Physics Group
Health and Safety Research Division
Oak Ridge National Laboratory
Oak Ridge, Tennessee 37831-6122 USA

Department of Physics
The University of Tennessee
Knoxville, Tennessee 37996 USA

INTRODUCTION

In this chapter we focus on interphase physics - a rather new and transdisciplinary field of science - which deals with the effects of the nature, density, and state of the medium on the microscopic and macroscopic properties of - and processes in - matter in the intermediate density range between the low-pressure gas and the condensed (liquid) state. Such studies traditionally begin either at the liquid density and move progressively to lower - but still high - densities or begin at low densities (low pressure gases) and gradually move to denser and denser gaseous media. The former (the chemical) approach has been amply discussed by others in this volume; the latter (the physical) approach is the one we adopted in this chapter. Both approaches contributed greatly to the bridging of the density gap between the low pressure gas and the liquid.

EXAMPLES OF MEDIUM-DEPENDENT PROCESSES

In emission (fluorescence) spectroscopy two classic examples of density-dependent processes are the phenomena of excimer and exciplex formation and decay, and the imprisonment (trapping) of resonance radiation in atomic (e.g., Hg, rare gases) gases.

---

* Research sponsored by the Office of Health and Environmental Research, U.S. Department of Energy, under contract DE-AC05-84OR21400 with Martin Marietta Energy Systems, Inc.

In absorption spectroscopy two examples of medium-dependent processes are the well-known phenomena of pressure broadening of spectral lines in gases and solvent shifts in liquids. In general, however, the distribution of oscillator strength in low-pressure gases is not too dissimilar from that in dense gases and liquids except perhaps for the postulated (e.g., see Williams et al., 1975) collective excitations (plasmons) in liquids, analogous to those in metals.

Low-lying valence states of molecules in dense gases and liquids generally retain their identity in dense media. This can be seen by the example shown in Fig. 1a where the $^1A_{1g} \rightarrow ^1B_{2u}$ transitions of benzene vapor are seen to be virtually identical with and without a high-pressure helium background gas. Similar spectra in liquid perfluoro-n-hexane show little change from the low-pressure data. In the condensed phase, low-lying processes may dominate over higher-lying ones because of the possibility of extensive energy relaxation. Quasi-charge-separated states such as high-n-Rydberg states (n is the principal quantum number) - in contrast to low-lying valence states - undergo profound changes and (except in rare cases) lose their identity. This is illustrated by the example in Fig. 1b where the 5p → 6s Rydberg transition of methyl iodide (CH_3I) is progressively asymmetrically broadened with increasing N_2 pressure and thus perturbation (see further discussion in the next section).

Charge-separated states (electrons, positive ions, negative ions) and physical quantities that describe their behavior and reactions are seriously affected by the density and nature of the medium. Thus, the electron energy ε, the quasifree (excess) electron "ground state" energy V_o, the electron drift velocity w, the electron mobility μ, the cross sections for electron scattering σ_{sc}, electron capture σ_c, electron attachment σ_a, dissociative attachment σ_{da}, autodetachment σ_{ad}, ionization σ_i, and the associated physical quantities [e.g., the electron affinity EA, the vertical attachment energy VAE, the vertical detachment energy VDE, the ionization threshold energy I, and the polarization energy of the positive (P^+) and the negative (P^-) ions] are all strong functions of the nature and density of the medium in which these elementary physical processes occur. While these physical quantities have well-defined values at low gas number densities, they assume a spectrum of values in dense media.

In this chapter we shall elaborate on the effect of the nature and density of the medium - in the intermediate density range between the low pressure gas and the liquid - focusing on the high-n-Rydberg states and the charge-separated states. In connection with the latter, we shall discuss the effect of the medium on the electron state, electron transport, negative ion states, electron attachment, and ionization.

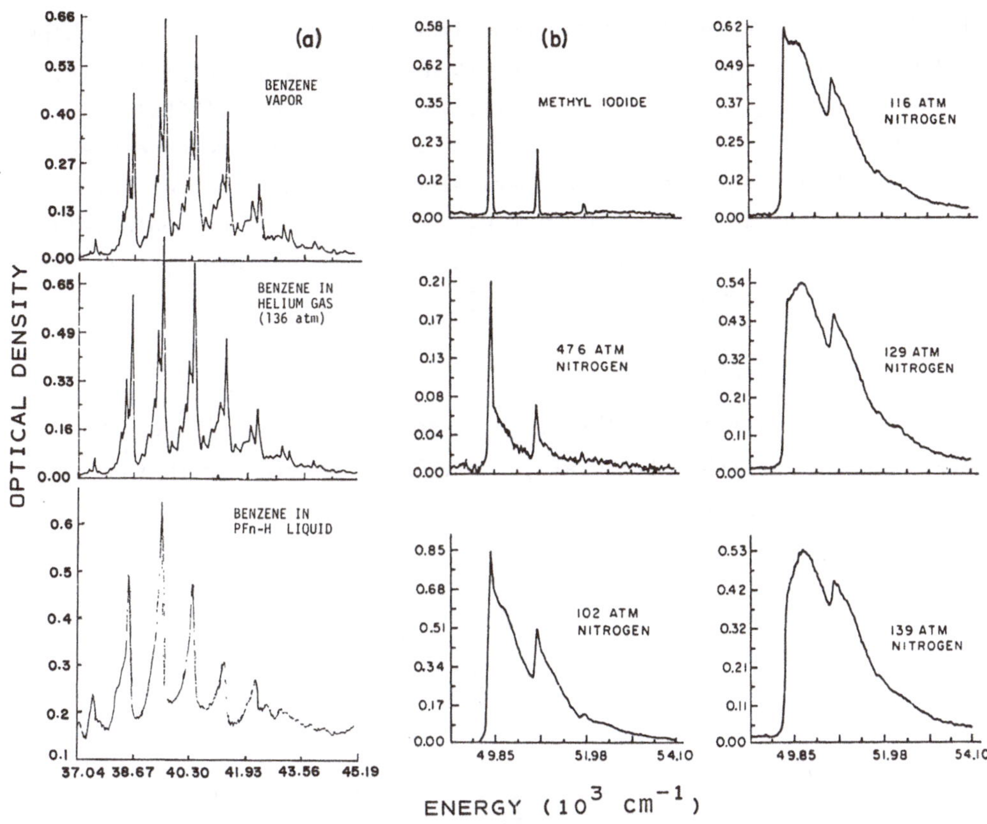

Fig. 1. (a) $^1A_{1g} \rightarrow {}^1B_{2u}$ transitions of benzene vapor before (upper) and after (middle) application of 136 atm of helium gas (Robin, 1974) and in liquid perfluoro-n-hexane (T≈295K; Kourouklis et al., 1982). The spectra are almost identical and are characteristic of valence-shell excitations. (b) Progressive asymmetric broadening of the 5p → 6s (B system) Rydberg transition in methyl iodide under nitrogen-gas perturbation (Robin, 1974). (From Christophorou and Siomos, 1984)

HIGH-LYING RYDBERG STATES

An isolated atom (molecule) in a Rydberg state of sufficiently high n ($\geq$ 10) is nearly hydrogenic. The radius r_n (= $n^2 a_o$; a_o is the Bohr radius) of the Rydberg orbit is large (≃ 53Å for n = 10) and the binding energy B_n(= R_H/n^2; R_H is the Rydberg constant) of the Rydberg electron is small (B_{10} ≃ 0.136 eV). The Rydberg electron is weakly bound and is, therefore, easily ionizable and transferable. Bound-electron capture reactions ($A^{**}_{Ryd} + M \rightarrow A^+ + M^-$; see Christophorou et al., 1984); for example, have shown that the Rydberg electron can roughly be treated independently of the cationic core and virtually as a free electron with sub-thermal kinetic energy (< 0.022 eV for n $\geq$ 25).

How, then do Rydberg states evolve as the gas number density N (or the gas density ρ) increases from a low pressure gas to the condensed matter[1]? The effect of the medium on high-n Rydberg states can perhaps be illustrated by the results in Fig. 1b and Fig. 2a on respectively, CH_3I in N_2 and CH_3I in Ar. The changes in the spectrum of CH_3I in N_2 (and Ar) with increasing pressure (say of N_2) would be profound when the density of N_2 is such that the Rydberg electron cannot complete its orbit due to scattering in collisions with N_2. If l is the electron's mean free path and $L = 2\pi r_n$ is the circumference of the nth orbit, the number density N_c of N_2 at which $l = [\langle\sigma_{sc}\rangle N_c]^{-1} = L$ is

$$N_c \simeq \left[2\pi r_n \langle\sigma_{sc}\rangle\right]^{-1} . \tag{1}$$

For n=10 and $\langle\sigma_{sc}\rangle(N_2) \simeq 10^{-16}cm^2$, $N_c \simeq 3x10^{21}$ molecules cm^{-3} which corresponds to $P_{N_2}(T{\simeq}300K){\simeq}120$ atm, - a value consistent with the observations in Fig. 1b.

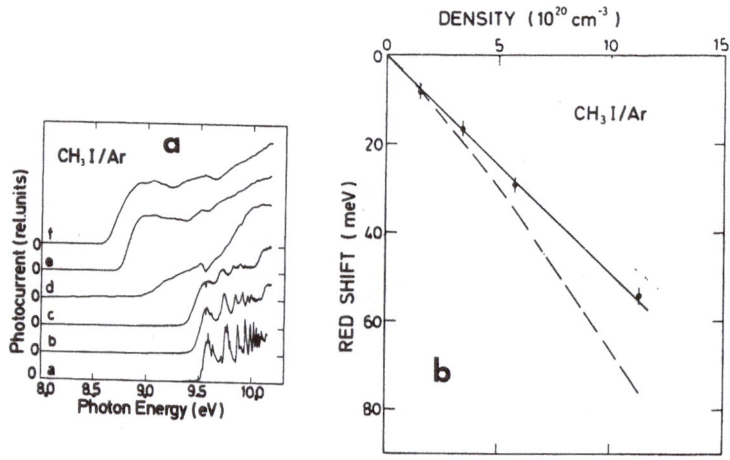

Fig. 2. (a) Photoconductivity excitation spectrum of methyl iodide CH_3I impurity in fluid argon for various argon densities: a) pure CH_3I, 0.1 Torr; b) 0.57; c) 1.1; d) 4.2; e) 8.0 and f) $16x10^{21}$ atoms cm^{-3}. (From Reininger et al., 1985).
(b) CH_3I d′ series energy shift versus Ar number density. • experimental points [the average (for n ≥ 10) d′ energy shift]; . . . . Δsc (ρ); -.-.-. Δρ [Eq. (2)]; ———Δ(ρ) [Eq. (5)]. (From Köhler et al., 1986)

---

1 In condensed matter the energy E_n of the nth Wannier impurity state is (e. g., see Messing et al., 1977) $E_n = E_g - G/n^2$, where $E_g = I_g + V_o + P^+$, I_g is the gas-phase ionization threshold energy, $G = 13.6\ m^*/\varepsilon^2$ is the effective Rydberg constant, m^* is the electron effective mass in the conduction band, and ε is the dielectric constant of the host medium.

The energy shift in E_n and the ionization onset I has generally been treated within the free-electron model of Fermi (1934). In this simple model, the shift $\Delta(\rho) \equiv E_n(\rho) - E_n(\rho=0)$ in E_n with increasing ρ consists of two terms: $\Delta_{sc}(\rho)$ due to the free-electron scattering and $\Delta_p(\rho)$ due to the polarization of the medium by the cationic core, viz.,

$$\Delta(\rho) = \Delta_{sc}(\rho) + \Delta_p(\rho) = \frac{2\pi\hbar^2}{m} a\rho - 10e^2 \alpha\rho^{4/3} , \tag{2}$$

where a is the scattering length and α is the polarizability of the medium. The shifts $\Delta_{sc}(\rho)$ and $\Delta(\rho)$ [see Eq. (2)] are shown in Fig. 2b for the CH_3I d′ series in Ar along with the experimental results of Köhler et al. (1986) on $\Delta(\rho)$. Köhler et al. found that the shifts (and their ionization limits) which they determined experimentally, decreased linearly with ρ, a finding not quite in accord with the Fermi model. Their measurements are in better agreement with the more general treatment of Alekseev and Sobel'man (1966) which gave the same expression for Δ_{sc} as the Fermi model, but a somewhat different expression for $\Delta_p(\rho)$, viz.,

$$\Delta_p(\rho) = -9.87 \left(\frac{\alpha e^2}{2}\right)^{2/3} (\hbar v)^{1/3} \rho , \tag{3}$$

where v is the average speed of the atoms of the medium. The total shift $\Delta(\rho)$ is [Alekseev and Sobel'man, 1966; Köhler et al., 1986].

$$\Delta(\rho) = \frac{2\pi\hbar^2}{m} \rho\, a_{ef} \tag{4}$$

where

$$a_{ef} = a - 9.87 \left(\frac{m}{2\pi\hbar^2}\right) \left(\frac{\alpha e^2}{2}\right)^{2/3} (\hbar v)^{1/3} , \tag{5}$$

is represented in Fig. 2b by the solid line. Köhler et al. (1986) found – in agreement with the theory – that for Ar and Kr for which the scattering length a is negative (<0) the shift is negative (to the red) and that for He for which a is positive (>0) the shift is positive (to the blue).

THE ELECTRON STATE IN THE DENSE GAS AND THE LIQUID

In low-pressure gases ($P \lesssim 1$ atm) the electron mean free path l is much larger than the electron de Broglie wavelength λ and the electrons in such gases are free. Their transport properties are treated within the Boltzmann transport equation and their kinetic energies can be well in excess of thermal depending on the value of the density reduced electric field E/N. Their steady-state energy distributions $f(\varepsilon, E/N)$ are non-Maxwellian (except at very low E/N) and - depending on the gas, temperature (T), and E/N - their peak energies range from 3/2kT to $\lesssim$10 eV (see Christophorou, 1971, 1984).

In dense gases ($\gtrsim$ 400 atm) and liquids, 1 < $\bar{\lambda}$ ($\lambda = 2\pi\bar{\lambda}$) and the (excess) electrons in such media are either quasi-free (e_{qf}) and/or localized (e_l) depending on the medium, N, and T. Electrons are generally localized in densed media whose V_o > 0 eV and are quasi-free in densed media whose V_o <0 eV; V_o is <0 eV for a <0 and V_o is >0 eV for a>0. The V_o itself can be a function of N as can be seen from the V_o(N) data in Fig. 3 for the heavier rare gases whose scattering length is negative [for He whose scattering length is positive the V_o increases with increasing N (Asaf and Steinberger, 1986)]. Quasi-free electrons have much higher mobilities than localized electrons as can be seen from the representative sample data on V_o and thermal electron mobilities for a number of liquids shown in Table 1 [for further discussion see Christophorou and Siomos (1984), Schmidt (1984), and Freeman (1987)]. During their drift, the electrons can, of course, be partly in the localized and partly in the quasi-free state.

The energies of the excess electrons in dense media are generally thermal. In dense media with negative V_o values, however, the electrons can - at high E/N - attain energies well in excess of thermal as can be seen from the data on liquid Ar in Fig. 4 (Christophorou, 1985). Similar situations have been reported for liquid Xe [Schmidt (1984), Gushchin et

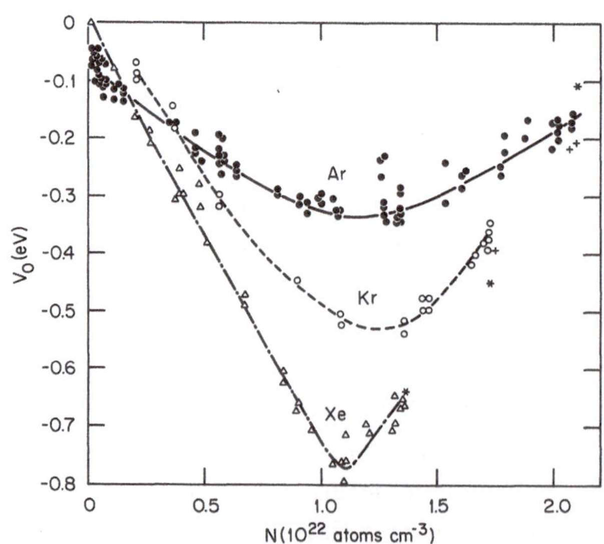

Fig. 3. V_o versus N for Ar, Kr, and Xe • (Ar; Reininger et al., 1983); o (Kr; Reininger et al., 1982); Δ (Xe; Reininger et al., 1982); * (Von Zdrojewski et al., 1980); + (Tauchert et al., 1977).

Table 1. Values of V_o and Thermal Electron Mobility for Some Liquids (From Christophorou and Siomos, 1984)

	LIQUID (T IN K)	ELECTRON ENERGY RELATIVE TO VACUUM V_o (eV)	ELECTRON MOBILITY (T) $cm^2\ V^{-1}\ s^{-1}$	(K)
	Xe (161)	- 0.67	2200	(163)
e_{qf}	Ar (87)	- 0.20	475	(85)
a<0	Neopentane (296)	- 0.43	55	(296)
	He (1.1)	+ 1.02	0.04	(4.2)
e_1	n-Hexane (292)	0.02	0.06	(294)
a>0	n-Hexane (193)	0.21	0.005	(229)

al., (1982), Warman and de Haas (1985)] and tetramethyl silane [Bakale and Beck (1986)].

The state of the electron and its energy crucially depend on the medium and profoundly affect the magnitude of the interaction cross sections of excess electrons in dense media.

EFFECT OF DENSITY ON ELECTRON TRANSPORT

Examples of Experimental Observations

Up until the early 1960's the electron transport coefficients w (electron drift velocity) and D_T (transverse electron diffusion coefficient) were considered to be independent of N. This is probably because the experimental studies were traditionally conducted at low N (gas pressures $\lesssim$ 1 atm). The transport coefficients w and D_T [actually D_T/μ where μ is the electron mobility ($w=\mu E$)] are functions of the gas, T, and E/N (e.g., see Hunter and Christophorou, 1984).

The electron drift velocity w (or μ or μN) has since been found[2] to decrease slowly (e.g., for H_2, N_2, C_2H_6, C_3H_8, and other species with small electron scattering cross sections) or to decrease strongly (e.g., for He, CO_2, 1-C_3F_6 for which permanent or transient anions are formed as N increases and for polar media for which the electron scattering cross section is large due to the electron-electric dipole interaction) with

[2] The author is not aware of any reports on the dependence of D_T/μ on N, although some small changes have been predicted (O'Malley, 1983).

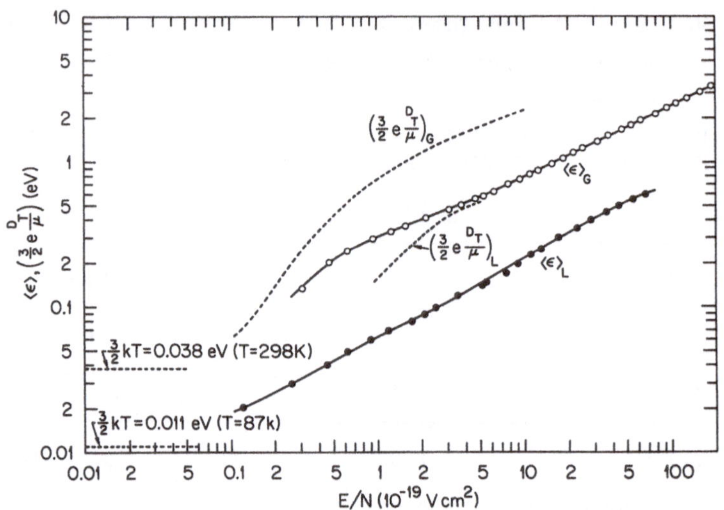

Fig. 4. Mean electron energy $\langle\varepsilon\rangle_G$ in gaseous Ar (Christophorou, 1984), characteristic energy $\left(\frac{3}{2} e \frac{D_T}{\mu}\right)_G$ of electrons in gaseous Ar (based on the $\frac{D_T}{\mu}$ data of Miloy and Crompton, 1977), characteristic energy $\left(\frac{3}{2} e \frac{D_T}{\mu}\right)_L$ of electrons in liquid Ar [T=87K; based on the $\frac{D_T}{\mu}$ measurements of Shibamura et al., 1979) and mean electron energy $\langle\varepsilon\rangle_L$ in liquid Ar estimated by Christophorou (1985), all as a function of E/N. (From Christophorou, 1985)

increasing N. For polar media the changes in w (or μ) begin at much lower values of N. On the other hand the w (or μ or μN) values were found to increase with N for the heavier rare gases and the spherical hydrocarbon dielectric liquids for which V_o<0eV.

In Fig. 5 are shown early examples of the variation of the ratio $\mu_p'/\mu_{<500}$ versus N for three representative cases: slow decreases with N (e.g., C_2H_6), large decreases with N (e.g., CO_2) and increases with N (e.g., CH_4). The μ_L/μ_G value shown in the figure is the ratio of the measured (μ_L) value of μ in the liquid and the thermal mobility in the gas μ_G adjusted for the difference in N between the gas and the liquid. These early analyses (Christophorou, 1975) indicated that an increase in μ (or w) with N was characteristic of media with a Ramsauer-Townsend minimum in their electron scattering cross section and that it required a decrease with increasing N of the scattering cross section and a shift of the cross section minimum to lower energy (see further discussion later in this section).

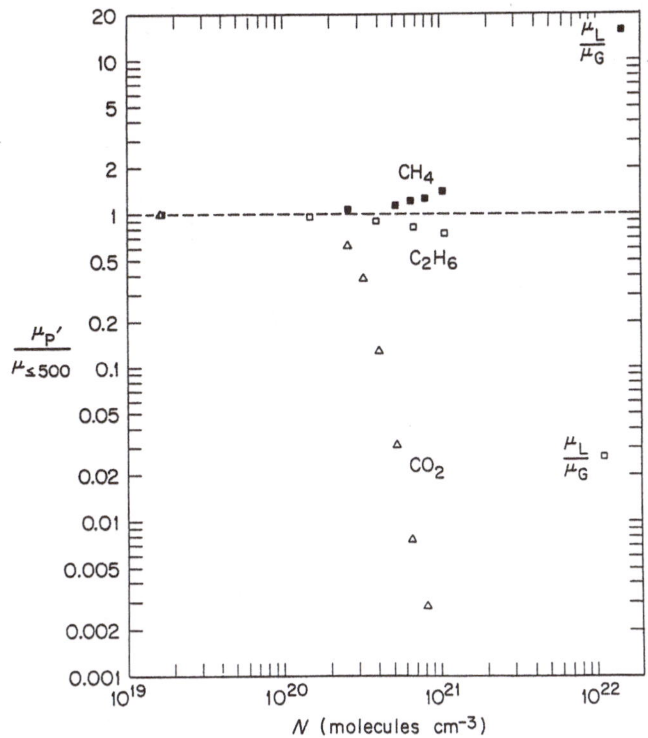

Fig. 5. $\mu_{p'}/\mu_{\leq 500}$ versus N for CH_4, C_2H_6, and CO_2; $\mu_{p'}$ is the measured mobility at a pressure P and $\mu_{\leq 500}$ that measured at low pressures ($\leq$ 500 Torr). (From Christophorou, 1975)

In Fig. 6a the dependence of w(E/N) on N(T=300K) and in Fig. 6b the dependence of $(\mu N)/(\mu N)_o$ on N (at various T) are given for the polar gas NH_3. Clearly w and μN decrease greatly with increasing N at relatively low N. This is due to the large electron-electric dipole scattering which effects a large scattering cross section and thus multiple scattering interference effects (causing a decrease in w) at relatively low values of N. For NH_3, the average value of the electron scattering cross section at thermal (T=300K) energies is ~1.2 x 10^{-13} cm^2 (Christophorou et al., 1982) and $l = \lambda\!\!\!^{-}$ for $N_c \simeq 5.5 \times 10^{19}$ molecules cm^{-3}. As N increases beyond the range of values in Fig. 6 electron localization or transient/permanent electron trapping occur and cause a rather sharp decrease in w or μ as can be seen from the data in Fig. 7. These latter processes are, obviously, a function of T; their effect decreases with increasing T.

An example of the opposite behavior, namely w or μ increasing with increasing N is shown in Fig. 8 for Xe. The increases in w up to ~3x10^{21} atoms cm^{-3} is associated with a reduced contribution to electron

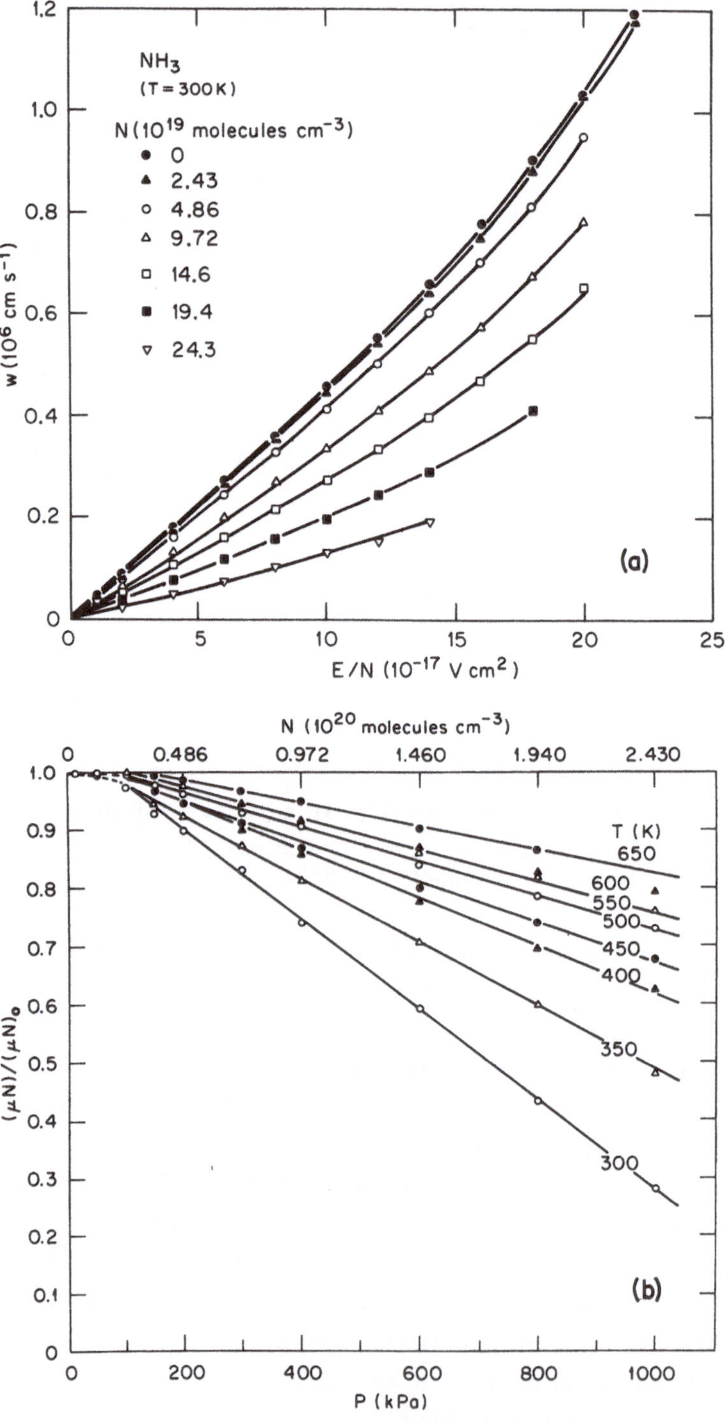

Fig. 6. (a) w vs. E/N for NH_3 at various values of N at T = 300K.
(b) $\mu N/(\mu N)_0$ vs. P (or N) for NH_3 at a number of T.
For a given value of E/N, the w and μN data were plotted as a function of N and extrapolated to $N \to 0$. These values are designed in (a) by the solid circles (N = 0) and in (b) by $(\mu N)_0$. (From Christophorou et al., 1982)

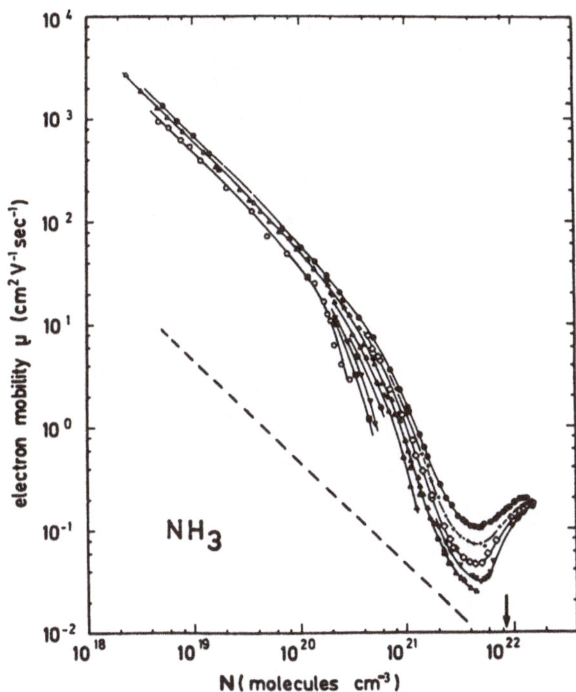

Fig. 7. μ vs. N in subcritical and supercritical NH_3 vapor at various T:300 (o), 320 (■), 340 (▼), 360 (•), 380 (▲), 400 (Δ), 410 (∇), 420 (◇), 440 (+), and 460K (●). The arrow indicates the critical density of NH_3 and the dashed line represents the averaged mobility of unidentified impurity ions (T < 400 K). (From Krebs and Heintze, 1982; see also Krebs, 1984)

scattering from the polarization component of the interaction potential by overlapping of the fields of adjacent atoms. The shape of and maximum in the μ versus E/N function and its modification with N are worth noting since they provide clues (see following section) as to the magnitude and energy dependence of σ_{sc} as a function of N and as to the $\sigma_{sc}(\varepsilon)$ in the liquid.

If one restricts himself to the low-field (thermal value) mobility and plots the product μN as a function of N (see Fig. 9) he can envision the evolution of the dependence of the electron mobility for this gas (Xe) on the gas number density over the entire density range from the low pressure regime to the liquid. At low N ($\lesssim 3 \times 10^{20}$ molecules cm^{-3}; see Fig. 9) μN is independent of N since in this range $N\sigma_{sc}$ is low enough for single atom scattering to prevail. Beyond this value, μN decreases, passes through a minimum, increases to a maximum and falls again. The decrease in the density range from ~3 to ~40 x 10^{20} atoms cm^{-3} was attributed (see discussion in Christophorou, 1984) to enhanced scattering due to multibody interactions and the large subsequent increase due to

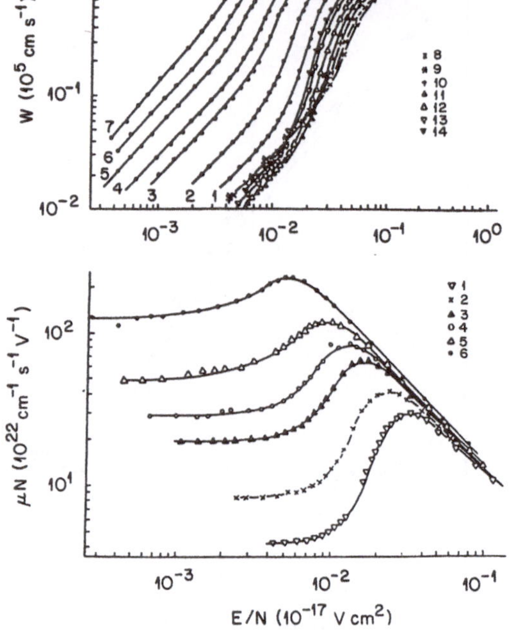

Fig. 8. w and μN versus E/N for Xe (T = 298K) at various values of N from Dmitrenko et al. (1983). (a) Curves 1 through 14 correspond to N (in units of 10^{21} atoms cm^{-3}) of: 4.24, 4.97, 5.38, 6.3, 6.97, 7.34, 7.75, 0.1, 0.438, 0.91, 1.92, 2.74, 3.54, and 3.92, respectively. (b) Curves 1 through 6 correspond to N (in units of 10^{21} atoms cm^{-3}) of: 4.24, 4.97, 5.58, 6.30, 6.97, 7.75, respectively. (From Dmitrenko et al., 1983)

interference effects. Certainly the structure of the medium needs to be considered along with the other possible processes at these high densities. These density ranges, however, have been covered by others elsewhere in this volume.

Theoretical Rationalizations

Low N: For sufficiently low N, $N\sigma\lambda_T \ll 1$ [$\lambda_T = \hbar/(2mkT)^{1/2}$] and w can be represented by the classical expression (e.g., see Hunter and Christophorou, 1984)

$$w = -\frac{4\pi}{3}\frac{e}{m}\frac{E}{N}\int_0^\infty \frac{v^2}{\sigma_m(v)}\frac{df_0}{dv}\,dv\ , \tag{6}$$

where $\sigma_m(v)$ is the momentum transfer cross section and f_o is given by

$$f_0 = \left(\frac{m}{2\pi kT}\right)^{3/2} e^{-mv^2/2kT}\ . \tag{7}$$

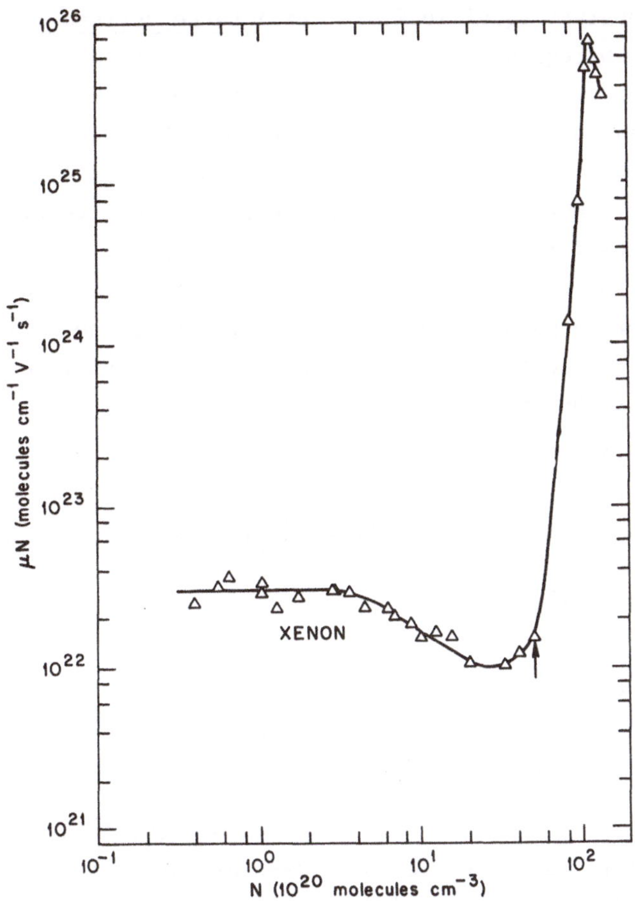

Fig. 9. μN versus N for Xe. (From Freeman, 1981)

From (6) and (7) we have for the thermal, low density value $(\mu N)_o$ of μN

$$(\mu N)_o = \frac{4}{3} \frac{e}{(2\pi mkT)^{1/2}} \frac{1}{\langle\sigma\rangle} , \tag{8}$$

where $\langle\sigma\rangle$ is an average of the scattering cross section at thermal energies.

High N: (i) Small decreases in w (or μ) with N. In the N region over which $N\sigma\lambda\!\!\!^{-}_T < 0.5$ (the electron is still "free" in this density range), the small decreases in w (or μ or μN) with increasing N for gaseous media with a > 0 and α small have been treated quantum mechanically considering the effects of multiple scattering on w (e.g., see Legler, 1970; O'Malley, 1980; Atrazhev and Yakubov, 1977, 1980; Braglia and Dallacasa, 1978, 1982; Yakubov and Polischuk, 1982; Polischuk, 1984, 1985). Most such treatments attempted to account for the effect of N on w by introducing phenomenological density corrections to the low-N mobility (Eq. 8). For example, Atrazhev and Yakubov (1977) gave

$$\mu N = (\mu N)_o \left(1 - \frac{\sqrt{\pi}}{2} F N \langle\sigma\rangle \lambda\!\!\!^{-}\right) , \tag{9}$$

where F is a constant equal to 2.09. (ii) Large decreases in w (or μ) with N. In the N region over which $N\sigma\lambda_T > 1$ the large decreases in w with N have been attributed to quasi or permanent electron trapping by single molecules, dimers, polymers, clusters or by the bulk medium. No adequate theoretical account of these changes exists. (iii) Increases in w (or μ) with N. Multiple scattering theories have attempted to account for these changes for gases with a < 0 and α large by considering the attenuation of the electron scattering by density effects due to the screening of the long-range polarization interaction potential. For example, Atrazhev and Yakubov (1980) gave for the density-dependent attenuation of α the expression

$$\left(1 + \frac{8\pi}{3}\alpha N\right)^{-1} = A , \tag{10}$$

so that the effective scattering cross section as a function of N is

$$\sigma_{ef}(N) = 4\pi a_{ef}^2(N) , \tag{11}$$

where $a_{ef}(N) = a + \frac{2}{3}\frac{\pi^2\alpha^2 N}{R_o\, a_o} A$ is the effective scattering length at N, a is the isolated molecule scattering length, a_o is the Bohr radius and R_o is the "atomic radius".

In general, the theoretical treatments of w(N) fail at high N especially in the transition region. Even at low N they fail to correctly predict the magnitude and functional dependence of w (or μ) on N over wide density ranges. Part of the difficulty lies in the fact that the theory often retains the atomic scattering picture and introduces corrections to the low-density scattering as the N is increased. Another serious problem is the proper description of the scattering potential and its screening by the medium. Bound electron effects present another outstanding problem.

Very High N: (Liquid; $N\sigma\lambda >> 1$). The electron under these conditions is in the field of strongly correlated scatterers and is thus strongly influenced by the structure of the medium. The thermal electron scattering cross section $(\sigma_L)_{th}$ and thermal electron mobility $(\mu_L)_{th}$ in the liquid are now expressed by

$$(\sigma_L)_{th} = \sigma_{ef}(N)S(0) = 4\pi a_{ef}^2(N)S(0) \tag{12}$$

$$(\mu_L)_{th} = \frac{2}{3}\frac{1}{N}\left(\frac{2}{\pi mkT}\right)^{1/2}\frac{e}{4\pi a_{ef}^2 S(0)} , \tag{13}$$

where $\sigma_{ef}(N)$ and $a_{ef}(N)$ are the effective scattering cross section and length, respectively, at a density N, $S(0) = NkT\chi$ is the structure factor at thermal energies ($K \to 0$), and χ is the isothermal compressibility. [$S(0) \simeq 0.03 - 0.05$ near the triple point of simple fluids; $S(K) \to 1$ for $\varepsilon \gtrsim 4$eV.] Electron motion and electron scattering in liquids have been the topics of other lectures at this institute and will not thus be discussed further. It should be emphasized, however, that the success of any theoretical treatment of electron motion in liquids depends on the assumed form of the scattering potential and the proper correction for its screening by the medium.

<u>Is there a Ramsauer-Townsend minimum in the electron scattering cross section for liquids with $a < 0$ ($V_o < 0$eV) and the heavier liquefied rare gases (Ar, Kr, Xe) in particular?</u>

We have seen earlier in this chapter that an increase in w with N was observed in dense media whose electron scattering cross sections have a Ramsauer-Townsend minimum at low energies [e.g., heavier rare gases (see Fig. 10), CH_4, $C(CH_3)_4$]; gases with a negative scattering length and negative V_o. In these media the scattering cross section at low energies (thermal and epithermal) decreases and the position of the Ramsauer-Townsend minimum shifts to lower energy as N increases. An early indication of this has been obtained by Christophorou and McCorkle (1976) from an analysis of the pressure dependence of w for CH_4 (Lehning, 1969) and Ar (Bartels, 1974) at pressures $\lesssim$ 100 atm (see Fig. 11). Consistent with Fig. 11 and the interpretation of Christophorou (1975, 1976) are the findings in Fig. 8 and also the results of a recent Boltzmann equation

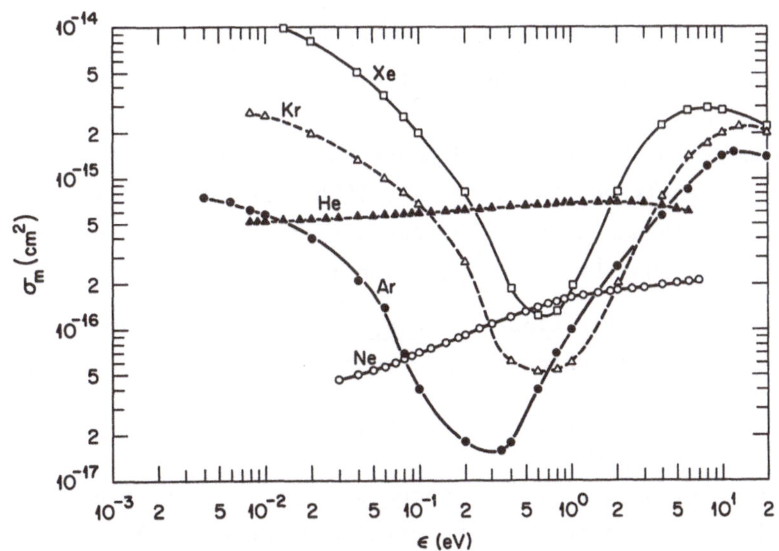

Fig. 10. Momentum transfer cross section σ_m versus electron energy ε for the rare gases. (From Christophorou et al., 1980)

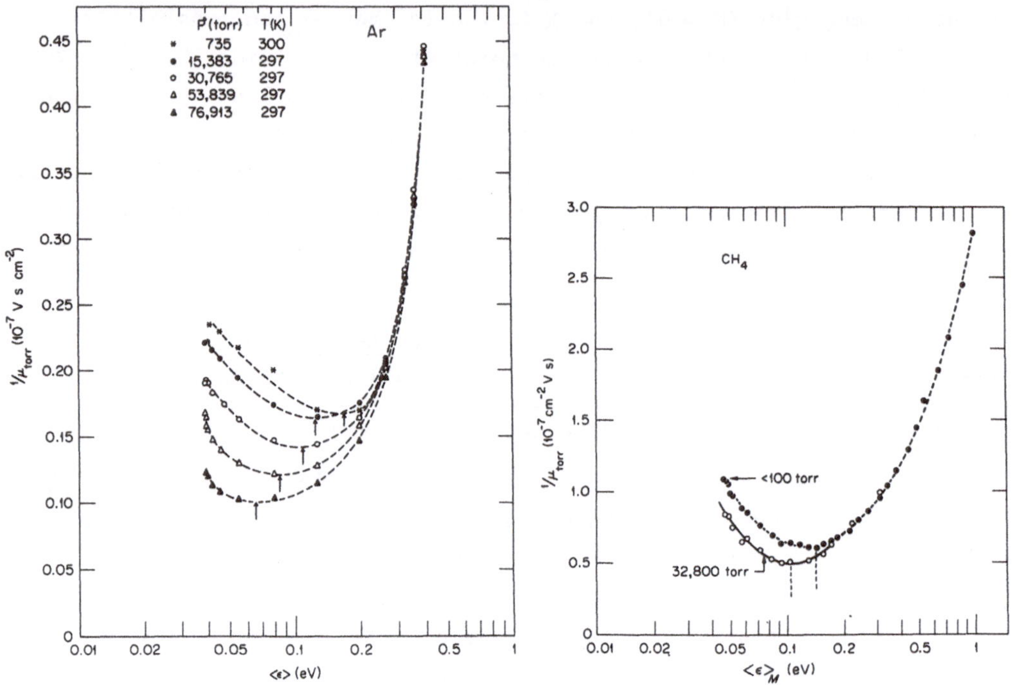

Fig. 11. $1/\mu_{Torr}$ (μ_{Torr} is the electron mobility at 1 Torr pressure) versus mean electron energy for Ar and CH_4 at the indicated pressures. (From Christophorou and McCorkle, 1976)

analysis of experimental data on the drift velocity of electrons as a function of E/N in liquid Ar, Kr, and Xe.

In Fig. 12 are shown the measured w(E/N) for gaseous (low pressure) and liquid Ar and Xe. These data can be used in a Boltzmann transport equation analysis to determine a set of scattering cross sections and an electron energy distribution function f(ε, E/N) consistent with the w(E/N) measurements. The liquid phase data (Miller et al., 1968) have recently been analyzed (Sakai et al., 1985; Nakamura et al., 1986) in this manner (this has originally been suggested by Cohen and Lekner, 1967) incorporating three cross sections:

-Cross section for <u>elastic energy loss</u> (energy transfer to acoustic phonons which is independent of the liquid structure)

$$\sigma_0(\varepsilon) = 2\pi \int_0^{\pi} (1-\cos\Theta)\ \sigma(\varepsilon,\Theta) \sin\Theta\, d\Theta\ , \tag{14}$$

-Cross section for <u>elastic momentum transfer</u> [which depends on the liquid structure through the structure factor S(K)]

$$\sigma_1(\varepsilon) = 2\pi \int_0^{\pi} (1-\cos\Theta)\ \sigma(\varepsilon,\Theta)\ S(K) \sin\Theta\, d\Theta \tag{15}$$

and

-Cross section for <u>inelastic electron scattering</u> σ_{in}.

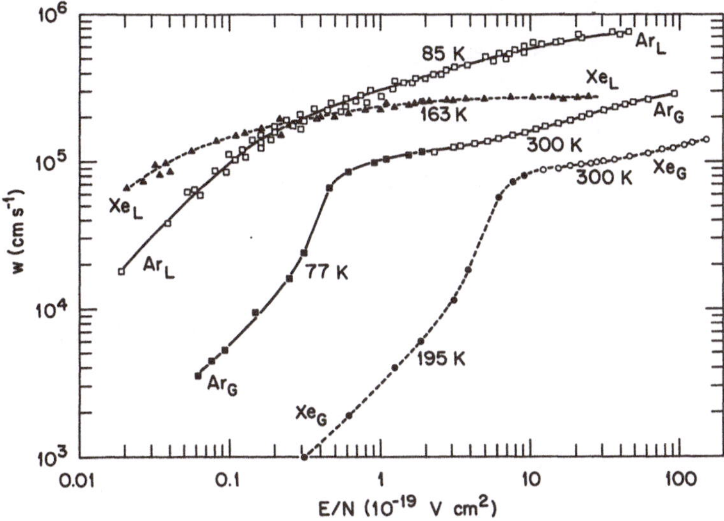

Fig. 12. w versus E/N for gaseous (low-pressure) (see Christophorou, 1971) and liquefied (Miller et al., 1968) Ar and Xe.

In these expressions $\sigma(\varepsilon,\Theta)$ is the differential scattering cross section. The cross sections $\sigma_0(\varepsilon)$, $\sigma_1(\varepsilon)$ and $\sigma_{in}(\varepsilon)$ obtained by Sakai et al. are compared in Fig. 13 with the low density gaseous momentum transfer cross section $\sigma_m(\varepsilon)$. Clearly $\sigma_0(\varepsilon)$ and $\sigma_1(\varepsilon)$ are lower than $\sigma_m(\varepsilon)$ below ~0.2 eV, and exhibit a shallower Ramsauer-Townsend minimum which is shifted to lower ε compared to that of $\sigma_m(\varepsilon)$. (At $\varepsilon \gtrsim 4$ eV, as expected, σ_0 and $\sigma_1 \rightarrow \sigma_m$). It would thus seem that a weak Ramsauer-Townsend minimum exists in dense and liquefied rare gases at lower energies than in the low density gas as predicted by the earlier analyses (Christophorou, 1975; Christophorou and McCorkle, 1976). The failure of the theory (e.g., Lekner, 1967) to clearly identify such a minimum may be due to an overcorrection for the screening of the electron scattering potential. Clearly further theoretical work is indicated.

EFFECT OF MEDIUM ON NEGATIVE ION STATES (TRANSIENT ANIONS)

To aid our discussion on this topic, let us focus on the negative ion state (NIS) of N_2 at ~2.3 eV. This is a shape resonance of $^2\Pi_g$ symmetry. The incident electron enters an orbital of the N_2 molecule and is temporarily retained in the N_2^{-*} complex by a centrifugal potential barrier; the lifetime $\tau_a = \hbar/\Gamma_a$ of N_2^{-*} ($^2\Pi_g$) toward autodetachment is ~5×10^{-15} s (Christophorou, 1978). This, and similar resonances for other molecules are both strong and abundant. With the aid of Fig. 14 let us see what is usually looked for in an electron scattering experiment employed for the study of such transient species. The energy ε_i of a

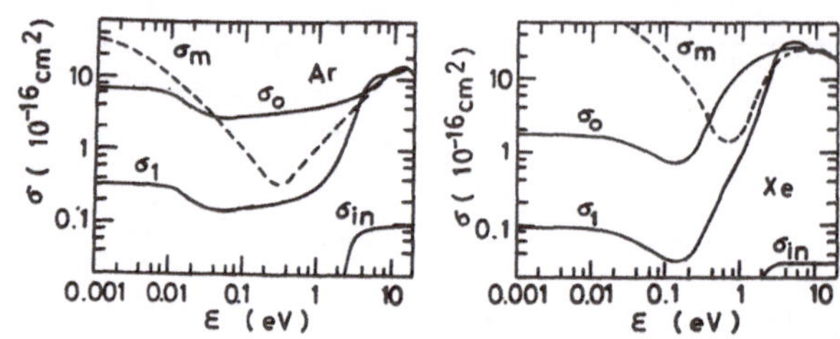

Fig. 13. Cross sections $\sigma_0(\varepsilon)$, $\sigma_1(\varepsilon)$, and $\sigma_{in}(\varepsilon)$ for liquefied Ar and Xe (see the text); σ_m is the low density gaseous momentum transfer cross section. (From Sakai et al. 1985)

monoenergetic incident electron beam is increased and the scattered electrons (in the forward direction or at an angle) are detected without or with energy analysis. When ε_i is high enough for the ν' =0 level of N_2^{-*} to be reached (see Fig. 14) a large enhancement is observed in the elastic and inelastic scattered electron current as a result of the decay of the N_2^{-*} to the various ν = 0 (elastic), ν = 1,2, ...(inelastic) vibrational levels of N_2. As the ε_i is increased, higher levels ν' of $N_2^-(^2\Pi_g)$ are reached. If, then, one is looking at those electrons which have been scattered indirectly via this resonance having excited the $N_2(X^1\Sigma_g^+)$ molecule in, for instance, the first vibrational level (i.e., having lost energy equal to $h\nu_1$), the cross section function for such a process will show maxima at those values of ε_i which coincide with the positions of the ν' = 1,2,3...levels of $N_2^-(^2\Pi_g)$. Generally, the temporal trapping of the incident electron alters the adiabatic potential governing the nuclear motion and this favors excitation of higher vibrations [indirect (via NISs) excitation far exceeds that for direct (without quasi electron trapping) scattering (see Christophorou, 1984)].

While such studies on the isolated N_2 molecule are abundant, none exists - to the author's knowledge - on liquids. However, such studies do exist on solid films of N_2 [Demuth et al., 1981; Sanche and Michaud, 1981b, 1983, 1984) and, also on the other systems such as H_2, CO, O_2, C_6H_6 (Demuth et al., 1981; Sanche, 1979; Sanche and Michaud, 1981a, 1984)]. The results of Sanche and Michaud (1981b, 1983) on the N_2^{-*} resonance in solid films and the principle of their experiment are shown in Fig. 15. They clearly show (i) strong vibrational excitation (ν_1, ν_2, and ν_3 channels, see Fig. 15); (ii) resonance shifts to lower energy compared to the isolated molecule (by ~0.8 eV for a N_2 or Ar matrix and by ~1.4 eV for a Xe matrix; this difference in the energy shift can be

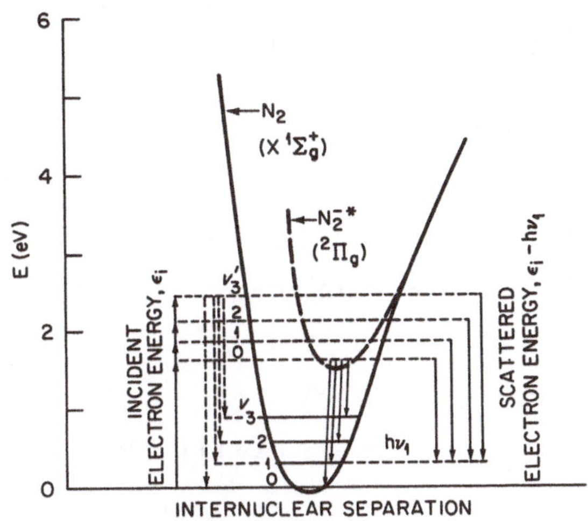

Fig. 14. Schematic potential energy curves for N_2 $(X^1\Sigma_g^+)$ and $N_2^{-*}(^2\Pi_g)$; incident and scattered electron energies (see the text).

accounted for by the polarization of the matrix by the temporarily localized electron); and (iii) contrary to the single-molecule scattering case where the elastic channel is the strongest - the resonance is undetectable in the elastic channel in the solid film.

In an attempt to rationalize the gaseous and the condensed phase behavior, Fano et al. (1986) noted that while the formation of the NIS in the condensed phase should not differ from that in the gaseous phase since it is dominated by the short-range forces, the decay of the NIS in the condensed phase should differ from that in the gas because it is affected by the medium due to polarization. The NIS in the dense medium changes the polarization potential and thus the intra and inter molecular motion. The intramolecular changes lead to excitation of discrete vibrations (as in the gas) while intermolecular changes lead to continuous acoustic excitations (background) which in effect replace the (low-pressure) elastic scattering. This process is very probable due to the large number of such low frequency vibrations.

Concerning the shift in the energy position of the NIS, it is instructive to refer to Fig. 16 which shows schematically the vertical attachment energy (VAE), the electron affinity (EA) and the vertical detachment energy (VDE) and to Fig. 17 which shows schematically the increase in the VAE (less negative) and the increase in the EA (more positive) in going from dilute gas to the liquid.

Concerning the possible changes in the autodetachment lifetime τ_a of the transient anions, the cross section for their formation and their decomposition by dissociative electron attachment in making the

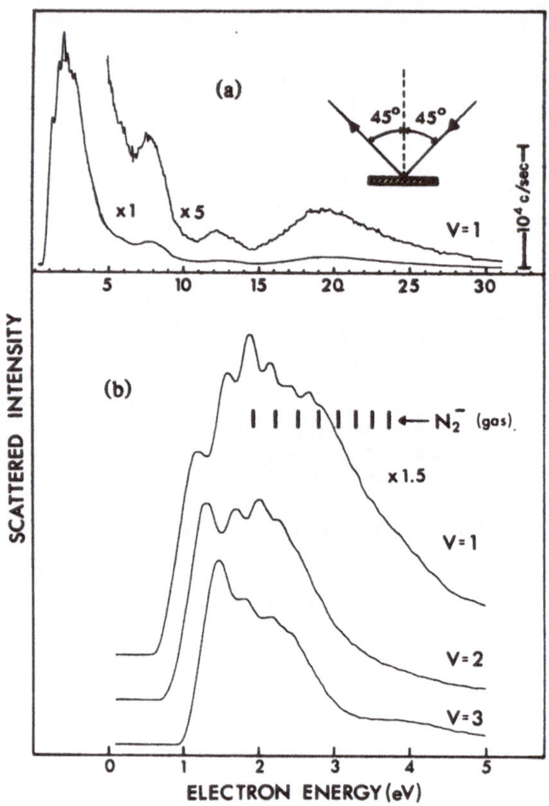

Fig. 15. (a) Energy dependence of the vibrational excitation of the $\nu = 1$ level of N_2 $(X^1\Sigma_g^+)$ in ~50-Å multilayer molecular N_2 film. The four maxima are attributed to various states of N_2^-.

(b) Excitation function for the $\nu = 1,2$ and 3 levels of N_2 below 5 eV. The structure reflects the vibrational states of $N_2^{-*}(^2\Pi_g)$ (see Fig. 14); the corresponding energy positions of the peaks in the gas phase are also shown. (From Sanche and Michaud, 1983.)

transition from the gas to the liquid, it is necessary to keep in mind that any such changes would depend on whether the electron affinity of the molecule in the gas $(EA)_G$ is negative (<0 eV) or positive (>0 eV) and on whether a negative $(EA)_G$ becomes positive in the condensed phase.

For $(EA)_G$ <0 eV, the results on the lowest NISs of N_2,CO, and H_2 (peaking, respectively at 2.3, 1.8 and 3.75 eV for the isolated molecules) on solid films have shown (Sanche, 1984; Demuth et al., 1981) that in the condensed phase the centrifugal barrier is greatly distorted due to symmetry changes which are effected by the medium. This, results

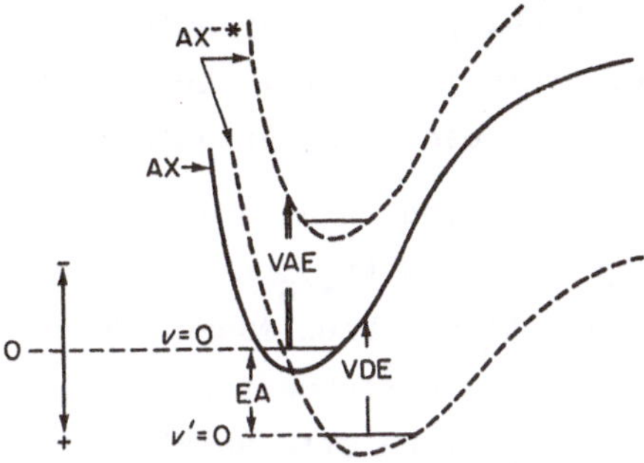

Fig. 16. Schematic potential energy diagrams for AX and AX^{-*} indicating the electron affinity EA, the vertical attachment energy VAE, and the vertical detachment energy VDE.

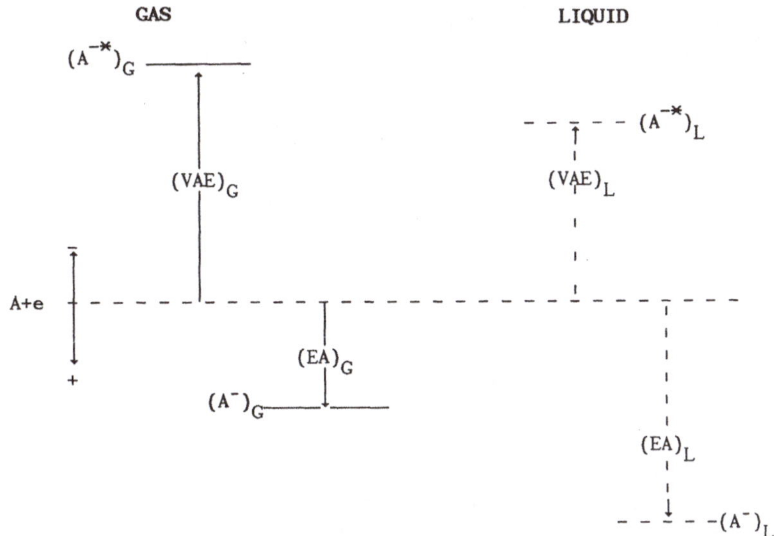

Fig. 17. Schematic illustration of the relative values of EA and VAE in a gas and a liquid.

in increased autodetachment, i.e., a decrease of the autodetachment lifetime of the NIS, in the solid compared to that $(\tau_a)_G$, in the low-pressure gas. One might, thus, expect $(\tau_a)_G$ to be longer than the lifetime $(\tau_a)_L$ of the NIS in the liquid. Because of this, the cross section for dissociative electron attachment σ_{da} for molecules such as O_2 ($O_2 + e \rightarrow O_2^{-*} \rightarrow O^- + O$) would be smaller in the liquid that in the gas due to the decrease of the survival probability (Christophorou, 1971, 1984). However, one should note that since the position of the resonance is lower in the liquid than in the gas and since the magnitude of the dissociative attachment cross section is larger the lower the energy position of the resonance (Christophorou, 1971, 1984), the cross section for a dissociative attachment process may actually be much larger in the liquid than in the gas. This is clearly the case for the process $e + N_2O \rightarrow N_2O^{-*} \rightarrow O^- + N_2$ as can be seen from the data in Fig. 18.

For $(EA)_G > 0$ eV (or when the EA of a molecule is negative in the gas but positive in the liquid), then $(\tau_a)_G < (\tau_a)_L$ due to the faster energy relaxation in the liquid. It has been shown (Christophorou, 1984) that when $(EA)_G > 0$ eV, the electron attachment cross section (for a dissociative or a nondissociative electron attachment process) in the liquid $(\sigma_a)_L$ is maximum when the captured electron is quasifree (e_{qf}) and that $(\sigma_a)_L$ is close to its diffusion controlled value when the captured electron is localized (e_l). The electron state in the liquid crucially determines both the magnitude of $(\sigma_a)_L$ and the relation of $(\sigma_a)_L$ to the corresponding value $(\sigma_a)_G$ in the gas. As will be shown in the next section, $(\sigma_a)_L$ (e_{qf}) >> $(\sigma_a)_L$ (e_l). It is only for liquids for which the electron is in the quasi-free state that a comparison with gaseous data is meaningful.

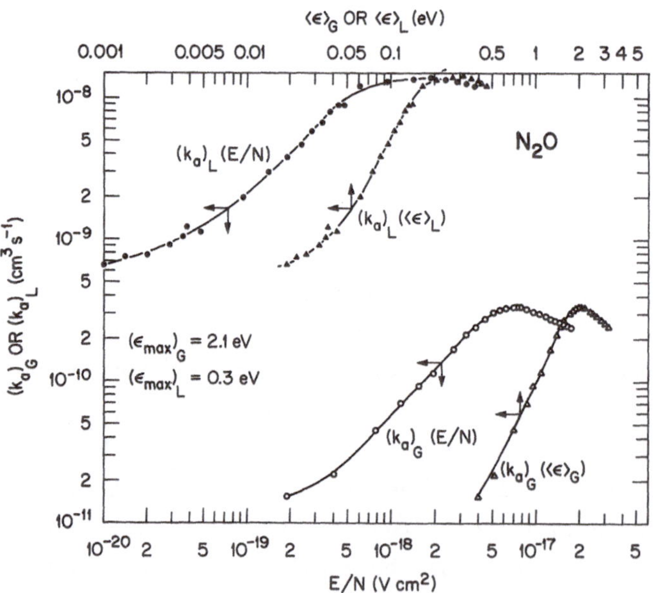

Fig. 18. Electron attachment rate constant for the reaction $e + N_2O \rightarrow O^- + N_2$ measured in a low pressure Ar buffer gas [$(k_a)_G$ vs. E/N or $\langle\varepsilon\rangle_G$, ($\langle\varepsilon\rangle_G$ = mean electron energy in the gas)] and in liquid Ar [$(k_a)_L$ vs. E/N or $\langle\varepsilon\rangle_L$ ($\langle\varepsilon\rangle_L$ = mean electron energy in the liquid)]. (Christophorou, 1985.)

It is beyond the scope of this lecture to review the electron attachment processes in dense gases and the many and varying effects of the density on them. The reader is referred to the work of the author elsewhere (e.g., Christophorou, 1976, 1984). It suffices here to present a few selected examples to illustrate the effect of the medium on electron attachment to molecules in gases and liquids and the gas to liquid transition.

In general, the effect of the medium on electron attachment is a function of the mode of electron attachment (dissociative or nondissociative), the anionic state involved in the attachment process and its energy position, as well as the nature and density of the medium. The effect of the medium and its density on dissociative electron attachment is insignificant at low N because the fragment anion(s) does not, as a rule, require collisional stabilization. However, at high N and in liquids where the anionic state (NIS) can be influenced by the medium and its position be shifted to lower energy, changes are expected and indeed observed (e.g., Fig. 18). The effect of the medium and its density on nondissociative electron attachment is a strong function of the lifetime of the transient anion which is initially produced (i.e., the lifetime τ_a of the NIS) and can be profound - even at low N - when the τ_a is short ($\lesssim 10^{-10}$s). The magnitude and - at high N - the energy dependence of the rate constant/cross section for nondissociative electron attachment are a function of N. To illustrate some of these factors, we shall refer to the electron attachment process in O_2 forming O_2^- via the short-lived ($<10^{-10}$s) O_2^{-*} and to the electron attachment process in SF_6 forming SF_6^- via the long-lived ($> 10^{-5}$ s) SF_6^{-*}, when these processes occur in dense gases and liquids.

$\underline{O_2}$: At low N, electron attachment to O_2 producing O_2^- is well understood. The process (see Fig. 19) is described as

$$O_2(X^3\Sigma_g^-;\ \nu{=}0) + e \xrightarrow{k_c} O_2^{-*}(X^2\Pi_g;\ \nu' \geq 4) \begin{cases} \xrightarrow{\tau_a^{-1}} O_2^{(*)} + e^{(\prime)} & \text{(16a)} \\ \xrightarrow[(+M)]{k_{st}N_M} O_2^-(X^2\Pi_g;\ \nu'<4) + M + \text{energy} & \text{(16b)} \end{cases}$$

where M is usually a buffer gas molecule which stabilizes O_2^{-*} in binary collisions. If, now, the e, O_2 system is situated in a buffer gas (e.g., N_2) whose number density is increased, the magnitude of the attachment rate constant for the production of O_2^- increases and its energy dependence changes (electron attachment becomes progressively larger at the lower energies) with increasing N_2 density (see Fig. 20). The electron

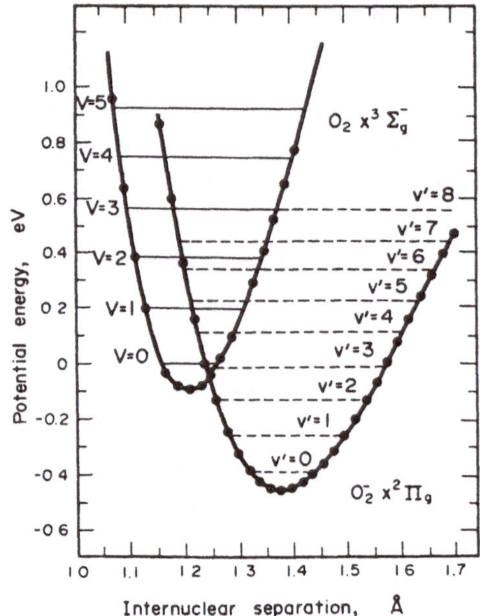

Fig. 19. Potential energy curves for O_2 $(X^3\Sigma_g^-)$ and $O_2^-(X^2\Pi_g)$. (Boness and Schulz, 1970; Christophorou et al., 1984)

attachment process is no longer described by (16), but it requires the involvement of more than one N_2 molecules. Moreover, for a fixed buffer gas number density, the attachment rate constant depends on the nature of the buffer gas itself as is exemplified by the data in Fig. 21. While the modeling of the data in Fig. 21 for electron attachment to O_2 in the buffer gas N_2 or C_2H_4 required the involvement of more than two buffer gas molecules, the modeling of O_2^- formation in the buffer gas C_2H_4 was consistent with reaction (16) over the entire density range in Fig. 21 (Goans and Christophorou, 1974; Christophorou, 1976). When these latter data were extrapolated to liquid density, the rate constant obtained was comparable with the values measured in dielectric liquids and liquefied rare gases. The rate constant for formation of O_2^- measured in liquefied rare gases (Bakale et al., 1976) decreased with E/N in a manner analogous to that in gases (Fig. 20). Other studies (e.g., see Hatano and Shimamori, 1981) contented that electron attachment to O_2 forming O_2^- in dense buffer gases M, is principally due to electron capture by Van der Waals molecules $[O_2.M]$.

$\underline{SF_6}$: This is a molecule which has been extensively studied both in the gaseous and in the liquid state. In gases, SF_6^- is formed at thermal and near thermal energies with a very large (close to the maximum s-wave value) cross section which decreases rapidly as the electron energy increases above thermal and which is pressure independent (Christophorou,

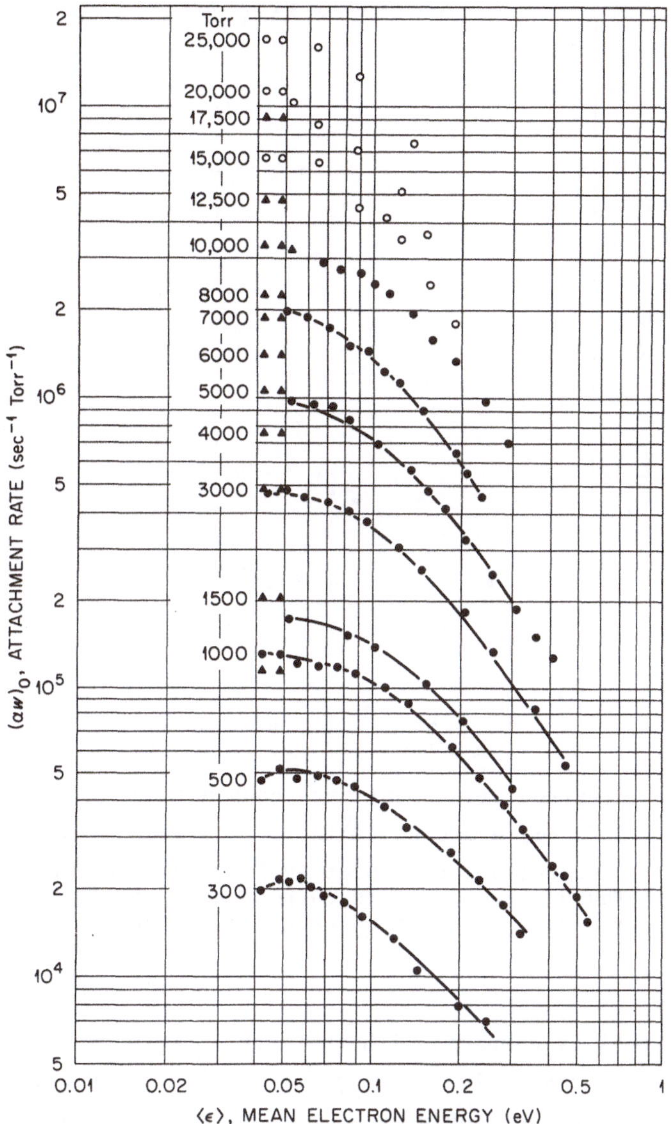

Fig. 20. Electron attachment rate constant $(\alpha\ w)_o$ versus the mean electron energy $<\varepsilon>$ for O_2 in N_2 at the indicated total pressures. Note that the attachment is due to the formation of O_2^- and that the rate constant is expressed in units of s^{-1} $Torr^{-1}$ ($T \simeq 298K$). (From Goans and Christophorou, 1974.)

1984). The lifetime of the isolated SF_6^{-*} at thermal energies is > 30 μs (Christophorou, 1978). In liquefied rare gases, the attachment of slow electrons to SF_6 forming SF_6^- is similar to that in gases, both in magnitude and energy dependence. This can be seen from Fig. 22 where the rate constant $(k_a)_G$ for electron attachment to SF_6 in low-pressure ($\lesssim 3$ atm) gaseous Ar ($T \simeq 300K$) as a function of E/N or $<\varepsilon>_G$ (the mean electron

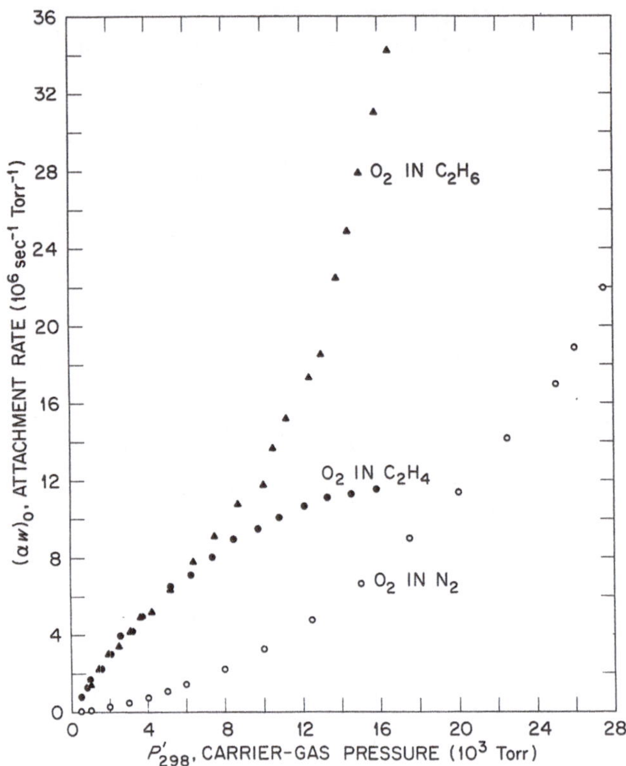

Fig. 21. Electron attachment rate constant for O_2 in N_2 (o), C_2H_4 (•) and C_2H_6 (▲) as a function of the pressure (corrected for compressibility) of these buffer gases. These rate constants correspond to a value of ⟨ε⟩ ≃ 0.05 eV. (From Goans and Christophorou, 1974)

energy in the gas) is compared with that, $(k_a)_L$ (E/N) in liquefied Ar (T = 87K). The thermal value of $(k_a)_G$ at 298K and at 87K (extrapolated) agree very well with the liquid $(k_a)_L$ data for the lowest E/N, showing that $(k_a)_G = (k_a)_L$ when the electron energy distribution is thermal. In general, the comparison of $(k_a)_L$ with $(k_a)_G$ and the analogy of electron attachment processes in liquids with those in gases is more appropriate for high mobility ($\mu \gg 1$ cm^2 V^{-1} s^{-1}; $V_o < 0$ eV) liquids. In these the electron is quasi-free and its attachment to molecules AX embedded in the liquid can be viewed - as in gases - as a vertical transition between the initial $(e + AX)_L$ and the final $(AX^{-*})_L$ state; the attachment process depends on the properties of AX and the medium (especially V_o). In many such cases - e.g., the case of SF_6 in liquefied rare gases just discussed - the $(k_a)_L$ is comparable to $(k_a)_G$. On the other hand, in liquids in which the electron is initially in a localized state, the rate determining step is the diffusive motion of the electron, and $(k_a)_L$ depends only weakly on the medium and varies little with AX. In such

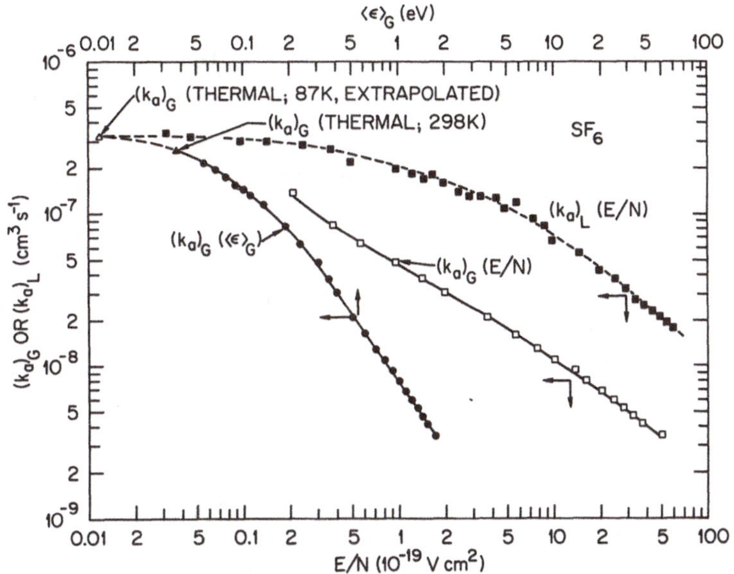

Fig. 22. $(k_a)_G$ versus $\langle\varepsilon\rangle_G$ or E/N for SF_6 in gaseous Ar and $(k_a)_L$ versus E/N for SF_6 in liquid Ar. ▲ , thermal value of $(k_a)_G$ at 298K; Δ , thermal value of $(k_a)_G$ extrapolated to 87K. (From Christophorou, 1985)

cases $(k_a)_L$ can attain diffusion-controlled values as long as a negative ion state of AX exists at thermal energies; this condition seems to be satisfied for most liquids when $(\varepsilon_{max})_G \lesssim 1$ eV (Christophorou, 1976). In these cases $(k_a)_L$ can be expressed as

$$(k_a)_L \simeq 4\pi R D_e \ , \tag{18}$$

where R is the encounter radius, and D_e is the electron diffusion coefficient. Since, moreover, $D_e = (kT/e)\mu$, $(k_a)_L$ is predicted to increase linearly with μ, a behavior observed experimentally in some instances (Bakale et al., 1976; Bakale and Schmidt, 1981). Finally, when the electron drifts part of the time as quasi-free and part of the time as localized, $(k_a)_L$ can be expressed as

$$(k_a)_L = (k_a)_1 \ p+(k_a)_f(1-p) \quad , \tag{19}$$

where $(k_a)_1$ and $(k_a)_f$ are, respectively, the attachment rate constants involving e_1 and e_{qf} and p is the probability of finding the electron in the localized state (see further discussion and references in Christoporou and Siomos, 1984).

While electron impact ionization cross section (or coefficient) measurements in gases are abundant (e.g., see Christophorou, 1971, 1984), no such measurements are known to exist in liquids or in high-pressure gases (the latter experiments are planned at the author's laboratory). Perhaps the only known measurements of this type are those of Derenzo et al. (1974) on electron avalanches in liquefied Xe from which they determined the "ionization coefficient" α (E/ρ) for liquefied Xe. These results are shown in Fig. 23 where they are compared with similar data in the gaseous phase. The α (E/ρ) function in liquefied Xe does not scale from the gaseous one by considering the density difference between the gas and the liquid.

In contrast to the situation on electron impact ionization, there have been some studies on the photoionization processes and energetics of molecules in dense gases [e.g., see Reininger et al. (1984) and references quoted therein] and in liquids [e.g., see Lipsky (1981), Schmidt (1984), Christophorou and Siomos (1984), and references quoted therein]. The ionization studies in dense gases have shown that the ionization onset (photoconductivity threshold) I(N) at a number density N can be represented by (Raz and Jortner, 1969)

$$I(N) = I_G + P^+(N) + V_0(N) + \Delta\, E(N) \quad . \tag{20}$$

In Eq. (20) I_G is the low-gas-density adiabatic ionization onset, P^+ (a negative quantity) is the polarization energy of the positive ion produced in the photoionization process, V_0 (a negative or positive quantity) is the ground state energy of the photoionization electron, and Δ E is a small quantity associated with the broadening of the valence levels of the isolated atom (molecule) in the dense gas or the liquid. The quantities V_0 (see Fig. 3), P^+ and Δ E are all functions of N.

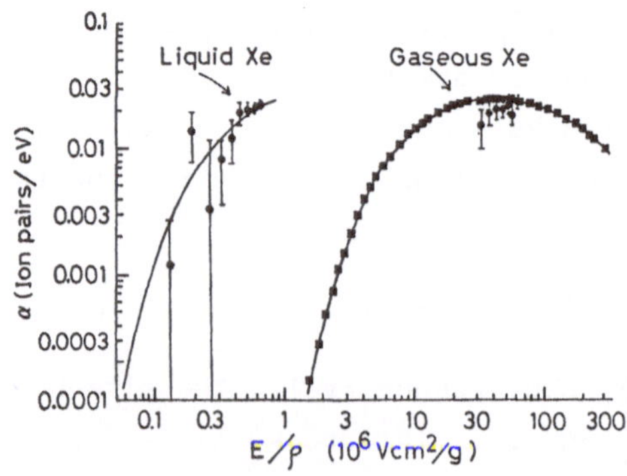

Fig. 23. "Ionization coefficient" α (ion pairs/eV) versus E/ρ for gaseous and liquefied Xe. (From Derenzo et al., 1974.)

Neglecting ΔE, we can see that I(N) differs from I_G by an amount equal to $P^+(N) + V_o(N)$. In Fig. 24a, I(N)($\equiv E^i_{pc}$ = photoconductivity onset) is shown (Reininger et al., 1984) for the molecules C_2H_6, C_3H_8, and n - C_4H_{10} embedded in Ar as a function of Ar density. Reininger et al. (1984) used the known adiabatic values of I_G for C_2H_6, C_3H_8, and n-C_4H_{10}, the values of I(N) they measured for these molecular additives in Ar as a function of the Ar density, their earlier measurements on the $V_o(N)$ of Ar (Fig. 3), and determined through Eq. (20) (with ΔE=0) values of $P^+(N)$. These are shown in Fig. 24b where they are compared with those (solid lines) they calculated from the Born equation

$$P^+ = -\frac{e^2}{2\sigma}\left(1 - \frac{1}{\varepsilon_{opt}}\right), \tag{21}$$

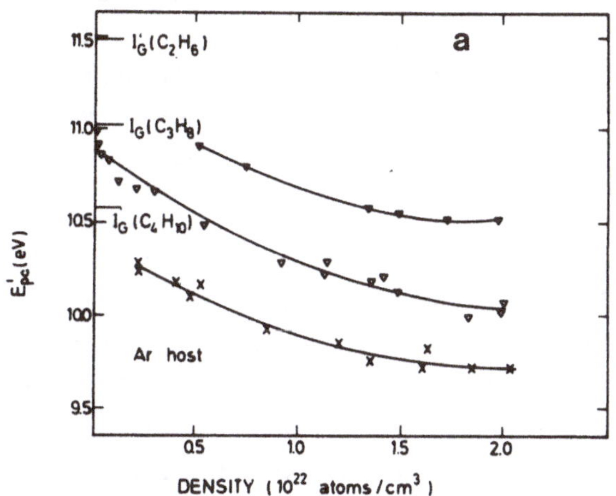

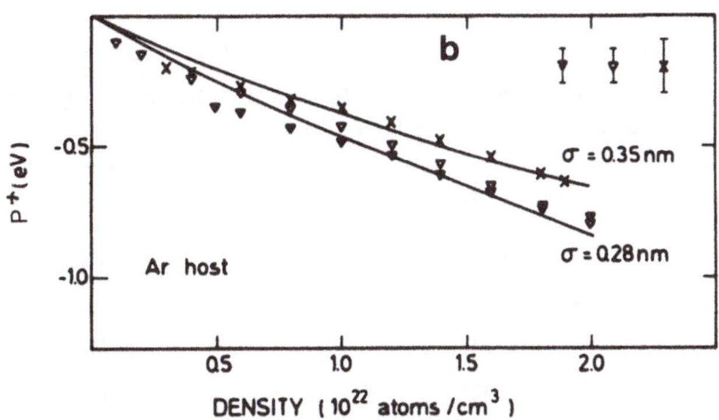

Fig. 24. (a) I(N) for C_2H_6, C_3H_8, and n-C_4H_{10} in Ar as a function of argon density.
(b) P^+ for C_2H_6 (▼), C_3H_8 (▽), and n-C_4H_{10} (x) in argon as a function of the argon density.

The solid lines are calculated values of P^+ using Eq. (21) for the two values of σ shown in the figure (see the text). (From Reininger et al., (1984.)

for two assumed values of the "effective radius" σ of the positive ion. The reasonable agreement between the experimental data on the density dependence of P^+ with that calculated from Eq. (21) for these systems would suggest that - in spite of the approximate nature of Eq. (21) - the density dependence of $P^+(N)$ is dominated by the density dependence of the optical dielectric constant ε_{opt}.

In liquids, Eq. (20) can be written as

$$I_L = I_G + P^+ + V_o \ . \tag{22}$$

For organic molecules in nonpolar liquids, Eq. (22) was generally found to be valid and I_L to be generally lower than I_G by 1 to 3 eV depending on the solute/liquid system. If for a particular solute dissolved in liquid media of differing V_o values, the quantity P^+ does not change significantly from one liquid to another, a linear relationship might be expected between I_L and V_o. Early work (e.g., Holroyd and Russell, 1974) has indicated such a relationship and recent multiphoton ionization conductivity studies on azulene in nonpolar liquids (Faidas and Christophorou, 1987) independently confirmed its validity for the solute/liquid systems in Table 2.

Finally in Fig. 25 a schematic is pictured of the variation of I(N) from its I_G to its I_L value. In the extreme low-density limit $I(N) = I_G$ and the discrete energy levels of the isolated species shown schematically converge to I_G. However, as N increases, I(N) decreases towards its liquid-phase value (I_L) and this downward shift of I(N) results in progressively more and more of the higher-lying states of the isolated

Table 2. Values of I_L, $I_L - I_G$, V_o, and P^+ for the Azulene Molecule in Various Nonpolar Liquids (from Faidas and Christophorou, 1988)

	I_L[a] (eV)	$I_L - I_G$[b] (eV)	V_o (eV)	P^+ (eV)
n-Tridecane	6.28	- 1.14	+ 0.21	- 1.35
n-Pentane	6.12	- 1.30	+ 0.01	- 1.31
Tetramethylpentane	5.70	- 1.72	- 0.33	- 1.39
Tetramethylsilane	5.45	- 1.97	- 0.55	- 1.42
Tetramethyltin	5.33	- 2.09	- 0.75	- 1.34

[a] $\pm$ 0.05 eV

[b] I_G = 7.42 eV

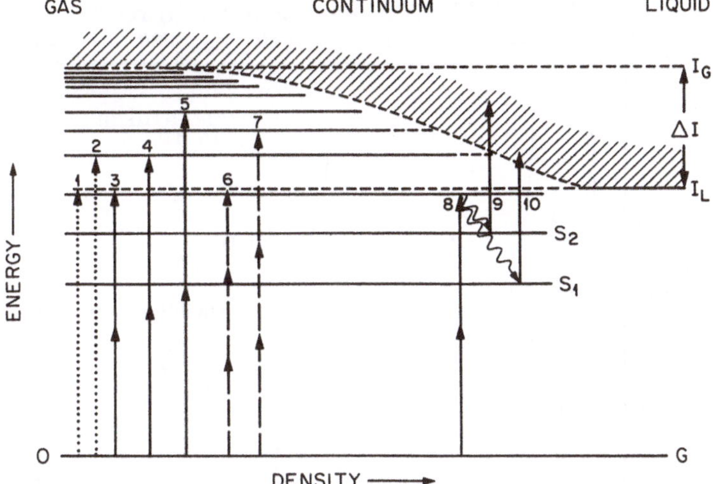

Fig. 25. Schematic illustration of the gradual lowering of I(N) from its low-density value I_G to its value I_L in the liquid and the progressive increase in the number of electronic states which become autoionizing as I_L is approached. The arrows indicate various possible resonant and nonresonant single and multiphoton ionization processes for a molecule in the liquid (see the text).

molecule becoming autoionizing (superexcited) converging at each value of N to the corresponding I(N) limit. Structure in the photoionization spectrum can result from a number of processes. In Fig. 26, for example, the structure below ~420 nm is due to one-photon resonant two-photon ionization, and that in the range 550 to 450 nm due to two-photon resonant three-photon ionization.

In condensed media, proper determination of the energetics of the photoionization process requires proper identification of the ionization mechanism(s). The latter can become complicated depending on the characteristics of the laser source used and the characteristics of the resonance states involved, especially their lifetimes and intramolecular relaxation pathways. To illustrate this we again refer to Fig. 25 where the arrows designate various photoionization mechanisms for a molecule embedded in a (nonpolar) liquid. (Note that we use the same energy levels as in the low-pressure gas although the position of these levels should be lowered in the liquid.) These include (i) direct nonresonant one-photon (process 1), two-photon (process 3), and three-photon (process 6) ionization; (ii) direct one-photon (process 2), two-photon (process 4), and three-photon (process 7) ionization resonant with a superexcited state; (iii) three-photon ionization which is two-photon resonant with an excited state below I_L. Concerning the last case (case iii), Faidas and Christophorou (1987, 1988) found that for aromatic molecules in nonpolar

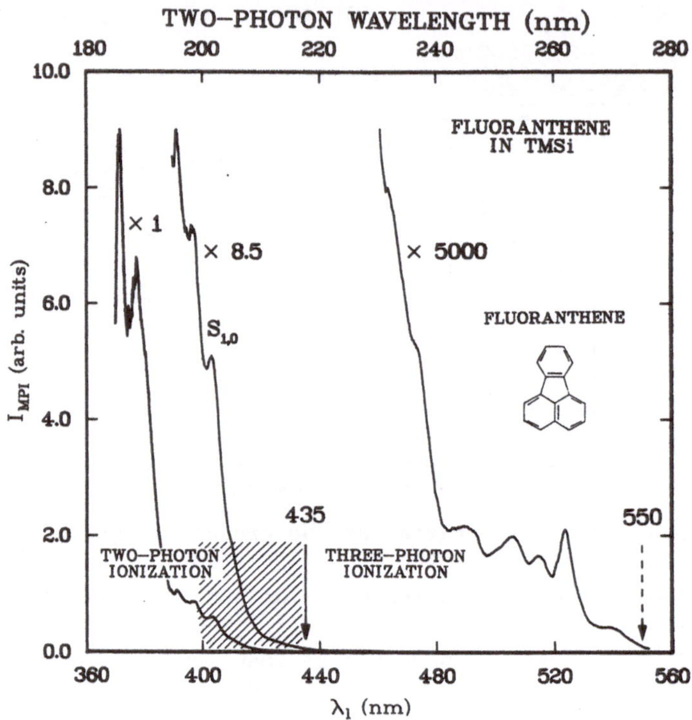

Fig. 26. The multiphoton ionization spectrum I_{MPI} vs. λ_{laser} of fluoranthene in tetramethylsilane. The solid arrow indicates the laser wavelength (435 nm) above which only three-photon ionization is observed. In the hatched region both two- and three-photon ionization processes are observed. The broken arrow indicates the onset of the three-photon ionization. The structure below ~420 nm is due to the one-photon resonant with S_1 state two-photon ionization. The structure between 460 < λ_{laser} < 550 nm is due to two-photon resonant three-photon ionization. (From Faidas and Christophorou, 1987.)

liquids the state from which ionization occurs by the absorption of a third photon generally is not the one reached by the two-photon absorption but rather one which lies energetically below and to which the upper state quickly internally converts. For example, for benzene and other typical aromatic hydrocarbons for which internal conversion to the lowest excited π-singlet state S_1 is fast (~10^{-12} s) compared to the duration (~10^{-8} s) of the laser pulse used by Faidas and Christophorou (1987, 1988) the transition to the ionization continuum by the absorption of the third photon occurs from the lowest excited π-singlet state S_1 (lifetime ~3×10^{-8} s; process 10), for azulene for which the second excited π-singlet state S_2 is much longer-lived (1.4×10^{-9} s) than the S_1 (2×10^{-12} s) the third photon is absorbed from S_2 (process 9), and for fluoranthene for which both the S_2 and the S_1 states are long-lived (lifetimes 2 to 5×10^{-8} s), the transition to the ionization continuum occurs by absorption of a third photon from either the S_1 or the S_2 states. Other mechanisms are, of course, possible in the condensed phase which do not occur in dilute gases (e.g., two-photon excitation just above I_L, followed by geminate recombination, relaxation to a lower excited state and ionization from this state with the absorption of a third photon).

The unraveling of these processes in liquids and dense gases and their relation to those in low-pressure gaseous media will greatly contribute to our understanding of the gas to liquid transition.

REFERENCES

Alekseev, V.A., and Sobel'man, I.I., 1966, Sov. Phys. JETP, 22:882.
Asaf, U., and Steinberger, I.T., 1986, Chem. Phys. Lett., 128:91.
Atrazhev, V.M., and Yakubov, I.T., 1977, J. Phys. D., 10:2155.
Atrazhev, V.M., and Yakubov, I.T., 1980, High Temp., 18:966.
Bakale, G., Sowada, U., and Schmidt, W.F., 1976. J. Chem. Phys., 80:2556.
Bakale, G., and Schmidt, W.F., 1981, Z. Naturforsch. 36a:802.
Bakale, G., and Beck, G., 1986, J. Chem. Phys., 84:5344.
Bartels, A.K., 1974, Ph.D. Dissertation, University of Hamburg, Hamburg, W. Germany.
Boness, M.J.W., and Schulz, G.J., 1970, Phys. Rev. A, 2:2182.
Braglia, G.L., and Dallacasa, V., 1978, Phys. Rev. A, 18:711.
Braglia, G.L., and Dallacasa, V., 1982, Phys. Rev. A, 26:902.
Christophorou, L.G., 1971, "Atomic and Molecular Radiation Physics," Wiley-Interscience, New York.
Christophorou, L.G., 1975, Intern. J. Radiat. Phys. Chem., 7:205.
Christophorou, L.G., 1976, Chem. Revs., 76:409.
Christophorou, L.G., and McCorkle, D.L., 1976, Chem. Phys. Lett., 42:533.
Christophorou, L.G., 1978, Adv. Electr. Electron Phys., 46:55.
Christophorou, L.G., Maxey, D.V., McCorkle, D.L., and Carter, J.G., 1980, Nucl. Instrum. Methods, 171:491.
Christophorou, L.G., Carter, J.G., and Maxey, D.V., 1982, J. Chem. Phys., 76:2653.
Christophorou, L.G. (Ed.), 1984, "Electron-Molecule Interactions and Their Applications," Academic, New York, volumes 1 and 2.
Christophorou, L.G., McCorkle, D.L., and Christodoulides, A.A., 1984, in: "Electron-Molecule Interactions and Their Applications," L.G. Christophorou (Ed.), Academic, New York, Vol. 1, Chap. 6.
Christophorou, L.G., and Siomos, K., 1984, in: "Electron-Molecule Interactions and Their Applications," L.G. Christophorou (Ed.), Academic, New York, Vol. 2, Chap. 4.
Christophorou, L.G., 1985, Chem. Phys. Lett., 121:408.
Cohen, M.H., and Lekner, J., 1967, Phys. Rev., 158:305.
Demuth, J.E., Schmeisser, D., and Avouris, Ph., 1981, Phys. Rev. Lett., 47:1166.
Derenzo, S.E., Mast, T.S., Zaklad, H., and Muller, R.A., 1974, Phys. Rev. A, 9:2582.
Dmitrenko, V.V., Romanyuk, A.S., Suchkov, S.I., and Uteshev, Z.M., 1983, Sov. Phys. Tech. Phys., 28:1440.
Faidas, H., and Christophorou, L.G., 1987, J. Chem. Phys., 86:2505.
Faidas, H., and Christophorou, L.G., 1988, Rad. Phys. Chem. (in press).
Fano, U., Stephens, J.A., and Inokuti, M., 1986. J. Chem. Phys., 85:6239.
Fermi, E., 1934, Nuovo Cimento, 11:157.
Freeman, G.R., 1981, in: "Electron and Ion Swarms," L.G. Christophorou (Ed.), Pergamon, New York, 93.
Freeman, G.R., (Ed.), 1987, "Kinetics of Nonhomogeneous Processes," Wiley-Interscience, New York.

Goans, R.E., and Christophorou, L.G., 1974, J. Chem. Phys., 60:1036.
Gushchin, E.M., Kruglov, A.A., and Obodovskii, I.M., 1982, Sov. Phys. JETP, 55:650.
Hatano, Y., and Shimamori, H., 1981, in: "Electron and Ion Swarms," L.G. Christophorou (Ed.), Pergamon Press, New York, 103.
Holroyd, R.A., and Russell, R.L., 1974, J. Phys. Chem., 78:2128.
Hunter, S.R., and Christophorou, L.G., 1984, in: "Electron-Molecule Interactions and Their Applications," L.G. Christophorou (Ed.), Academic, New York, Vol. 2, Chap. 3.
Köhler, A.M., Reininger, R., Saile, V., and Findley, G.L., 1986, Phys. Rev. A, 33:771; Private Communications.
Kouroukiis, G.A., Siomos, K., and Christophorou, L.G., 1982, J. Molec. Spectr., 92:127.
Krebs, P., 1984, J. Phys. Chem., 88:3702.
Krebs, P., and Heintze, M., 1982, J. Chem. Phys., 76:5484.
Legler, W., 1970, Phys. Lett., 31A:129.
Lehning, H., 1969, Phys. Lett. A, 29:719.
Lekner, J., 1967, Phys. Rev., 158:130.
Lipsky, S., 1981, J. Chem. Ed., 58:93.
Messing, I., Raz, B., and Jortner, J., 1977, Chem. Phys., 25:55.
Miller, L.S., Howe, S., and Spear, W.E., 1968, Phys. Rev., 166:871.
Milloy, H.B., and Crompton, R.W., 1977, Australian J. Phys., 30:51.
Nakamura, S., Sakai, Y., and Tagashira, H., 1986, Chem. Phys. Lett., 130:551.
O'Malley, T.F., 1980, J. Phys. B, 13:1491.
O'Malley, T.F., 1983, Phys. Lett., 95A:32.
Polischuk, A. Ya, 1984, J. Phys. B, 17:4789.
Polischuk, A. Ya, 1985, J. Phys. B, 18:829.
Raz, B., and Jortner, J., 1969, Chem. Phys. Lett., 4:155.
Reininger, R., Asaf, U., and Steinberger, I.T., 1982, Chem. Phys. Lett., 90:287.
Reininger, R., Asaf, U., Steinberger, I.T., and Basak, S., 1983, Phys. Rev. B, 28:4426.
Reininger, R., Saile, V., Laporte, P., and Steinberger, I.T., 1984, Chem. Phys., 89:473.
Reininger, R., Saile, V., Findley, G.L., Laporte, P., and Steinberger, I. T., 1985, in: "Photophysics and Photochemistry Above 6 eV," F. Lahmani (Ed.), Elsevier Science Publishers, Amsterdam, 253.
Robin, M.B., 1974, "Higher Excited States of Polyatomic Molecules," Volume 1, Academic, New York, 78.
Sakai, Y., Nakamura, S., and Tagashira, H., 1985, IEEE Trans. Electr. Insul., EI-20:133.
Sanche, L., 1979, J. Chem. Phys., 71:4860.
Sanche, L., and Michaud, M., 1981a, Phys. Rev. Lett., 47:1008.
Sanche, L., and Michaud, M., 1981b, Chem. Phys. Lett., 84:497.
Sanche, L., and Michaud, M., 1983, Phys. Rev. B, 27:3856.
Sanche, L., and Michaud, M., 1984, Phys. Rev. B, 30:6078.
Schmidt, W.F., 1984, IEEE Trans. Electr. Insul., EI-19:389.
Shibamura, E., Takahashi, T., Kubota, S., and Doke, T., 1979, Phys. Rev. A, 20:2547.
Tauchert, W., Jungblut, H., and Schmidt, W.F., 1977, Can. J. Chem., 55:1860.
Von Zdrojewski, W., Rabe, J.G., and Schmidt, W.F., 1980, Z. Naturforsch., 35A:672.
Warman, J.M., and de Haas, M.P., 1985, IEEE Trans. Electr. Insul., EI-20:147.
Williams, M.W., Hamm, R.N., Arakawa, E.T., Painter, L.R., and Birkhoff, R.D., 1975, Int. J. Radiat. Phys. Chem., 7:95
Yakubov, I.T., and Polischuk, A. Ya., 1982, J. Phys. B, 15:4029.

CALCULATIONS OF V_o AND THE ENERGY DISPERSION OF ELECTRONS IN RARE GAS LIQUIDS

G. Ascarelli

Physics Department
Purdue University
West Lafayette, IN 47907

During this lecture I wish to discuss how V_o has been calculated in the case of liquid Argon (Lekner, 1967; Plenkiewitz et al., 1986). When the energy of states of non-zero wave vector is calculated the effective mass is obtained in the same way as for electrons in a crystalline solid (Kittel, 1963). The effect of density fluctuations can be easily estimated in the framework of the effective mass theory.

The existing calculations of V_o in liquids are calculations of the bottom of the conduction band in a crystal. The "unit cell" is the Wigner-Seitz sphere whose radius is $r_{ws} = (3/4\pi n)^{1/3}$, where n is the atomic density. The potential that is seen by an extra electron in the liquid is a pseudo-potential used to describe the Coulomb interaction to which is added a term arising from the dipole moment induced by the electron, both directly on the atom at the origin and through the polarization of the remaining atoms of the fluid (Lekner, 1967). The radial distribution function of the fluid arises only through this last term. The "crystal structure" that is assumed is that of a close packing of spheres. The closest approximation to the spherical symmetry characteristic of a liquid will be a f.c.c. crystal.

The difference between the calculations carried out by Lekner (1967) and Plenkiewitz et al. (1986) lies in the choice of the pseudo-potential. The pseudo-potential used by the latter is the same used for the band calculations in the corresponding rare-gas solids.

The solution of the quantum mechanical problem is carried out as in the now classical case of Na by imposing that the radial derivative of the wave function corresponding to zero pseudo-momentum, k, is zero at the surface of the Wigner-Seitz cell. When one identifies V_o with the energy of this state one implicitly assumes that the minimum of the band is at the center of the Brillouin Zone (Raimes, 1961).

It is to be stressed that this is a crystal-like theory. The detail of the band anisotropy will arise from the assumption of a specific crystal structure. Since the wave function we are considering is S-like, a band dispersion calculated for a f.c.c. crystal should not differ too much, at least near the band minimum, from that of a fluid if the effect of disorder is completely taken into account by the use of the radial distribution function in the calculation of the dipole moment induced on the atom at the center of the Wigner-Seitz cell.

Although it had not been evaluated, the effective mass tensor can be calculated using the $\vec{k} \cdot \vec{p}$-method (Kittel, 1963) (this is nothing but second-order perturbation theory):

$$\left(\frac{m_o}{m^*}\right)_{\mu\nu} = \delta_{\gamma\mu} + \frac{2}{m_o} \sum_{\delta}{}' \frac{\langle \gamma 0 \mid \pi_\mu \mid 0\delta \rangle \langle \delta 0 \mid \pi_\nu \mid 0 \gamma \rangle}{E_{\gamma 0} - E_{\delta 0}} \quad . \tag{1}$$

Here δ and γ are band indices and the zeros indicate that the calculation is carried out at the center of Brillouin Zone (B.Z.). In the absence of spin-orbit interaction the operator π reduces to the momentum operator.

In the case of rare-gas crystals where the conduction band has Γ_6^+ symmetry, the effective mass tensor is diagonal with all elements equal. It reduces to a scalar effective mass whose value is, however, in general, different from a free electron mass. For reasons of symmetry, if the conduction band of a liquid has a minimum at the center of the B.Z., it must be spherical.

We must, however, ask for what range of energies a parabolic dispersion relation, $E+(\hbar k)^2/2m^*$, is valid. In the rare-gas solids, both theory (Rossler, 1970) and experiment (Kessler et al., 1987) indicate the existence of significant non-parabolicities due to a band gap about 2 eV above the conduction-band minimum. Since the potential seen by the electron in a liquid is close to the potential acting on the electron in the solid, similar band gaps are expected in the liquid.

The effective mass that is measured from exciton spectra in crystal Xe is 0.33 m_o, while that in the liquid is 0.27 m_o (I. Steinberger, 1988). The values of V_o are also close in both states of Xe (Kessler et al., 1987). Although comparable data for the effective masses is not known in other rare gas liquids, no large differences of these parameters are to be expected between the liquid and the solid.

Both in the solid as well as in the liquid, the electron in the conduction hand has a charge density that is large between the atoms because of the repulsion of the electrons making up the closed atomic shell. Near the triple point a large fraction of the volume of the Wigner-Seitz cell is taken up by this repulsive potential. The kinetic energy of the conduction electron (which has nothing to do with its thermal energy) is largely due to this confinement when the atomic density is large. When the atomic density initially decreases, the kinetic energy of the electron decreases rapidly, but its attractive potential energy, due to the interaction with the nuclear cores, does not change much because the average distance of the electron from the nuclei has not significantly changed (Ascarelli, 1985). Thus, upon decreasing the density the magnitude of V_0 initially increases because the electron is more tightly bound. However, when the atomic density is decreased further, the average potential energy decreases as well, because the average distance of the electron from the nuclei increases. As a result the density dependence of V_0 must have a minimum, as is effectively observed. Clearly the binding energy is zero in the dilute gas limit.

At this point it might be worth connecting a little further with Professor Steinberger's lecture.

The lowest excited state of the system is by definition the exciton state at the bottom of the exciton band. If this state is not degenerate, when the sample contains N atoms, there will be N states in the exciton band. In the case when this lowest exciton is p-times degenerate, there will be pN states in the exciton band. In general, the degeneracy of the exciton states will be removed when the exciton wave vector is not zero. Similar arguments apply to each of the exciton states so that the total number of exciton states is equal to the number of excited states that exist in a system of N non-interacting atoms. This is only an enumeration of the states and does not imply that an exciton can be described by the same wave function as that of an excited atomic state. Professor Steinberger showed us data indicating that the energy of an exciton is different from that of a broadened excited atomic state.

The electron and the hole are not necessarily localized on the same atom and the electron hole-pair can be localized on a series of equivalent sites of the system. Both possibilities give rise to additional terms in the Hamiltonian that do not exist in the atomic case.

From the point of view of the enumeration of the states, the exciton series limit, that defines the bottom of the conduction band, corresponds to the vacuum. This does not, however, indicate that the wave functions, and, thus, e.g., the respective scattering probabilities, have necessarily any similarity. In the case of an electron in the conduction band, the wave function is represented by a plane wave, while in the case of an electron in the conduction band the wave function is a product of a plane wave and another function that oscillates rapidly within each unit cell. The latter has a large influence when one considers scattering by an entity whose dimensions are comparable to an individual atom. The energies of the electrons are also quite different: in the condensed phase, there are significant contributions from the polarization of all the atoms in the system, a repulsion by all the electrons of all the atoms, the attraction by all the atomic nuclei, as well as the additional repulsion due to the exclusion principle and the possibility of exchange betweem all the electrons of the system. By way of contrast, in the atomic case there exists only one ion. These interactions give rise to a bottom of the conduction band that can be either above or below vacuum. From an "average" point of view these interactions can be characterized by a value of V_o, an effective mass, and a dielectric constant.

All the above arguments are based on a "rigid" lattice, i.e., a system in which there are no large atomic displacements that give rise to self-trapping like, e.g., in the case of the electron in a "bubble" characteristic of liquid He.

In the case of electrons, such self-trapping is particularly important when the bottom of the conduction band is above the vacuum level. The energy of the system may then be decreased by localizing the electron in an empty region, the bubble. The price to be paid is associated with the compression of the fluid and the kinetic energy associated with the localization of the electron.

We can now make a guess on the effect of large density fluctuations capable of localizing an electron.

For simplicity we shall assume a spherically symmetric square well due to a spherically symmetric uniform density fluctuation. In this case, the first bound state appears (with zero binding energy) when the radius a, and depth ΔV_o, of the density fluctuation, are related by

$$\Delta V_o = \frac{\pi^2 h^2}{8m^* a^2} \quad . \tag{2}$$

In the effective-mass approximation, i.e., when the volume $4\pi a^3/3$ contains many atoms (a few hundred is probably a good guess), the value of ΔV_o cannot be any larger than the difference between the value of V_o at the average density and that corresponding to the minimum in the V_o-vs.-n curve. With $m^* = m_o$, a is $\simeq 80\text{Å}$ for $\Delta V_o \sim 0.1$ eV.

Provided the coupling between the electron and the lattice is sufficiently strong, the electron may create a density fluctuation and be trapped in it, as in the case of liquid helium. However, unlike the case of He where the bottom of the band is significantly above the vacuum and that energy can be used to create the "bubble", this is not the case in the heavier rare gases where V_o is negative and ΔV_o is small. In the case of argon near the triple point, such a "bubble" can be created by an electron whose energy is about 1 eV above the minimum of the band at the equilibrium density, i.e.,

$$E = \Delta F_o - \Delta V_o = \frac{(\Delta n)^2 \Omega}{2n^{-2} \chi_T} - \Delta V_o , \tag{3}$$

where $\Omega = 4\pi a^3/3$ and χ_T is the isothermal compressibility. If the electron-density fluctuation coupling is sufficiently strong this will flatten the electron dispersion and create a small energy gap as in the case of the electron-optical phonon coupling in polar crystals, or even in semiconductors, e.g., InSb.

During this lecture I hope I impressed you with the fact that an electron in a fluid like Ar is bound, its wave function is extended, and, in a "frozen" liquid (without thermal motion of the atoms), the electron in the conduction band would not scatter, i.e., it would be in a stationary state. Scattering corresponds to a transition from one stationary state to another. It is not the result of the interaction with a single atom but instead with a change of potential brought about by the displacement of the atoms of the fluid. This can be described by phonons, if we consider their time dependence (usually in the GHz range), or "static" if we consider a much slower time dependence so that the electron wave packet (whose dimensions are of the order of the thermal wave length of the electron, $\Lambda = \left(\frac{2\pi \, mkT}{h^2} \right)^{-1/2}$) has moved far from the region of the disturbance during the life of the density fluctuation.

The quantity we consider in transport theory and call the electron momentum is the pseudo-momentum of the center of mass of the wave packet made up of eigenstates of the Schroedinger equation. I wish to remind you that in the case of a crystalline solid this pseudo-momentum is the eigen value of the operator that describes the translational symmetry of the lattice. In the case of a solid, the range of pseudo-momentum ($\hbar k$), used in making the wave packet, is small in comparison with the size of the Brillouin Zone. It is a consequence of the effective-mass approximation that the pseudo-momentum of this wave packet behaves like the momentum we use in transport theory. In calculating transport properties we must substitute the free electron mass with the effective mass. This is a way of taking into account the energy dispersion relation of the electron.

It is not clear what the equivalent of this pseudo-momentum would be in a disordered system like a liquid. Presumably, if the potential changes associated with the disorder are small, as it appears to be in the case of liquids similar to the rare gases heavier than Ne, the concepts developed in the case of the crystals are a good approximation. In the case of molecules that are neither spherical nor have a shape represented by one of the cubic groups, the fluctuations of the potential resulting from changes of molecular orientation can be large.

We may guess that if the scattering due to "static" density fluctuations is comparable to that due to phonons, the main difference between transport in a solid and the corresponding liquid would be that the range of pseudo-momentum we must use to make up the wave packet in a liquid is somewhat larger than in the corresponding crystal with phonons.

Near the bottom of the conduction band of a rare-gas liquid the energy momentum relation is probably parabolic and may be characterized by an effective mass. At higher energies, possibly as low as ~1 eV above the band minimum, a significant non-parabolic behavior is likely. This may arise, as in the solid, due to band structure effects, or alternatively due to the interaction with density fluctuations in a way reminiscent of the formation of the "bubble" in liquid helium.

REFERENCES

Ascarelli, G., 1985, The motion of electrons injected in classical nonpolar insulating liquids, Comments Solid State Phys., 11:179.

Kessler, B., Eyers, A., Horn, K., Muller, N., Schmiedskamp, B., Schonhence G., and Heinzmann, V., 1987, Determination of X_e valence and conduction bands by spin-polarized photo emission, Phys. Rev. Lett., 59:331.

Kittel, C., Wiley, J., 1963, "Quantum Theory of Solids," New York, 186.

Lekner, J., 1967, Motion of electrons in liquid argon, Phys. Rev., 158:130.

Plenkiewitz, B., Jay-Gerin, J. P., Plenkiewitz, P., and Bachelet, G. B., 1986, Conduction band energy of excess electrons in liquid argon, Europhysics Letters, 1:455.

Raimes, S., 1961, "The Wave Mechanics of Electrons in Metals," North Holland, Amsterdam (particularly Chapter 9).

Rossler, U., 1970, Electron and exciton states in solid rare gases, Phys. Stat. Sol., 42:345.

Steinberger, I., 1988, these Proceedings.

INTERFACIAL PHENOMENA

ELECTRICAL ASPECTS OF LIQUID/VAPOR, LIQUID/LIQUID, AND LIQUID/METAL INTERFACES

B.E. Conway

Chemistry Department, University of Ottawa
32 George Glinski Way, Ottawa, KIN 6N5
Ontario, Canada

INTRODUCTION: TYPES OF LIQUID INTERFACES

Manifestation of electrical behavior of interfaces depends in important ways on the type of liquid. Three principal types can be distinguished:

a) metallic liquids, e.g., Hg, its amalgams, Ga;

b) dipolar liquids, including H-bonded fluids; and

c) non-polar liquids of spherical atoms or molecules, e.g., Ar, Ne, Ch_4, etc.

The origin of electrical aspects of liquid interfaces is very different in the three cases, as will be indicated later.

Several categories of liquid interfaces having characteristic properties (Table 1) and electrical behavior must be distinguished:

(a) pure liquids adjacent to their vapors, or virtual vacuum, as with Hg or involatile fluids;

(b) metallic liquids, e.g., the Hg electrode, adjacent to a polar liquid such as water, or non-aqueous solvents;

(c) metallic liquids, e.g., the Hg electrode, adjacent to an electrolyte solution of a salt in a polar solvent; and

(d) interfaces between two virtually immiscible liquids, e.g., water/nitrobenzene; water/hydrocarbons; water lipid-bilayer membranes, or between two immiscible electrolyte solutions between which, however, certain kinds of ions may be transferred, e.g., Bu_4N^+,Cs^+ (Samec et al., 1979a, 1979b).

Amongst liquid metals, Hg is the ideal one, both for metal/vapor and metal/electrolyte (electrochemical) studies, because of its relative unreactivity with many vapors and solvents, and because experiments can be conducted over a convenient temperature range around 298 K, as well as down to its melting point and up to ca. 629.9 K. Hg can also be obtained in a state of very high purity.

Ga is less satisfactory owing to its much more basic character and consequent reactivity with water and acids, tending to suffer self-polarization through corrosion, with H_2 evolution. Similar problems arise with base-metal amalgams. Studies with other metals in the liquid state are much more difficult but electrochemical measurements have been made with some low-melting point alloys, and with Sn and Pb in molten salts at elevated temperatures.

Table 1. Types of liquid interfaces and factors in the molecular state of the liquid near such interfaces.

Factors in the state of a liquid at its interface	Types of interface
Number-density of its molecules	Free liquid against its vapor
Structure of the liquid near its interface	Free liquid against another, immiscible free liquid (oil/water, water/Hg)
Orientation of its molecules	
Resulting surface-potential	
Vibration-rotation behavior of its molecules	Free liquid against a liquid-like membrane (e.g., lipid membrane)
Relative adsorption (surface excesses) of solvent and solute, if the liquid is a solution	Liquid against an insulating solid, e.g., an oxide
State of hydration of ions, if the liquid is an electrolyte solution	Liquid against a semi-conducting solid, e.g., some oxides, sulfides (including minerals), Ge, Ga-As, etc.
Distribution of cations and anions in the case of an electrolyte solution	Liquid against a metallic conductor (Hg and other metal electrodes)
Effects of the ions on the orientation of the solvent molecules at the interface	
Effects of an applied external field, in the case of electrode/solution interfaces, on solvent orientation, ion distribution, and liquid structure near the interface.	

The extent of knowledge of the interfacial electrochemistry at Hg enormously exceeds in quantity, precision, and reliability that for all other liquid-metal systems (for reviews, see, e.g., Grahame, 1947; Parsons, 1954; Conway, 1977).

A progression of complexity of the interface of liquids arises in the several cases referred to above, especially in the case of liquid metal electrode/electrolyte-solution boundaries, as indicated below:

Liquid Metal/Electrolyte Interface: Levels of Complexity

(a) Liquid metal/gas (vacuum) interface - primary surface.

(b) Liquid metal/vacuum interface + monolayer of solvent molecules.

(c) Liquid metal/vacuum interface + oriented monolayer and bulk solvent molecules.

(d) Liquid metal/liquid interface (c) with ions in bulk solvent at interface.

(e) Liquid metal/liquid interface (d) with ions and submonolayer of monolayer oxide films present.

Figure 1 illustrates the increasing complexity of liquid interfaces when the liquid is a metal as the metal/vacuum interface (a) is replaced by metal interface with adsorbed solvent dipoles (b), bulk dipolar solvent (c), and bulk electrolyte solution (d).

ORIGIN OF ELECTRICAL BEHAVIOR OF LIQUID INTERFACES

Electrical properties of liquid interfaces arise because of the non-homogeneous environment of atoms or molecules in the liquid phase, at its interface, in the direction normal to its surface.

The interface of any type of liquid involves a discontinuity of the atoms and molecules of the material along one direction normal to the surface. The local distribution function at or near the surface is different from that in the bulk and introduces an inhomogeneity in the distribution of electron and nuclear charges in the case of metals, and of dipole partial charges and their orientation in the case of surfaces of polar liquids. This non-uniform distribution of electron or partial charges gives rise to a local charge separation at the interface*, i.e., a surface dipole layer is generated together with a corresponding interfacial field.

---

* Since the modified distribution of particles and associated partial charges arises over a finite distance of ca. 0.2 to 0.5 nm, it is useful to refer to this boundary region of finite dimensions as the "interphase", a term first employed by Guggenheim (1932) in dealing with the thermodynamics of surfaces and adsorption.

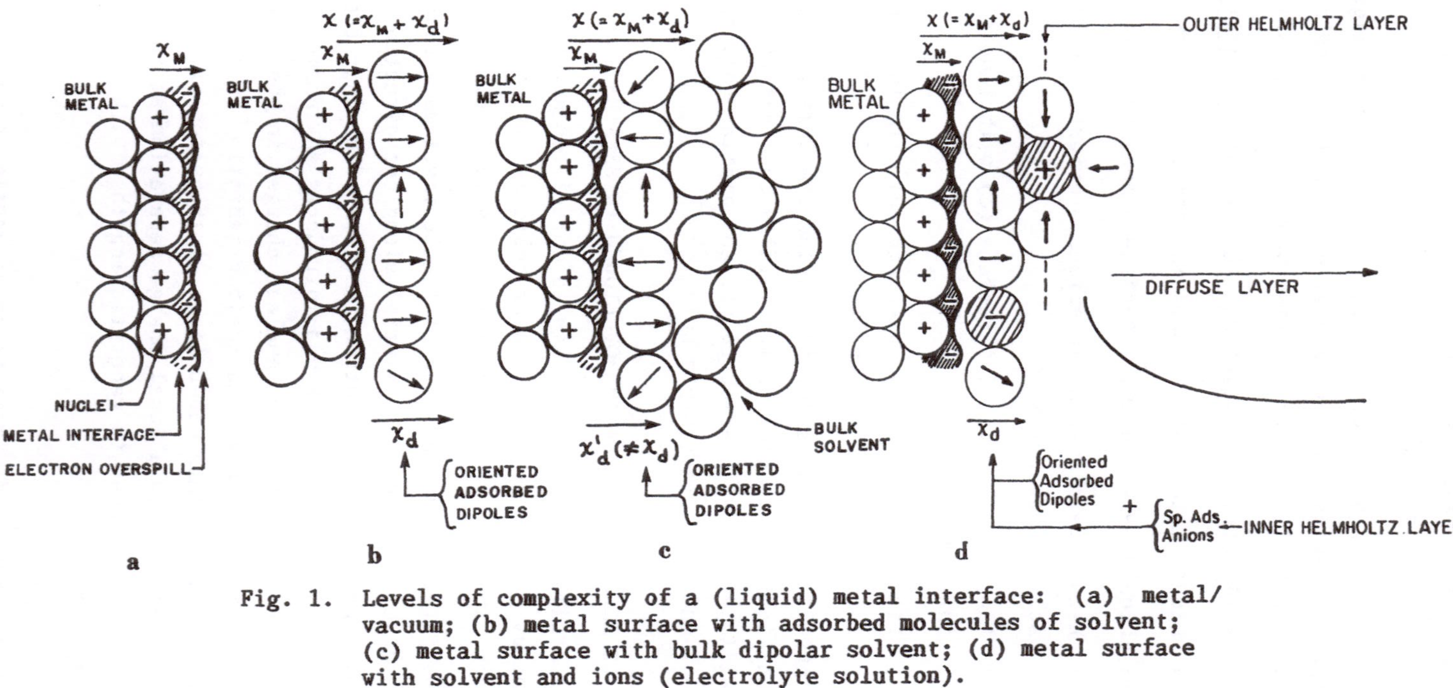

Fig. 1. Levels of complexity of a (liquid) metal interface: (a) metal/vacuum; (b) metal surface with adsorbed molecules of solvent; (c) metal surface with bulk dipolar solvent; (d) metal surface with solvent and ions (electrolyte solution).

Even with non-polar liquids, the local environment of, and interaction between, the outer electron shells of atoms or molecules in the surface is different from that in the bulk phase, so that an electrical inhomogeneity can still arise, but is less marked than with dipolar or free electron fluids.

At electrodes and in oriented dipole layers, the field can be on the order of 3×10^7 V cm^{-1} (10^5 e.s.u.). In the metal electrode/solution interface case, as at Hg, this field can be modulated by external control of the metal/solution potential difference or the corresponding surface charge density, q_M.

In the case of metallic liquids, as with solid metals, the manifestation of electrical properties is first of all determined by the distribution of conduction electrons (the electron plasma) at the metal's surface with respect to the plane of nuclei of exposed surface atoms. This so-called electron "overspill" effect (Fig. 1a) depends on the orientation of the crystal face at solid metals and, in the case of liquid metals, on the mean surface density of metal atoms. The overspill effect determines the intrinsic electric surface potential of the metal, χ_o.

In the case especially of dipolar liquids or of dipolar liquids adjacent to a liquid or solid metal (electrode case), the electrical properties of the interphase are intimately connected with (a) the structure of the interphase, i.e., the extent of dipole orientation as measured by the component resolved in the normal to the interface, and (b) any degree of charge-transfer that arises from donor/acceptor interaction between the dipoles of the polar liquid or adsorbable components dissolved in it and the metal, e.g., in the case of water, alcohols, amines, heterocyclic bases such as pyridine, etc.

At unpolarized metal interfaces, spontaneous dipole orientation, leading to a dipole surface potential and corresponding field, can arise when specific donor/acceptor interaction arises between the metal and one end or the other of the dipole of the liquid or of a solute component dissolved in it. A dipole orientation effect will also arise spontaneously when a dipole, existing asymmetrically with the molecule, interacts with its own image (see the section on IMAGE CHARGES AT LIQUID INTERFACES) in the metal phase, as, e.g., with H_2O and other polar solvents. Thus, at the potential of zero charge of Hg in contact with liquid water (containing 0.1 M NaF as electrolyte), the water dipoles are oriented with their negative ends towards the Hg, and a net charge of $q_M \approx -2$ μC cm^{-2} is required to bring about zero net H_2O dipole orientation.

At electrode metals (see the section on THE MERCURY ELECTRODE/ SOLUTION INTERFACE), dipole orientation, if not already spontaneous, is induced by the polarization field determined by the net charge density, q_M, that can be imposed in a controllable way in an electrochemical experiment, e.g., at Hg. Further details of this effect are treated in THE MERCURY ELECTRODE/SOLUTION INTERFACE.

LIQUID OR SOLID METAL/VACUUM INTERFACE AND SURFACE DIPOLE POTENTIAL

Unfortunately, unlike the situation at the Hg/electrolyte solution interface, independent variation of metal surface charge at metals cannot be made at the metal/vacuum interface, but the electron work function Φ of the metal can be determined accurately <u>in situ</u> (e.g., by means of the u.v. photoelectric effect), and <u>changes</u> of Φ due to adsorption of polar and nonpolar molecules can be measured, giving the net surface dipole potential χ of the adsorbed film of atoms or molecules. For low coverages (Θ), χ is usually linear in Θ but increases less rapidly than linearly as $\Theta \to 1$, due to mutual depolarization amongst similarly oriented surface dipoles:

$$\chi_\Theta = \chi_{\Theta = 0} /(1 + k\Theta) \; . \qquad (1)$$

The formal basis of the relation between the work function Φ for liquid (and solid) metals and the χ component, is as follows: a quantity ϕ called the "inner potential" of the phase is determined by the sum

$$\phi = \psi + \chi \; , \qquad (2)$$

where ψ is the so-called "outer potential" of the phase due to free charge density (referred to as q_M in this paper) residing on its surface (for a sphere of radius r, $\psi = q/r$; if q is the total charge, $= 4\pi r^2 q_M$). In terms of the model of a layer of oriented dipoles each of moment m, at a surface density of n cm^{-2},

$$\chi = 4\pi \, n \, m \; , \qquad (3)$$

according to a relation of von Helmholtz.

Hence

$$\phi = \psi + 4\pi \, n \, m \; , \qquad (4)$$

omitting here the possible depolarization of the dipoles due to lateral interaction, i.e., m usually decreases with n, as in Eq. (1).

As discussed by Lange (1930, 1951) (cf. Parsons, 1954), since ψ is experimentally measurable, it is useful to define another, so-called "real potential" in terms of the chemical potential μ and the surface potential, χ:

$$\alpha = \mu + ze\chi \; , \qquad (5)$$

or the electrochemical free energy, $\bar{\mu}$, is

$$\bar{\mu} = \alpha + ze\psi = \mu + ze\chi + ze\psi \ , \qquad (6)$$

for the electrical energy of a charge ze. When the phase is uncharged, $\psi = 0$, so the real potential $\alpha \equiv \mu$ for such a condition. For an uncharged metal, α is the electronic work function denoted earlier as Φ, i.e., the work involved in bringing a particle of charge e from infinity through the surface of the phase (where the surface potential χ resides) to its interior when its surface bears no net charge: $q_M = 0$, $\psi = 0$. Then

$$\Phi \equiv \bar{\mu} + \mu + e\chi \quad (z = 1) \ , \qquad (7)$$

i.e., the electron work function of a metal has an internal chemical potential component (μ) and a surface component, $e\chi$. Hence, for solid metals, unlike those in the liquid phase, Φ is dependent on the crystallographic face through which the electron charge exits or enters the phase. The differences between the values for the principal-index planes of f.c.c. metals are quite appreciable (Hamelin, 1986), viz. up to ca. 1 V, and follow differences of potential of zero charge when the single-crystal materials are employed as electrodes. Of course, for liquid metals, an average value arises similar to that for the close-packed (111) plane, although liquid metals, like Hg, at appreciable temperature do not have ideally smooth surfaces due to Boltzmann fluctuations (see Watts-Tobin and Mott, 1961) for the same reasons that liquids contain "holes" at finite temperatures.

When chemisorption occurs, e.g., of Na on W or I^-, or S^{--} on Hg, appreciable charge-transfer usually takes place and the changes of Φ are then usually quite large. However, it is interesting that even the physical adsorption of noble gases, having zero permanent moment, give rise to quite substantial χ values (Mignolet, 1950, 1953), and this must be due to an influence of the outer s and p orbitals of the adsorbate atoms on the electron distribution at the metal surface, and, reciprocally, of the metal's intrinsic surface field, E, on the otherwise spherical electron distribution of noble atom adsorbates through the atom's polarizability, α.

That the intrinsic surface potential at the vacuum/metal interface is substantial is indicated by the measured surface potential changes that are associated with physical adsorption of noble gases which, by themselves, could contribute no molecular surface dipole moment. The changes $\Delta\chi$ of χ, due to noble gas adsorption, are on the order of 0.5 ~ 0.8 V and thus correspond to the polarizing influence of a substantial surface field. Taking the polarizability of say Ar as 1.63×10^{-24} cm^3, the surface field is seen to be ca. 4×10^7 V cm^{-1}. Alternatively, the screening effect of the noble gas electron shells can be regarded as eliminating effectively the intrinsic χ_M so that $\Delta\chi = -\chi_M$. This surface field will tend to exert a considerable polarizing influence on solvent molecules when a metal, e.g., Hg, is placed in a solution in the zero charge condition.

The density of free electrons near a surface of a metal such as Hg or a semiconductor determines the distance within the solid phase, from its interface, to which an external field vector can penetrate and modulate the electron distribution. This distance, called the Thomas-Fermi screening length, is only of the order of 0.1 nm at metals and has a significance analogous to the Debye ionic atmosphere screening distance, $1/\kappa$, in electrolyte solutions or ionic double-layers where ion charge rather than electron charge distribution is involved. More closely analogous effects to those in ionic solutions arise near semiconductor interfaces and the screening length is then normally much larger than at a metal, depending on the electron or hole density. The Thomas-Fermi screening distance is important in electric modulated reflectivity studies at metal electrode interfaces (McIntyre and Peck, 1973).

ADSORPTION AND ELECTRICAL PROPERTIES OF LIQUID METAL/VAPOR INTERFACES

At the interface of Hg with various gases or vapors at controlled but variable fugacity, f, it is possible to measure changes of surface tension and derive the surface excess, Γ, of molecules adsorbed from the vapor phase:

$$(\partial\gamma/\partial\mu) = -\Gamma \ , \tag{8}$$

with μ, the chemical potential of the adsorbate in the gas phase, being given in the usual way by:

$$\mu = \mu^\circ + RT \ln f \ , \tag{9}$$

in relation to its standard value, μ°.

Coupled measurements of the work function at various f values (for not too large pressures) and hence corresponding Γ's through Eq. (11), enable $\Delta\chi$ to be related to Γ, so that some idea of the (changes of) electrical state of the interphase as a function of adsorbate coverage can be deduced. Evaluation of Γ for alcohols and water at Hg has been made by Kemball (1947), using surface tension measurements, while $\Delta\Phi$ measurements have been made on various metal surfaces in relation to physical and chemisorption of a variety of gases, especially O_2, H_2, CO, Cl_2, H_2O, Ne, and Ar.

The surface dipole potential that is measured in such experiments represents a <u>change</u>, Δ, in the value of χ, and is not a simple quantity to interpret since it can be comprised of components due to:

(i) modification of the intrinsic surface potential of the metal associated with electron distribution at its interface;

(ii) induction of a dipole in the adsorbate atom or molecule by the metal's intrinsic surface field, as with Ne, Ar (Mignolet, 1950);

(iii) specific orientation of permanently polar adsorbate molecules through electron-pair donor/acceptor interactions with the metal, as with H_2O, CH_3OH, and H_2S at Hg, or CO at Pt or Ni; and

(iv) development of net charges at the metal interface when specific chemisorption, with charge-transfer, occurs, as with I^- at Pt ($I^- + Pt \rightarrow Pt(e)I$), or Cs at W ($Cs\cdot + W \rightarrow W(e)^-Cs^+$). Direction of charge transfer can usually be known from the <u>sign</u> of change of Φ. The adsorbate film, in these cases, is a compact double-layer of + and - charges, unlike the situation at the Hg electrode where the "double-layer" has a complex structure (Fig. 1) comprising: (a) a compact inner charged (+ -) layer of ions and metal excess charge, q_M; (b) a diffuse layer of thermally distributed cations and anions from 1 to several hundred nm away from the metal surface; and (c) associated with the compact region of separated charges (Helmholtz layer), a region 0.3 ~ 0.5 nm thick of solvent dipoles oriented to a mean extent determined by the population of ions in the solution side of the Helmholtz layer and the net charge density, q_M, on the metal.

THE MERCURY ELECTRODE/SOLUTION INTERFACE

Charge-Potential Relations

The distribution of electrons at metal interfaces has become a topic of major current interest in electrochemistry and metal physics. In electrochemistry, in particular, the possibility arises, especially at Hg, of varying the surface excess electron density, q_M, in a known and controllable way (e.g., see Grahame, 1947) which modifies the electron overspill relative to the plane of surface-atom nuclei, displacing it outward under negative polarization or inward under positive polarization (Lang and Kohn, 1973, Feldman et al., 1986). Experimental access to this effect requires contact between the metal, e.g., Hg, and an electrolyte solution, e.g., aqueous NaF, HCl, with a three-electrode arrangement: an Hg, or Pt, counter electrode for application of a polarizing voltage and a reversible reference electrode against which a varied potential, E, at the Hg can be measured. Then q_M, as a function of E, can be experimentally measured either from the variation of surface tension, γ, with E, according to the Gibbs-Lippmann equation (Grahame, 1947)

$$-(\partial\gamma/\partial E)_{\mu_{i,j}} = q_M \, , \tag{10}$$

for constant composition, $\mu_{i,j}$ of the solution, or from double-layer capacitance (C) measurements made by means of a.c. modulation

$$C = (\partial q_M / \, E) \, . \tag{11}$$

The modulation of surface electron density at Hg implied by Eq. (10) is manifested not only through changes of surface tension, γ, but may be indirectly observed in the change of relative specular reflectivity (McIntyre and Peck, 1972; Gottesfeld et al., 1973) with electrode potential at Hg and at other metals under conditions not restricted to liquid state surfaces.

At the Hg electrode/electrolyte-solution interface, the electrical and structural situation is particularly complex (Fig. 1d) and can be thought of as the result of combining the electrical anisotropy of the metal interface with that of a dipolar liquid, together with the distribution of dissolved cations and anions that arises, depending on the net charge, q_M, on the metal, plus any specific chemisorption affinity cations, or especially anions, may have for the metal surface.

The complexity of the situation was illustrated schematically in Fig. 1, especially 1d for the Hg/electrolyte interface.

<u>Adsorption as a Function of Potential</u>

Using Eq. (8) in a form appropriate (Grahame, 1947; Parsons, 1954) for a liquid metal electrode, viz.

$$\left(\partial\gamma/\partial\mu_i\right)_{\mu_{j,k,T,E}} = \Gamma_i(E) \, , \tag{12}$$

enables the surface excess Γ_i of any component, i, of the system to be determined as a function of its chemical potential, μ_i, and of electrode potential, E. Similar results can be derived from interfacial capacitance (Eq. 11) measurements as $f(\mu_{i,E})$. Since the relation between q_M and E is also known (Eq. 10), application of electrochemical thermodynamics is able to give, especially for Hg, a very detailed account of the electrical and adsorption behavior of this liquid/liquid (electrolyte) interface. For Hg in contact with various electrolyte solutions, and others containing, in addition, non-electrolyte adsorbates, a great deal is now known about electrical effects in adsorption behavior not only of ions, but also of dipolar and non-polar uncharged solute molecules, and indirectly also the behavior of solvent sipoles (see following section).

Surface tension measurements required for Eqs. (10) or (12) are made either by the electrocapillary technique or by means of drop-time measurement as a function of potential. Analysis of the shape of a recumbent Hg drop (Young, 1805) is also possible (Kemball, 1947) as also in the analysis of meniscus rise at a plate, as used by Morcos (1971, 1972).

SURFACE ELECTRON DENSITY DISTRIBUTION AT METAL INTERFACES

One of the most important manifestations of electrical behavior of liquid interfaces arises at the liquid metals (the behavior of Hg and Ga has attracted special interest) and also correspondingly at solid metals, on account of asymmetry of the electron distribution that arises at the metal's surface (Fig. 1a) over a distance of 0.1 ~ 0.3 nm (the Thomas-Fermi screening distance). While the net excess charge q_M (plus or minus) can be exactly (and thermodynamically) evaluated at Hg (Eq. 10) when the latter is set up as an electrode exposing an interface between itself and an electrolyte solution, it should also be noted that, along with the absolute value of q_M, the corresponding asymmetry of electron distribution is also variable with electrode potential or q_M, but is not susceptible to direct evaluation. Much less is known about this variation of electron distribution than that of q_M itself but, in recent years, its role in determining, in part, the electrical properties of the double-layer (e.g., its capacitance) has become recognized (Badiali et al., 1981; Schmickler, 1979, 1983; Schmickler et al., 1984) and much theoretical work has been carried out. From the point of view of the physics of metal surfaces, Lang and Kohn (1973) have applied the "jellium model" (see below) of the degenerate electron system in metals to representation of the situation of electrons at or near metal surfaces, especially in relation to the induction of charge by a neighboring ion. A recent review by Feldman et al., (1986) covers the whole question of electron distribution and screening theory in relation to metal/electrolyte interfaces in a thorough way, so this matter will not be treated in detail again here. Also, the papers of Badiali et al., (1981) and of Schmickler (1983) deal with this factor in connection with the double-layer behavior at electrode/situation interfaces.

Apart from the intrinsic physical interest in the role of potential-dependent free-electron distribution in determining the electrical behavior of metal/electrolyte interfaces, work on this problem has been stimulated by the experimental observation that the compact component of the metal/electrolyte interfacial capacitance (Grahame, 1947) is metal-dependent, e.g., for Hg relative to (liquid) Ga (Frumkin and Damaskin, 1974) in the supposed absence of specific chemisorption of solvent dipoles or solute ions. In the liquid state, of course, the specific lattice arrangements of atoms in various single-crystal planes of different metals that lead to metal- and surface-specific interfacial behavior (Hamelin, 1986) are absent and liquid metal surfaces acquire, uniformly, virtually hexagonal close-packing, albeit with some thermal fluctuations in the longer-range order.

In most treatments of the surface electron distribution, the so-called "jellium model" has been used, for which the atomic cores are smeared into a planar uniform background. The jellium model has been found, e.g., Lang and Kohn (1973), to work well for simple metals and, with regard to the surface electron response, has been employed to treat the behavior of some of the noble metals (Equiluz, 1984).

The first treatment of electrostatic screening (of an external field) by electrons of a metal near/at its surface was given by O.K. Rice (1928) as long ago as 1928, almost at the same time (1926, 1927) as the statistical theory of Thomas (1926) and Fermi (1927). The main deficiencies of these early treatments, which involved a sharp-boundary model for the electron distribution, arose from neglect of the screening of the external field by electrons, distributed outside the jellium edge plane. Already Ku and Ullmann (1964) and Tsong and Muller (1969) had discussed in the '60's the effect of significant penetration of a field into a metal on the capacitance of capacitors having a microscopic gap, a problem closely related to the behavior of metal electrolyte double-layer capacitance (except, in the latter case, the compact layer "gap" is only some 0.3 ~ 0.5 nm).

An interesting metal-dependent result derived from the equations of Partenskii et al. (1976) is the value of the critical field $E_{o,cr}$ for which $\bar{x}$ ($E_{o,cr}$) = 0, where $\bar{x}$ measures the position of the boundary of the electron distribution (the boundary $\bar{x}$ approaches the bulk metal as E_o increases; note that E_o is $-4\pi\, q_M$, assuming a local permittivity of 1). Some results for $E_{o,cr}$, based on several types of calculation, are given in Table 2.

In the case of an idealized metallic conductor, the screening electron charge is regarded as being located at the atomic surface of the material and no potential drop or corresponding field can exist within the metal. The recent treatments of the interfacial electron distribution recognize that the ideal conductor metal is unrealistic, so that there is a fall of potential over a small, finite distance within the metal, creating a contribution to the overall double-layer capacitance that is in series with the compact and the diffuse-layer capacitance components.

Table 2. Critical Values, $E_{o,cr}$, of the Electrostatic Field for Several Metals Having Electron Densities n

METAL	n 10^{-3} a.u.	$E_{o,cr}$ /V A^{-1}			
		TF	TFD	TFDH	HK
Li	6.92	6.92	2.65	1.78	
Na	3.8	4.2	0.72	0.33	1.5
Cu	12.6	11.4	5.1	4.6	
Ag	8.7	8.6	3.3	2.8	4.1
In	17	14.6	7.3	6.7	
Zn	19.5	16.4	8.6	8.0	

TF ≡ Thomas, Fermi; TFD ≡ Thomas, Fermi, Dirac; TFDH ≡ Thomas, Fermi, Dirac, Hambosh; HK ≡ Hokenberg, Kohn

* Data quoted from Feldman et al. (1986).

In the treatment of Badiali et al. (1981) the jellium model of the metal electron system is used with the jellium edge being assumed to be a plane passing through the centers of surface atoms of the metal. Solvent molecules then lie in contact with the surface at a distance equal to the radii, τ, of surface metal atoms and hence are separated from the jellium edge by a distance τ_1. This is not an altogether realistic model and, in fact, does not take into account the "overspill" effect associated with the wave function of the metal's electrons at the surface. Another problem is that the solvent is represented by an electron-repulsive dielectric continuum, little related to the properties of water dipoles which are involved at the Hg surface in aqueous systems that have mostly been experimentally studied in double-layer capacitance works.

The plane of the edge of electron density approaches the (bulk) metal side with increasing q_M, i.e., the metal electron capacity contribution decreases. Agreement with experiment has been claimed with regard to the experimentally known appreciable difference between the compact-layer capacitance at Hg and Ga at $q_M = 0$. However, Feldman et al. (1986) regard this apparent agreement as an artifact of the choice of the jellium edge positions at these two metals.

It should be noted that the intrinsic surface potential at metal electrode interfaces, due to the discontinuity of distribution of electrons and atomic nuclei at metal surfaces, becomes modified by any adsorption of solvent dipoles, ions of the electrolyte or electrodeposited ad-atoms. Along these lines a more realistic treatment of the

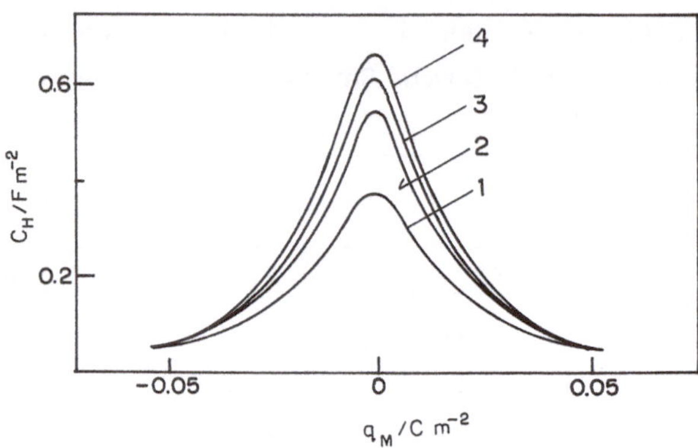

Fig. 2. Calculations of the Helmholtz layer capacitance at a metal/water interface (273 K) for various values of the background e density: (1) n = 0.01 a.u.; (2) n = 0.02 a.u.; (3) n = 0.03 a.u.; (4) n = 0.04 a.u. (Schmickler, 1983).

electron/solvent dipole system at the metal interface was made in the calculations of Schmickler (1983) in terms of a lattice of point dipoles. The results for the following values of the parameters of the calculation, viz., $N_s = 10^{15}$ solvent molecules cm^{-2}, dipole moment of solvent molecules = 6.12×10^{-20} cm, radii = 0.3 nm, width of Helmholtz layer = 0.64nm, as evaluated by Schmickler (1983), give the capacitance curves shown in Fig. 2.

In further developments, with Schmickler et al. (1984), two models of the solvent layer at the metal interface were considered. These self-consistent calculations of charge-induced electron relaxation predict in one form or another the well known hump in the compact-layer capacitance and introduce a dependence of the capacitance behavior on the properties of the metal electron system, that is, of course, not indicated in previous, purely molecular treatments of the metal/electrolyte interface. In general, metal-specific behavior (apart from that associated with specific orientation of solvent dipoles due to donor-acceptor interaction with the metal) is related to the free electron density of the metal. For further details, readers are referred to the review mentioned earlier (Feldman et al., 1986).

IMAGE CHARGES AT LIQUID INTERFACES

General Aspects

Another important aspect of the electrical properties of liquid interfaces is the so-called image charge induced by a real charge near the surface of an insulating liquid or near the surface of an electronically conducting liquid (or solid) metal such as Hg. This problem is of importance in (a) adsorption of ions at Hg, (b) negative adsorption of ions at the interface of a dipolar dielectric fluid, and (c) the energy profile for escape of electrons (field emission) from a liquid or solid metal into vacuum or solution.

The image charge formally represents the equivalent electrical effect of a continuous 2-dimensional charge distribution induced by the real charge lying above or below the liquid interface. Since the treatment and representation of image charges arises from a concept in classical electrostatics (e.g., see Slater, 1939), problems arise when image changes are considered in terms of the microscopic structure of liquid interfaces, viz., the finite sizes of, and local charge distribution in, real atoms or molecules of the phase, and when the real inducing charge is near, i.e., within fractions of a nm from the surface, as in double-layers at charged liquid interfaces, especially the Hg and other metallic electrodes.

The image potential is of importance in three significant ways in the electrochemistry of interfaces: (a) in contributing to the adsorption energy of ions at metal interfaces and modifying the electron work function of the metal when ion adsorption takes place; (b) in determining, in part, the negative adsorption of lyophilic ions at liquid/vapor (vacuum) interfaces; and (c) in determining the electron-exit barrier for metal/vacuum interfaces, especially in cold field emission.

These two situations correspond formally to the generation of an image charge at metals which is <u>opposite</u> in sign to that of the inducing charge while, at the liquid/vapor interface, the image charge is of the <u>same</u> sign as that of the inducing charge; hence the repulsion effect leading to negative adsorption in the latter case. These two cases arise from dielectric theory where, for the metal case, the medium has effectively infinite permittivity while for the liquid/vapor (vacuum) case, the permittivity of the exterior medium is 1, or very near 1, and normally less than that of the medium in which the inducing charge resides, e.g., for water $\varepsilon = 78$.

The Question of Charge-Image Interactions at Metal Surfaces: Statement of Problems

When ions or partially charged ad-atoms are adsorbed at an electrode surface, e.g., Hg, the free energy of their adsorption is made up of several recognizable, but not easily determinable, components: (a) an electron overlap or charge-transfer energy, e.g., for I^- on Pt or SH^- and I^- on Hg; (b) an energy associated with partial displacement or deformation of the ion's solvation shell, including the cosphere/coplane overlap (Conway, 1975) of water interacting with the metal surface and the ion; (c) a polarization energy of the ion due to the electrode surface field;

and (d) an energy U(r) of interaction of the ionic charge q with its image, classically given by $-q^2/4d$ where d is the distance of closest approach to the electrode and q the ion's charge.

The image energy should provide an important contribution to the energy of specific adsorption of ions at the potential of zero charge and of solvent dipoles at electrode surfaces. It is usually felt that for distances corresponding to contact adsorption (0.1 ~ 0.3 nm, depending on solvation effects), the classical image law will not hold at all accurately due to the discreteness of atomic surface structure and distribution of surface electron density referred to earlier, as well as by screening effects by solvent molecules at the surface and near the ion. Quantitative and even qualitative uncertainties about this problem lead to serious difficulties in formulating a treatment for the energy of adsorption of ions and solvent dipoles at metal electrode interfaces.

Three principal problems arise in representing the interaction of a charge with a metal electron distribution at or near the surface of a metal, other than by the classical image law which will not be expected to hold very near (< 0.5-1.0 nm) to a surface: (a) the quantum properties of electrons in the metal, especially near it surface in the Thomas-Fermi screening range; (b) atomic nuclear discreteness of the surface structure and charge distribution at the interface of a metal; and (c) the wave-mechanical penetration of the surface barrier, associated with "electron overspill". At electrodes, the overspill and surface electron density will additionally be potential dependent (cf., the Gibbs-Lippmann relation, Eq. 10).

From an electrochemical point of view, the significant result of theoretical calculations (see below) is that the deviations from the classical image energy are from 0.5 to 1 eV near the surface, within 0.02 to 0.03 nm. This is a large energy, comparable with the ranges of energy of adsorption of ions at metal electrodes, e.g., in specific adsorption of anions. Of course, the total classical image energy to within 0.1 nm of the metal would be on the order of 3.44 eV or 330 kJ mol^{-1}. Most of the physical calculations for electrons go to significantly closer distances to the surface which are not realistic either for contact adsorption of ions (0.06 ~ 0.18 nm) or for adsorption of hydrated ions (0.2 ~ 0.36 nm) which have distances of closest approach not smaller than the approximate figures listed above in the brackets.

A further major factor in this comparison with electron image interactions, which, as we have said, are important in vacuum-surface science, particularly in cold field emission of electrons and resonance tunneling, is the <u>state</u> of the charge for larger r from the surface: near electrodes in solution, the charge is <u>solvated</u> so that it is screened up to quite short distances from the surface by the electron distributions and nuclear charges of the solvating dipoles, especially when water and similar solvents are the solvent for the ion. At a "vacuum"-metal interface, electrons or ions feel the image interaction more strongly. Also, the dipoles of the solvent, including particularly those adsorbed and oriented at the electrode, contribute their own large charge distributions and screen the ion's charge from interaction with the metal electron distribution near the surface. Upon close or contact adsorption of an ion, usually an anion, at a metal surface, the chemisorption or electron-exchange interactions (Schmickler, 1979) become predominant and very difficult to distinguish from the image charge interaction. For example, in many chemisorptions of ions at metals, a partial charge transfer occurs, characterized by the electrosorption valency (Schultze and Vetter, 1973).

As was illustrated schematically in Fig. 1, the electrical situation at or near an electrode interface in solution is very complicated on a microscopic scale, as is also the metal surface itself when atomic or nuclear discreteness on the scale of 0.1 - 0.2 nm is taken into consideration in a realistic view of the structure of the interphase.

<u>Treatments for the "Vacuum"/Metal Interface</u>

Some early attempts to provide a solution to this question for the simpler case of a free charge in vacuum near a metal surface were made by Bardeen (1940), Sachs and Dexter (1950), Cutler and Davis (1964), and most recently by Lang and Kohn (1973), but mainly for the interaction of a charge in vacuum with electrons at or near a metal surface.

An empirical relation, suggested by Slater (1939) gives the classical image energy $U(r) = -e^2/4r$ at long distances r from the surface, and limitingly approaches a "non-electrostatic" chemisorption energy at distances comparable with the ion's size. A treatment of Cutler and Davis (1964) for the problem of the energy of an electron interacting with its image in an electron emission process gives another result for U(r), more specifically expressed in a paper by Cutler and Gibbons (1958).

The quantum limitations of the electrostatic image potential function, $U(r) = -e^2/4r$, were treated by Sachs and Dexter (1950). They derived a correction term which gives the order-of-magnitude deviation from the classical image energy due to quantum-mechanical effects in the metal. For distances $r > 0$, outside the metal, the form of their function is

$$\Delta E = U(r) + \text{constant } \kappa^{1/2} r^{-1} \cdot |U(r)| , \tag{13}$$

where U(r) is the image term and the other term represents quantal corrections. The constant depends on some assumed forms of the perturbed charge distribution and κ is a screening parameter characteristic of the quantum properties of metal electrons. It can be treated in terms of the Thomas-Fermi statistical model which depends only on free electron density.

A more recent improvement is that of Cutler and Gibbons (1958) which incorporates the important features of Bardeen's (1940) and Sachs and Dexter's (1950) treatment. Cutler and Gibbon's expression has the forms

$$U(r) = -e^2/4r + \eta\, e^2/4r + \eta\, e^2/4r^{\,2}$$

$$= W(r) + \eta\, r^{-1} |W(r)| (r > r_s > 0) , \tag{14}$$

and

$$U(r) = -W_a \qquad (r < r_s) . \tag{15}$$

As in earlier work on surface potentials, a free-electron model of the metal was used with a structureless surface and a surface potential barrier normal to the interface. Several forms of the "image" potential are shown in Fig. 3.

The Bardeen potential, calculated quantum-mechanically, approaches classical image behavior at large distances, r, and goes (as required) to a constant value for the potential in the interior of the metal. In the surface region, there is a minimum in the potential due to a maximum in the exchange charge density, arising from an electrostatic contribution. The general features of the Bardeen potential functions have been verified by Juretschke (1953). This function exhibits the following properties:

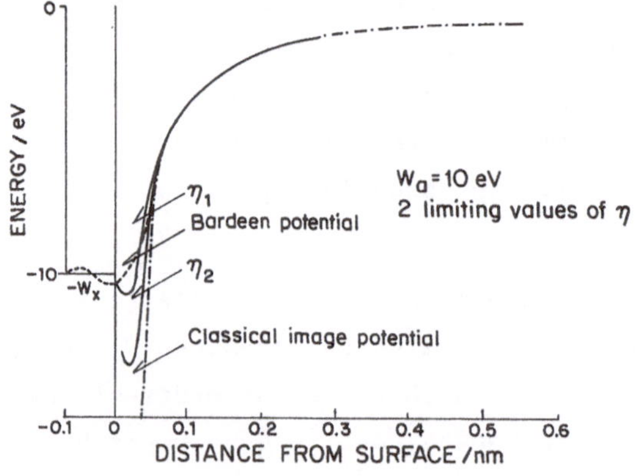

Fig. 3. Forms of the image potential according to Cutler and Davis (1958).

(a) image behavior outside the surface;

(b) a potential minimum near the surface; and

(c) a smaller force on the charge than that given by the classical image formula, except at large r.

The parameter η is determined by the surface properties of the metal. The bounds on η are determined by

$$r_s = 1/2r_1 \left(1 \pm \left[1 - 4\eta(r_1)^{-1}\right]^{1/2}\right) , \tag{16}$$

so that $0 < \eta < r_1/4$. In numerical evaluation, η was taken as 0.00679 nm. The form of the function for two values of η is shown below. Minima in W(r) arise at <u>ca</u>. 0.02 ~ 0.025 nm from the surface. Different η's cause the curve to be raised or lowered relative to the image curve. The energy is periodic in r for r < 0, in the metal, as expected.

Analytical approximations to the Bardeen potential were written as

$$V_-(r) = -W_a \left[1 + (A/r)\left(-e^{\lambda r}\right) \cos(\kappa r + \delta)\right] \text{ for } r < 0 , \tag{17}$$

$$V_+(r) = -\left(e^2/4r\right)\left(1 - e^{\lambda r}\right) \quad \text{for } r > 0 , \tag{18}$$

where W_a, A, λ, κ and δ are adjustable parameters.

It will be noted that a weakness of these treatments is the assumption of a smoothed continuous surface. The atomic discreteness of real surfaces, with periodicicity in screened positive nuclear charges, is one of the factors that makes the classical image law, arising from distributed induction of charge on a 2-dimensional surface, inapplicable at distances from the surface comparable with the scale of discreteness of surface atoms and interatomic "holes". This effect is well demonstrated, for example, by the microscopic inhomogeneity of the electron work-function, physically visible in electron emission pictures of surfaces and of surfaces bearing adsorbed Cs or Na atoms, e.g., on W electron emitters.

More recently, Lang and Kohn (1973) gave a detailed discussion of this problem, including the question of chemisorption, e.g., of alkali metal atoms on transition metal surfaces such as those of W, Ta, Re. The problem is approached first by considering the profiles of the charge induced by a uniform external electric field for metals of different bulk electron densities. At electrodes, it is to be noted, an additional factor is the potential-dependent surface electron density, q_M, given, in the case of liquid metals of surface tension γ, by $q_s = -(\partial\gamma/\partial E)_{\mu_{i},T,P}$, where E is the electrode potential and μ_i the chemical potential of species in the solution. Solvent adsorption is another problem.

A quantity of interest in the profiles of induced charge density ρ is the position of the center of mass, d_o, of these profiles. Positions of this "center of mass" of the induced charge distribution relative to the effective edge d_b of a uniform smeared-outpositive charge background (the charges of the "ions" of the bulk metal structure in the electron jellium) can be calculated. Some values of $d_o - d_b$ in atomic units (= 0.0529 nm) are listed below for three values of a parameter r_s which characterizes the bulk electronic densities. Thus r_s is defined through the relation

$$\frac{4}{3} \pi r_s^3 = -\rho_e^{-1} , \tag{19}$$

where ρ_e is the bulk electronic charge-density expressed in atomic units:

r_s (a.u.)	$d_o - d_b$ (a.u.)
2	1.6 ± 0.05
4	1.3 ± 0.2
6	1.2 ± 0.2

The results in this table allow d_o to be located for a metal relative to its lattice planes. Following various calculations on the polarizing effect of an external field, normal to the metal surface lattice, the useful conclusion was reached by Lang and Kohn (1973) that the plane through d_o is to be regarded as the effective location, electrically speaking, of the metal surface. Profiles of induced surface charge density $\rho(r,o)$ in atomic units were calculated for the three values of r_s given in the table above.

Schematically, the background of atomic nuclei can be represented by a plane of "uniform background charge" and an effective metal surface at d_o. This is shown in Fig. 4, together with the situation for adsorption of an alkali metal as a cation on a metal of suitably high function, e.g., W, Ta.

Distances a and s characterize the adsorption geometry, viz., the distance of the adion from the electronically effective metal surface and from the outer plane of atomic nuclei of the surface structure.

The general conclusion is that image energies are to be represented in terms of a function that expresses the distance from the "surface" relative to the effective metal surface plane at r_o, i.e., distances are written relatively as $r-r_o$, then the image energy in the simplest analysis is written $W(r) = e^2/(r-r_o)$. r_o introduces characteristic properties of the metal, related for example to the surface potential.

Other connections to surface potential and atomic chemisorption behavior have been given by Lang (1981), and Lang and Williams (1978).

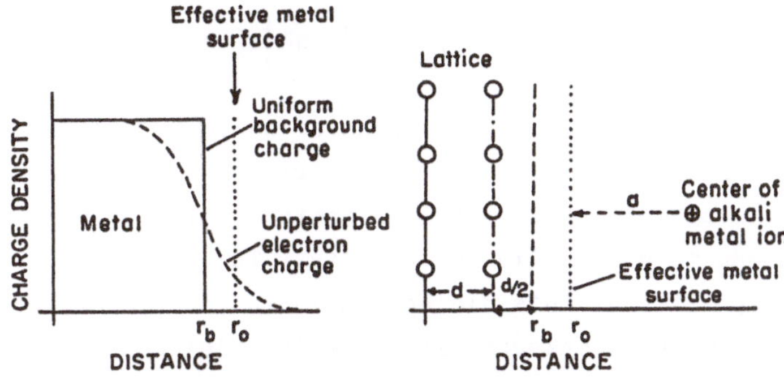

Fig. 4. Model for a metal interface with a charge near to it at r_1 (from Lang and Kohn, 1973).

Treatment by Kornyshev et al. (1977)

Basically the image law derives from the interaction of a charge in a dielectric medium with an array of induced charges, or ideally a continuous 2-dimensional distribution of charge induced on the surface of the dielectric discontinuity -- the metal "surface". At short distances from the metal surface, problems arise with the local distribution of electrons and ions of the metal structure, the quantal behavior of the metal "electron gas," and the overspill of the electron plasma and screening effects of this plasma around the closely associated ion in the solution dielectric at the surface. Also, at the surface, is an array of solvent dipoles, usually partly oriented (see MOLECULAR MODEL TREATMENTS OF WATER ORIENTATION AT CHARGED INTERFACES), which themselves will induce image dipole charge distributions at the interphase.

One of the main problems in the treatment of the image potential relation for charges near metallic surfaces is the way the screening effects of the metal electron plasma are handled.

An important treatment of the image potential problem for charges near a metal interface was given recently by Kornyshev, Rubinshtein and Vorotyntsev (1977). In most work, the non-local screening effect near a surface results in a significant decrease in the image potential U(r) with respect to the classical image law value: $U(r) = -q^2/4r$. In particular, the value lim U(r) is found to be finite and negative in accordance with the intuitive estimate of $r \rightarrow 0$ screening effects:

$$U(r) = -q^2/4\ (a + \kappa^{-1})\ , \tag{20}$$

where κ^{-1} is an effective screening length (Gomer and Swanson, 1963).

Calculations were based on the screening properties of a dielectric plasma-like medium (PLM) interface for which the image potential takes the form

$$\frac{1}{q^2} W(r) = - \frac{1}{4\varepsilon r} \left[1 - S(r)\right] = - \frac{1}{4\varepsilon r} + \int_0^\infty dK \left[\varepsilon + T(K)\right]^{-1} e^{-2rK} , \quad (21)$$

where $S(r) > 0$, $T(K)$ are functions depending on the model for screening in the PLM half-space, and ε is the dielectric constant. The behavior of a test charge in a dielectric medium in contact with the metal was considered and leads to physically new results. The treatment is developed for various properties of this dielectric medium characterized by a parameter $\xi = \varepsilon/\varepsilon_0$ which measures its relative dielectric or electrical screening behavior. The "ionic skeleton" of the metal is taken to have a dielectric constant $\varepsilon_0 = 1$.

Two approaches are made to the evaluation of the screening function S(r) (Eq. 22). First, that the potential in the PLM half-space follows a Poisson-Boltzmann function as in Debye-Hückel electrolyte theory, but with a screening length κ^{-1} corresponding to the Thomas-Fermi screening length for a degenerate electron plasma. This model, (i), leads to an image potential function giving for S(r) the expression

$$S(r) = 4r\kappa\xi \, A(2r\kappa, \xi) , \quad (22)$$

with

$$A(p, \xi) = \int_0^\infty dx, x \left[(1 + x^2)^{1/2} + \xi \, x\right]^{-1} \exp\left[-px\right] . \quad (23)$$

The approach involves a non-local constitutive relation between electric field and induction. Another formula for S(r) [model (ii)] is obtained as

$$S(R) = 4a\kappa\xi \, B(2r\kappa, \xi) , \quad (24)$$

with

$$B(p, \xi) = 8 \int_0^\infty dx \, \left(3 + 2x\left[\left(1 + x^2\right)^{1/2} + x\left(4\xi + 3\right)\right]\right)^{-1} x^2 \exp\left[-px\right] . \quad (25)$$

An elucidation of the results obtained is achieved by replacing the PLM in the half-space by a dielectric slab and an impenetrable metal. The image potential in the complementary electrolyte solution half-space is then given with S(r) taken as

$$S(r) = 4\xi \frac{r}{d} \, c\left(\frac{2r}{d} , \xi\right); \text{ with } C(p, \xi) = \int_0^\infty dx \left(\xi + \coth x\right)^{-1} \exp\left[-px\right], \quad (26)$$

where $d = \kappa^{-1}$, the screening distance.

Deriving the asymptompic expansions of this integral leads to

$$U(r) - \frac{q^2}{4\xi r}\left[1 - \xi\frac{d}{r} + \xi^2\frac{d^2}{r^2} + 0\left(\frac{\xi^3 d^3}{r^3}, \frac{\xi d^3}{r^3}\right)\right], \quad r >> \xi\, d/2,\ d/2, \tag{27}$$

$$U(r)\ \frac{q^2}{4\xi r}\left[\frac{\xi-1}{\xi+1} - \frac{r}{d}\frac{4}{\xi^2-1}\ln\frac{\xi+1}{2} + \frac{r^2}{d^2(\xi+1)^2}\frac{4\xi}{}(1 + 0\left(\frac{r}{d}\right))\right], \tag{28}$$

$$r << \frac{d}{2}, \quad \frac{\xi d}{2},$$

$$U(r) = \frac{q^2}{4\xi r}\left[1 - \frac{4r}{\xi d}\left(\ln\frac{\xi d}{4r} - 1\right) + 0\left(\frac{r^2}{\xi^2 d^2}\ln\frac{\xi d}{r}, \frac{d}{r_\xi}\right)\right], \tag{29}$$

$$1 << \varepsilon,\ \frac{d}{2} << r << \frac{\xi d}{2},$$

where 0 is a function of the indicated quantities. Again, attraction to the metal surface arises at large distances r. The superposition of "effective" interactions with the interface of two semi-infinite dielectrics placed at $z = 0$ and with the metal surface placed at $z = -d$ arises at small distances. If ξ is not too close to unity, the first term in Eq. (28) is the leading one. It results in a repulsion from the boundary when $\xi > 1$, and U(r) has a minimum at $r \simeq \xi\, d/2$; for $\xi >> 1$, an approximate formula, $U(r) = q^2/4\xi\, r$, is valid in the range of $r << \xi\, d/2$.

The general conclusions are as follows: for each of the models (i) or (ii) for S(r), and consequently U(r), qualitatively the same profile of the image potential in the solution dielectric as a function of r is obtained. At large r, the charge is attracted to the surface as in the classical image law. When $\varepsilon < \varepsilon_o$, U(r) is monotonic and $\to -\infty$ as $r \to 0$. However, when $\varepsilon > \varepsilon_o$, $U(r) \to \infty$ as $r \to 0$ with a minimum. If $\varepsilon >> \varepsilon_o$, this minimum stands away from the boundary at a distance much larger than the PLM screening length. Here an adsorption energy minimum could arise from purely electrostatic considerations rather than, or in addition to, the usually invoked balance of bonding and repulsive forces in adsorption. Thus, e.g., with a singly charged ion in a dielectric with $\kappa^{-1} = 0.1$ nm and ε_o taken as = 1, U at the minimum is found to be > 1 eV. Here ε_o is taken as the permittivity of the "ionic lattice skeleton" of the metal, usually assumed to be unity. ε_o is also involved in κ. Thus, for a 1-component Maxwell plasma with n charge carriers cm^{-3},

$$\kappa = \left(4\pi\, ne^2/\varepsilon_o kT\right)^{1/2}, \tag{30}$$

or, for the degenerate electron gas case,

$$\kappa = \left(6\pi\, ne^2/\varepsilon_o E_F\right)^{1/2}, \tag{31}$$

where E_F is the Fermi level energy.

The first terms both in Eq. (27) and (29) are identical with the image-potential energy in a dielectric (ε) near its boundary with another dielectric (ε_o). The second terms represent the interaction of the test charge with the space charge in the PLM. The first term is the leading one, if ξ is not too close to unity. For $\xi < 1$ the charge is attracted to the boundary and the whole curve is monotonic. An effective repulsion of the charge from the surface arises when $\xi < 1$. In this case, the U(r) curve has a minimum between the "repulsive" branch at small, and the "attractive" branch at large, distances. For $\xi < 1$, $r_{min} \sim (2\kappa)^{-1}$ and $U(r_{min}) \sim -q^2\kappa/\varepsilon$. For $\xi >> 1$, the minimum is situated in the range of $\xi/2\kappa$ [model (i)] or $\sqrt{\xi/2\kappa}$ [model (ii)]. Its depth is of the order of $q^2\kappa/\varepsilon\xi$ and $q^2\kappa/\varepsilon\sqrt{\xi}$, respectively. Furthermore, if $\xi >> 1$, an approximate formula, $U(r) = q^2/4\varepsilon r$, is valid in the region of $r << \xi/2\kappa$ or $r << \sqrt{\xi/2\kappa}$ for each model, respectively. For each model, the first term in the image energy equations is $q^2/4\varepsilon r \left(\frac{\xi - 1}{\xi + 1}\right)$ so that when the dielectric screening parameter $\xi = 1$, U(r) at small distances is negative and finite so that the following limiting cases appear:

$$U(r) \sim -\frac{q^2\kappa}{\varepsilon}\left(\frac{1}{3} + \frac{1}{4}\kappa r \ln 2\kappa r\right) \text{ for the model (i) ,} \tag{32}$$

and

$$U(r) \sim -\frac{q^2\kappa}{\varepsilon}\left(0.36 + \frac{1}{4}\kappa r \ln 2\kappa r\right) \text{ for the model (ii) .} \tag{33}$$

For ξ close to unity, U(r) at $r >> r_o \sim |\xi - 1|/\kappa$ is monotonic and coincides with the curve for the case $\xi = 1$. At $r << r_o$, W(a) tends to $+\infty$ or $-\infty$, depending on the sign of $\xi - 1$. In the limit $(\xi - 1) \to +0$ the position of the minimum coincides with $r \to +0$ and its depth is equal to $-q^2\kappa/3$ and $-0.36\ q^2\kappa$ for the model (i) and model (ii), respectively. The curves for U(r) in the whole range of r are plotted in Figs. 5a and 5b from the work of Kornyshev et al. (1977).

For electrode-solution interface problems, the treatment of Kornyshev et al. (1977) seems to be the most promising and explicit for elucidation of short-range electrostatic interaction of a charged or partially charged adsorbate with a metal-electrode surface.

THE SURFACE POTENTIAL OF THE AIR (VAPOR)/WATER INTERFACE AND OF AQUEOUS ELECTROLYTE SOLUTIONS

The surface potential of a pure liquid would give useful information on the surface structure of the fluid and the electric behavior arising from orientation of its molecules at its interface. Unfortunately, absolute values of the surface potentials of phases cannot be measured, only differences due to adsorption of surface-active neutral molecules or

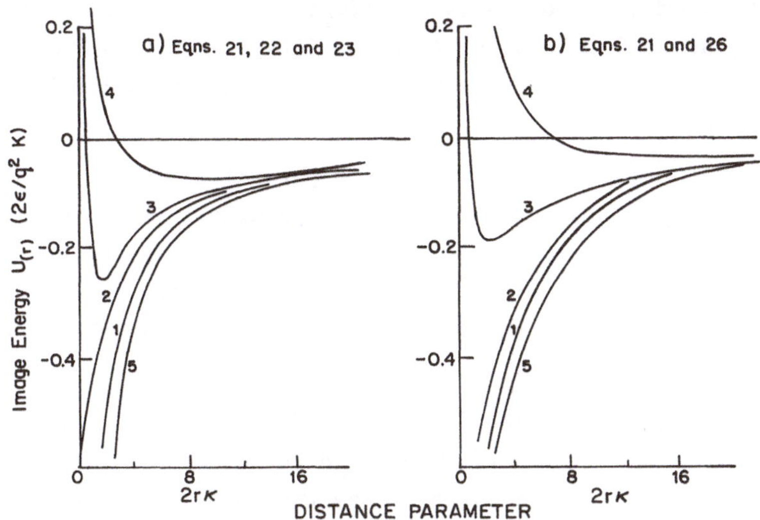

Fig. 5a and 5b. Image potential energy profiles U(r) in a dielectric near the boundary of the PLM calculated for various ξ values (From Kornyshev et al., 1977): (1) 0.5; (2) 1; (3) 1.5; (4) 10 and (5) 0. ($\xi = \varepsilon/\varepsilon_0$)

ions. In the latter case, the surface potential change $\Delta\chi$ is due to any double-layer distribution of cations and anions in the interphase, as well as any changes of solvent dipole orientation at the interface caused by the ions, i.e.,

$$\Delta\chi = \Delta\chi_{\text{solvent}} + \Delta\chi_{\text{ions}} . \tag{34}$$

Surface-potential changes are usually measured by one of the following methods:

(a) a d.c. technique involving direct measurements across an air (vapor) gap with an ionizing probe (Po or an artificial radioisotope).

(b) Kelvin's method using a vibrating condenser plate above the liquid surface as the second plate, with a.c. null-point detection and a controllable d.c. bias which, at zero a.c. current in the vibrating plate capacitor circuit, gives the surface potential.

(c) Kenrick's method (1896) using flowing surfaces of electrolyte solutions.

Usually $\Delta\chi$ measurements are coupled with measurements of the surface excess Γ of adsorbate which causes the change of χ; Γ data are derived from surface tension measurements by means of Gibb's adsorption equation. The derivative $\partial(\Delta\chi)\partial\Gamma$, or the integral slope $\Delta\chi/\Delta\Gamma$, then gives a measure of the average surface potential change per molecule of adsorbate in the interface. For the case of ionic adsorption, Γ measures the total surface excess quantity of ions over a diffuse-layer region near the liquid surface. For neutral molecules, $\Delta\chi/\Delta\Gamma$ or $\partial(\Delta\chi)\partial\Gamma$ can be interpreted in terms of molecular models of adsorption, and adsorbate and solvent orientation in a more compact dipolar layer.

The absolute value of solvent dipole potential at pure liquid interfaces, we have mentioned, cannot be thermodynamically measured. The sign of χ is of general importance for it indicates which is the preferred direction, e.g., of water dipoles at air or vapor/water interfaces. This is of interest in relation to structure and H-bonding in the air/water interfaces and theories of χ.

It can be argued from the data of Randles and Whiteley (1956) (for the interface) that the sign of χ can be deduced* from the direction of change of χ with temperature since it may reasonably be assumed that any numerical value of χ of either sign will decrease with increasing temperature due to the tendency for thermal disorientation. The observed temperature coefficient is negative. A complicating factor is, however, that the number of density of dipoles in the surface will also decrease with temperature due to the thermal expansion of the interphase, and lateral interaction effects will have diminished effectiveness in reducing χ as temperature is increases. The direction of decrease can, however, be measured so that the orientation of water at the liquid water/air interface can be deduced (Randles, 1963) as that with the O atom of H_2O molecules directed outwards. This is also the direction favored at the Hg/water interface.

Data are available for neutral molecule monolayers at the air/water interface. Together with the results for the temperature coefficient of surface potential of the air/water interface, these data indicate a potential drop of ca. 0.13 V. At the Hg electrode/water interface, shifts of potential at constant q_M due to neutral molecule adsorption indicate a surface potential at $q_M = 0$ of ca. 0.07 to 0.10 V. Of course, there is no a-priori reason why the direction of orientation of water dipoles should be the same at the air/water interface as at Hg since: (a) in the latter case, chemisorption effects involving the O lone-pair orbitals can be significant (more so at other metals such as Ga, Pt, etc.,), and (b) the H-bonding situation can be different at the two

---

* This conclusion turns on the point that if a surface potential $\Delta\chi$ of a certain sign exists at a given potential, its numerical value must diminish as the temperature is increased and become eventually zero at a sufficiently high temperature. The sign of the temperature coefficient can be determined, hence the sign of the surface potential itself at a given temperature can be deduced.

interfaces due to the dipole-image interactions in the case of metal interfaces with water, as well as the van der Waals water-metal interaction which can be asymmetric.

An attempt to make an a priori calculation of the air/water surface potential was made by Stillinger and Ben-Naim (1967). The water molecule was idealized as a point dipole and point quadrupole encased in a spherical shell representing its excluded volume. Classical electrostatics was applied to the determination of the electric field surrounding such a molecule in the interphasial region of the liquid and vapor phases. From this evaluation, the mean torque on the molecule in the interphase can be deduced and leads to a spontaneous orientation polarization at the interface. The resulting p.d. at the interface was derived for several temperatures and corresponded to a tendency for the O atoms of water dipole/quadruples to be oriented outwards from the bulk as is indicated indirectly from the various experimental approaches mentioned above.

The physical basis of this treatment is that the polar water molecules will tend to orient themselves in such a direction as to maximize the interaction of their fields with the bulk of the liquid where the dielectric constant is larger than in the interphase or in the vapor (a similar tendency governs the expulsion of ions from the air/water interphase into the bulk). If water molecules had only a dipole moment, the local fields due to either limiting orientation of dipoles in the interphase would interact similarly with the bulk, so no net preferential orientation would result. However, in the case of the quadrupole, the quadrupole moment leads to a displacement of field lines either toward the front or the back of the polar H_2O molecule, depending on its sign, and the consequent symmetry breaking leads to a surface dipole layer with a preferred orientation, giving rise to a finite, intrinsic surface potential. The effect is analogous to that in the preferred binding of water dipole/quadrupoles to anions rather than cations of the "same" radii which leads to higher hydration energies of anions than cations.

The applicability of the treatment of Stillinger and Ben-Naim (1967) to the practical experimental situation is somewhat restricted since the calculations were made for temperatures just below the critical temperature. This was done on account of several simplifications which result for such conditions, principally because the interphasial zone becomes very wide as the critical point is approached so that molecules at the "surface" reside in a region of slowly varying dielectric constant. The statistical-mechanical expression for mean orientation probability may then be linearized to give convenient explicit results in elementary form. Readers are referred to the original paper for details of the mathematical treatment.

At lower temperatures, it would be expected that the surface potential is determined to an appreciable extent by orientations which are favored by H-bonding in one direction rather than the other. Hitherto this aspect of the problem has not been considered.

TREATMENTS FOR THE DIELECTRIC LIQUID/VAPOR (VACUUM) INTERFACE AND THE ROLE OF IONIC SOLVATION

Nature of Behavior

The approach of an ion to an interface of a liquid or a metal can have several effects on the hydration or solvation of the ion:

(a) At a metal or at an oxide interface, water may already be intrinsically oriented, e.g., as at Hg, Ga, Pt, SiO_2 etc (see Two-State Dipole Orientation Treatments, and Three-State Orientation and Cluster Models in following section). Hence, approach of an hydrated ion to the interface will locally involve cosphere interaction of the hydration shell of the ion and the hydrated layer of the interface; positive or negative free energy changes may accompany such an approach (within 0.5 ~ 1 nm) depending on the relative orientation of water dipoles at the ion and at the surface, and the degree of "structure-making" or "structure-breaking" that either the ion or the interface causes in the water.

(b) If the dielectric constant of the medium beyond the water interface differs from that of water (or the solvent), then the ion will experience an image interaction with the interface due to induction of charges at that interface. If the medium beyond the water interface has a dielectric constant greater than that of water (e.g., in the case of a metal where ε is effectively infinite), an attractive image interaction arises, pulling the ion to the interface. This is opposed by the reluctance of the ion to suffer displacements of its coordinating solvent molecules. If the medium beyond the bulk solvent phase is vacuum, air, or its vapor, then an image of like sign to that of the ion arises and the hydrated ion is repelled from the surface, giving a negative surface excess. The electrostatic theory connected with these effects was first treated by Wagner (1924) and by Onsager and Samaras (1934).

(c) Double-layers may be set up near the surface if differential adsorption of cations or anions arises (Bell and Rangecroft, 1971) due to their different extents of hydration and surface activities (see previous sub-section).

In either case, near a metal or at an air/water interface, the extent of hydration of an ion tends to be restricted (Conway, 1975) due to limited accessibility of the solvent to the ion. This may lead to an increase or a decrease in energy dependent on whether the image, or equivalent polarization interactions on each side of the interface, are repulsive or attractive, respectively. At electrodes, the latter situation usually arises, because of the almost infinite dielectric constant of a metal, while at air/water interfaces the former arises, giving negative adsorption.

The experimental behavior relations between adsorption of various species at the air/water and Hg-electrode/water interfaces have been explored in a number of papers by Frumkin (1926, 1929), and by Kaganovich and Gerovich (1966). Jarvis and Schieman (1968) and Frumkin et al. (1929) have shown that ionic adsorption at the air/water interface generally decreases in the order $CNS^- > I^- > NO_3 > Br^- > Cl^- > OH^- > F^-$ with Pr_4N^+ salts, i.e., in a similar (but not identical) order to that for strength of adsorption at a Hg electrode. Relations between ionic adsorption, hydration, and surface potential at the air/water interface have also been studied by Frumkin et al. (1926) and Frumkin (1924).

It is generally found for simple inorganic ions that their adsorption or surface excess Γ_i is negative at air/water interfaces. For sufficiently hydrophobic ions, however, e.g., alkyl- and aryl-ammonium-type ions, or fatty acid anions, the Γ_i is positive.

Surface potentials for a number of electrolytes in aq. solution were measured by Jones and Ray (1937, 1941) and more recently by Jarvis and Schieman (1968) using the radioactive probe method. Changes in surface potential $\Delta\chi$ were found to vary from 64 mV for Na_2SO_4 at 1.8 m to -180 mV for NaSCN at 7.5 m. The group Ia chlorides in water gave $\Delta\chi$ values that decreased in the order $K^+ = NH_4{}^+ > Na^+ > Li^+$, while the values for group IIa cations decreased in the order $Ba^{2+} > Sr^{2+} > Mg^{2+}$. At a constant anion concentration of 2 m, $\Delta\chi$ due to the sodium salts were in the order $SO_4{}^{2-} > CO_3{}^{2-} > CH_3COO^- > Cl^- > NO_3{}^- > Br^- > I^- > SCN^-$. In general, anions with the smaller hydration energies give greater decreases in surface potential. The magnitude of the respective surface-potential changes, however, does not appear to be a simple function of the hydration energy of the ions. The surface-potential changes must also involve the orientation and structure of the water molecules at the water/air interface, which may be only partially dependent upon the ionic properties exhibited in bulk solution.

Electrostatic Treatment in Terms of Image Interactions

The electrostatic image treatment of Wagner (1924) was improved by Onsager and Samaras (1934), taking account of Debye-Hückel screening effects (the κ parameter) in the ion distribution resulting from repulsive image interactions. They derived the relation for the relative ion concentration $c(l)/c_o$ at a distance l from the interface as

$$\frac{c(l)}{c_o} = \exp -e^2 \cdot e^{2\kappa l}/4\epsilon kT \ . \tag{35}$$

The screening effect greatly diminishes the distance from the surface over which significant ion distribution occurs (cf., the Gouy - Chapman theory of the diffuse double-layer). Much more recently, Bell and Rangecroft (1971) gave a fuller treatment for 2:1 electrolytes of the ion distribution problem and the resulting surface potentials generated.

Near to the surface, i.e., for distances of ionic approach that are comparable with hydration radii and with structural discontinuities in the water solvent, the image potential cannot be expected to apply as a basis for calculating the ion distribution corresponding to negative adsorption; more specific ion hydration changes must be considered, taking account of the structure of the hydration shells.

Electrostatic Treatment of Restricted Ion Hydration at Air/Water Interfaces

The origin of negative adsorption of simple inorganic ions is easy to understand: the surface phase, being 2-dimensional or semi-3-dimensional, provides a region where an ion can only be partially hydrated. It is, hence, from a free-energy point of view, in a less favorable situation than in the bulk, so that its extent of adsorption will be negative. At finite salt concentrations, the factor mentioned above connected with the asymmetry of the ionic atmosphere imposed by the discontinuity introduced by the presence of the surface, also arises.

Here we give the elements of a calculation (Conway, 1975) for evaluating the change of electrostatic hydration energy due to an ion as the ion is transferred from the bulk solvent medium to a position near the interface of the solvent with another dielectric medium (vacuum or vapor). It is convenient to evaluate the hydration effects in ion adsorption at infinite dilution to eliminate complications due to ionic atmosphere effects which screen the ion/ion-image repulsion at finite concentrations.

Case (i): Born model for total hydration energy

The first and simplest case is that where the ionic solvation is treated according to Born's model (1920). This provides only a preliminary basis for discussing the model since it is well known that the Born equation does not apply at short distances from an ion without corrections for dielectric saturation or introduction of a change of model in this region.

The Born energy, G_σ, of charging an ion limitingly in the surface (Fig. 6a) is given by

$$G_\sigma = 1/2\ \frac{(ze)^2}{2r_i}\left(\frac{1}{1}\right) + 1/2\ \frac{(ze)^2}{2r_1}\left(\frac{1}{\varepsilon}\right) = \frac{(ze)^2}{4r_i}\left(1 + \frac{1}{\varepsilon}\right), \qquad (36)$$

taking half the charging energy in the gas phase ($\varepsilon = 1$) and half in the liquid (ε). The energy of negative adsorption due to hydration energy change is therefore the difference of energy of the ion in the surface and in the bulk, i.e.,

$$\Delta G_{ads} = \Delta G_{b\to\sigma} = G_\sigma - G_b = \frac{(ze)^2}{4r_i}\left(1 + \frac{1}{\varepsilon}\right) - \frac{(ze)^2}{4r_i}\frac{1}{\varepsilon} = \frac{(ze)^2}{4r_i}\left(1 - \frac{1}{\varepsilon}\right). \qquad (37)$$

It is seen that this is half of the desolvation energy of the ion, a result that is expected from this model (Fig. 6b). The energy of transfer of the ion from the bulk to a position in the surface is hence accompanied by an expected positive free energy change which will lead to the commonly observed negative surface excess of the ion in the surface.

It is obvious that this model corresponds only to an extreme case. A distribution of positions of the ion will be taken up corresponding to a range of hydration energies near the surface, on the solution-side of the interface. This distribution will be treated below.

Case (ii): Ion below liquid interface

A more realistic situation is that where the ion is centered at some position, l, below the liquid interface (Fig. 6c).

The hydration energy of the ion can be treated in terms of the ion's self-energy in the bulk dielectric medium diminished by the energy that would have been associated with the region of bulk solvent unavailable to the ion beyond the nearby surface of liquid, but to which is added the charging energy associated with the same volume in the free space above the liquid where $\varepsilon = 1$.

The model in Fig. 7 is considered. Spherical volume elements above the liquid are centered on the ion but at distances x from the surface (Fig. 7). The polarization energy due to the ionic field E is calculated below and above the dividing surface, where the dielectric constants are ε and ε' (=1) respectively.

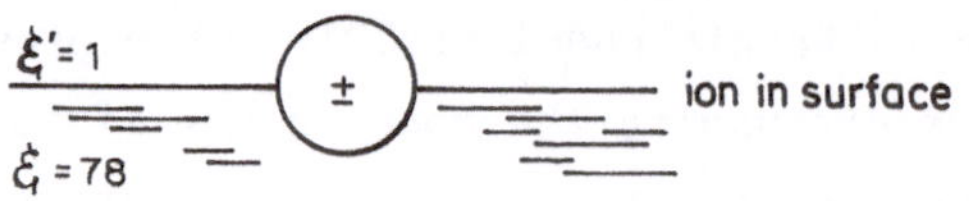

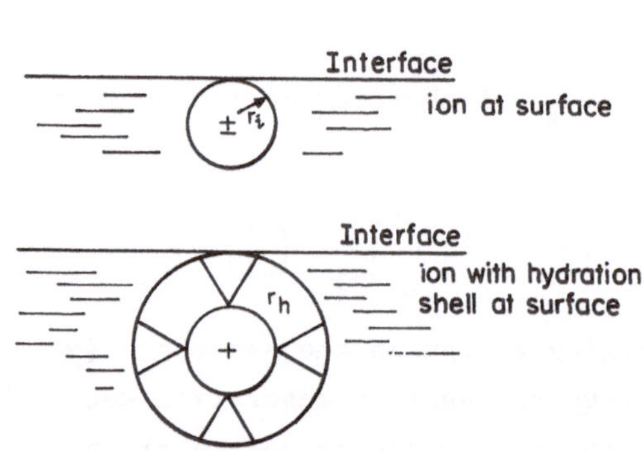

Fig. 6. Models for situations of an ion in or near an air/water interface: (a) Ion in the interface; (b) Ion below interface; and (c) Ion with primary hydration shell.

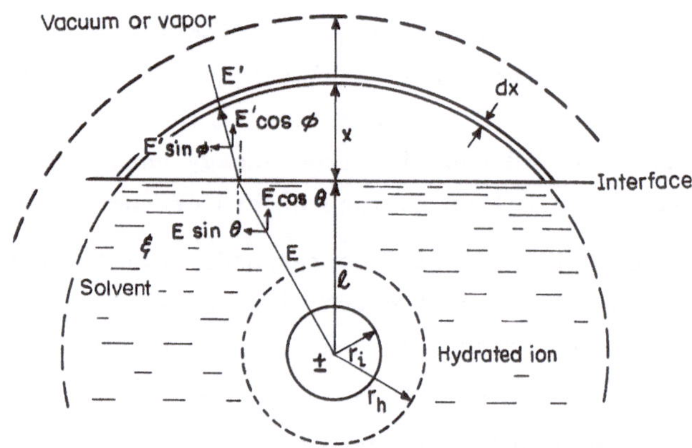

Fig. 7. Working diagram for calculation of restricted Born hydration energy of an ion near an air/water interface (from Conway, 1975).

At the interface of two dielectrics across which a field exists, three conditions apply: (a) the potential has a common value at the interface; (b) the normal components of the dielectric displacements, D, just within and just outside the interface are equal; and (c) the tangential components of the field on each side of the interface are equal. At the dielectric discontinuity, the direction of field vectors also changes from an angle Θ to an angle ϕ to the normal to the surface (see Fig. 7).

The conditions give

$$\varepsilon E \cos \Theta = \varepsilon' E' \cos \phi \text{ and } E \sin \Theta = E' \sin \phi \, . \tag{38}$$

The field vector outside the liquid surface, E′ is hence, since $\varepsilon'=1$,

$$E' = E\left[\varepsilon^2 \cos^2\Theta + \sin^2\Theta\right]^{1/2}, \tag{39}$$

or since $\varepsilon^2 >> 1$ for water,

$$E' = E\varepsilon \cos\Theta . \tag{40}$$

The work of charging the ion is $\varepsilon \int (E^2/8\pi)\, dV$, where E is the ionic field and dV is the volume element in which polarization due to the ionic field arises. For the volume above the surface, the charging energy contribution, if it were still the fluid of dielectric constant ε, is seen from Fig. 7 to be

$$G = \frac{\varepsilon\, e^2}{8\pi} \int_0^\infty \frac{2\pi(1+x).dx}{\varepsilon^2(1+x)^4} = \frac{\varepsilon\, e^2}{8\pi}\int_0^\infty \frac{x.dx}{\varepsilon^2(1+x)^3}, \tag{41}$$

since the area of the spherical sector (Fig. 7) is $2\pi(1+x)x$ and its volume is $2\pi(1+x)x.dx$. Then

$$G = \left[\frac{e^2}{4} - \frac{1}{1+x} + \frac{1}{2(1+x)^2}\right]_0^\infty = \frac{e^2}{8l\varepsilon}, \tag{42}$$

for an ion of charge e.

The calculation of the remaining polarization energy due to E′ in the spherical shells dx in the free space above the interface ($\varepsilon = 1$) presents substantial mathematical difficulties since it involves integration over x and various solid angles determined by $\phi = \arc\tan\left[\frac{E}{E'}\tan\Theta\right]$. If, however, the main polarization effect of the ion were attributed to a spherical angle for which cos Θ is near 1 (which is equivalent to neglecting the refraction of lines of force), $E' = \varepsilon E$ so that for an ion of radius r_i at the position $l = r_i$, $G = e^2/8r_i$ above the surface and its self-energy in the bulk is $e^2/8\varepsilon r_i$. If the bulk extended above the dividing interface, the charging energy contribution in the medium above the surface would be $e^2/8\varepsilon r_i$ from Eq. 42. Hence the net charging energy for the ion in the situation shown in Fig. 7 would be very approximately*

$$G_\sigma = \frac{e^2}{2\varepsilon r_i} - \frac{e^2}{8\varepsilon r_i} + \frac{e^2}{8r_i} = \frac{e^2}{8r_i}\left(\frac{3}{\varepsilon} + 1\right). \tag{43}$$

---

* It is interesting that the terms $\frac{e^2}{2\varepsilon r_i} - \frac{e^2}{8\varepsilon r_i} = \frac{3}{8}\frac{e^2}{\varepsilon r_i}$ in Eq. 43, which arise from the Born charging relation and Eq. 40 without any approximations, and correspond to the charging energy with respect to the solvent fluid available to the ion below the dividing surface, give G_σ already larger than does the image-energy calculation.

The self-energy in the bulk is, as before, $(e^2/2\varepsilon r_i)$, so that

$$G_{ads} = G_\sigma - G_b = \frac{e^2}{8r_i}\left(\frac{3}{\varepsilon} + 1\right) - \frac{e^2}{2\varepsilon r_i} = \frac{e^2}{8r_i}\left(1 - \frac{1}{\varepsilon}\right) , \tag{44}$$

a quantity which is always positive, corresponding to negative adsorption.

Case (iii): Ion with Primary Hydration Shell near the Interface

Equation 43 would predict a large positive ΔG_{ads} if r_i is taken as the crystal ionic radius. However, it is much more probable that ions in the surface region of the liquid interface do not lose their primary hydration layer. This is because the hydration of small ions is characterized by a high-energy ion-solvent interaction region corresponding to the primary hydration shell, and a region of weaker ion-solvent interaction beyond this shell where the polarization energy is determined to a good approximation by Born's relation, using the bulk value of the dielectric constant ε. Hence, because of the strong interaction within the primary hydration shell of radius r_h, almost no ions will be able to approach the liquid surface to distances less than r_h where they would have to suffer some loss of interaction with their primary coordinating water molecules. It is therefore reasonable to suppose that the restriction on extent of hydration which ions near a liquid surface must suffer arises principally (but not entirely; see below) from loss of the long-range Born polarization energy beyond r_h in comparison with the total-range polarization energy the ion, with its primary hydration shell, would have in the bulk solvent.

The model (Fig. 6b), in which a fixed primary hydration shell (radius r_h) is assumed, allows a Born treatment to be pursued more reliably since it is applied only to the solvent polarization beyond the primary hydration shell. Under such a limitation, ε for the solvent can be taken as the normal bulk value and the molecular structure of the solvent is then also much less important than it is within one solvent molecular diameter from the ion.

Then with $l = r_h$, Eq. 42 gives, at the surface,

$$\Delta G_{ads} = \frac{e^2}{8r_h}(1 - 1/\varepsilon) . \tag{45}$$

For a sodium ion, r_h is approximately 0.095 + 0.276 nm which gives ΔG_{ads} = 42 kJ mol^{-1}. This energy is sufficiently large to more or less completely exclude the ion from the surface region at $l \sim r_h$ at ordinary temperatures.

Distribution of Hydrated Ions Near a Liquid/Vapor Interface

The previous calculation applies to a particular situation of the ion at the surface. It is evident, however, that a distribution of ions will be generated (cf., Onsager and Samaras, 1934) near the surface according as ΔG_{ads} becomes smaller with increasing l as $l > r_h$. The net surface excess must then be calculated by integration of this distribution with respect to l away from the surface.

Let c_o be the bulk concentration of a particular kind of ion in ions cm^{-3}. Then the local number $dn_{(l)}$ of ions in a 1 cm^2 lamina of thickness dl will be given by

$$dn_{(l)} = c_o \exp - \Delta G_{ads(l)}/RT \cdot dl \ , \tag{46}$$

so that the surface excess Γ_i is

$$\Gamma_i = n - n_o = c_o \int_{r_h}^{\infty} \left[\exp - \frac{\Delta G_{ads(l)}}{RT} - 1\right] dl \ . \tag{47}$$

Γ_i can be evaluated approximately within the limitations of $\Delta G_{ads(l)}$ being given by Eq. 44, with r_i taken as the variable distance l. Then, substituting ΔG_{ads} from Eq. 44 for the ion at any position $l > r_h$,

$$\Gamma_i = c_o \int_{r_h}^{\infty} \left[\exp - \frac{Ne^2}{8\ RT} (1 - 1/\varepsilon) - 1\right] dl \ , \tag{48}$$

where N is Avogadro's number and ε can be considered almost constant with l for $l > r_h \cdot n$ since f(l) involves an exponential integral function and must be evaluated numerically, so that Γ_i can be calculated.

Mathematical problems of convergence arise with Eq. (48) with ∞ as the upper limit of the integral unless the screening factor κ is introduced. However, if the integral is taken from ∞ to a distance L at which the Born co-sphere overlap energy with the interface is < kT, i.e., when $L = e^2 (1 - 1/\varepsilon)/8kT$, which has a value of 7 nm for a univalent ion in water at 298 K, a finite effective Γ_i can be evaluated.

In concluding this section, we stress that the local structure of solvent near its interface and the structure of the hydration shells of ions introduce significant specificities in ion adsorption at liquid interfaces, especially in H-bonded solvents. Thus, both the structure of the primary hydration shell of ions and the "structure-broken" region (Frank and Evans, 1945) just outside the primary shell will also be important in ion adsorption since the properties of water in this region are known to be ion-specific, but these factors will be significant only near the interface.

Most real liquids, especially water, have an interface in which some net degree of solvent orientation arises, corresponding to minimization of surface free energy. In the case of water, the H atoms of H_2O molecules in the surface are oriented inwards towards the bulk. For hydrated ions near the interface, this will introduce some ion-specific "co-sphere" interaction type of effect (analogous to that between ions themselves in the bulk at appreciable concentrations) which will either locally diminish or increase the ionic concentration within one or two molecular diameters of the liquid interface in comparison with the situation in the absence of such "co-sphere" structure-interactions (for the interphase, this could be termed the "co-plane" interaction effect). The effect will arise because the structure-changed region near the ion will overlap with any structure-changed region at or near the surface of the liquid itself and lead to an additional positive or negative free energy contribution to ΔG_{ads} for the ion. By comparison, with hydration co-sphere structure interactions between ions in the bulk of water solutions, or in the double-layer, it may be anticipated that this effect could amount to at least $\pm$ 4-6 kJ mol^{-1}, due to H-bond breaking or reorientation effects, e.g., from the "A_{ij}" factors for electrolyte solution activity coefficients.

MOLECULAR MODEL TREATMENTS OF WATER ORIENTATION AT CHARGED INTERFACES

Basis of Treatments

Since one of the principal sources of electrical effects at liquid metal and non-metal interfaces is orientation of molecular solvent and/or solute dipoles (the other source is differential accumulation of free electric charges in a double-layer), it is important next to review this question, especially for the case of electrodes where net charge q_M, causing orientation, is controllable.

Treatments of solvent dipole orientation as a function of electrode surface charge, q_M, seek to evaluate a reciprocal capacitance contribution arising from the dependence of the surface-dipole layer potential difference, χ_d, on q_M. At electrode interfaces, $\Delta\chi_d$ is a function of potential and hence q_M; then the dependence of $\Delta\chi_d$ on q_M can be written as a differential capacitance, $1/C_d$, due to dipole orientation. It combines in a series relationship with other capacitance contributions of the overall double-layer due to free charge accumulation. Therefore, only if C_d is sufficiently small will its effect on the overall measured capacitance be important. Hence its correct evaluation is a critical matter in modern theories of the double-layer which properly take into account the solvent layer. For example, the two-state Watts-Tobin model

(1961) gives rise to anomalous behavior with respect to this capacitance (see below).

One of the main questions is whether the capacitance contribution due to the dependence of orientation on q_M can account for the capacitance hump usually observed near the potential of zero charge (at Hg).

Two-state Dipole Orientation Treatments

A useful molecular model treatment of solvent orientation polarization at an electrode interface was given by Watts-Tobin and Mott (1961). Two orientation states of the solvent dipoles, up ↑ and down ↓, aligned with the electrode field E arising from net surface charge density q_M, were envisaged. The polarization in the interphase at the electrode surface was calculated in terms of the relative population $N\uparrow/(N\uparrow + N\downarrow)$ and $N\downarrow/(N\uparrow + N\downarrow)$ of the two states of orientation. $N\uparrow$ and $N\downarrow$ are determined (a) by the field, (b) by temperature, and (c) by any lateral interaction forces between the oriented and unoriented dipoles. Interaction effects were not, however, taken into account in the original treatment.

The Boltzmann distribution function is employed to calculate the relative populations $N\uparrow$ and $N\downarrow$ and, hence, the local polarization per cm^2 may be calculated. Thus

$$\frac{N\uparrow}{N\uparrow + N\downarrow} = e^{U/kt}; \quad \frac{N\downarrow}{N\uparrow + N\downarrow} = e^{-U/kt} , \tag{49}$$

where U is the net energy of the oriented dipole in the field. When lateral interactions are significant, U has the form

$$U = \mu E + Wn\, N\downarrow/N - Wn\, N\uparrow/N , \tag{50}$$

where $N = N\uparrow + N\downarrow$, W is the pairwise lateral interaction energy between oriented dipoles, and 2n the coordination number in the interphase. An orientation distribution function R can be defined (Bockris et al., 1963) as

$$R = \frac{N\uparrow - N\downarrow}{N\uparrow + N\downarrow} = \tanh[U/kT] , \tag{51}$$

i.e.,

$$R = \tanh\left[\frac{\mu E}{kT} - RWn/kT\right] . \tag{52}$$

R measures the orientation polarization per unit area in the interphase at the charged surface since $N\uparrow + N\downarrow$ equals the total number of orientable dipoles per cm^2, assuming electrostriction in the double-layer does not change this number. The variation of R towards a saturation orientation value of $\pm$ 1, i.e., when $N\uparrow \equiv N\downarrow + N\uparrow$ or $N\downarrow \equiv N\uparrow + N\downarrow$ occurs over range of E dependent on the magnitude of nW. $N\uparrow + N\downarrow$, i.e., $\Theta\uparrow$ or $\Theta\downarrow \rightarrow 1$, corresponds to dielectric saturation in the double-layer. E can be related to the surface charge q_m by $E = 4\pi\, q_M/\varepsilon$ so that R as $f(a_M)$ can be obtained. A more developed treatment by Levine et al. (1969) took into account (a) the induced dipole moment, μ_i, contribution due to neighboring dipoles of permanent moment μ_p; (b) the effect of neighboring dipoles on the field at a given site, viz.,

$$E = 4\pi\, q_M + \left(n_e/d^3\right) \left(\mu_p + \mu_i\right) , \tag{53}$$

where n_e is an effective coordination number for dipoles with distance d between each other, and n is taken as 11 according to a relation due to Topping; and (c) the possibility that one orientation is intrinsically preferred to another due to chemisorption or image forces (this is found experimentally, e.g., from the non-zero charge at which maximum absorption of a symmetrical organic adsorbate of zero net dipole moment arises).

The main weakness of the treatments of Watts-Tobin and Mott (1961), and Levine et al. (1969), is the neglect of H-bond structural effects in the interphase, i.e., correlated orientation effects. Such effects are not really taken account of by the lateral interaction term WnR, since this has no general angular dependence on orientation, only extreme orientations being considered. The limiting 2-state situation is undoubtedly a serious oversimplification since molecules of the water dielectric in the interphase will be oriented at <u>various</u> angles and some average polarization orientation angle Θ will be a function of field and Wn_e, and will also be determined by the angular dependence of H-bond energy between neighboring molecules.

It should be mentioned that on the basis of the 2-state model (Conway et al., 1973) using the non-dipolar probe, 1,4-pyrazine at Hg, derived the dipole orientation function R for water as a function of potential from experimental electrocapillary results at the Hg electrode.

<u>Three-state Orientation and Cluster Models</u>

It should be mentioned that a 3-state model of solvent dipole orientation at electrodes, with configurations ↑ → or ← ↓, was considered in detail by Fawcett (1978) and gives a better account of inner layer capacitance behavior and organic molecule absorption than the Watts-Tobin 2-state model. This model was shown to be the favored one, energetically, by Parsons, with predominantly "flat" (→ or ←) orientations of H_2O dipoles at the potential of zero charge of metal surfaces.

More generalized extensions of the 2-state or 3-state models have been investigated in terms of H-bonded clusters of H_2O dipoles at Hg/H_2O interfaces, e.g., by Parsons (1975) and Frumkin and Damaskin (1974).

Unfortunately, the greater the complexity of models developed, the more empirical becomes the testing of fit to the experimental behavior.

Models are compared in Table 3.

Catastrophe Situation With the 2-State Model

For certain conditions, the two-state model represented by Eqs. (49), (51), and (52) leads to either a negative capacitance of the dipole layer or a corresponding imaginary dielectric constant. The origin of this effect, referred to as the "Cooper-Harrison catastrophe", was treated in some detail by Marshall and Conway (1984) who showed how it originated from neglect of the dipole's own reactive polarization so that the interfacial field should be represented not by $-4\pi q_M$ but rather by $-4\pi(q_M-q_{dipole})$, where q_{dipole} is the net reactive charge per cm^2 arising from ends of oriented dipoles. The detailed consequences of this revision are to be found in Marshall and Conway's paper (1984).

Table 3. Models for inner-layer solvent behavior at charged interfaces.

Dielectric Polarization Models	Dipole Orientation Models	Cluster/Dipole Models
Continuum dielectric theory with normal dielectric constant	Free dipole orientation	Free dipole + dimer orientation
	Electrostatically interacting dipole orientation in 2 or 3 states	Free dipole + cluster orientation
Continuum dielectric theory with dielectric saturation	H-bonded dipole orientation*	
(a) Debye-Langevin model (b) Kirkwood-Onsager (Booth-Grahame)	Partial orientation in two-dimensional H-bonded network with varying degrees of H-bond bending	

* In some ways, the case of electrostatically interacting dipoles is similar to that for H-bonded dipoles. However, directional factors and some electronic delocalization become important in the H-bonded model, e.g., for H_2O.

LIQUID/LIQUID INTERFACES

While the boundary between liquid Hg and electrolytes is a special and important case of a liquid/liquid interface, we consider here the situation of non-metallic liquid/liquid interfaces, i.e., between two immiscible liquids, e.g., H_2O/nitrobenzene, H_2O/CCl_4, H_2O/hydrocarbons. Such systems are significant biologically where boundaries between aqueous cytoplasmic fluid and hydrophobic lipid membranes arise.

For immiscible polar fluids in contact, the dipole orientation and structure on each side of the interface will be somewhat modified from that at each liquid/vapor interface separately, depending on the interaction between the dipoles of one fluid and the other, as with H_2O/nitrobenzene. Selective orientation can occur.

Because the dielectric constants of the two liquids in contact will normally be different, one greater than the other, a special dipole imaging situation arises: in the component of higher ε, image charges of unlike sign will arise while in the lower ε liquid, like-sign charges will appear, giving rise to dipole repulsion between one liquid and another. This will be a component of the overall interfacial interaction energy at the boundary (polarization and van der Waals forces).

In the case of each liquid containing an electrolyte, ion-charge imaging will arise in each medium together with a double-layer associated with negative or positive adsorption of the cations and anions related to their solvation which, when significant, will be comprised of a contact region at the interface and a diffuse-layer region extending into the bulk of each medium as with electrolytes at the air/water interface (see Bell and Rangecroft, 1977).

Considerable significant work has been done by Samec (1979) and by Samec et al. (1977, 1979b) on the characterization of the electrochemical behavior of interfaces of immiscible liquids (e.g., H_2O/nitrobenzene) containing electrolytes whose ions can be transferred selectively between one liquid and the other, as with Cs^+ or $(n\text{-}Bu)_4N^+$ (cf., the situation with phase-transfer catalysis). A four-electrode system is used with two potentiostats and the interfacial behavior can be followed then by means of cyclic-voltammetry.

Ion transfer between water and a non-aqueous medium, through a liquid-liquid interface, will normally involve a change of state of solvation of the transferred ion which provides a kinetic barrier (at least in one direction) to transport. The situation is a little like intercalation of Li^+ ions from a molecular solvent, e.g., THF or CH_3CN, to a host lattice like TiS_2, at the interface of which desolvation has to occur.

At a liquid/liquid junction, an interphase potential difference, E_j, can arise in addition to that caused by differential diffusion of cations and anions (the Nernst diffusion potential).

Alfenaar et al. (1967) have pointed out that when two different solvents form a boundary, then there may also be a significant contribution ($E_{j,\ solvent}$) to the liquid junction potential from the transport of solvent molecules across the interface, especially if they interact strongly. Following the procedure of Cox et al. (1973), the liquid junction potential due to solvent transport is then given, by analogy with the diffusion potential $E_{j,\ ion}$, as

$$E_{j,solvent} - \frac{1}{F}\int_{H_2O}^{S} \Sigma t_s' d\mu_s \ , \tag{54}$$

where $d\mu_s$ is the change in chemical potential of each solvent on crossing its junction with the other solvent, and the coefficient, t_s', represents the total number of moles of solvent carried across that junction per Faraday of charge passed.

If both the transport numbers and the solvation numbers, e.g., K^+ and Cl^- ions, through the junctions are approximately constant for the two solvents involved, then the junction potential is given by

$$E_{j,\ solvent} = (t'\ H_2O/F)\ {}^{H_2O}\Delta^{S}\mu_s\ (H_2O) + (t_s'/F)\,{}^{S}\Delta^{H_2O}\mu_s\ (S)\ , \tag{55}$$

where ${}^{H_2O}\Delta^{S}\mu_s(H_2O)$ represents the Gibbs energy per mole of solution of H_2O in solvent, S, and ${}^{S}\Delta^{H_2O}\mu_s(S)$ represents the Gibbs energy per mole of solution of S in H_2O at $t_s' = (t_{K+} \times n_{K+}) + (t_{Cl-} \times n_{Cl-})$.

Table 4 shows the Gibbs energies of solution of some solvents in other solvents together with calculated solvent junction potentials,

Table 4. Solvent junction potentials calculated from free energies of some solvents S in other solvents (H_2O) at 298 K.

(Here the solvents are miscible)

Solvent S	$^a\Delta G_s$/kJ mole^{-1} ($H_2O \rightarrow S$)	$^a\Delta G_s$/kJ mole^{-1} ($S \rightarrow H_2O$)	$E^*_{j,\ solvent}$/V
CH_3OH	-2.09	1.26	-0.009
$HCONH_2$	-0.80	7.95	0.074
DMF	-0.84	-2.09	-0.030
AN	4.81	6.28	0.115

* $E_{j,\ solvent} = (t'_{H_2O}/F)\ G_s\ (H_2O\ S) + (t_s'/F)\ G_s\ (S\ H_2O);\ t'_{H_2O} = t_S$

a From Cox et al. (1973)

$E_{j, \text{solvent}}$, assuming equal ionic transport numbers ($t_{K+} = t_{Cl-} - 0.5$) and equal solvation numbers (taken here as $n_{K+} = n_{Cl-} = 1$) for K^+ and Cl^-.

Appreciable liquid junction potentials are found which are consistent with, e.g., the experimental observations of Cox et al. (1973), where variations up to 100 mV in some cell potentials were found to arise from one solvent to another when the bridge solvent was changed.

REFERENCES

Alfenaar, M., de Ligny, C., and Remijnse, 1967, Rec. Trav. Chim. P.B., 86:986.

Badiali, J.P., Rosinberg, M.L., and Goodisman, J., 1981, J. Electroanal. Chem., 130:31.

Bardeen, J., 1940, Phys. Rev., 58:727.

Bell, G.M., and Rangecroft, P.D., 1971, Trans. Faraday Soc., 67:649.

Bockris, J.O'M., Devanathan, M.A.V., and Muller, K., 1963, Proc. Roy. Soc., London, A274:55.

Conway, B.E., 1975, J. Electroanal. Chem., 65:691.

Conway, B.E., 1977, Adv. Coll. Interface. Sci., 8:91.

Conway, B.E., and Dhar, H.P., 1973, Croatica Chem. Acta, 45:173.

Cox, B.G., Parker, A.J., and Waghorne, W.E., 1973, J. Amer. Chem. Soc., 95:1010.

Cutler, P.H., and Gibbons, J.J., 1958, Phys. Rev., 111:394.

Cutler, P.H., and Davis, J.C., 1964, Surface Sci., 1:194.

Equiluz, A.G., 1984, Phys. Rev., 30:4366.

Fawcett, W.R., 1978, J. Chem. Phys., 82:1385.

Feldman, V.J., Partenskii, M.B., and Vorob'ev. M.M., 1986, Progress in Surface Science, 23:3.

Fermi, E., 1927, Rend. Acad. Lincei, 6:602.

Frank, H.S., and Evans, M., 1945, J. Chem. Phys, 13:507.

Frumkin, A.N., 1924, Zeit. Phys. Chem, 109:34.

Frumkin, A.N., 1926, Koll. Zeit., 39:8.

Frumkin, A.N., and Williams, J.W., 1929, Proc. Natl. Acad. Sci, U.S.A., 15:400.

Frumkin, A.N., and Damaskin, B.B., 1974, Electrochim. Acta, 19:173.

Gomer, R., and Swanson, L.W., 1963, J. Chem. Phys., 38:1613.

Gottesfeld, S., Conway, B.E., and Dhar, H.P., 1973, J. Coll. Interface, Sci., 43:303.

Grahame, D.C., 1947, Chem. Rev., 47:447.

Guggenheim, E.A., 1940, Trans. Faraday Soc., 36:397.

Hamelin, A., 1986, Chapter I in Modern Aspects of Electrochemistry, Eds. B.E. Conway, J.O'M. Bockris, and R. White, Plenum Publ. Corp., N.Y.

Jarvis, N.J., and Schieman, M.A., 1968, J. Chem. Phys., 72:74.

Jones, G., and Ray, W.A., 1937, J. Amer. Chem. Soc., 59:181.

Jones, G., and Ray. W.A., 1941, J. Amer. Chem. Soc., 63:288.

Juretschke, H., 1953, Phys. Rev., 92:1140.

Kaganovich, R.I., and Gerovich, V.M., 1966, Elektrokhimiya, 2:977.

Kemball, C., 1947, Proc. Roy. Soc., London, A188:117.

Kenrick, F.B., 1896, Zeit. Phys. Chem., 19:625.

Kornyshev, A.A., Rubinshtein, A.I., and Vorotyntsev, M.A., 1977, Phys. Stat. Solid, b84:125.

Kornyshev, A.A., Partenskii, M.B., and Schmickler, W., 1984. Zeit. Naturfonch., 39a:1122.

Ku, N.Y., and Ullmann, F.G., 1964, J. Appl. Phys., 35:265.

Lang, N.D., and Kohn, W., 1973, Phys. Rev., B7:3541.

Lang, N.D., and Williams, A.R., 1978, Phys. Rev., B18:616.

Lang, N.D., 1981, Phys. Rev., 46:842.

Lange, E., and Miscenko, S., 1930, Zeit, Phys. Chem., 149:1.

Lange, E., 1951, Zeit. Elektrochem., 55:76.
Levine, S., Bell, G.M., and Smith, A.L., 1969, J. Phys. Chem., 73:3534.
Marshall, S., and Conway, B.E., 1984. J. Chem. Phys., 81:923.
McIntyre, J.D., and Peck, M., 1973, Disc. Faraday Soc., 56:122.
Mignolet, J.C.P., 1950, Disc. Faraday Soc., 8:105; 1953, J. Chem. Phys., 21:1298.
Morcos, I., 1971, J. Chem. Phys., 55:4125.
Morcos, I., 1972, J. Chem. Phys., 56:3996.
Onsager, L. and Samaras, W.N.T., 1934, J. Chem. Phys., 2:528.
Parsons, R., 1954, Chapter 3 in Modern Aspects of Electrochemistry, Ed. J.O'M. Bockris, Butterworths, London.
Parsons, R., 1975, J. Electroanal. Chem., 59:229.
Partenskii, M.B., Porov, E.I., and Kuzema, V.E., 1976, Fiz, Metalov i Metalovedenie, 41:280.
Randles, J.E.B., 1963, Adv. Electrochem. and Electrochem. Eng., Eds., P. Delahay, and C. Tobias, 3:1 John Wiley and Sons, N. Y.
Randles, J.E.B. and Whiteley, K.W., 1956, Trans. Faraday Soc., 52:1509.
Rice, O.K., 1928, Phys. Rev., 31:1051.
Sachs, R.G., and Dexter, D.C., 1950, J. Appl. Phys., 21:1304.
Samec, Z, Marecek, V., Koryta, J., and Khalil, M.W., 1977, J. Electroanal. Chem., 83:393.
Samec, Z., 1979, J. Electroanal. Chem., 99:197.
Samec, Z., Marecek, V., and Weber, J., 1979a, J. Electroanal. Chem., 96:245.
Samec, Z., Marecek, V., and Weber, J., 1979b, J. Electrochem. Chem., 100:841.
Schmickler, W., 1979, J. Electroanal. Chem., 100:533.
Schmickler, W., 1983, J. Electroanal. Chem., 150:19.
Schmickler, W. and Henderson, D., 1984, J. Electroanal. Chem., 176:383.
Schultze, J.W., and Vetter, K., 1973, J. Electroanal. Chem., 44:83.
Slater, J.C., 1939, Introduction to Chemical Physics, McGraw Hill, N. Y.
Stillinger, R.H., and Ben-Naim, A., 1967, J. Chem. Phys., 47:4431.
Thomas, L.H., 1926, Proc. Cambridge Phil. Soc., 23:542.
Tsong, T.T., and Muller, E.W., 1969, Phys. Rev, 181:530.
Wagner, C., 1924, Phys. Zeit., 25:474.
Watts-Tobin, R.J., and Mott, N.F., 1961, Electrochem. Acta., 4:79; Phil. Mag., 6:133.
Young, T., 1805, Phil. Trans. Roy. Soc., London, p.84.

SPACE CHARGE EFFECTS IN DIELECTRIC LIQUIDS

Markus Zahn

Massachusetts Institute of Technology
Department of Electrical Engineering and Computer Science
Laboratory for Electromagnetic and Electronic Systems
High Voltage Research Laboratory, Cambridge, MA 02139

ABSTRACT

Insulating liquid dielectrics used in power apparatus, such as transformers and cables, and in pulse power technology, such as Marx generators and pulse forming lines, have their performance affected by injected space charge. The space charge distorts the electric field distribution and introduces the charge migration time between electrodes as an additional time constant over the usual dielectric relaxation and fluid transport times. Pipe flows and a Couette flow system of coaxial cylinders where the inner cylinder can rotate are described which measure the streaming current for flow electrification where the mobile part of the electrical double layer is entrained in the flow. Charge injection from high voltage stressed electrodes is measured by this Couette apparatus as well as by Kerr electrooptic field and charge mapping measurements. Such optical measurements are described between parallel plate and coaxial cylindrical electrodes for nitrobenzene, ethylene carbonate, highly purified water, water/ethylene glycol mixtures, transformer oil, liquid and gaseous SF_6, and high voltage stressed and electron beam irradiated polymethylmethacrylate. Specific attention is directed to highly purified water because Kerr electrooptic measurements, voltage-current terminal measurements, and electrical breakdown tests have shown that the magnitude and polarity of injected charge and the electrical breakdown strength depend strongly on electrode material combinations and voltage polarity.

APPLICATIONS OF LIQUID INSULATION

For greater efficiencies in ac and dc power generation and transmission, and for larger energies in such pulsed power applications as lasers, inertial confinement fusion, charged particle beam devices, and directed energy devices, it is necessary to operate at the highest voltage levels with minimum volume. These technologies are thus limited by the electrical breakdown strength of available materials.

The study of electrical insulation and dielectric phenomena is mainly empirical, with most measurements made at electrical terminals of voltage and current, thus providing no information on the electric field distribution throughout the volume between electrodes. These terminal measurements cannot distinguish among a wide range of models of pre-breakdown and breakdown behavior. The electric field distributions cannot be calculated from knowledge of system geometries alone because of space charge effects. This volume charge is due to flow electrification, injection from electrodes and/or interfacial space charge layers, or to dielectric ionization and depends on the electric field, which in turn through Gauss's law self-consistently depends on the charge distribution.

Electric Power Apparatus

The most common high voltage insulant is petroleum mineral oil, usually called "transformer oil," but which is used as insulation in all high voltage devices. Because transformer oil usage began with the growth of the utility industry, its general physical, chemical, and electrical properties have been well characterized. However, because transformer oil is a blend of various hydrocarbon types, specific properties differ between samples. In the past, mineral oil flammability was reduced by adding polychlorinated biphenyls (PCB's). However, later research has shown the non-biodegradability of PCB's to be a health hazard so that current standards do not allow the use of PCB oil. This has resulted in a search for new natural or synthetic oils that are biologically and environmentally safe and yet still have high electric strength, low dielectric dissipation, good chemical stability, low volatility and high flash point, good arc quenching properties, and are nonflammable, non-toxic, and inexpensive.

Silicone oil is one candidate for replacement for PCB-contaminated oil with a number of utilities in the midst of a retrofilling program. Trichlorotrifluoroethane ($C_2Cl_3F_3$) (trade name, Dupont Freon 113) is liquid at room temperature and is also used for high voltage insulation and cooling in high voltage transformers and compact dc valves.

Flow Electrification - The importance of charge effects is demonstrated by the flow electrification of such insulating liquids through insulating or conducting pipes or filters. Here, the mobile part of the electrical double layer that forms at the interface between dissimilar materials is swept away by the flow and can accumulate on insulating surfaces leading to strong electric fields and ultimately to spark breakdown. This was an old problem with jet fuels but has recently re-surfaced in power apparatus as the need for more compact, high efficiency equipment has resulted in new material combinations and increased flow speeds. Current efforts are using anti-static additives and charge relaxation volumes to provide leakage so that charge cannot significantly accumulate (Gasworth, 1985; Gasworth, 1986; Gasworth et al. 1986).

Electric field generation can be analyzed in terms of the four basic processes of charge generation, transport, accumulation, and leakage. In addition to charge injection from high field stressed surfaces, charge generation can be due to entrainment of double layer charge near duct walls or from upstream sources such as pumps which greatly agitate the fluid. Other generation factors include obstacles such as spacer blocks, protuberances, and bends leading to secondary flows that increase turbulence even at small Reynold's numbers. The flowing fluid can then transport this generated charge to other regions where it can accumulate on insulating walls or on electrically isolated conductors. This charge can then build up until the leakage currents just equal the rate of charge accumulation or until the insulation fails and spark discharges occur due to the high electric field generation. In the duct region of a transformer, these four processes generally overlap, making it difficult to sort out causes and effects. Autopsies of failed transformers have often found discharge tracks on the pressboard at the bottom of the windings where the turbulence and oil flow velocity are greatest as the entrance duct narrows. However, the primary failure is often due to large discharges in the upper oil space near the high voltage output (Crofts, 1986). Laboratory measurements have shown that for paper/oil systems, positive charge is entrained in the oil, leaving behind negative charge on the pressboard wall (Oommen, 1986). The small surface tracking discharges on the pressboard are due to local accumulation of negative surface charge, while the large breakdown in the upper tank oil volume is due to significant positive volume charge accumulation.

Table 1 tries to organize the four electrification mechanisms in terms of the factors observed to influence the electrification process in transformer oil (Crofts, 1986).

Most electrification measurements use pipe flow like that shown in Fig. 1a where the current entering the pipe plus the current generated within the pipe are measured by electrometer E_1 and equals the current measured at the outlet by electrometer E_2. For insulating tubes, the radial and axial electric field components are measured by non-contacting field probes in the wall of a grounded surrounding metallic enclosure as

Table 1. Observed factors on electrification and their effect on charging processes in transformers.

FACTORS	MECHANISMS			
	GENERATION	TRANSPORT	ACCUMULATION	LEAKAGE
TEMPERATURE	Debye length, Laminar sub-layer, Arrhenius	Viscosity (Reynold's #)		Oil and paper conductivity, Paper/moisture equilibrium
MOISTURE	Source of ions at paper/oil interface			Increase of conductivity at paper, paper/oil interface, and oil volume
FLOW RATE	Streaming current	Reynold's #, Flow profile, Laminar sub-layer thickness		
TURBULENCE	Laminar sub-layer thickness	Wall shear stress		
CONTAMINANTS or SURFACE ACTIVE AGENTS	Zeta potential, Adsorption		Selective adsorption	Enhanced conduction
OIL ELECTRICAL CHARGING TENDENCY (ECT)	Charge build-up on duct walls, charge build-up in oil volume	Stripping off of mobile double layer charge into oil flow	Accumulation of entrained charge on isolated components or duct walls	
SURFACE CONDITION	Roughness, Turbulence	Drag on wall		
ENERGIZATION	Charge injection	Electro-convection		
CONFIGURATION	Inlet current from upstream sources (pumps), ducting, paper/oil interface	Pipes	Upper plenum paper/oil interface, charge precipitation from upstream injection	Paper/oil interface, paper bulk, oil bulk
DIELECTRIC STRENGTH	External discharges due to potential build-up			Tracking, surface discharges, bulk discharges

in Fig. 1b. The axial electric field component which drives surface leakage can be greatly minimized at the expense of increasing the radial field by making the enclosure a tightly fitting sleeve as in Fig. 1c. These laboratory models simulate physical processes at work in high voltage apparatus and can be used as sensors and monitors of the electrification process to determine such important factors as the generation and distribution of volume charge entrained in the liquid, the distribution of surface charge accumulating at insulating interfaces, bulk and

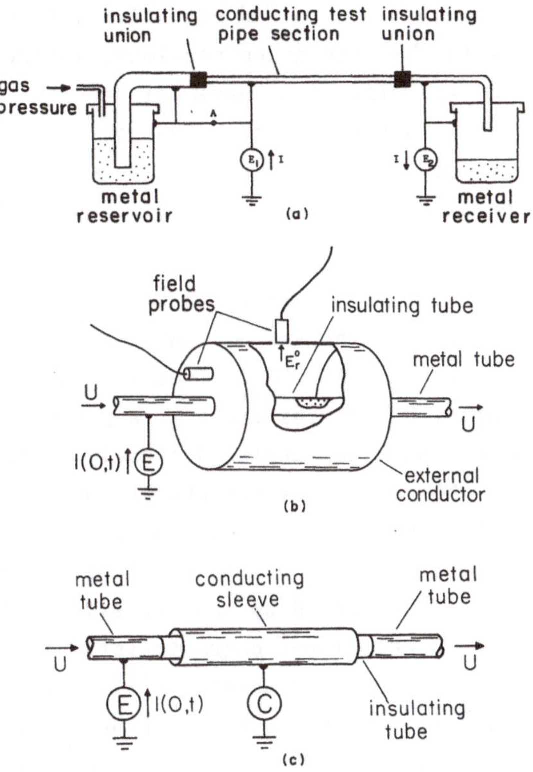

Fig. 1. (a) Experimental arrangement for measuring streaming currents in conducting pipes. Gas pressurizes a metal reservoir to drive liquid from left to right. The test pipe section is joined at both ends by insulating unions. Electrometer E_1 monitors currents entering and generated in the test section while electrometer E_2 measures the outlet current collected by the metal receiver. As a check, both electrometers should measure the same current I.
(b) For insulating pipes, the radial and axial electric field components are measured by noncontacting field probes in the wall of a grounded metallic enclosure.
(c) The axial electric field component which drives surface leakage can be greatly minimized at the expense of increasing the radial field by making the enclosure a tightly fitting sleeve.

surface conductivity including the effects of additives, and the configuration of nearby conductors (Gasworth et al., 1986; Gasworth 1985).

Basic Electrokinetics - Electrification problems in high voltage equipment stem from the entrainment of diffuse electrical double layer charge in the circulating liquid coolant as it flows past metallic or insulating surfaces and the accumulation of this charge in the volume or deposition of this charge on insulating or isolated surfaces so that the electric potential rises. The process is analogous to operation of a Van de Graaff generator. Metallic sections are often adequately characterized by the readily measured streaming currents they generate, but because of their distributed nature and sensitivity to external conditions, processes in the insulating sections, including charge convection, migration, and diffusion, have not been so experimentally accessible. Under the influence of image charges on external conductors, these processes are responsible for potentially hazardous charge accumulations and the associated electric field generation across the insulation.

Because of slight fluid ionization and presence of trace impurities, and perhaps by design through the use of additives, insulating liquids carry positive and negative ions. These ions try to neutralize each other in the bulk, but at boundaries there is a preferential adsorption of one species with the opposite carrier diffusely distributed over a thin boundary region. The degree of net charge and the depth to which it penetrates into the liquid volume are related to the balance of ion diffusion, migration, and convection. In equilibrium, diffusion due to concentration gradients is balanced by the electric field induced by the separated charges. The double layer characteristic thickness, called the Debye length λ, is related to fluid permittivity ε, ohmic conductivity σ, and molecular diffusion coefficient D as $\lambda = [\varepsilon D/\sigma]^{1/2}$.

In aqueous electrolytes, $\sigma \sim 10^{-1}$ mhos/m, $D \sim 10^{-10}$ m^2/s, $\varepsilon \approx 80\varepsilon_0$ $\sim 7 \times 10^{-10}$ farads/m so that $\lambda \sim 8 \times 10^{-10}$ m is extremely small. A thermal voltage drop at room temperature, $kT/q \approx 25$ mV, across this double layer results in a very large internal electric field $E \approx (kT/q)/\lambda \approx 3 \times 10^7$ volts/m.

Because of the large conductivity in aqueous electrolytes, externally applied fields cannot approach such magnitudes and thus hardly disturb the electrical double layer equilibrium. Similarly, convection has a negligible effect as is seen by comparing the electrical relaxation time $\tau_e = \varepsilon/\sigma \sim 84$ µs to a liquid transport time $\tau_{trans} \approx l/v$ where l is a characteristic travel length of a fluid moving at a velocity v. For representative values of $l = 1$ cm and $v = 1$ m/s, $\tau_{trans} = 10^{-2}$s is much greater than τ_e. The ratio of these times is called the electric Reynold's number $R_e = \tau_e/\tau_{trans} = \varepsilon v/\sigma l \sim 84 \times 10^{-4} \ll 1$. Because $R_e \ll 1$, the effect of convection in electrolytes on the charge is small.

Highly insulating dielectric fluids with their lower conductivities $\sigma \sim 10^{-10}$ mhos/m have much larger Debye lengths and smaller internal electric fields $\lambda \sim 4 \times 10^{-6}$m, $E \approx kT/\lambda q \sim 6 \times 10^3$ volts/m where we take $\varepsilon \sim 2\varepsilon_0$ and $D \sim 10^{-10}$ m^2/s. More insulating oils as used in transformers will further increase λ, and thus decrease the internal electric field.

With such low internal electric fields, reasonable externally applied fields are significant in determining the charge distribution in the layer. With a long relaxation time, τ_e~0.18 s, the typical electric Reynold's number is also large, R_e ~ 18. Thus fluid convection also strongly influences the distribution of charge in the double layer. Electrification occurs when the mobile part of the double layer is entrained in the flow. This charge transport leads to charge build up on charge collecting surfaces, eventually causing electrical breakdown if leakage processes are slower than the rate of charge collection.

Standing voltage Wave Interaction with Imposed Flow in Tube - Double layer charge can be sensed by external windings and thus be a non-contacting monitor of electrification. A turbulent fluid flow is imposed, but now a helical winding around a section of the insulating tube replaces the external conductor as shown in Fig. 2. The winding is excited with a standing-wave distribution of potential which perturbs the original distribution of volume charge in the flowing liquid. The extent to which the perturbed volume charge is displaced relative to the winding by the flow is reflected as an imbalance in currents drawn by two of the virtually grounded conductors.

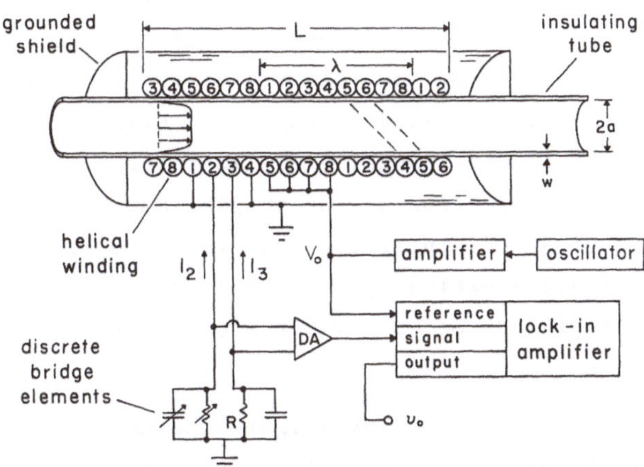

Fig. 2. An eight-conductor helical winding around an insulating tube imposes a standing-wave voltage distribution. Conductors 5-8 are connected to an ac high voltage for which electric fields are capacitively coupled to the inside of the tube. Conductors 1 and 4 are directly connected to ground to provide shielding for the sensing windings 2 and 3 which are connected to ground through current viewing resistors. In the absence of flow, the capacitive currents in 2 and 3 are equal ($I_2 = I_3$) so that there is no output $v_o \propto (I_2 - I_3)R$ from the differential amplifier (DA). The bridge is used to compensate for any slight imbalances in geometry. With flow, the perturbed volume charge is displaced, causing an imbalance in currents ($I_2 \neq I_3$) giving an output v_o that depends on applied voltage amplitude V_o and frequency, double layer parameters, and the flow velocity.

The sensed differential voltage increases linearly with applied standing wave ac high voltage for a fixed flow velocity and signal frequency and has a resonance that broadens and shifts towards higher frequencies with increasing flow rate. This standing-wave interaction can be pursued both as a research tool to probe the properties of the interface and as a real-time monitor of the entrained volume charge in operating equipment (Gasworth et al, 1986; Gasworth, 1985).

Voltage Traveling-wave Pumping of the Liquid - The standing-wave technique is useful as a means for studying the natural processes in the volume because the induced charge constitutes only a perturbing influence on the original charge distribution within the sub-layer. Thus, the applied fields are typically much smaller than those that can be generated by the flow or inject charge. To investigate the injection process, much higher potentials are applied when the helical winding is exploited again in the configuration of Fig. 3a. However, now to help distinguish injected charge from that induced by the flow, the liquid is initially stationary, and here it is the mechanical displacement d elicited by an imposed traveling-wave of potential that is of interest.

The observation that the liquid is pumped at all indicates the presence of a net charge in the neighborhood of the liquid wall. With no net flow and a static head d, the velocity distribution returns in the center of the pipe as illustrated in Fig. 3b. Evidence that this charge is injected from the liquid-insulating solid interface is the finding that reversals in the direction of the pumping shown in Fig. 3c occur for excitation frequencies and amplitudes such that the distance traveled by an ion in one excitation period approximates the diameter of the tube with the voltage amplitude necessary for flow reversal increasing linearly with frequency.

Couette Mixer - To further investigate the physics of the electrification process in transformer oil/cellulosic systems, a compact Couette flow system of coaxial cylinders for electrification measurements is shown in Fig. 4, where the inner cylinder can rotate at speeds giving laminar and turbulent flows as well as the transition cellular convection regime (Melcher et al., 1986; Lyon, 1987). The cylinder walls on either side of the oil gap can be bare metal or can be covered with oil-impregnated paper used in transformers.

By using an inlet expansion region, oil slowly flowing into the inlet will be uncharged, but normal molecular self-dissociation will provide the small amount of ionized positive-negative pairs necessary for electrification. The channel flow between cylinders will entrain the mobile part of the electrical double layer to give a measured current

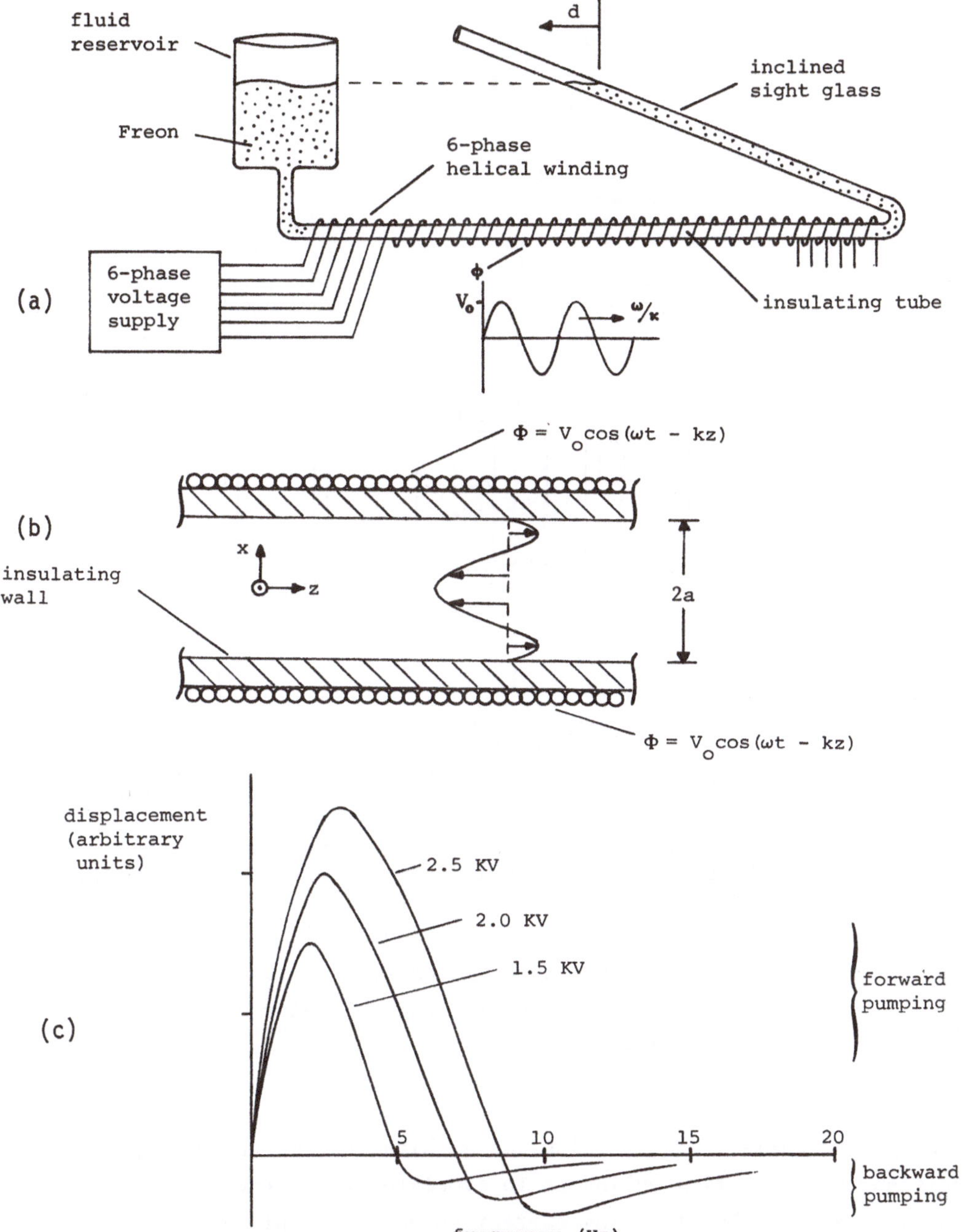

Fig. 3. (a) High applied voltages result in charge injection through the insulating tube into the liquid. The 6-phase helical winding imposes a traveling wave of potential that exerts a force on charge near the liquid/tube interface causing the initially stationary liquid to be displaced by a distance d.
(b) The flow profile for no net flow is shown with pumping occurring within the charge layer near the wall and return flow in the central region.
(c) For a given voltage amplitude, at low frequencies the fluid is displaced in the same direction as the traveling wave (forward pumping), but at higher frequencies the fluid is displaced in the opposite direction (backward pumping). The cross-over frequency of no displacement increases linearly with voltage and corresponds to the distance traveled by an ion in one excitation period that approximates the tube diameter.

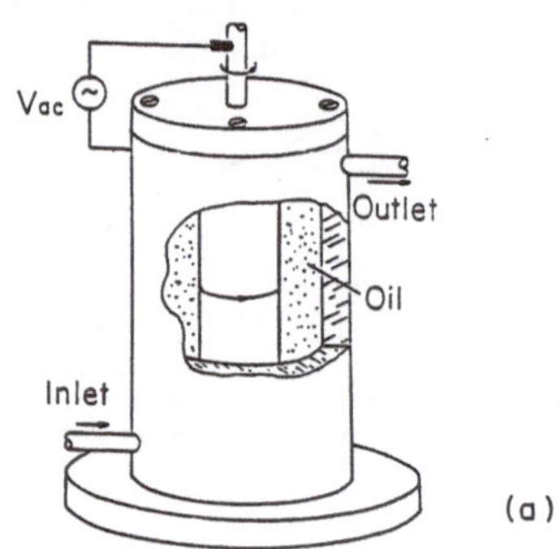

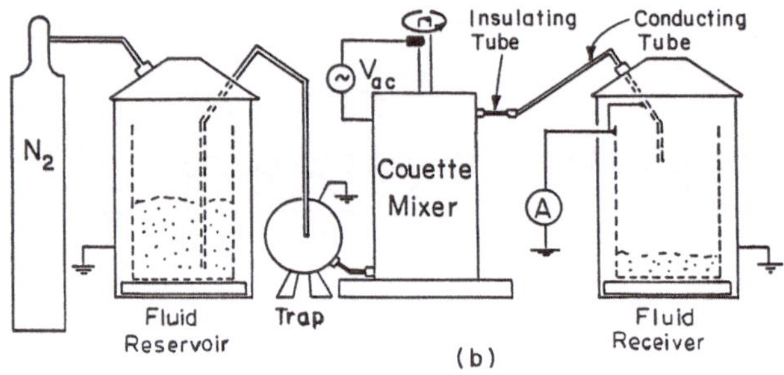

Fig. 4. (a) Couette charging apparatus of coaxial cylinders showing capability of applying radial electric field; (b) Couette electrification system showing dry nitrogen pressure drive, fluid reservoir and receiver, charge trap, Couette mixer where inner cylinder rotates, and measurement of current exiting receiver. For our apparatus the inner cylinder radius was 2.54 cm, the outer cylinder radius was 3.81 cm, and the height was 22.9 cm. Typical oil properties at room temperature were dielectric permittivity $\varepsilon = 2.2\varepsilon_o$, ohmic conductivity σ was 1.2×10^{-12} mhos/m, ion mobility $\mu = 10^{-9}$ $m^2/(V\text{-}s)$, kinematic viscosity $\nu = 1.8 \times 10^{-5}$ m^2/s, mass density $\rho_m = 900$ kg/m^3, and molecular diffusion coefficient was $D = \mu kT/q = 2.5 \times 10^{-11}$ m^2/s where kT/q is the thermal voltage.

through the outlet. In addition, an ac high voltage can be applied between the conducting cylinders to simulate transformer energization. The increase in the streaming current shows that applied fields lead to increased charge injection at interfaces between solid and liquid insulation.

This compact apparatus allows for relative flexibility in testing liquids, trace impurities, and insulating paper without involving large amounts of material. By virtue of the reentrant flow, it provides for the study of equilibrium electrification even in highly insulating systems, where the electrical development length $\varepsilon v/\sigma = v\tau_e$ for a pipe flow system is likely to be longer than the practical length of a test section.

Typical steady state streaming measurements at inner cylinder speed Ω=1400 rpm and sampling flow rate Q = 7 ml/s are ~50 pA with both bare metal cylinders and 125 pA with either the inner cylinder alone covered with kraft paper or both cylinder walls covered with kraft paper. The positive currents indicate that positive charge is entrained in the oil flow, leaving negative charge on the paper. If the paper is installed into the system without a drying treatment, the oil moisture content rises from ~ 15 ppm to 50 ppm, eventually settling at 35 ppm. Measurements during this few-day transient interval are not reproducible including polarity reversals. If the paper is first treated in a heated vacuum oil-impregnation system, the results are immediately consistent from run to run. For treated paper the moisture content in the oil was ~7 ppm. If the system was exposed to air, the oil moisture content rose to 12 ppm. During a one-day transient interval as the moisture content between paper and oil came to equilibrium, the results were not reproducible. However, once moisture equilibrium is reached, the streaming current measurements are the same as for dry paper.

Figure 5 shows the time evolution of representative streaming current measurements in the cellular convection and turbulent flow regimes. The periodic variations at lower speeds has period corresponding to the frequency $f = U/\lambda_c$ where $\lambda_c = 2\pi d/3.12$ is the spacing of flow cells at the onset of instability where $d = R_2-R_1$ is the annulus gap and U is the average upwards axial flow (Taylor, 1923; Chandrasekhar, 1961). These oscillations get washed out by high speed turbulent flow. With the rotor

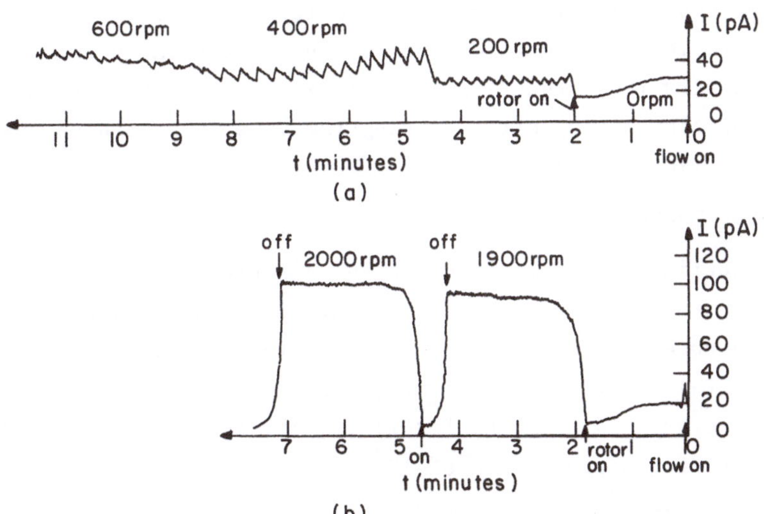

Fig. 5. Streaming current strip chart recordings in the (a) transition cellular flow regime showing periodic variations about a dc level due to flow instability and (b) in the turbulent flow regime for Q~7 ml/s.

stationary and the sampling flow on, a small streaming current of order 5-10 pA flows due to inlet charge to the Couette mixer and charge generation within the Couette mixer and associated plumbing. This background current is generally small compared to the typical currents with the inner cylinder rotating.

Energization - A typical equilibrium electric field in the double layer is of order $E_{DL} \approx (kT/q)/\lambda$ where kT/q is the thermal voltage and $\lambda = [\varepsilon D/\sigma]^{1/2}$ is the Debye length with D the molecular diffusion coefficient. For our experiments with the Debye length $\lambda \sim 20$ µm [$\sigma \sim 1.2 \times 10^{-12}$ mhos/m, $\varepsilon = 2.2\varepsilon_0$, $D = 2.5 \times 10^{-11}$ m^2/s], at room temperature $E_{DL} \sim 1.25$ kV/m. When applied fields exceed this low value, diffusion and self-migration processes can no longer maintain this equilibrium. Mobile charges then migrate in this externally imposed field.

We postulate that net charge is injected from boundaries either because one boundary injects a larger magnitude charge or the mobilities of the species differ, and that turbulent diffusion carries this charge into the core flow. For an applied electric field $E_0 \cos\omega t$, charges with mobility μ have ion velocity $dx/dt = \mu E_0 \cos\omega t$, so that the peak migration distance is $x_p = 2\mu E_0/\omega$. Figure 6 shows representative streaming current measurements with ac high voltage applied at various frequencies. The average current as well as the amplitude of the sinusoidal oscillations about the dc level increase with decreasing frequency because more of the charge from the wall crosses the diffusion sub-layer near the boundaries into the turbulent core as x_p increases with decreasing ω. Note that the measurement even picks up the sinusoidal variations about the dc level because the conducting tube in Fig. 2b entering the receiver is electrically tied to the receiver, together acting as a Faraday cage. As soon as charge enters the conducting tube, it is measured by the ammeter, and thus the flow mixing after exiting the Couette apparatus does not wash out the fine details.

Pulsed Power Technology

For inertial confinement fusion and directed energy devices, high peak power at the terawatt level is needed for short times of order 100 ns. Pulsed power technology collects and stores electric energy at a low input power (~1 kW) for a long time (~1 s) and then delivers this 1 kilojoule energy in a much shorter time (~100 ns) at a much higher power level (10 GW) for a power gain of 10^7.

Figure 7 illustrates a representative pulsed power machine used until 1986 in inertial confinement fusion experiments at Sandia National Laboratory. It is one module of the Particle Beam Fusion Accelerator I (PBFA-I). The complete machine consisted of 36 modules with ratings of

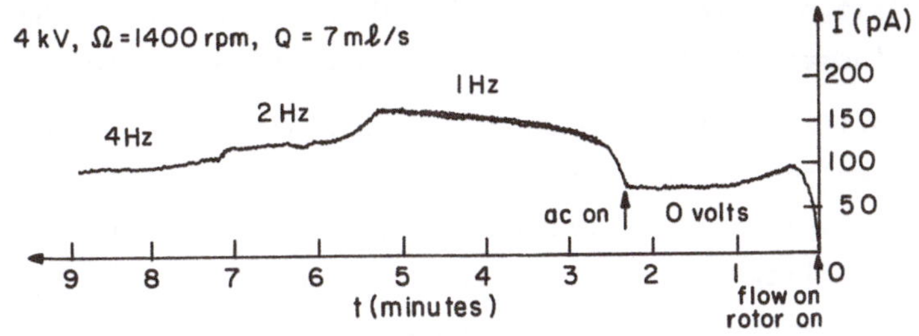

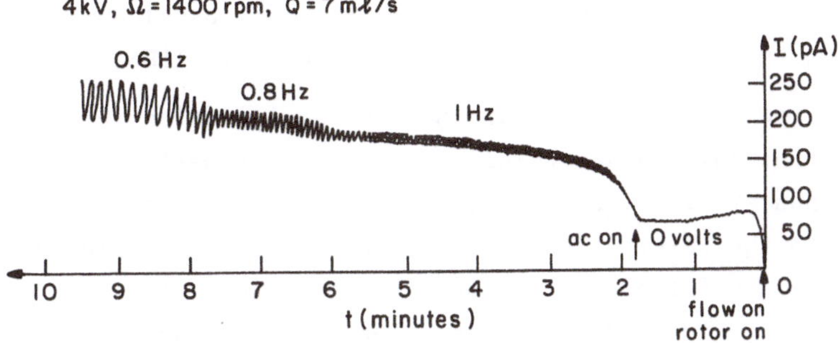

Fig. 6. Streaming current strip chart recordings in the turbulent flow regime with the inner cylinder rotating at 1400 rpm with a 4 kV peak ac high voltage applied across the cylinders and flow rate of 7 ml/s. The dc level of the streaming current and the amplitude of the sinusoidal oscillations about the dc level increase with decreasing frequency of the high voltage because more of the charge injected from the wall crosses the diffusion sub-layer into the turbulent core.

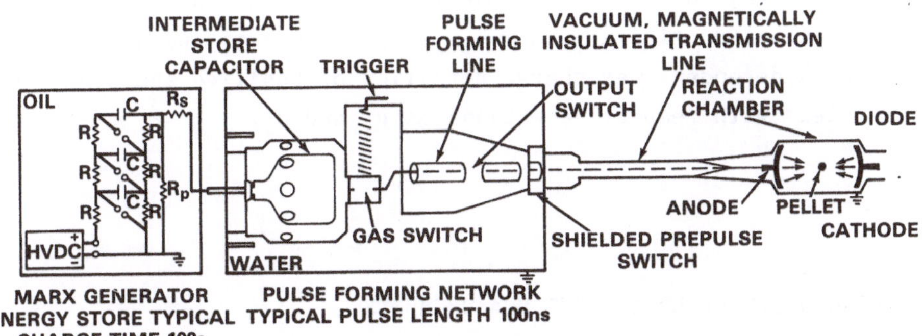

Fig. 7. A representative pulse power machine used for inertial confinement fusion is the Sandia National Laboratory Particle Beam Fusion Accelerator which consists of a slowly charged primary energy storage Marx capacitor bank under insulating oil, a water dielectric intermediate energy store capacitor, a switched pulse forming line which compresses the waveform in time, and a magnetically insulated transmission line which delivers the power pulse to a vacuum diode to produce energetic electrons or light ions, which are accelerated and focussed by electric and magnetic fields onto a target.

2 MV, 15 MA delivered for 35 ns with 30 TW peak power and 1 MJ energy. Inertial confinement fusion research continues with the 100 TW PBFA-II.

Highly purified water is used as the dielectric in pulse forming lines of most pulsed power machines because its high dielectric constant (~80) and high resistivity (>18MΩ-cm) allow short and efficient low impedance high voltage lines for pulse durations less than 100 μs. Water/ethylene glycol mixtures have higher resistivity and allow low temperature operation with negligible loss for longer millisecond time-scales, so that rotating machinery, rather than the usual Marx generator, can be used as the primary energy store [Zahn et al., 1986].

However, recent work has shown that charge injection into water can play an important role in determining the limits of performance of the dielectric (Zahn et al., 1985). Injection of space charge from the electrodes into the water affects the charging and discharging circuit characteristics and introduces another time constant: the time of flight for injected charge to migrate between electrodes. High voltage open circuit decay curves of a 20% water/80% ethylene glycol mixture by weight show a dielectric relaxation time that depends on electrode spacing and that decreases with increasing initial voltage at room temperature where the nominal low voltage relaxation time is ~1.4 ms, while at -10°C where the low voltage dielectric relaxation time is ~25 ms, the open circuit decay curve has a negative second derivative with time (Zahn et al., 1983). This is in contrast to the expected exponential decay, which always has a positive second time derivative and with dielectric relaxation time that is independent of electrode geometry and voltage. This anomalous behavior is due to injected charge that increases the effective ohmic conductivity σ to $\sigma + q_+\mu_+ - q_-\mu_-$, where q_+ and q_- are the time and space dependent net charge densities of positive and negative charge carriers with respective mobilities μ_+ and μ_-.

KERR ELECTRO-OPTIC FIELD MAPPING MEASUREMENTS

Governing Equations

High voltage stressed materials are usually birefringent where the refractive indices for light of free space wavelength λ polarized parallel, $n_{\parallel}$, and perpendicular, $n_{\perp}$, to the local electric field are related as:

$$n_{\parallel} - n_{\perp} = \lambda B E^2 , \quad (1)$$

where B is the Kerr constant and E is the applied electric field magnitude.

The simplest model assumes that molecules are anisotropic with two different polarizabilities along two perpendicular molecular axes. In addition, for the polar molecules such as water, the molecule also has a permanent dipole moment p at an angle χ to the long molecular axis. An applied electric field exerts a torque on the molecule so that the net dipole moment tries to align with the field. This alignment is opposed by thermal motion so that the probability of finding a dipole at an angle θ to the field is described by Boltzmann statistics. As long as the exponential Boltzmann factor is small, it is possible to relate the Kerr constant B for dipoles of number density N as [Coelho, 1979]:

$$B = \frac{N\Delta\alpha}{30kT\lambda\varepsilon_0}\left[\Delta\alpha + \frac{p^2}{kT}\left(1 - \frac{3}{2}\sin^2\chi\right)\right], \tag{2}$$

where $\Delta\alpha$ is the difference in polarizabilities between the long and short molecular axes. A symmetric molecule with $\Delta\alpha = 0$ exhibits no Kerr effect no matter how large the permanent dipole moment p. Typically $\Delta\alpha > 0$, $\chi = 0$, and $p^2/kT \gg \Delta\alpha$ so that the Kerr constant B is positive, indicating that light polarized along an applied electric field travels slower than light polarized perpendicular to the electric field. Large permittivity liquids such as water usually have large positive Kerr constants because polar molecules have large $\Delta\alpha$ and large p.

Some molecules, particularly alcohols, have a negative Kerr constant, even with $\Delta\alpha > 0$ because χ is near $\pi/2$.

Measurement Methods

The phase shift ϕ between light field components propagating in the direction perpendicular to the plane of the applied electric field along an electrode length L is:

$$\phi = 2\pi BE^2L = \pi\left(\frac{E}{E_m}\right)^2 \; ; \; E_m = \frac{1}{\sqrt{2BL}}. \tag{3}$$

The birefringent medium converts incident linearly polarized light to elliptically polarized light. This effect is very similar to photoelasticity in which mechanical stresses rather than electrical stresses cause the birefringence.

Plane Polariscope - We assume that the electric field $\vec{E}$ in the birefringent medium is at an angle θ_K from the vertical y axis. A polarizer whose transmission axis is at an angle θ_{p1} from the vertical is placed on one side of the test cell while another polarizer with transmission axis at angle θ_{p2} from the vertical is placed on the other side. The light transmitted through this system is:

$$\frac{I}{I_0} = \cos^2\left(\theta_{p1} - \theta_{p2}\right) - \sin 2\left(\theta_K - \theta_{p1}\right)\sin 2\left(\theta_K - \theta_{p2}\right)\sin^2(\phi/2), \tag{4}$$

where I_0 is the peak light amplitude.

Crossed Polarizers ($\theta_{p1}-\theta_{p2} = \pi/2$) - If the transmission axes of the polarizers are perpendicular, then (4) reduces to:

$$\frac{I}{I_o} = \sin^2 2\left(\theta_K - \theta_{p1}\right) \sin^2(\phi/2) \ . \tag{5}$$

Light transmission minima then occur when:

$$\frac{\phi}{2} = n\pi \qquad n = 0,1,2,\ldots$$

$$2\left(\theta_K - \theta_{p1}\right) = m\pi \qquad m = -1,0,+1 \ . \tag{6}$$

The latter condition depends only on the directions of the incident polarized light and the applied electric field. Independent of the electric field magnitude, there will be dark fringes wherever the light polarization is parallel or perpendicular to the applied electric field. These field directional lines are called isoclinic fringes. Superimposed on this pattern are light minima wherever $\phi/2 = n\pi$ is satisfied. These field magnitude dependent lines are called isochromatic lines.

When $E = 0$, $\phi = 0$ and the light pattern is uniformly dark. As the electric field is increased, some light is transmitted. It is convenient in (3) to define the electric field magnitude necessary to reach the first light maximum when $\phi = \pi$ as:

$$E_m = \frac{1}{\sqrt{2BL}} \ , \tag{7}$$

so that (5) can be rewritten as:

$$\frac{I}{I_0} = \sin^2 2(\theta_K - \theta_{p1}) \sin^2\left[\frac{\pi}{2}\left(\frac{E}{E_m}\right)^2\right] . \tag{8}$$

Light maxima and minima will then occur for:

$$E = \sqrt{n}\, E_m \qquad \begin{array}{l} n = 1,3,5,\ldots \text{ n odd maxima} \\ n = 0,2,4,\ldots \text{ n even minima} \ , \end{array} \tag{9}$$

when n is odd for maxima and n is even for minima.

Aligned Polarizers ($\theta_{p1} = \theta_{p2}$) - If the polarizers are aligned, then (4) reduces to:

$$\frac{I}{I_0} = 1 - \sin^2 2(\theta_K - \theta_{p1}) \sin^2\left[\frac{\pi}{2}\left(\frac{E}{E_m}\right)^2\right] . \tag{10}$$

We then have light transmission maxima with the latter condition of (6) whenever the light polarization is parallel or perpendicular to the applied field. Similarly, light maxima also occur when the condition of

$\phi/2 = n\pi$ applies. Light minima can only occur when both $\theta_K - \theta_{pl} = \pm(\pi/4)$ and $\phi = (2n + 1)\pi$, n = 0,1,2,... Thus, the isochromatic field magnitude dependent light minima with aligned polarizers occur at the maxima for crossed polarizers and vice-versa.

Circular Polariscope - The isoclinic lines can be removed if crossed quarter wave plates are placed on either side of the test cell but between the polarizers with the incident polarization at an angle of $\theta = \pm 45°$ to either of the quarter wave plate axes, with the second quarter wave plate axes perpendicular to the first as shown in the optical section of Fig. 8. This configuration is called a circular polariscope. The necessary fluid handling and high voltage systems are also shown in Fig. 8, specifically for Kerr electro-optic field mapping measurements in highly purified water.

A quarter wave plate acts as a birefringent medium with phase shift $\phi = \pi/2$ between perpendicular field components along the plate axes. The incident linearly polarized light is converted to circularly polarized light by the first quarter wave plate. In the absence of birefringence, the second quarter wave plate aligned perpendicular to the first converts the incident circularly polarized light back into linearly polarized light. With birefringence, the light transmission intensity for either crossed or aligned polarizers with crossed quarter wave plates is then:

$$\frac{I}{I_0} = \begin{cases} \sin^2\left[\frac{\pi}{2}\left(\frac{E}{E_m}\right)^2\right] & \text{Crossed Polarizers} \\ \cos^2\left[\frac{\pi}{2}\left(\frac{E}{E_m}\right)^2\right] & \text{Aligned Polarizers .} \end{cases} \tag{11}$$

Thus, there are no field directional isoclinic fringes and the minima and maxima are interchanged when the polarizers are crossed or aligned. Aligned polarizers offer a slight advantage as the first minima when $E=E_m$ occurs at a lower field value than for crossed polarizers where the first minimum occurs when $E = \sqrt{2}\,E_m$. Thus, for a given voltage, there is generally one more dark fringe with aligned polarizers.

Examples with Parallel Cylindrical Electrodes - To illustrate the various plane and circular polariscope cases, we consider a two-wire line of parallel cylindrical electrodes stressed by voltage V. In the absence of space charge in the dielectric volume, the electric field distribution outside the cylindrical electrodes obeys Laplace's equation and can be found using the method of images [Zahn et al., 1983]:

$$\vec{E}(x,y) = \frac{V}{2\cosh^{-1}(D/2R)}\left[\frac{-4axy\vec{i}_y + 2a(y^2 + a^2 - x^2)\vec{i}_x}{[y^2 + (x+a)^2][y^2 + (x-a)^2]}\right], \tag{12}$$

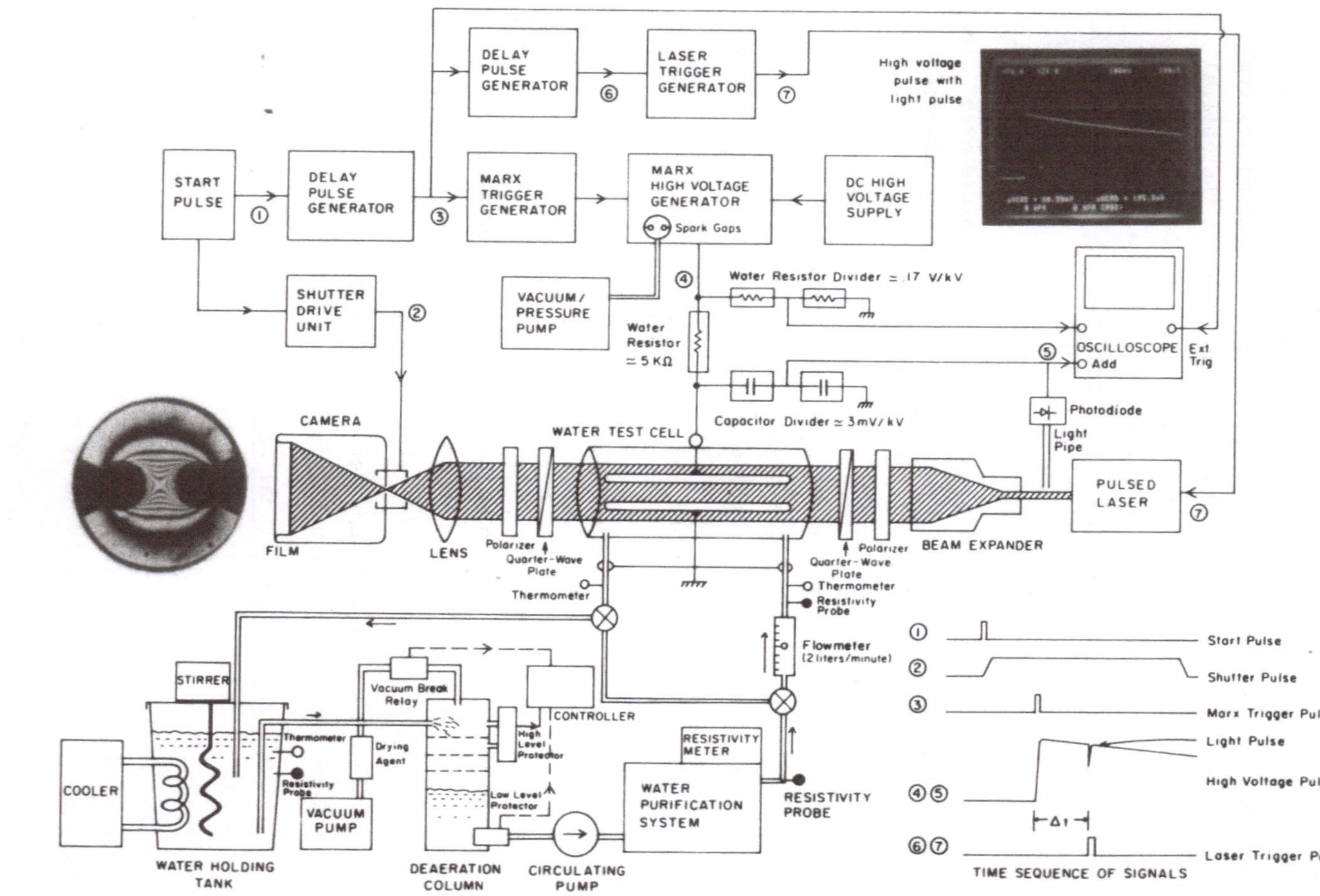

Fig. 8. Schematic of Kerr electro-optic experiments in highly purified water showing the water purification, cooling, and circulation system; circular polariscope optics, high voltage supplies, triggering electronics, and recording systems. The representative aligned polarizer Kerr measurement photograph is for space charge free parallel cylindrical electrodes stressed by a high voltage pulse of 136 kV with $E_m \simeq 36.4$ kV/cm.

where x is the coordinate direction on the line connecting the image charges which are a distance 2a apart. If the cylinders have identical radius R with their centers a distance D apart, then:

$$a = \sqrt{\left(\frac{D}{2}\right)^2 - R^2} \ . \tag{13}$$

Lines of constant electric field magnitude and direction are shown in Fig. 9 and should be compared to the circular polariscope field magnitude dependent isochromatic lines in Fig. 8. Figure 10 shows the measured plane polariscope Kerr electro-optic isochromatic and isoclinic fringe patterns with crossed and aligned polarizers for various angles of incident light polarization. The isoclinic minima for crossed polarizers

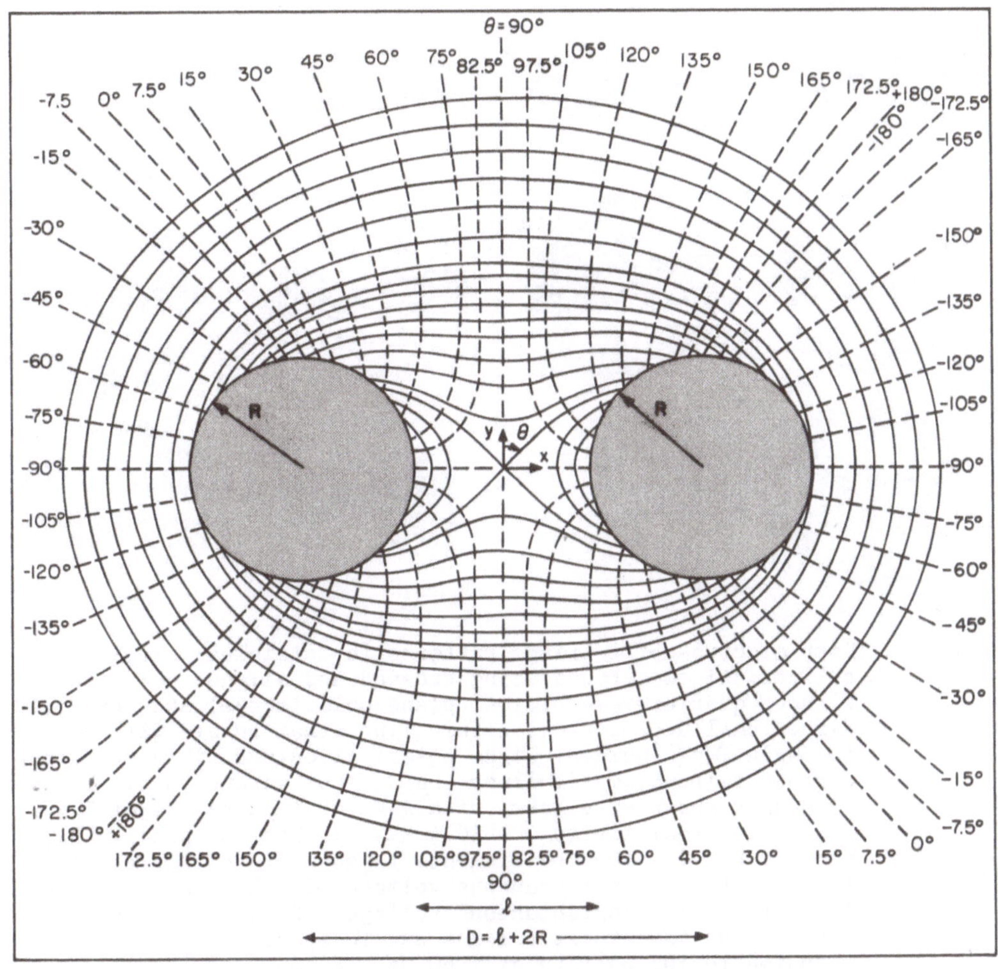

Fig. 9. Lines of constant electric field magnitude (solid) and direction (dashed) for the electric field distribution between parallel cylindrical electrodes. The proportions of center-to-center spacing D, gap l, and radius R correspond to the test cell used in Figs. 8 and 10 where D = 2.27 cm, l = 1 cm and R = 0.635 cm.

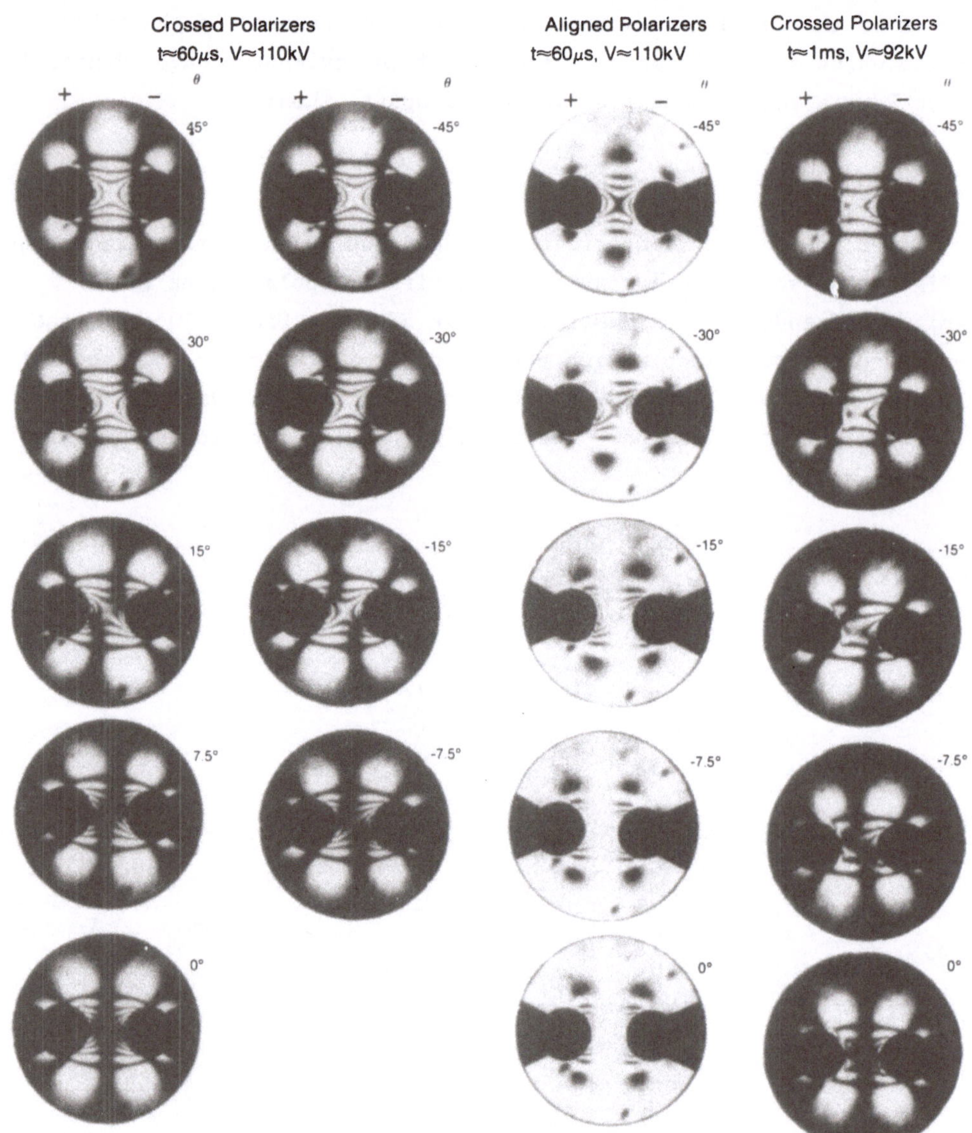

Fig. 10. Kerr electro-optic fringe patterns for stainless steel parallel cylindrical electrodes using crossed and aligned polarizers without quarter-wave plates (plane polariscope) showing field directional dependent isoclinic lines that rotate with changes in angle Θ of incident light polarization and field magnitude dependent isochromatic lines that do not depend on Θ. The measurements shown were taken with a water temperature of T = 26.4°C and resistivity ρ = 16.4 MΩ-cm. The charging voltage was 120 kV with the measurements taken near voltage crest 60 μs from start with instantaneous voltage ~110 kV, and t = 1 ms from start with instantaneous voltage ~92 kV. Note the distortion of the isochromatic lines due to space charge at t = 1 ms compared to the pattern at t≈60 μs. E_m≈36.4 kV/cm.

and maxima for aligned polarizers rotate with changes in direction Θ of incident light polarization, while the field magnitude dependent isochromatic lines remain unchanged as Θ is varied. The isoclinic minima

(Crossed Polarizers) and maxima (Aligned Polarizers) which occur along lines where the applied electric field is either parallel or perpendicular to the light polarization agree with the dashed lines shown in Fig. 9 and provide a measure of electric field direction. For instance, the crossed polarizer dark isoclinic lines in Fig. 10 for $\theta = +15°$ agree with the lines of Fig. 7 with $\theta = 15°$, $\theta = 105°$, and $\theta = -165°$. The lines at $\theta = -75°$ are obscured by the electrode holders. Similarly, the isoclinic lines $\theta = -15°$ agree with Fig. 9 lines for $\theta = -15°$, $\theta = 75°$, and $\theta = 165°$, with the $\theta = -105°$ lines obscured by the electrode holders. However, these isoclinic lines are broad and obscure the field magnitude dependent isochromatic lines, so it becomes preferable to remove them using the circular polariscope configuration as shown in Fig. 8.

The measurements in the first three columns of Fig. 10 were near voltage crest, 60 μs from the start of the high voltage pulse. The last column was taken at time t = 1 ms. Note the distortion of the isochromatic lines as compared to the pattern at t~60 μs.

On the short time scale, injected space charge does not have time to propagate into the dielectric volume so that the Kerr measurements agree with the space charge free solution shown in Fig. 9. However, for longer times, space charge does significantly accumulate in the dielectric volume and distorts the electric field distribution. The Kerr pattern becomes significantly non-symmetric with the x-coordinate for times longer than 500 μs after the start of the high voltage pulse. The electric field near the positive electrode drops relative to the electric field at the negative electrode, indicating positive charge injection. To verify that this positive injection is not unique to the particular characteristics of each electrode, polarity was reversed, with positive charge injection now from the opposite electrode.

In this parallel cylindrical geometry, it is difficult to calculate the space charge density q, related to the electric field through Gauss's law, where now the charge density depends on spatial derivatives of both (x,y) field components. Since the Kerr measurements cannot accurately separate electric field components, the spatial derivatives required cannot be accurately measured. For this reason, most future measurements in this paper will use parallel plane or coaxial cylindrical electrode geometries where the electric field is predominantly in one direction and only depends on that coordinate, allowing easy computation of the charge density.

Space Charge Distortion of the Electric Field

We consider the case of planar electrodes with an x directed electric field E which is distorted by net space charge with density q(x) only dependent on the x coordinate. Gauss's law requires that the slope of the electric field distribution be proportional to the local charge density:

$$\nabla \cdot \vec{E} = \frac{q}{\varepsilon} \rightarrow \frac{\partial E}{\partial x} = q/\varepsilon \ . \tag{14}$$

For no volume charge shown in Fig. 11a, the electric field is uniform given by $E_0 = V/d$ for a voltage V across a gap of d. The electric field drops at a charge injecting electrode, but its average value E_0 remains constant at V/d. Charge injection thus causes the electric field to increase above the average value at other positions. In Fig. 11b, unipolar injection has the electric field maximum at the non-charge injecting electrode, thus possibly leading to electrical breakdown at lower voltages. The strongest distortion due to unipolar space charge occurs for space charge-limited injection at one electrode where the injected charge density is infinite so that the electric field at the injecting electrode is zero. The electric field at the noninjecting electrode is then 50 percent higher than the average field value. For bipolar homocharge injection in Fig. 11c, positive charge is injected at the anode and negative charge is injected at the cathode, so that the

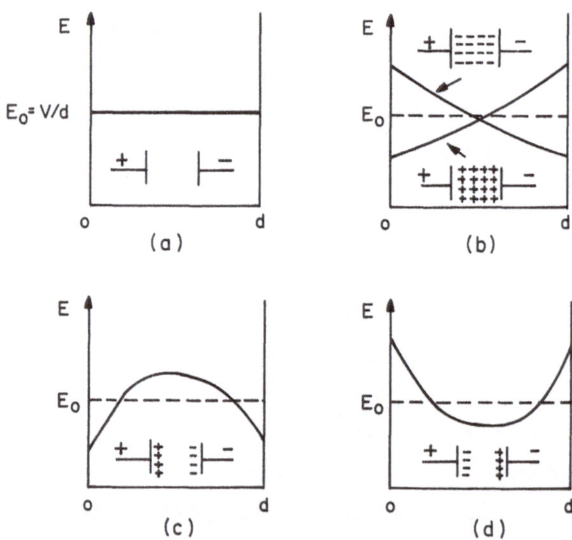

Fig. 11. Space charge distortion of the electric field distribution between parallel plate electrodes with spacing d at voltage V so that the average electric field is $E_0 = V/d$. (a) No space charge so that the electric field is uniform at E_0. (b) Unipolar positive or negative charge injection so that the electric field is reduced at the charge injecting electrode and enhanced at the non-charge injecting electrode. (c) Bipolar homocharge injection so that the electric field is reduced at both electrodes and enhanced in the central region. (d) Bipolar heterocharge distribution where electric field is enhanced at both electrodes and depressed in the central region.

electric field is lowered at both electrodes and is largest in the central region. Since electrical breakdown often initiates at the electrode/dielectric interface, this case can allow higher voltage operation without breakdown, being up to ~40% higher in highly purified water (Zahn et al., 1985). If the voltage suddenly reverses, the electrical field must also instantaneously reverse, but the charge distribution cannot immediately change because it takes some time for the volume charge to migrate. Thus, for early times with voltage reversal, positive charge is near the cathode and negative charge is near the anode, enhancing the fields near the electrodes as shown in Fig. 11d. Such a bipolar heterocharge configuration leads to breakdown at lower voltages as has been found in HVDC polyethylene cables when the voltage is instantaneously reversed to reverse the direction of power flow. A similar configuration to Fig. 11d also occurs when the dielectric is ionized so that free ions are attracted to their image charges on the electrodes.

Nitrobenzene - An example of a homocharge bipolar conduction process is nitrobenzene, which has the highest known Kerr constant, $B \approx 3 \times 10^{-12}$ m/V^2. When stressed by a 30-kV step high voltage with ~1 μs risetime, applied to stainless steel parallel plate electrodes with 1-cm spacing and length L = 10 cm, Fig. 12 shows selected frames taken from a high speed movie at ~5000 frames per second of Kerr electro-optic field mapping measurements using crossed polarizers, where $E_m \approx 12$ kV/cm. At t = 0 and for t > 16 ms, the field is essentially uniform, evidenced by the lack of fringes indicating no net charge density. However, the bipolar conduction process is shown during the transient interval by the propagation of fringe lines from the positive and negative electrodes. The field distributions plotted in Fig. 12 have a slowly propagating positive slope near the positive electrode, indicating positive space charge, and a faster moving negative slope near the negative electrode, indicating more mobile negative space charge. At about t = 2.33 ms, the two charge fronts meet and the charges recombine to a charge-neutral steady state.

Ethylene Carbonate - Another highly polar fluid is ethylene carbonate [$C_3H_4O_3$, glycol carbonate; dioxolone-2, $(-CH_2O)_2CO$], which is solid at room temperature with dielectric constant ~5.4 but which melts at about 36°C and has dielectric constant of ~90 at 40°C (Lee, 1976).

This high dielectric constant with its ease of handling, low corrosivity, and low toxicity have made ethylene carbonate suitable as a solvent for specialized applications and for the preparation of binary mixed solvents with water, methanol, acetonitrile, sulpholane, and acetone. It is also used as an electrolyte in nonaqueous batteries, as a plasticizer, as a monomerer or reagent for the synthesis and modification

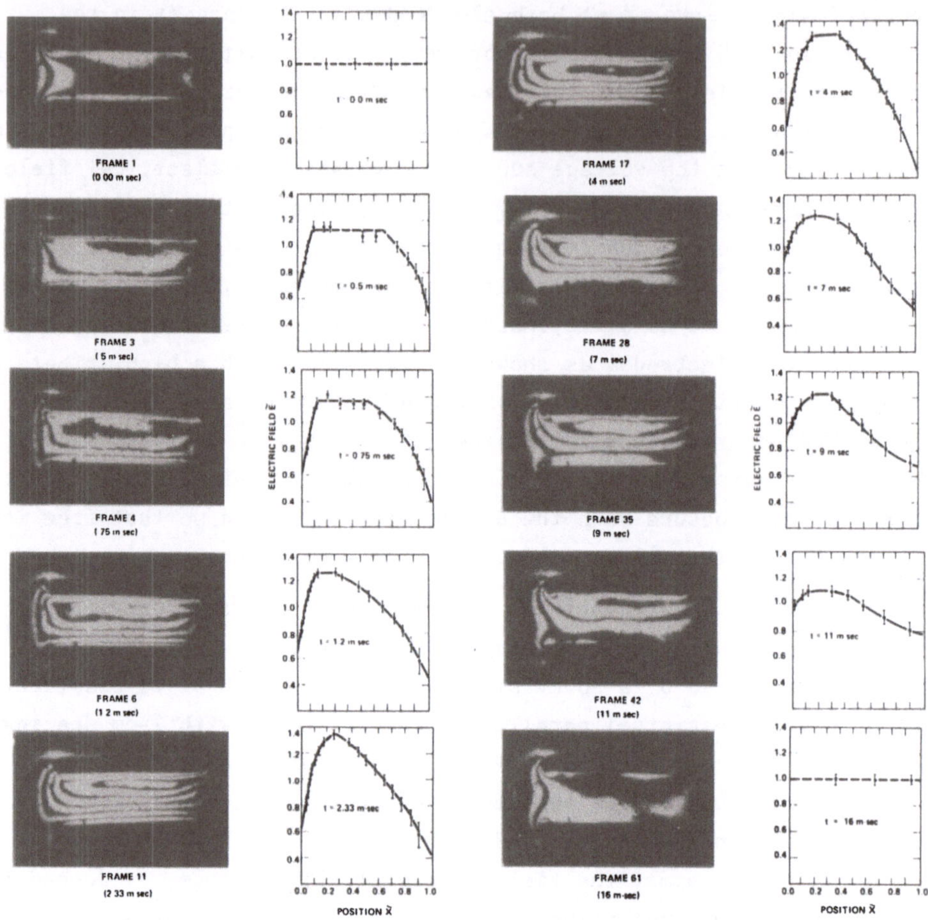

Fig. 12. Transient Kerr electro-optic field mapping measurements at ~5000 frames/second with crossed polarizers in nitrobenzene ($E_m \approx 12$ kV/cm) showing bipolar conduction when a stepped 30 kV high voltage is applied across a 1-cm gap. To the right of each photograph is a plot of $\tilde{E} = Ed/V_0$ vs $\tilde{x} = x/d$ where x is measured from the upper positive electrode. Data points are shown for each light transmission maxima and minima. The space-charge density is proportional to the slope of the electric field distribution, so we have positive charge injected from the upper ($\tilde{x} = 0$) positive electrode and negative charge injected from the lower ($\tilde{x} = 1$) negative electrode. The negative charge moves faster than the positive charge.

of a wide variety of polymers, and as a reactant in many chemical processes [Dow Chemical Co.].

For our purposes, ethylene carbonate was of interest because of its reported high Kerr constant, close to that of nitrobenzene, the usual fluid used in electro-optic field mapping measurements. However, unlike nitrobenzene, ethylene carbonate is not a health hazard and is thus easier to handle. Another motivation for using ethylene carbonate is the possibility of performing future Kerr measurements when the dielectric is solid.

Dow Chemical Co. experimental grade ethylene carbonate with a purity of 99% was used. From the bottle, the resistivity near 40°C was ρ = 0.02MΩ-cm, giving a dielectric relaxation time of $\tau = \varepsilon\rho \simeq 0.16$ μs. Space charge would be neutralized on this fast time scale and thus not easily observed. Another difficulty was that the low resistance of the test cell, ~190 Ω, would quickly discharge the 132 nF Marx generator with an RC time constant of ~25 μs. The voltage would thus decay before the charge could migrate a measurable distance.

The resistivity could be increased to 0.10 - 0.15 MΩ-cm by passing the ethylene carbonate through a silica-gel column to dry it. Heating tape was wrapped around the column to maintain the temperature above 36°C to prevent freezing. The resistivity was further increased up to 5 MΩ-cm by distillation. Because ethylene carbonate decomposes below its 238°C boiling point at atmospheric pressure, the boiling point was lowered to 100°C by distilling at 8 mm Hg pressure. A completely closed system was used to minimize contamination. Recontamination of the ethylene carbonate was prevented by pumping out the distiller and then refilling with dry air.

A Marx impulse generator had typical pulse risetimes of order 20 μs with decay times of order 6 ms. At early times the electric field is uniform, given by V/d where V is the instantaneous voltage across electrodes of spacing d. The transmitted light distribution in the interelectrode region is then also uniform as shown in Fig. 13 for λ = 633 nm. In Fig. 13, at the lowest voltage the fringe number is n~19, while at the highest voltage n≃83, where as given in (9) and (10), n is odd for light minima with aligned polarizers. From this data at T≃42.9°C, we find that our test cell has $E_m \simeq 8.3$ kV/cm and $B \simeq 2.06 \times 10^{-12}$ m/V^2 [Zahn and Ericson, 1984]. This high Kerr constant is near that of nitrobenzene at λ = 633 nm, which is $B \simeq 3.89 \times 10^{-12}$ m/V^2 at 13°C, decreasing to $B \simeq 2.61 \times 10^{-12}$ m/V^2 at 41°C [Hebner et al., 1974]. The Kerr constant of water at 590 nm is much smaller, $B \sim 3.65 \times 10^{-14}$ m/V^2 at ~10°C (Zahn and Takada, 1983).

Figure 14 shows the Kerr fringes at later times for various voltages. At low voltages, the interelectrode light intensity is close to uniform, indicating a uniform electric field. As the voltage is increased, numerous dark lines appear between the electrodes, indicating a highly non-uniform electric field with a small field near the positive electrode and large field at the negative electrode, corresponding to

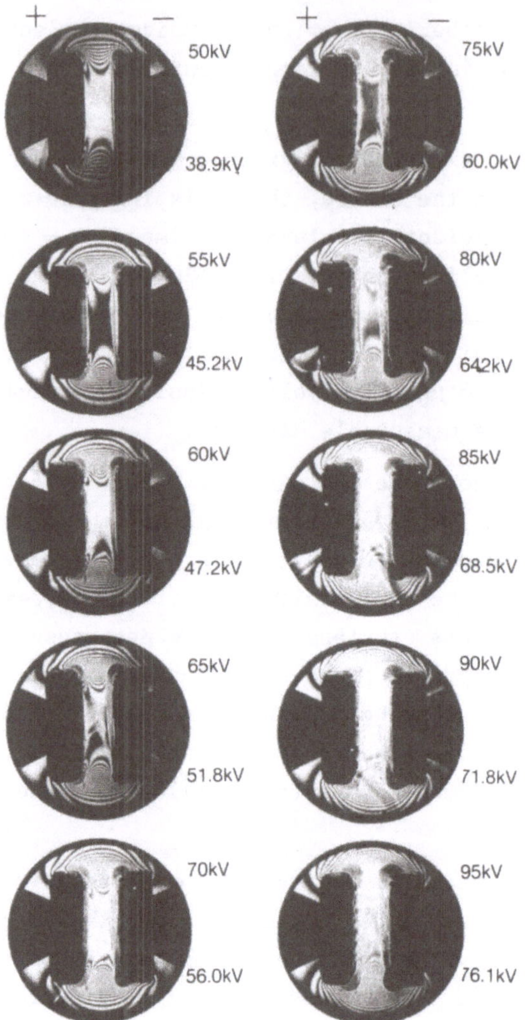

Fig. 13. At early times with high voltage stressed ethylene carbonate between stainless steel electrodes, the uniform light intensity (either bright, gray, or dark) in the central interelectrode region indicates the space charge-free uniform electric field. Aligned polarizers: T = 42.9°C, ρ = 4.0 - 4.2 MΩ-cm, and t ~20μs.

positive charge injection as in Fig. 11b with maximum charge density of order 1 C/m^3.

Coaxial Cylindrical Electrodes - Space charge shielding can also cause a stress inversion in non-uniform field electrode geometries as shown in Fig. 15 for coaxial cylindrical electrodes (Zahn and McGuire, 1980). In the absence of space charge we have the usual maximum field at the inner cylinder of radius R_i decreasing as 1/r to the outer cylinder of radius R_o. With unipolar injection from the inner cylinder, the field drops at the inner cylinder and increases at the outer cylinder, the area under the curve remaining constant at the applied voltage V. When the

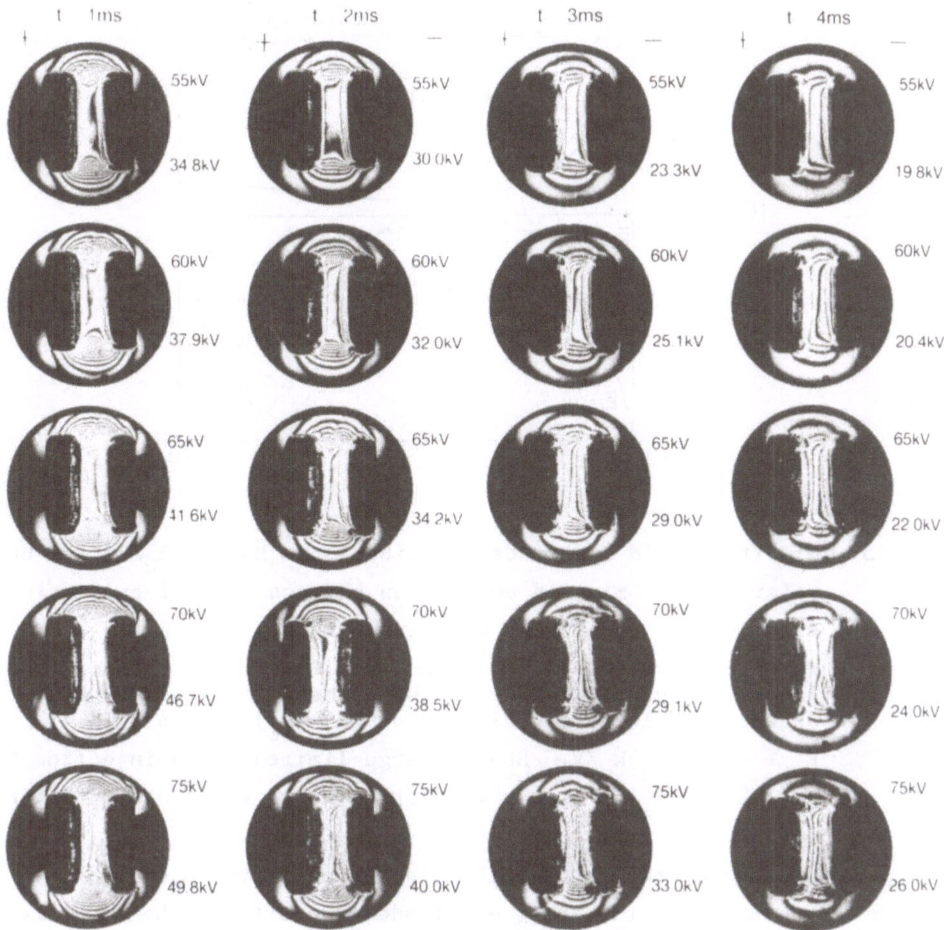

Ethylene Carbonate
Stainless Steel Electrodes

T = 42.9°C $\rho = 3.8 - 4.5$ MΩ-cm

Crossed Polarizers

Fig. 14. The light minima between electrodes with field decrease at the left positive electrode and field increase at the right negative electrode at long times and high voltages indicates significant positive charge injection in ethylene carbonate.

electric field E_i at the inner cylinder drops to $E_iR_o/V = (1-R_i/R_o)$, the electric field is constant across the gap. Any further lowering of the electric field at the inner cylinder results in a stress inversion as the electric field is least at the inner cylinder and increases to the outer cylinder. This occurs when the injected charge is so large that the space charge shielding reduces the electric field due to the sharp geometry. If the unipolar injection is from the outer cylinder, the field drops at the outer cylinder and further increases the field at the inner cylinder. Figure 16 shows Kerr electro-optic field mapping measurements

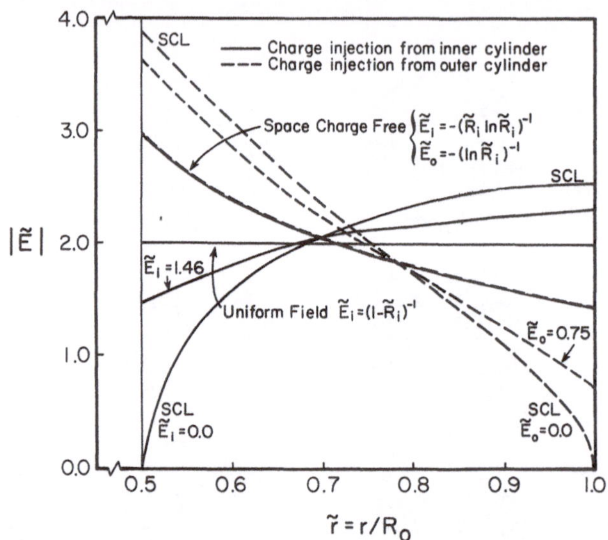

Fig. 15. Electric field magnitude distributions $|\tilde{E}| = |ER_0/V|$ for unipolar drift dominated conduction between coaxial cylindrical electrodes at voltage V with inner radius $\tilde{R}_i = R_i/R_o = 0.5$ with positive charge injection from the inner cylinder (solid lines) or outer cylinder (dashed lines) for various values of emitter electric fields $\tilde{E}_i = E(r = R_i) R_o/V$ or $\tilde{E}_o = E(r = R_o)R_o/V$. Space charge limited (SCL) injection has $E_i = E(r = R_i) = 0$ at $r = R_i$ or $E_o = E(r = R_o) = 0$ at $r = R_o$.

in highly purified nitrobenzene with unipolar positive injection with dc high voltages. When the inner electrode is positive, the electric field drops at the inner cylinder and increases at the outer cylinder. When the outer electrode is positive, the field drops at the outer cylinder and increases at the inner cylinder. For less purified nitrobenzene, Fig. 17 shows that there is an increase in charge injection so that there is a stress inversion with the field least at the positive charge injecting inner cylinder.

Figure 18 shows Kerr effect measurements over the course of a sinusoidal voltage cycle. The left half of photographs in Fig. 18 show the half cycle where the outer cylinder is positive and the field distribution corresponds closely to the 1/r charge-free case. The right half of photographs in Fig. 18 shows the opposite half cycle where the inner cylinder is positive. The electric field is highly uniform due to significant positive charge injection. Such asymmetrical behavior over the course of a sinusoidal cycle leads to harmonic generation.

Weakly Birefringent Dielectrics

High Kerr constant materials such as nitrobenzene or water for reasonable electric fields larger than E_m result in numerous maxima and minima, allowing easy determination of the electric field distribution.

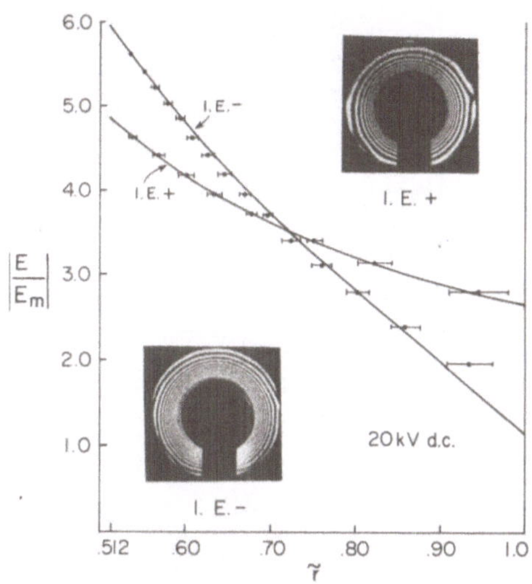

Fig. 16. DC electric field distributions with Kerr effect data showing effects of polarity reversal with inner electrode positive (I.E.+) and inner electrode negative (I.E.-) at 20 kV with highly purified nitrobenzene. Stainless steel cylindrical electrodes have inner radius 0.512 cm and outer radius 1 cm with length 10 cm. The solid lines are obtained from unipolar analysis with a best fit to the data chosen for the electric field at the positive electrode. Note that the weaker electric field regions, evidenced by broader, less densely spaced fringes, are near the outer cylinder. For both polarities increasing the voltage across the cylinders causes the fringes to move outwards and disappear into the outer cylinder. $E_m \approx 9.8$ kV/cm.

If B is so small that E_m exceeds the applied field, only a gray scale pattern results and measurement sensitivity is reduced. Many of our measurements use long electrode lengths L>1 meter to reduce E_m defined in (3) and (7).

With numerous maxima and minima, we used an expanded light beam with photographic film as the light distribution detector. For weak Kerr constant materials we improve measurement sensitivity by using a narrow light beam with a photomultiplier tube (PMT) for field measurement as a function of time at a single position x.

Water/Ethylene Glycol Mixtures - Rotating machine energy storage is being researched as compared to capacitive storage because of the possibility of much higher energy density. However, the slower response of rotating machines requires pulse forming line charging times on the order of milliseconds rather than microseconds, so the line dielectric must have a relaxation time of many milliseconds. This can be accomplished with mixtures of water and ethylene glycol because they maintain a high permittivity ($\varepsilon_{water} \sim 80$, $\varepsilon_{glycol} \sim 40$) with high resistivity especially at low temperatures ~-35°C allowed by the freezing point depression due to glycol (Fenneman, 1982). The dielectric relaxation time of pure water at

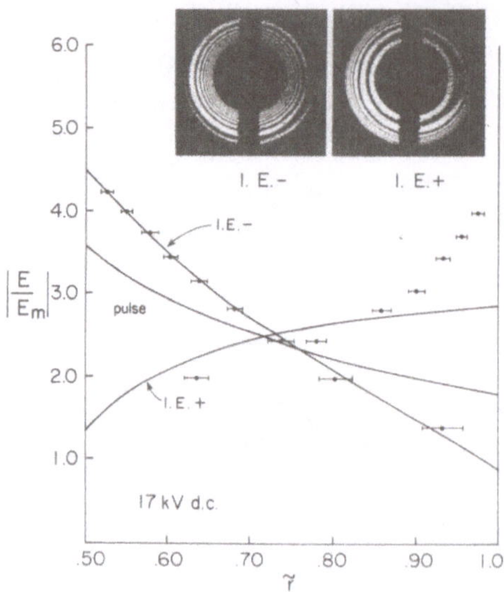

Fig. 17. Electric field distributions with Kerr effect data showing effects of polarity reversal with inner electrode positive (I.E.+) and inner electrode negative (I.E.-) at 17 kV with less purified nitrobenzene. Stainless steel cylindrical electrodes have inner radius 0.5 cm and outer radius 1 cm with length 8.5 cm. The solid lines are obtained from unipolar analysis with a best fit to the data chosen for the electric field at the positive electrode. Note that when the inner cylinder is positive (I.E.+), the region of weaker field strength is near the inner cylinder so that increasing the voltage across the cylinders causes the fringes to collapse upon the inner cylinder. There is a poor fit between theory and measurement when the inner electrode is negative (I.E.-), where the fringes move outwards as the voltage is increased. $E_m \approx 11.5$ kV/cm.

room temperature is about 150 μs which increases to about 650 μs near 0°C. THese values can be increased to as much as 65 ms at -35°C with a 60% glycol/40% water by weight mixture.

However, ethylene glycol has a negative Kerr constant so that water/glycol mixtures have a reduced Kerr constant from pure water. Because these mixtures have a much smaller Kerr constant than pure water, photographic measurements as in Fig. 10 were much less accurate than for pure water alone because of the decreased number of dark fringes. To increase sensitivity, a narrow beam from a cw He-Ne laser at 633 nm wavelength was aimed down the center of a stainless steel parallel electrode geometry system with a photomultiplier tube (PMT) as a detector (Zahn et al., 1984). When a high voltage pulse with about 50 μs risetime was applied, the PMT output went through a series of maxima and minima related by the integer n as given by (9) and (11) and shown in Fig. 19. Because the measurements are made at early time, space charge does not have time to migrate into the dielectric volume to distort the electric field distribution from its uniform value V/d.

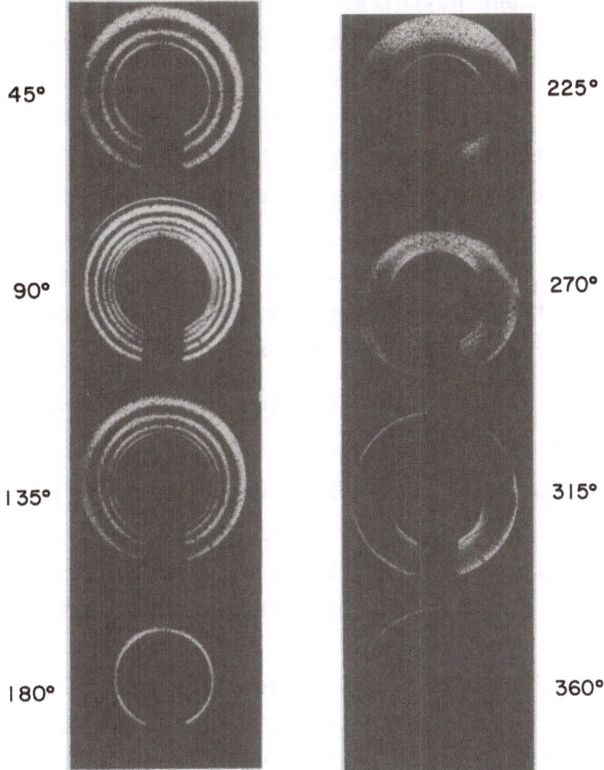

Fig. 18. Kerr effect data taken over the course of a 40 Hz, 14 kV peak sinusoidal voltage using the cylindrical electrodes described in Fig. 16. From 0° to 180° the inner cylinder is negative with very little charge injection, while from 180° to 360° the inner cylinder is positive with significant charge injection causing an essentially uniform electric field as evidenced by the lack of fringes. $E_m \approx 9.8$ kV/cm.

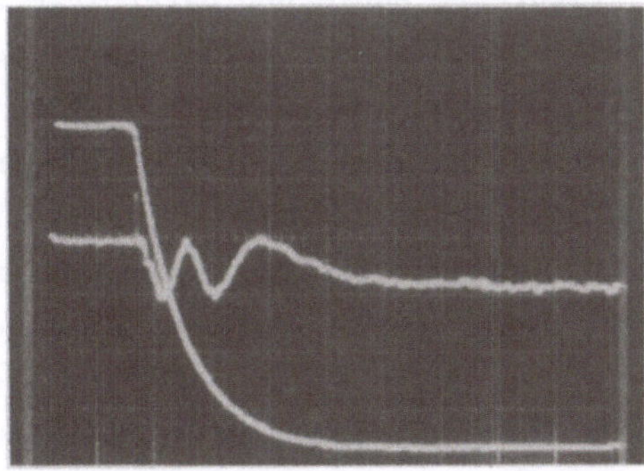

Fig. 19. The output of the photomultiplier tube goes through a series of maxima and minima (n = 1 to 4) when a 100 kV negative high voltage pulse is applied to 1.1 m long parallel plate stainless steel electrodes here shown for 77% water/23% ethylene glycol by weight at 2.7°C with crossed polarizers. Voltage waveform 17 kV/cm, PMT waveform 100 mV/cm, 20 μs/cm. From this measurement, the Kerr constant of the mixture at 633 nm wavelength is $B \approx 2.2 \times 10^{-14}$ m/V^2 with $E_m \approx 45.6$ kV/cm.

For various ethylene glycol/water mixtures, the applied field was increased from 60 kV/cm to 120 kV/cm in 5 kV/cm steps using aligned and crossed polarizers. The large number of maxima and minima described by integers were each related to E_m using (9) and then B is calculated using (7). Table 2 lists E_m and B and their standard deviations σ for all measured concentrations and temperatures. Because the temperature range was from -2.1°C to 20.0°C, Table 2 also lists B values normalized to temperature $B(T/T_o)^2$ using $T_o = 273°K$ as a reference and plotted in Fig. 20. This temperature dependence is used because it is the dominant effect in (2) from simple molecular theory.

Note in Fig. 20 that the Kerr constant is about zero with a mixture 79% glycol/21% water by weight.

Transformer Oils - Figure 21 shows representative high voltage pulse waveforms across bare brass electrodes of length L = 1.06 m with a 1-cm gap of mineral transformer oil and Silicone Transformer Fluid and the resulting PMT output for aligned polarizers with λ = 633 nm light from a cw He-Ne laser. For zero voltage, the light intensity is maximum so the PMT signal is at minimum negative voltage. Increasing voltage decreases the transmitted light intensity so the PMT signal decreases towards ground. At V≃111 kV, the PMT signal is zero for mineral oil so that $E_m \simeq 111$ kV/cm and $B \simeq 3.8 \times 10^{-15}$ m/V^2. For higher voltages, the PMT signal reaches ground during the voltage rise and then increases again towards the next maxima. A proposed retro-fill fluid for PCB-contaminated transformers is silicone oil. The small PMT signal in Fig. 21b should be compared to transformer oil. The Kerr constant of silicone oil is about 1/8 that of transformer oil. A single PMT measurement only allows measurement of the electric field at one point and not the electric field spatial distribution. Figure 22 shows photographic measurement of the electric field distribution with mineral oil. Here the right brass electrode is covered with creped paper held on with Vetak glue, materials used as insulation between transformer conductor windings. At the highest voltage before breakdown (~157 kV/cm), we see two fringes of light minima. We note that the light intensity in the central region between electrodes is uniform at early time indicating the electric field is also uniform. Space charge effects create a non-uniform electric field and thus a non-uniform light distribution as shown after 40 ms in Fig. 22.

Sulfur Hexafluoride (SF_6) - Sulfur hexafluoride (SF_6) is the primary insulating gas used in compact high voltage apparatus. The Kerr cell and optical measurement system with photomultiplier tube (PMT) detector are presented schematically in Fig. 23 (Carreras and Zahn, 1986). The Kerr cell is a cylindrical stainless steel pressure chamber with an internal

Table 2. Calculated values of electric field magnitude E_m necessary for the first minima with aligned polarizers or the first maxima with crossed polarizers and the resulting values of the Kerr constant B for various water/ethylene glycol mixture ratios at various temperatures.

Ratio by Weight H_2O %Wt	Glycol %Wt	Temp T(°C)	$E_m \pm \sigma$ (10^6V/m)	$B \pm \sigma$ (10^{-15}m/V^2)	$B(T/T_o)^2$ (10^{-15}m/V^2) T_o=273°K
100	0	+10.3	4.06±0.07	+27.8±1.0	+29.9±1.1
90	10	+3.7 to +4.1	4.20±0.06	+26.0±0.7	+26.7±0.7
82	18	+3.3 to +4.3	4.44±0.07	+23.2±0.8	+23.9±0.8
77	23	+2.5 to +2.8	4.56±0.06	+22.0±0.6	+22.4±0.6
63	37	-0.7	5.05±0.07	+18.0±0.5	+17.9±0.5
42	58	+0.8 to +0.9	6.37±0.10	+11.3±0.3	+11.4±0.3
40	60	+0.2 to +0.3	6.58±0.08	+10.6±0.2	+10.6±0.2
34	66	-1.4 to -0.5	7.89±0.11	+7.4±0.2	+7.3±0.2
30	70	+1.3 to +1.7	9.44±0.15	+5.1±0.2	+5.2±0.2
27	73	-1.7 to +0.1	12.49±0.54	+3.0±0.3	+2.9±0.3
24	76	-0.1	15.30±0.20	+2.0±0.1	+2.0±0.1
21	79	+0.7 to +1.4	>28	<|0.6|	<|0.6|
18	82	-2.1 to -0.3	~20.0	~1.2	~-1.2
15	85	-0.1 to +1.2	12.35 to 12.48	-2.9 to -3.0	-2.9 to -3.0
9	91	-0.5 to +1.1	9.14±0.11	-5.5±0.1	-5.5±0.1
4	96	-1.1 to +1.7	7.77±0.19	-7.6±0.4	-7.6±0.4
4	96	+9.8 to +10.5	8.36±0.23	-6.6±0.4	-7.1±0.4
4	96	+19.4 to +20.0	8.79±0.08	-5.9±0.1	-6.8±0.1
0	100	+10.5 to +11.7	7.40±0.20	-8.4±0.5	-9.1±0.5

coaxial arrangement. The inner diameter of the outer cylinder is 2.54 cm. The inner electrode is a solid stainless steel rod glued to a half round solid length of PMMA. The rod is 1.27 cm in diameter for liquid SF_6 measurements and 0.79 cm for gaseous SF_6 measurements. Both the inner electrode and PMMA are L = 1.2 meters long, to be used in (3) to calculate the Kerr constant B. At one end is an insulating high voltage bushing constructed of polycarbonate. This plastic bushing must withstand the high pressures of the SF_6 and insulate the pulsed voltage to 150 kV that is fed to the inner electrode. At each end of the chamber are half-inch thick quartz windows. The complete pressure cell was hydrostatically tested at ~600 psi and operates at ~300-350 psi. The SF_6 is obtained from a cylinder that is inverted to facilitate the access of liquid SF_6. The He-Ne beam is initially linearly polarized at 45° from

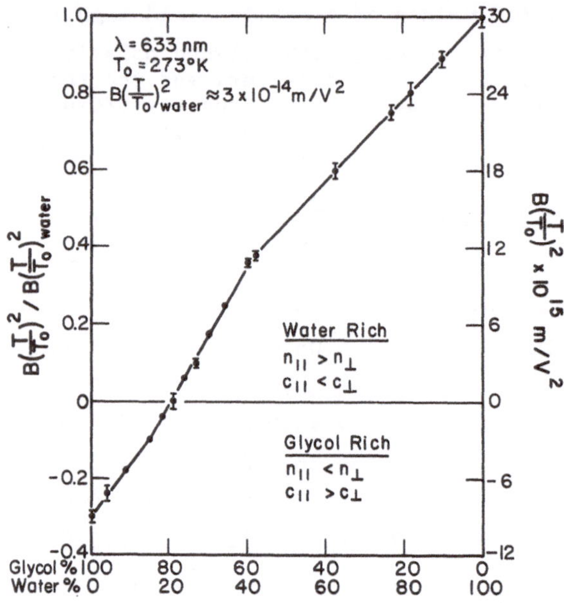

Fig. 20. Kerr constant at λ = 633 nm of ethylene glycol/water mixtures as a function of mixture ratio by weight. Note that water has a positive Kerr constant while glycol has a negative Kerr constant. The Kerr constant is zero at 79% glycol/21% water by weight.

the local electric field at the beam spot and the analyzer is crossed 90° from the initial laser polarization. Mechanical birefringence is introduced by the high pressure stressing the quartz windows so that the light phase is increased over (3) by a value δ_m. The light intensity for crossed and aligned polarizers is then given by

$$\frac{I}{I_m} = \begin{cases} \sin^2\left[\frac{\pi}{2}\left(\frac{E}{E_m}\right)^2 + \frac{\delta_m}{2}\right] & \text{Crossed Polarizers} \\ \cos^2\left[\frac{\pi}{2}\left(\frac{E}{E_m}\right)^2 + \frac{\delta_m}{2}\right] & \text{Aligned Polarizers .} \end{cases} \quad (15)$$

A quarter wave plate is placed before the analyzer to compensate for the mechanical birefringence by introducing a phase shift of $-\delta_m$ by placing its fast axis at angle of 45° + $(\delta_m/2)$ with respect to the incident polarization. A photomultiplier tube, connected directly to the 1 MΩ input resistance of the scope, is used as the detector. A Marx generator is used to supply the high voltage as a step pulse input. All of the presented measurements in Fig. 24 were conducted with crossed polarizers and a negative high voltage input step pulse on the inner electrode with He-Ne laser beam placement shown in Fig. 23.

The Kerr constants of gaseous SF_6 for low densities, < 0.071 g/cm^3 have been reported and are on the order of 10^{-18} - 10^{-19} m/V^2 [Buckingham

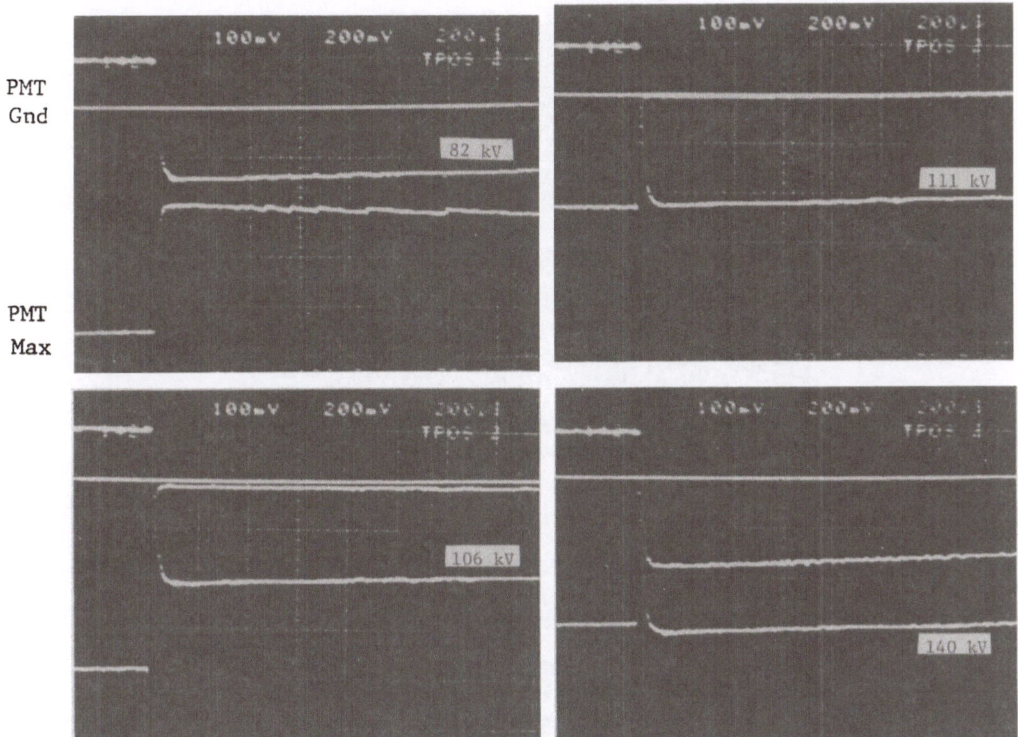

(a) Transformer Oil

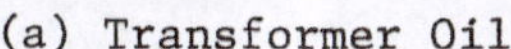

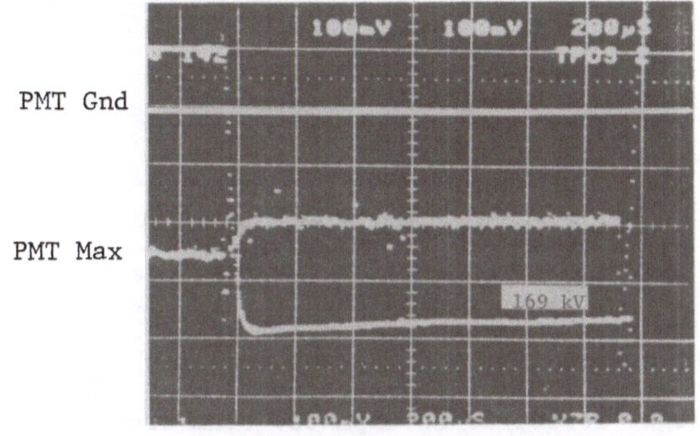

(b) Silicon Oil

Fig. 21. (a) PMT measurements with aligned polarizers in Gulf Transcrest H transformer oil using L = 1.06 mm long bare brass electrodes with 1 cm gap for a narrow beam (0.9 mm diameter) cw He-Ne laser light at 633 nm down the center. The upper waveforms show the pulsed negative high voltage with the lower waveforms being the PMTs output heading towards PMT ground shown as the upper solid line in each oscilloscope picture. Time scale is 200 µs/div. Measurements show that E_m~111 kV/cm and B~3.8 x 10^{-15} m/V^2.

(b) PMT measurement with aligned polarizers in Dow Corning 561 Silicone Transformer Fluid using L = 1.06 m long bare brass electrodes with 1 cm gap for cw He-Ne laser light at 633 nm. The upper waveform shows the large negative high voltage with the lower waveform being the PMT output heading towards PMT ground shown as the upper solid line. Time scale is 200 µs/div. Measurements show that E_m~300 kV/cm and B~0.5 x 10^{-15} m/V^2.

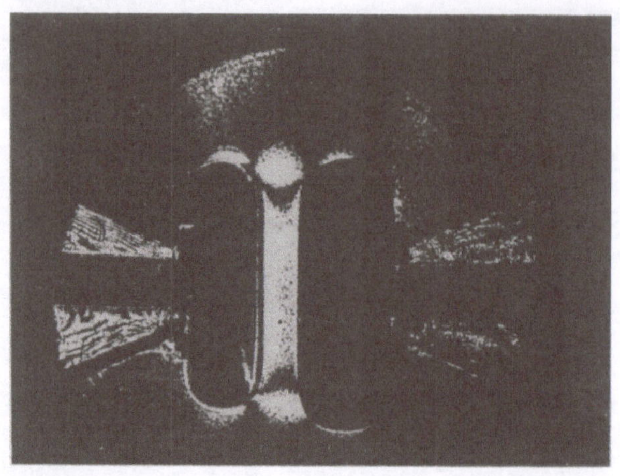

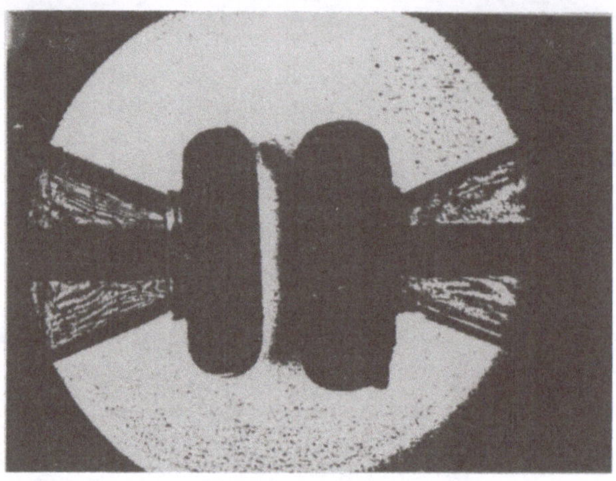

$t=40$ ms, $v=59$ kV, $v_m=87$ kV

Fig. 22. Kerr electro-optic field mapping measurements using photographic film with aligned and crossed polarizers in Gulf Transcrest H transformer oil using L = 1.06 m long brass electrodes of ~6mm gap with the right positive electrode covered with creped paper held down with Vetak glue. The light source is an expanded beam from a pulsed tunable dye laser of ~200 ns time duration at free space wavelength of λ = 590 nm. The light intensity in the central inter-electrode region is uniform at early time, but becomes non-uniform at later times due to charge injection. The maximum pulse voltage is v_m while the instantaneous voltage is v.

and Dunmur, 1968; Buckingham et al., 1970]. To get the strongest signal possible, liquid SF_6 was first used which had Kerr constant $B \approx 10^{-15}$ m/V^2. By lowering the pressure, the SF_6 became gaseous with reduced Kerr constants as pressure decreased: B(295 psi, density = 0.1885 g/cm^3) = 4.4×10^{-16} m/V^2; B (270 psi, density = 0.1686 g/cm^3) = 1.4×10^{-16} m/V^2, and B (240 psi, density = 0.1456 g/cm^3) = 7.1×10^{-17} m/V^2. PMT signals of these measurements are shown in Fig. 24.

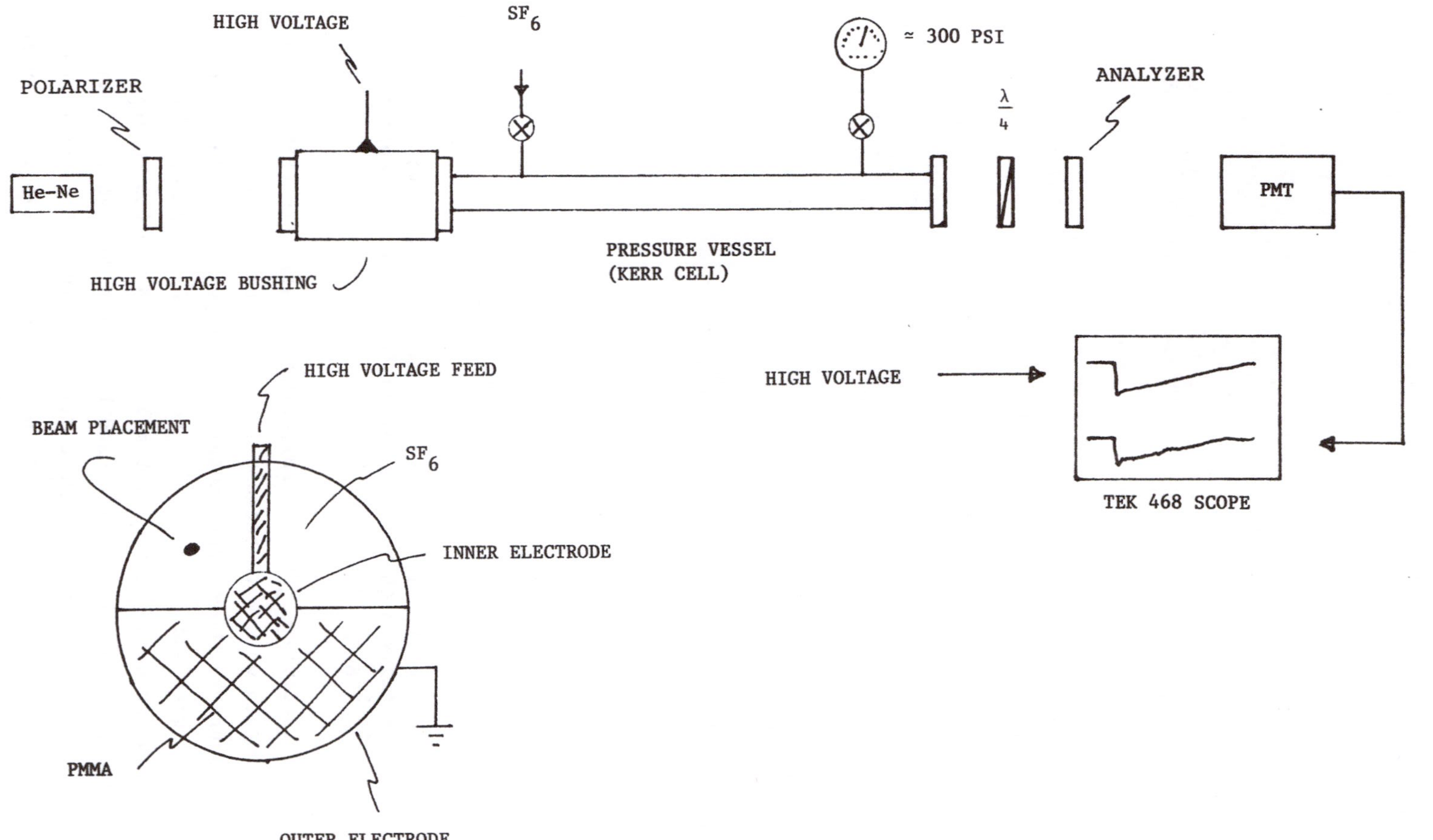

Fig. 23. Kerr electro-optic measurement system used with liquid and gaseous SF_6. The single quarter wave plate is used to compensate for the mechanical birefringence in the windows.

0.1 V/DIV

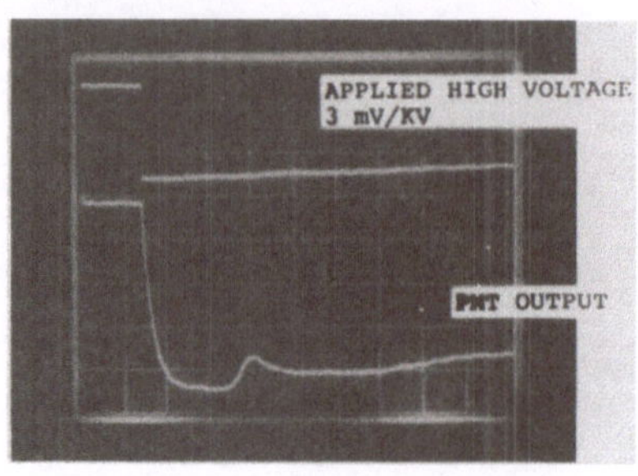

5.0 V/DIV T = 1 ms/DIV

LIQUID SF_6 @ ROOM TEMP.

-71 KV APPLIED VOLTAGE

0.2 V/DIV

5.0 V/DIV T = 1 ms/DIV

GASEOUS SF_6 @ 295 PSI., 18.9 °C 188.5 g/l

-130 KV APPLIED VOLTAGE

0.2 V/DIV

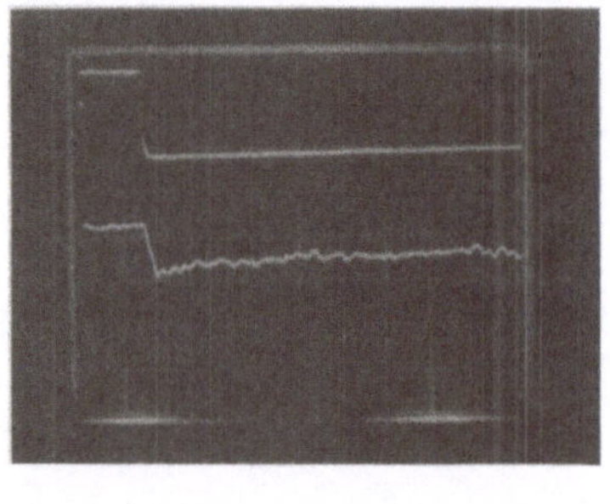

5.0 V/DIV T = 1 ms/DIV

GASEOUS SF_6 @ 270 PSI., 18.6 °C 168.6 g/l

-128 KV APPLIED VOLTAGE

0.2 V/DIV

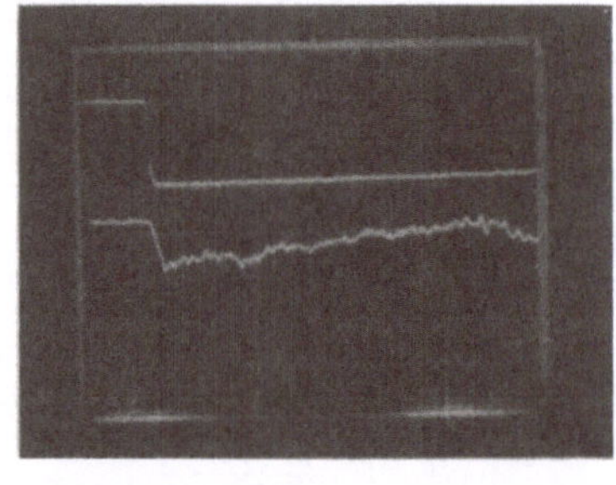

2.0 V/DIV T = 1 ms/DIV

GASEOUS SF_6 @ 240 PSI., 18.4 °C 145.6 g/l

-128 KV APPLIED VOLTAGE

TEK 468 SCOPE TRACES OF APPLIED HIGH VOLTAGE AND PHOTO MULTIPLIER TUBE OUTPUT FOR LIQUID SF_6 AND THREE DENSITIES OF GASEOUS SF_6.

Fig. 24. Photomultiplier tube responses (lower traces) for negative high voltage pulses (upper traces) applied to SF_6 at various pressures.

Polymethylmethacrylate (PMMA) - Polymethylmethacrylate (PMMA) is a convenient polymer to work with because it is transparent at room temperature. Figure 25 shows representative photomultiplier tube (PMT) responses with aligned and crossed polarizers with He-Ne laser light at 633 nm wavelength for a sample L = 1.09 m long and d = 0.635 cm thick between parallel-plane electrodes stressed by a high voltage pulse V(t). With no electric field (E = 0), separate measurements with aligned and crossed polarizers allowed determination of I_m and δ_m in (15). Early in the pulse V(t), space charge effects are negligible and the electric field in the central region is uniform at E = V(t)/d. With crossed and aligned polarizers we then measured the voltage V_m required for the first

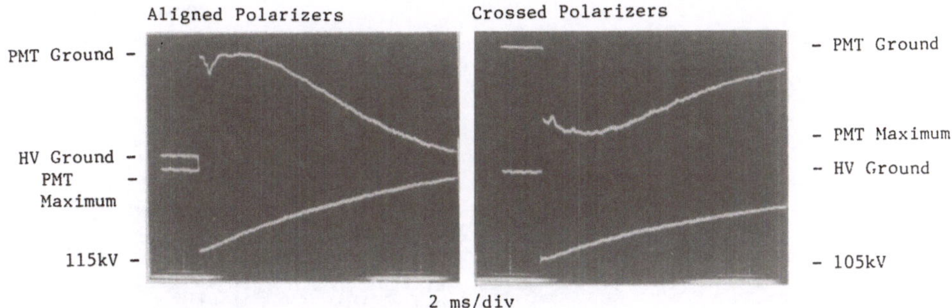

Fig. 25. Photomultiplier tube (PMT) outputs (upper traces) for 633 nm He-Ne light with negative high voltage pulses (lower traces) applied to a PMMA sample 1.09 m long and 0.635 cm thick with aligned and crossed polarizers. From these measurements, $E_m \simeq 152$-156 kV/cm so that $B \simeq 2 \times 10^{-15}$ m/V^2.

light maximum (crossed polarizers) or minimum (aligned polarizers) so that the total phase δ for either aligned or crossed polarizers for the trigonometric functions in (15) is $\pi/2$. For this long sample, we found V_m to be ~96.5 to 99 kV so that $E_m \simeq V_m/d \simeq 152$ to 156 kV/cm. From (7) we then obtained $B \simeq 2 \times 10^{-15}$ m/V^2 (Zahn et al., 1987). Note in Fig. 25 that the PMT ground is at the top of the oscilloscope picture and that increasing light gives a more negative signal. The high voltage has negative polarity. In these pictures mechanical birefringence has been compensated for using a quarter wave plate so that the net δ_m is negligibly small. Before voltage is applied, the PMT response is zero for crossed polarizers and maximum for aligned polarizers. Since the peak voltage amplitude exceeds V_m, with aligned polarizers the PMT response goes through minimum and starts to increase to the next maximum before the collapsing voltage again reaches V_m, returning the PMT output to a broad zero before returning to maximum as the applied voltage goes to zero. With crossed polarizers, the PMT output is zero before voltage is applied, rising briefly past the first maximum towards a minimum as $V(t) > V_m$ and then quickly returning to a broad maximum as the collapsing voltage passes through V_m again towards zero. Longer time electric field measurements to see space charge effects were taken in PMMA at the top, middle, and bottom of the sample. The PMT data in Fig. 26 for aligned and crossed polarizers with copper electrodes shows different light signals at the electrodes and in the middle of the sample. Figure 27 uses (15) to calculate the electric field from the measured data in Fig. 26. Figure 27 shows an electric field reduction at the positive electrode and field enhancement at the negative electrode with small changes in the middle. These results are consistent with positive charge injection as in Fig. 11b. The steady state electric field distribution

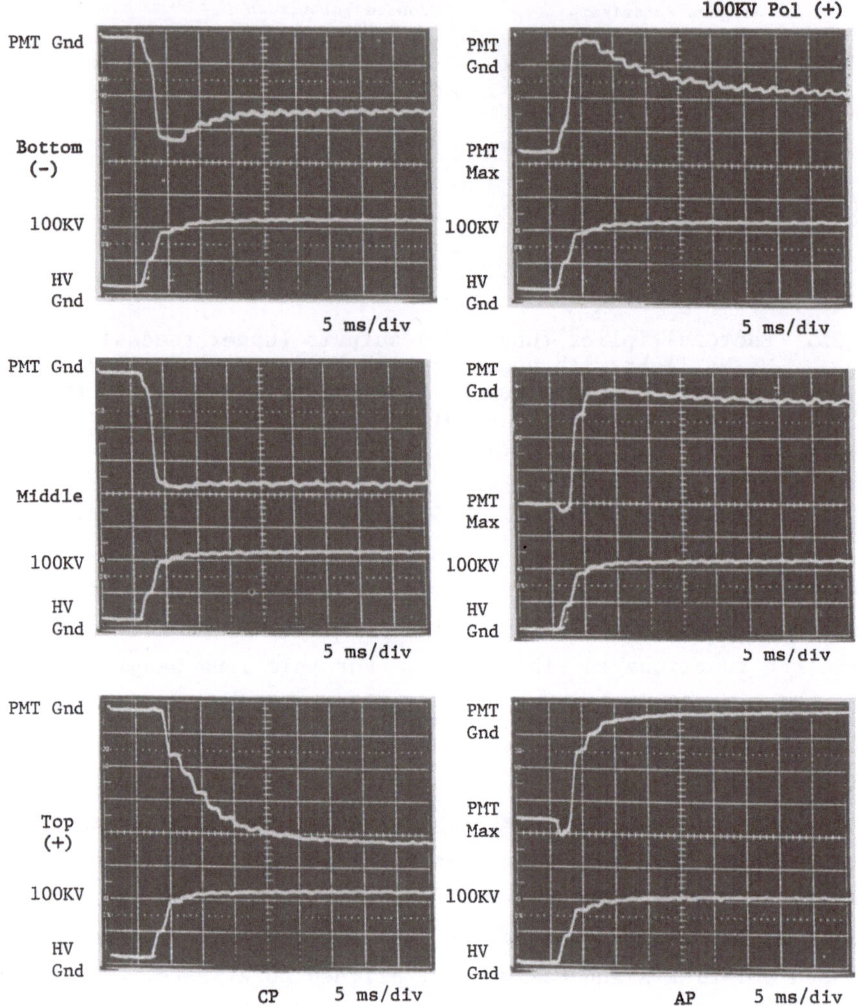

Fig. 26. Oscilloscope traces of PMT output and stepped positive high voltage for PMMA Kerr effect measurements with 100 kV charging voltage across copper electrodes with d = 0.635 cm gap. Measurements are shown for crossed and aligned polarizers to be different at the top (+), middle, and bottom (-) of the PMMA dielectric because of positive space charge injection.

can be approximated by a linear equation for a uniform volume charge distribution q_o:

$$E = q_o(x - d/2)/\varepsilon + V_o/d , \tag{16}$$

where the average field value is V_o/d. Figure 27 shows the electric field at the x = 0 positive injecting electrode, E(x = 0), is of order 90% of the average field value, and the electric field at the x = d non-injecting negative electrode, E(x = d), is about 110%. These field values at x = 0 and x = d can be used to solve for q_o:

$$q_o = \varepsilon[E(x = d) - E(x = 0)]/d \simeq 1.6 \times 10^{-2} \ C/m^3 , \tag{17}$$

where the relative dielectric constant for PMMA is 3.7.

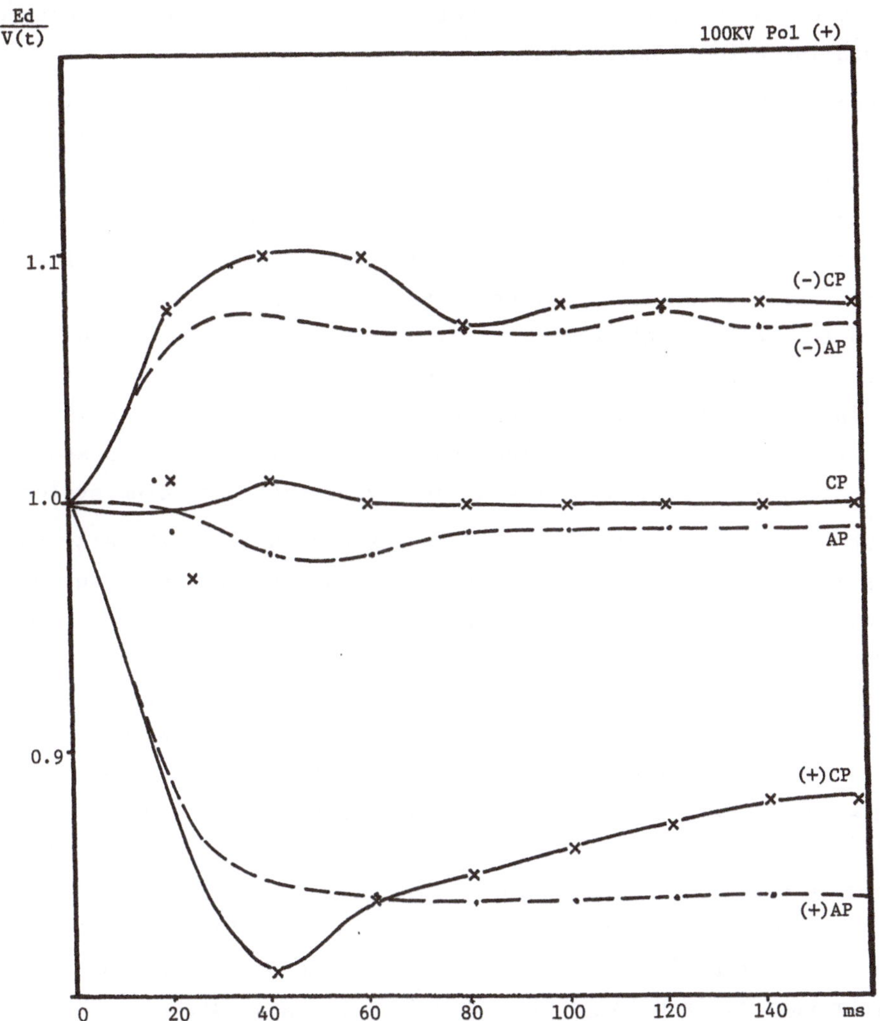

Fig. 27. Computed electric field time variations at the top (+), middle, and bottom (-) of the PMMA dielectric as calculated from the PMT measurements in Fig. 26.

Electron Beam Irradiation - These Kerr electro-optic field mapping measurements in high voltage stressed PMMA had a maximum applied electric field of the order of 160 kV/cm. The measurement required the use of a photomultiplier tube as a light detector because of the small Kerr constant of PMMA. Other work uses an electron beam to irradiate PMMA samples where the accumulated trapped charge leads to higher internal self-electric fields up to 3.5 MV/cm so that for a 10 cm-long PMMA sample, numerous light minimum and maximum arise, making photographic and videotape measurements possible (Zahn et al., 1987).

The motivation for this work is in part due to understanding electron-caused discharges along insulating surfaces on spacecraft. It is hoped that Kerr electro-optic field mapping measurements will allow the development and verification of models leading to a better understanding of radiation effects on solids.

Once we have determined that $B \approx 2 \times 10^{-15}$ m/V^2 from the measurements of Fig. 25, we find E_m for our shorter electron beam irradiated samples to be $E_m \approx 0.5$ MV/cm for L = 10.2 cm and $E_m \approx 0.44$ MV/cm for L = 12.7 cm. The linear polariscope configuration in Fig. 28 has incident light in the z-direction polarized at ±45° to the direction of the electric field E with an analyzing polarizer placed after the sample either crossed or aligned to the incident polarization. The He-Ne laser at 633 nm wavelength has its beam expanded to ~7.5 cm to allow measurements of the light intensity distribution over the entire sample cross section. Beam splitters allow simultaneous measurements for aligned and crossed polarizers using Polaroid cameras as well as a videotape recording system.

The electron beam is generated by a Van de Graaff generator and exists from the accelerator tube through a thin (76 μm) aluminum window and passes through ~50 cm air to the PMMA sample which is short-circuited through current monitors at the top and bottom surfaces. The energy loss in the window and intervening air is about 160 keV. Figure 28 shows representative data near the cameras for an accelerator beam energy of 2.6 MeV and a current density of 20 nA/cm^2 after ~60 s of beam irradiation for aligned and crossed polarizers for a d~1.27 cm thick PMMA sample uniformly irradiated over its width 5.1 cm and length 10.2 cm. The sample had E_m~0.5 MV/cm. The sample is shown before irradiation to have some mechanical birefringence. Also shown in Fig. 28 is a representative frame of light intensity distribution of maximum and minimum from our computerized image digitizer for a larger current density of 110 nA/cm^2 after 11 s of irradiation. This sample had a length of 12.7 cm so that $E_m \approx 0.44$ MV/cm.

Figure 29 shows the non-uniform optical pattern on film due to mechanical birefringence with no field and the Kerr measurement after the short-circuited sample is irradiated with high-energy electrons. Whereas the peak electric field in Fig. 25 is ~160 kV/cm, the peak field due to the trapped electrons in Fig. 27 is ~1.5 MV/cm, which led to the spark discharge treeing pattern shown.

Figure 29 also shows how we manually read the photographs to determine the electric field distribution. We plot the mechanical birefringence $\delta_m(x,y = 0)$ before irradiation which for simplicity we take to be box-like, either 0° or 90°, neglecting the gray scale smooth transition between maximum and minimum. We also plot the total birefringence from the Kerr effect photograph using the center of the dark minimum lines and the center of the light region between dark lines. To

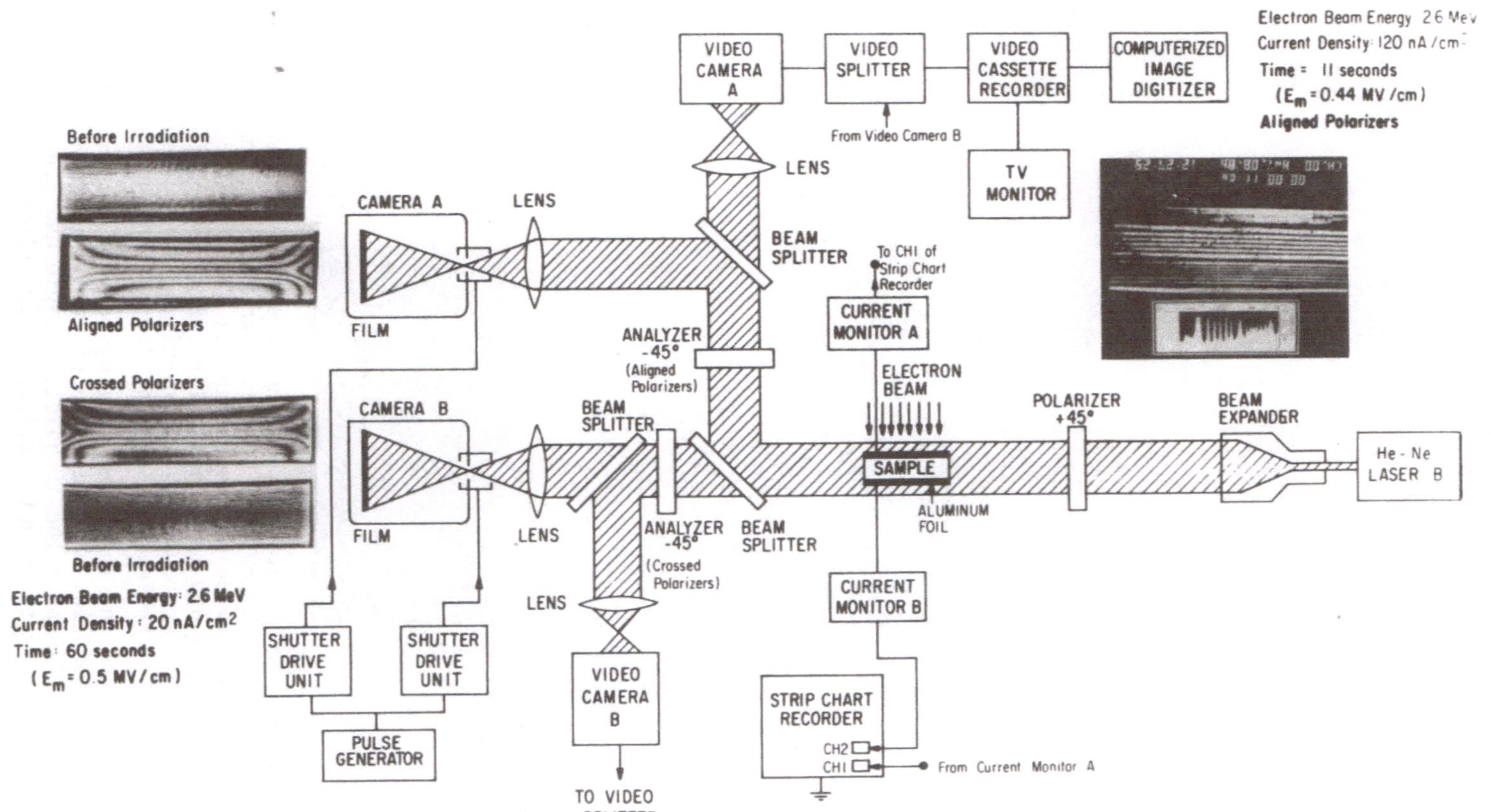

Fig. 28. Apparatus and representative data for Kerr electro-optic field mapping measurements with simultaneous aligned and crossed polarizers in electron beam irradiated samples using photographic film and a computer interfaced videotape recording system as detectors.

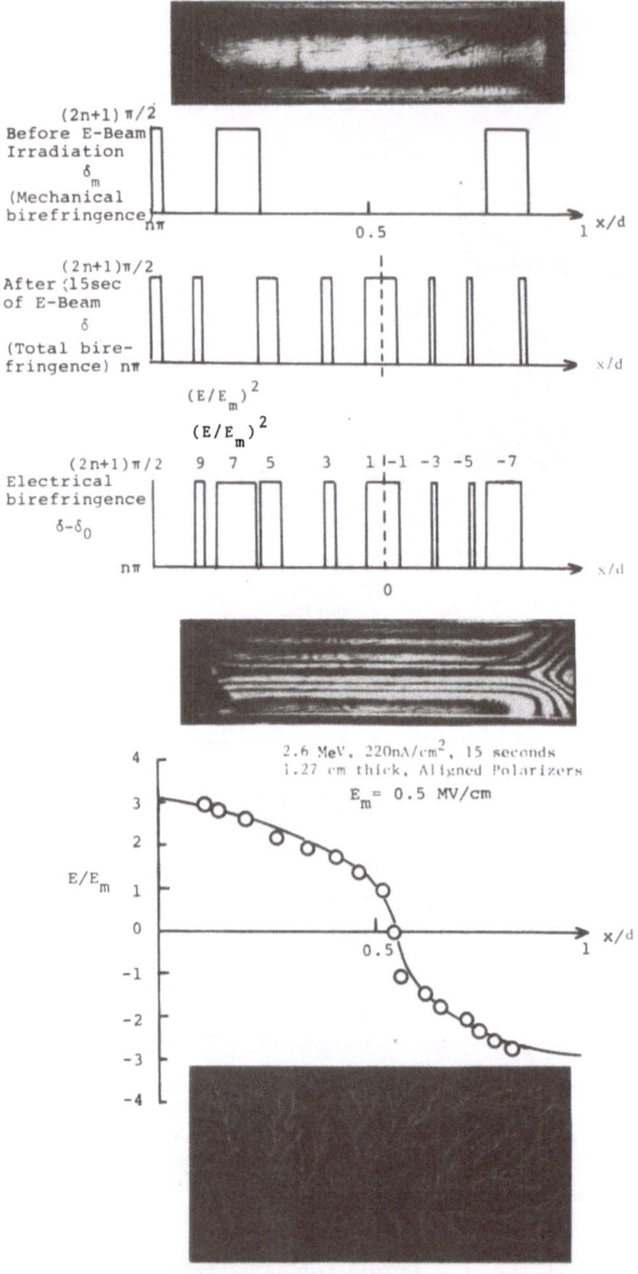

Fig. 29. Kerr electro-optic field mapping measurements in electron-beam irradiated short-circuited PMMA, 1.27-cm thick, using aligned polarizers with cw He-Ne laser light at 633 nm. Shown are the optical patterns due to mechanical birefringence before application of the electron beam, the resulting Kerr effect pattern after 15 s of irradiation at energy 2.6 MeV and current density 220 nA/cm^2, and the resulting treeing pattern after electrical breakdown. Box-like approximation of the total birefringence for the dark lines in the irradiated sample and the graphical subtraction to get the net electrical birefringence when $E/E_m = \sqrt{n}$, with $E_m \simeq 0.5$ MV/cm is shown. The resulting electric field distribution has zero average field with peak fields at the boundaries.

obtain the electric birefringence, we graphically subtract. We then calculate the electric field distribution as plotted. Note that the short circuit imposes the constraint of zero average electric field. The electric field is thus oppositely directed on either side of the zero-field point.

In the Kerr effect pattern of Fig. 29, the $n = 0$ and $n = \pm 1$ lines have coalesced in the central region. This obscures the $n = 0$ bright line for a zero field with aligned polarizers. Within the sensitivity of the optical measurement, the electric field is discontinuous across these coincident lines, so the charge at this position is effectively a sheet of surface charge with charge density $|\sigma/\varepsilon E_m| = 2.0$.

SPACE CHARGE EFFECTS IN HIGHLY PURIFIED WATER

Metallic electrodes differ in the magnitude and sign of injected charge where for highly purified water 304 stainless steel and ETP copper (with 0.04% oxygen) electrodes generally inject positive charge, 2024 aluminum injects negative charge, while standard brass can inject either positive or negative charge. Thus, by appropriate choice of electrode material combinations and voltage polarity, it is possible to have uncharged liquid, unipolar charged negative or positive, or bipolar charged liquid (Zahn et al., 1985).

Uniform Electric Field

All electrode material combinations have no space charge at early time after a high voltage pulse is applied, as it takes about 500 μs for the charge to migrate a significant distance from the electrodes as shown in Fig. 30. Kerr effect measurements then have a uniform electric field and thus uniform light distribution in the central gap as in Fig. 11a with no dark lines.

However, even at long times, Fig. 30 also shows no space charge with the negative electrode either stainless steel or copper with aluminum as the positive electrode.

Unipolar Charge Injection

Gauss's law of (14) and Fig. 11b shows that unipolar positive charge injection results in an electric field that everywhere has a positive slope while unipolar negative charge has an electric field everywhere with negative slope. For unipolar injection, the electric field is thus decreased at the charge injecting electrode, but is increased at the non-charge injecting electrode because the average electric field must always be V/l.

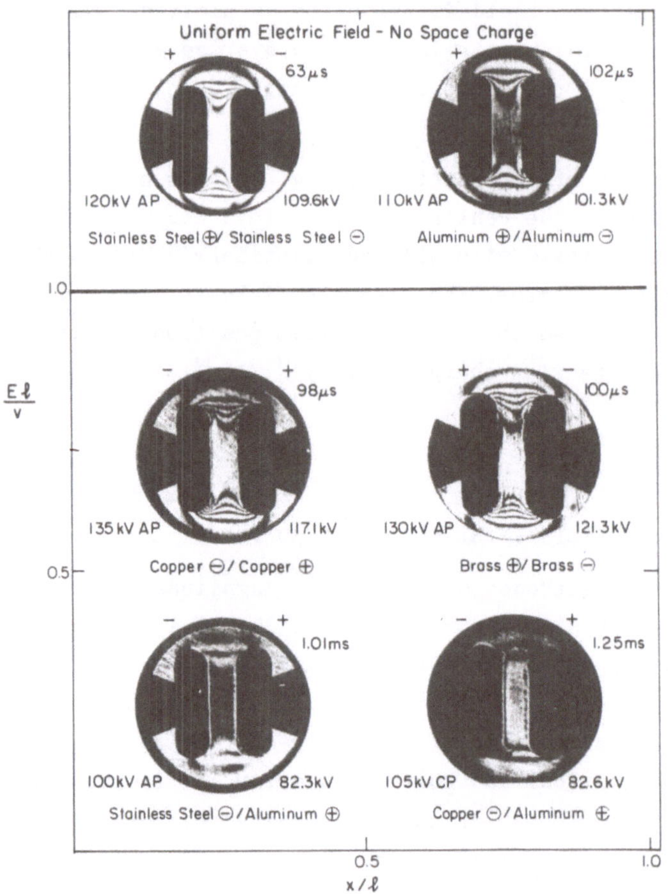

Fig. 30. For no space charge, the electric field distribution is uniform at V/l, where V is the instantaneous voltage and l = 1 cm is the electrode spacing, so that the Kerr effect shows uniform light transmission in the interelectrode gap region for all electrode combinations in highly purified water. $E_m \simeq 35\text{-}36$ kV/cm. In each photograph of Figs. 30 through 32, the initial charging voltage and whether aligned (AP) or crossed (CP) polarizers are used are given at the lower left, the instantaneous voltage is given at the lower right, and the time after high voltage is applied is given in the upper right.

Positive Charge Injection - Positive charge is injected from the positive electrode from a pair of stainless steel or copper electrodes as in Fig. 31. The same is true if copper is the positive electrode with stainless steel negative. However, there is an anomaly when stainless steel is positive and copper is negative. From our first measurements, we would expect this configuration to also have positive injection from the stainless steel electrode, but in fact it is bipolar homocharge, as shown in Fig. 32. There, the positive charge injection

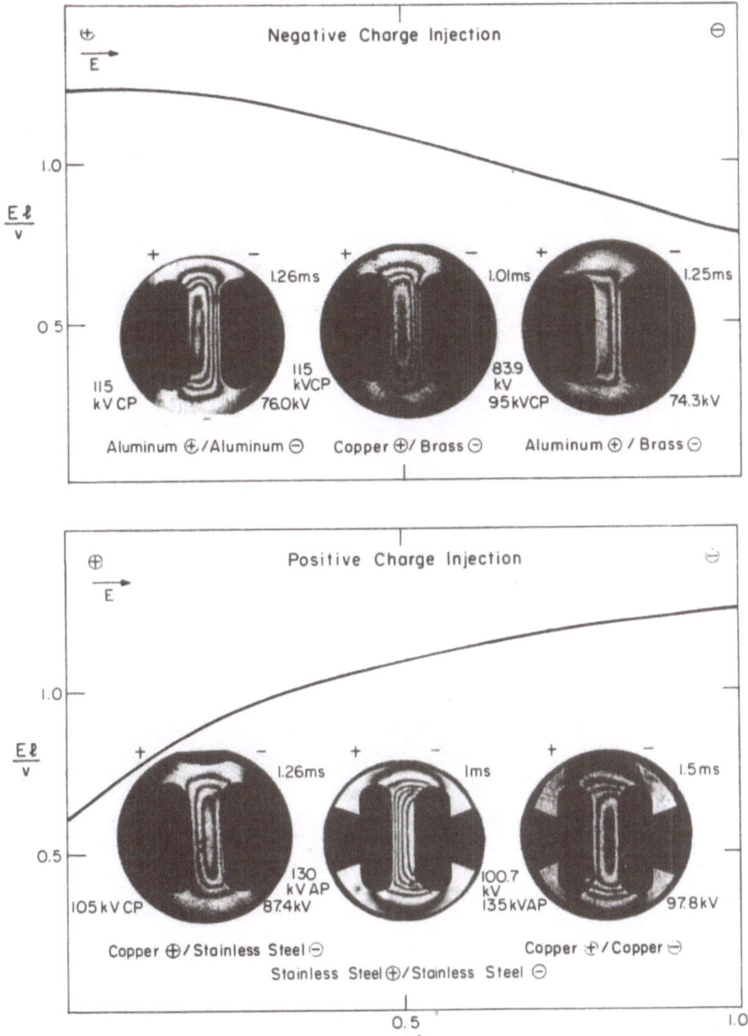

Fig. 31. The slope of the electric field distribution is the same as the sign of the injected charge. The electric field near the charge injecting electrode is decreased because of the space charge shielding while the field is increased at the non-charge injecting electrode, keeping the average field value constant at V/l, shown for various metallic electrode combinations in highly purified water.

from the stainless steel electrode gives clear fringes, while the negative charge injection from the copper electrode causes electrohydrodynamic motions that blur and smear the fringes together. Injected charge sometimes leads to this fluid turbulence, which for planar electrodes often persists only for a short time interval and then disappears.

The positive space charge injection from a pair of parallel-plate stainless steel electrodes as a function of time is shown in Fig. 33 to not exhibit any fluid turbulence. At early times, $t < 250\ \mu s$, the

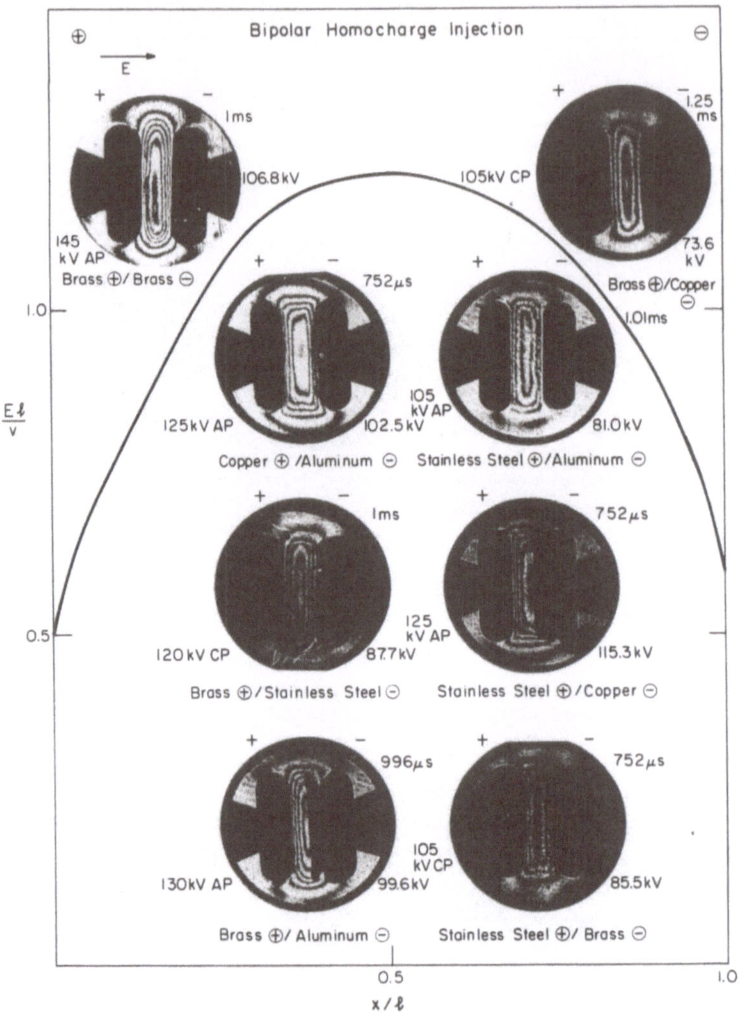

Fig. 32. With bipolar homocharge injection, the electric field is decreased at both charge injecting electrodes, with the peak field in the central gap region with average electric field V/l. The space charge shielding at the electrodes increases the voltage breakdown strength by up to 40% in water.

interelectrode light intensity distribution is uniform, indicating a uniform electric field. At t~500 μs, a number of dark fringes appear near the positive electrode. The light distribution is uniform in front of the fringes with the electric field decreasing back to the positive electrode. A weak field near the positive electrode and stronger field near the negative electrode from Fig. 11b indicates a net positive space charge distribution near the positive electrode with zero space charge in the uniform field region. At later times, the charge front moves towards the negative electrode. The electric field distribution at the various times in Fig. 33 is plotted in Fig. 34.

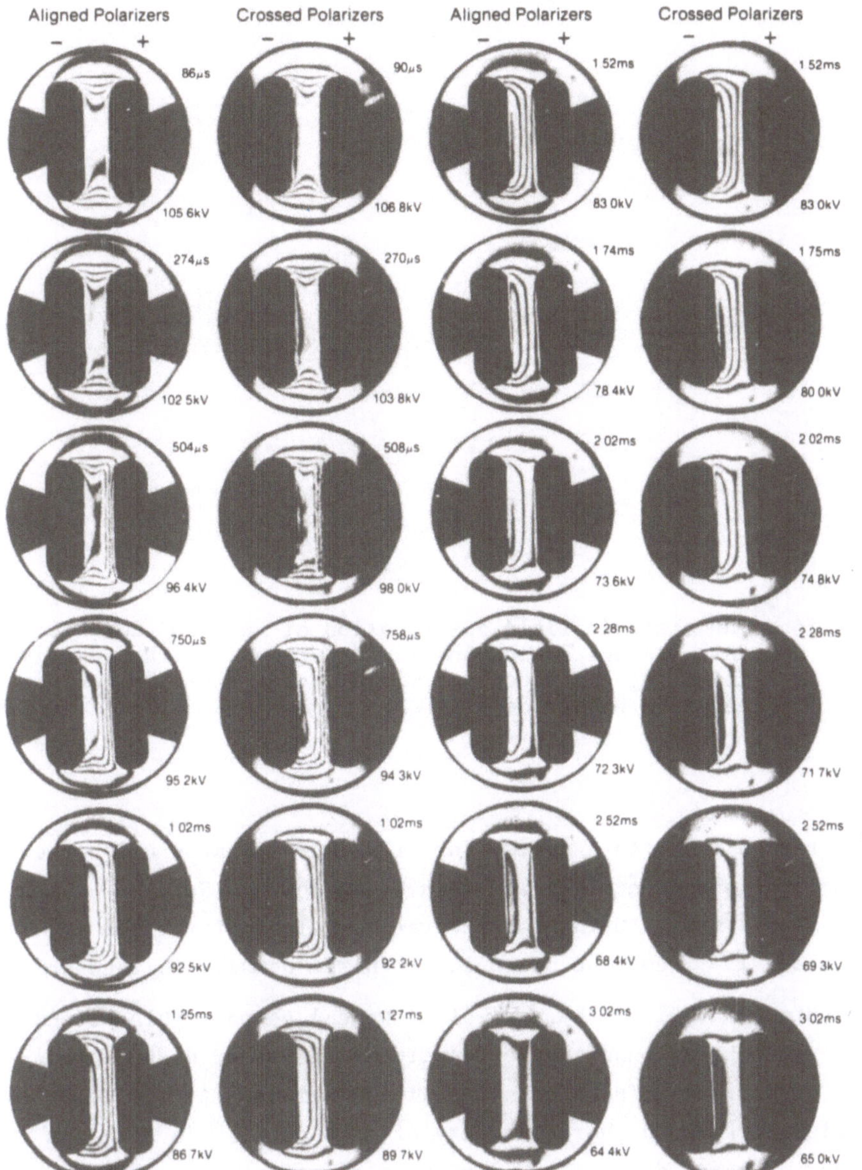

Fig. 33. Kerr effect measurements using 304 stainless steel electrodes with aligned and crossed polarizers at various times after the start of a high-voltage pulse with initial charging voltage of 115 kV showing positive charge injection (T = 8.5 - 10.4°C; ρ = 41.5 - 42.9 MΩ - cm; charging voltage 115 kV).

Note in Fig. 33 that the terminal voltage v listed at the lower right of each photograph decays with time after it has reached crest. In the time interval 0 to 500 μs, the nondimensional electric field El/v at the x = 0 electrode (anode) drops from 1.0 to ~0.4, while at the opposite x = l electrode (cathode), the nondimensional field rises from 1.0 to >1.1. For later times, the injection field at x = 0 remains about constant at El/v~0.5, while the electric field at x = l smoothly increases

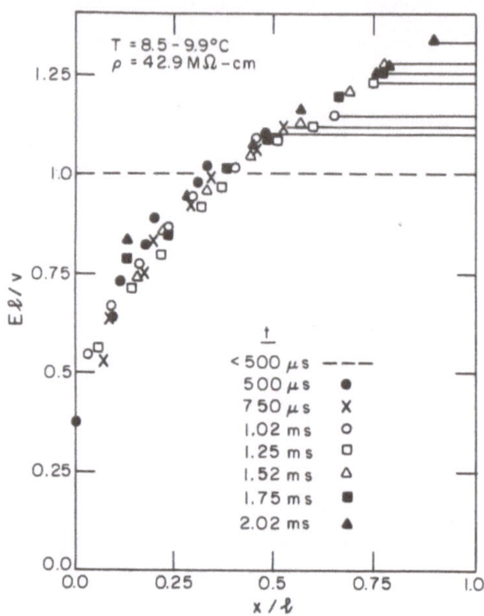

Fig. 34. The electric field distribution at various times for the data in Fig. 33 showing field decrease at the positive injecting electrode and field increase at the noninjecting negative electrode at x = 1.

to greater than 1.3. As a check, for all times the area under the nondimensional electric field curves must be unity. Near x = 0, the slope is about three. Using representative values of v = 100 kV, $\varepsilon = 80\varepsilon_o$, and l = 1 cm, the dimensional charge density is then $q \approx 2\ C/m^3$.

To obtain a feel of the size of q, we can compare this net charge density to the background charge density of the water dissociation products of hydronium (H^+) and hydroxyl (OH^-) ions. At T = 10°C, the equilibrium background charge density of each carrier is 5.21 C/m^3. Thus the measured charge density $q \approx 2\ C/m^3$ is a significant fraction of the background charge.

<u>Negative Charge Injection</u> - Figure 31 also shows negative charge injection from the negative aluminum electrode of a pair of aluminum electrodes, and from the brass electrode when it is negative with either a positive copper or aluminum electrode. Again, we have an anomaly with the copper electrode because we earlier found it to be a positive charge injector and thus expect a negative brass/positive copper electrode combination to inject bipolar homocharge.

The velocity v of migrating charges with mobility μ in a field E is v=μE. Examining Fig. 31 for the cases of positive charge injection from stainless steel electrodes and negative charge injection from aluminum electrodes, we find that the negative charge front has propagated about 35% across the gap in time $t_- = 1.26$ ms in an initial field of $E_- = 115$ kV/cm while the positive charge front has traveled about 70% across the gap in time $t_+=1$ ms in an initial field of $E_+ = 130$ kV/cm. The ratio of velocities and fields for the two cases then lets us calculate the ratio of high voltage charge mobilities:

$$\frac{v_-}{v_+} = \frac{0.35l/t_-}{0.7l/t_+} = 0.40 = \frac{\mu_- E_-}{\mu_+ E_+} = 0.88 \frac{\mu_-}{\mu_+} , \tag{18}$$

to be $(\mu_-/\mu_+) \approx 0.45$. The aluminum measurements were taken at ~5°C. At this temperature the low voltage hydroxyl ion mobility is about $\mu_- \approx 1.3 \times 10^{-7}$ $m^2/(V\text{-}s)$. The stainless steel measurements were taken at about 10°C where the low voltage hydronium mobility is about $\mu_+ \approx 2.9 \times 10^{-7}$ $m^2/(V\text{-}s)$. The ratio of low voltage mobilities is then $(\mu_-/\mu_+) = 0.45$, in agreement with the measured high voltage mobilities. Figure 35 shows the Kerr effect measurements for a pair of aluminum electrodes for aligned and crossed polarizers as the initial voltage is increased at time t = 1.25 ms after high voltage is applied.

Bipolar Homocharge Injection

The bipolar homocharge Kerr effect patterns in Fig. 32 generally show a symmetric set of oval lines. Here, the fields are weaker near the electrodes, increasing to a maximum near the center with positive charge near the positive electrode and negative charge near the negative electrode.

Figure 36 shows the bipolar homocharge Kerr effect patterns with aligned polarizers for a pair of brass electrodes at times 750 μs and 1 ms after high voltage is applied. The fluid turbulence near the positive electrode shown in Fig. 36 generally but not always occurs near either or both electrodes and seems to be associated with charge injection.

Electrohydrodynamic Effects

The Coulombic force on net space charge in the fluid gives rise to fluid motions causing convection currents in addition to conduction currents. The viscous diffusion time $\tau_v = \rho_d l^2/\eta$ determines whether fluid inertia with mass density ρ_d or fluid viscosity η dominates fluid motions over a characteristic length l. Water, with a room temperature fluid density of $\rho_d \approx 10^3$ kg/m^3 and viscosity of $\eta \approx 10^{-3}$ N · s/m^2, has $\tau_v \approx 100$ s over a characteristic length of l = 1 cm. Since this viscous diffusion time is very large compared to the dielectric relaxation time $\tau = \varepsilon/\sigma \approx 600$ μs (at T = 0°C) and representative charge migration times $\tau_{mig} \approx l/\mu E \approx 4$ ms (for hydronium ion at 0°C) based on a field of 100 kV/cm, fluid motions are essentially limited by their own inertia.

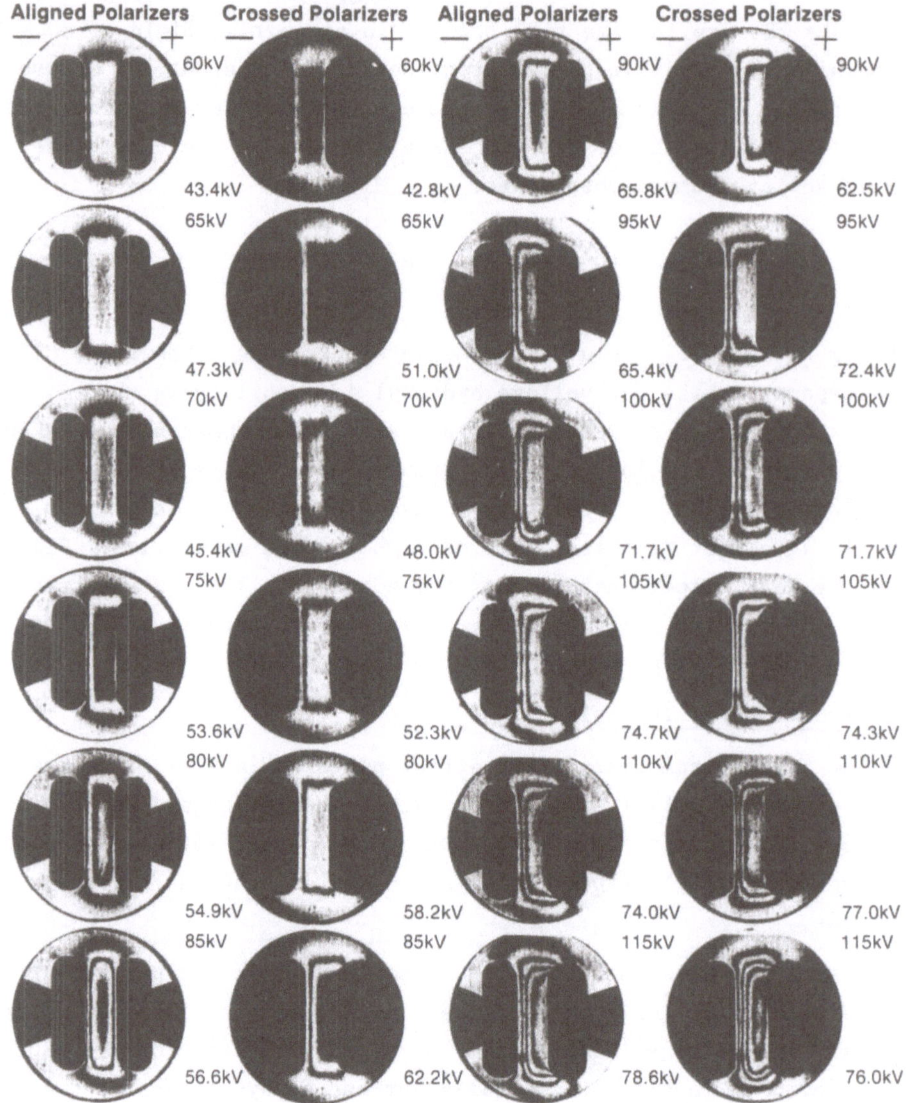

Fig. 35. Kerr effect measurements using 2024 aluminum electrodes with aligned and crossed polarizers at t = 1.25 ms after various high voltages are applied. The fringes propagating out from the left negative electrode indicate negative charge injection.

The ion mobility may also be enhanced by electrohydrodynamic motions. If the change in fluid kinetic energy equals the electrostatic field energy

$$\frac{1}{2}\,\rho_d v^2 = \frac{1}{2}\,\varepsilon\,E^2\,, \tag{19}$$

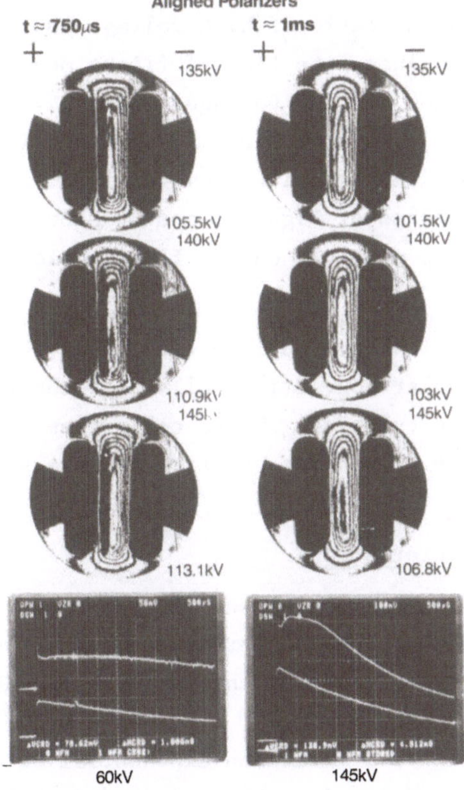

Fig. 36. Kerr effect measurements using standard brass electrodes with 1 cm gap using aligned polarizers at t = 750 μs (left) and t = 1 ms (right) after high voltages of 135 kV (top), 140 kV (middle), and 145 kV (bottom) are applied. The fringes propagating out from both electrodes indicate bipolar homocharge injection. Note the fluid turbulence at the positive electrode (left) at t = 750 μs and the non-exponential current trace (upper) due to space charge effects even though the terminal voltage (lower trace) decays approximately exponentially. Brass Electrodes T = 5.0 - 5.4°C, ρ = 63.7 - 64.9MΩ-cm.

the electrohydrodynamic mobility is [Felici, 1972]:

$$\mu_{EHD} = \frac{v}{E} = \sqrt{\varepsilon/\rho_d}\ . \tag{20}$$

For water [$\varepsilon = 80\varepsilon_o$, $\rho_d = 1000\ \mathrm{kg/m^3}$], $\mu_{EHD} \approx 8.4 \times 10^{-7}\ \mathrm{m^2/(V\text{-}s)}$.

The electroinertial time constant τ_{EI} is then the migration time based on the electrohydrodynamic mobility

$$\tau_{EI} = l/(\mu_{EHD}E) = l\sqrt{\rho_d/\varepsilon E^2}\ , \tag{21}$$

which at E=100 kV/cm for a 1-cm gap is $\tau_{EI} \approx 1.2$ ms. Because μ_{EHD} exceeds ion mobilities, it is expected that fluid turbulence like that shown in Fig. 36 might lead to faster charge migration times. For the unipolar charge injection analysis of (18) for positive and negative charge carriers, even though the ratio of high voltage mobilities agrees with low voltage mobility values, the time of flight measurement is faster than that predicted from the low voltage mobility value. For example, a hydronium ion with mobility $\mu_+ \approx 2.9 \times 10^{-7}$ $m^2/(V\text{-}s)$ in an average field $E \approx 100$ kV/cm would move across a $l = 1$ cm gap in a time $\tau_{mig} \approx l/(\mu_+ E)$ ~3.4 ms, while the measurements of Figs. 33 and 34 show a time of flight ~2 ms. The faster time is probably due to fluid motion during the ion transit time.

For the coaxial steel cylindrical electrodes shown in Fig. 37, fluid motions result in regular convection cells near the positive electrode, whether inner or outer cylinder. These cells grow with time away from the positive electrode. When the polarity is reversed, these cells also reverse.

Voltage-Current Characteristics

Space charge effects are also seen at the electrical terminals. In Fig. 36 both current waveforms (upper traces) show the initial impulse due to the fast rising voltage but then a current peak due to the ion's time of flight before current decay. This is especially noticeable at the high 145 kV voltage. Similar non-exponential decay reverse curvatures were seen in open circuit voltage decays from high resistivity water/glycol mixtures at T = -10°C as shown in Fig. 38. Unlike the expected exponential decay of a lossy dielectric with dielectric relaxation time $\tau = \varepsilon/\sigma$, independent of voltage amplitude and gap spacing, the results in Fig. 38 have non-exponential shape which depends on voltage V and gap spacing d because of the ion time of flight $\tau_{mig} = d^2/(\mu V)$ as an additional time constant.

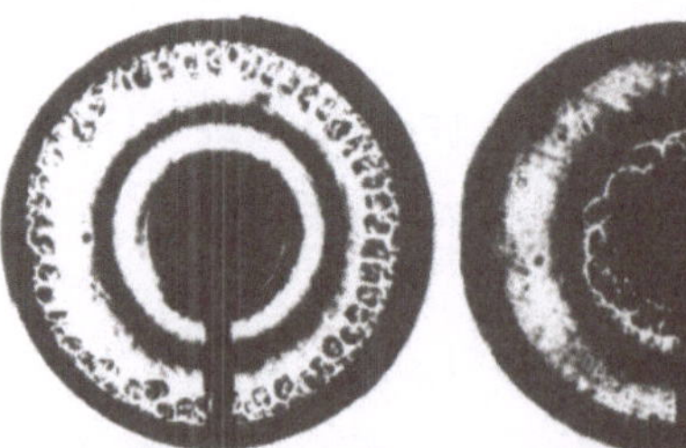

Fig. 37. Kerr electro-optic measurements in highly purified water with crossed polarizers showing electrohydrodynamic instability emanating from the positive electrode for stainless steel coaxial electrodes with initial charging voltage of 85 kV. E_m = 37.3 kV/cm.

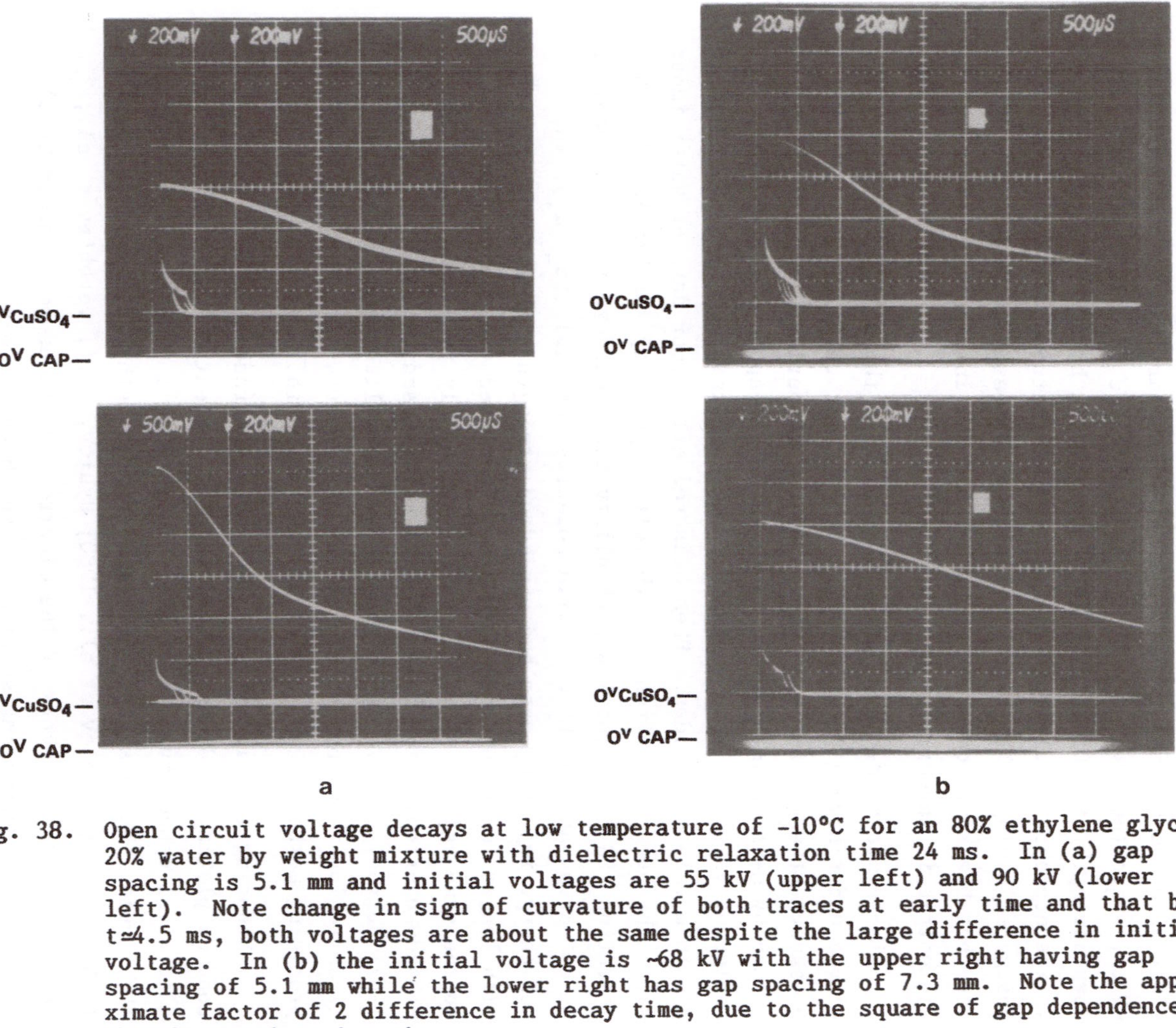

Fig. 38. Open circuit voltage decays at low temperature of -10°C for an 80% ethylene glycol/20% water by weight mixture with dielectric relaxation time 24 ms. In (a) gap spacing is 5.1 mm and initial voltages are 55 kV (upper left) and 90 kV (lower left). Note change in sign of curvature of both traces at early time and that by $t \simeq 4.5$ ms, both voltages are about the same despite the large difference in initial voltage. In (b) the initial voltage is ~68 kV with the upper right having gap spacing of 5.1 mm while the lower right has gap spacing of 7.3 mm. Note the approximate factor of 2 difference in decay time, due to the square of gap dependence of the charge migration time.

Electric Breakdown

Breakdown strengths in highly purified water are generally higher with bipolar injection. For example, with brass/aluminum electrodes, the polarity for bipolar injection had a breakdown strength of ~125-135 kV/cm, while the reverse polarity had negative charge injection with breakdown strength ~90-95 kV/cm. Similarly, stainless steel/aluminum electrodes had a breakdown strength with bipolar injection of ~125-140 kV/cm, while the reverse polarity had no charge injection with a breakdown strength of ~105 kV/cm. This increase in breakdown strength is due to the decrease in electric field at both electrodes due to the space charge shielding. The electric field is increased in the center of the gap, but breakdown does not occur because the intrinsic strength of the dielectric is larger than at an interface.

These measurements were taken soon after the metal electrodes were placed under highly purified water in our test cell. The improvement in breakdown strength for bipolar homocharge-injecting electrodes was confirmed by measurements at the Naval Surface Weapons Center (NSWC), Dahlgren, VA. They used a mixed set of 304 stainless steel (SS) and 2024 aluminum (Al) electrodes. When measured after four days of water immersion, the breakdown strength with voltage polarity for bipolar homocharge injection (SS_+/Al_-) the 10% probability for breakdown for electrodes of 81 cm^2 area was ~130 kV/cm. With polarity reversal (SS_-/Al_+), the non-charge injecting case had breakdown strength ~100 kV/cm. However, with longer immersion time both cases increased their breakdown strength to the same values. After 23 days under highly purified water, the breakdown strength for both polarities increased to 150 kV/cm (Gehman, 1987; McLeod, 1987). These effects are most likely due to the formation of corrosion films on the electrodes and deserves further research as a promising method to improve electrical strength.

Reducing the electric field at the electrodes has been previously done by placing conducting fluid layers adjacent to the electrodes whose electrical conductivity decreases smoothly as a function of depth into the liquid. In this way, the breakdown strength was increased fourfold to 1.5 MV/cm (Vorob'ev et al., 1980; Ryutov, 1974; Vorob'ev et al., 1974). Diffusion layers at the electrode surfaces were formed by slow extrusion of conducting solutions through porous electrodes. Gravitational instability was avoided by placing less dense liquids above higher density liquids. The lower electrodes used aqueous $CuSO_4$ solution with density slightly higher that water, while the upper electrode used $FeCL_3$ in ethyl alcohol with density slightly lower than water.

Such an involved process with attendant mixing problems is appropriate for laboratory testing but is not easily applied to a working pulsed power machine. The same electrode shielding effect for long charging times is more easily achieved by injecting space charge.

Electrochemical Effects

The injection charge polarity relationships we have shown for various metal electrodes are consistent with the ordering of the Galvanic series of alloys, proceeding from most negative to positive as zinc, aluminum, active steel, brass, copper, and passivated steel. To verify this ordering for highly purified water (resistivity 26 MΩ-cm at 20°C, 42 MΩ-cm at 5°C), we measured the open circuit voltage and short circuit current using a Keithley 616 electrometer for the 16 electrode material combinations, taking a pair at a time from pairs of 303 stainless steel, 1100 series aluminum (99% aluminum, 0.12% copper), type 110 copper alloy (99.9% copper, 0.04% oxygen), and C36000 brass (61.5% copper, 35.5% zinc, 3.0% lead). The plexiglas test cell containing the highly purified water and the electrodes had no extraneous metals in contact with the water. Screws and nuts used to support each electrode were made from the same metal stock as the electrode itself. Open circuit voltages and short circuit currents are listed in Table 3 for stationary water at 20°C and 5°C. These values changed with a flow rate of ~2 liters/minute because of streaming electrification effects and because charge profiles near the electrodes become more uniform with flow, thereby decreasing diffusive contributions (Rhoads and Zahn, 1987).

We believe that the initial charge leading to electrical breakdown can come from this electrochemical double layer and that reducing the charge or reversing its polarity just prior to a high voltage pulse can lead to greater voltage strength. By applying the potential of zero charge, pzc, which is generally less than 1 volt, the double layer charge can be brought to zero (Bockris and Reddy, 1970).

Another mixed pair of stainless steel (SS)/aluminum (Al) electrodes had an open circuit voltage of ~0.686 volts with SS positive with respect to Al. Fig. 39 shows a prestress of opposite polarity -2V applied for 96 seconds. If the electrodes are then open circuited, the open circuit voltage takes a quick jump toward the steady state voltage in a time ~150 μs corresponding to the water dielectric relaxation time $\tau = \varepsilon/\sigma$, and then a much longer time of a few minutes to asymptotically approach the steady-state value of 0.686 V. The long time corresponds to a diffusion time $\tau_{Diff} = L^2/D$ where L corresponds to the electrode gap spacing. The quick time constant corresponding to the dielectric relaxation time also obeys the diffusion time if L is the Debye length

Table 3. Measured open circuit voltages (volts) and short circuit currents (μA) for all possible electrode pair combinations using a Keithley 616 electrometer. The voltage polarity is defined as electrode A with respect to electrode B while the current polarity is defined as flowing from electrode A to electrode B in the external short circuit. The left and right upper entries correspond to open circuit voltages at 20°C and 5°C, while the lower entries correspond to short circuit currents.

V 20°C V 5°C / I 20°C I 5°C	Electrode A: Aluminum		Brass		Copper		Stainless Steel	
Electrode B: Aluminum volts	0.054	0.042	0.530	0.424	0.351	0.452	0.302	0.318
Aluminum μA	1.51	0.22	13.34	1.20	4.77	1.89	3.10	0.84
Brass volts	-0.522	-0.419	0.021	0.012	-0.027	-0.030	-0.107	-0.037
Brass μA	-13.4	-1.00	0.36	0.05	-0.48	-0.05	-1.63	-0.05
Copper volts	-0.354	-0.455	0.027	0.033	0.049	0.032	-0.029	-0.044
Copper μA	-5.02	-1.98	0.48	0.22	1.59	0.21	-0.50	-0.14
Stainless Steel volts	-0.298	-0.324	0.103	0.038	0.029	0.039	0.037	0.014
Stainless Steel μA	-3.30	-1.00	1.75	0.15	0.40	0.26	0.64	0.06

$\lambda = [\varepsilon D/\sigma]^{1/2}$. It appears that a low voltage prestress alters the electrical double layer equilibrium and that diffusion times in the double layer as well as the neutral bulk govern the time to return to equilibrium. Current research is examining the effects on electrical strength of applying high voltage pulses at various times after a low voltage prestress, such as when the open circuit voltage passes through zero.

Coaxial Cylindrical Electrodes

High voltage pulses up to ~100 kV are applied to a 1.1 m long coaxial water capacitor with inner radius $R_i \simeq 0.5$ cm and outer radius $R_o \simeq 1.9$ cm.

Figure 40 shows representative Kerr patterns for various times relative to the start of a 126 kV (3 stage Marx generator at 42 kV/stage) high voltage pulse. An unexpected optical phenomenon--the refraction of light away from the inner electrode--interferes with field mapping

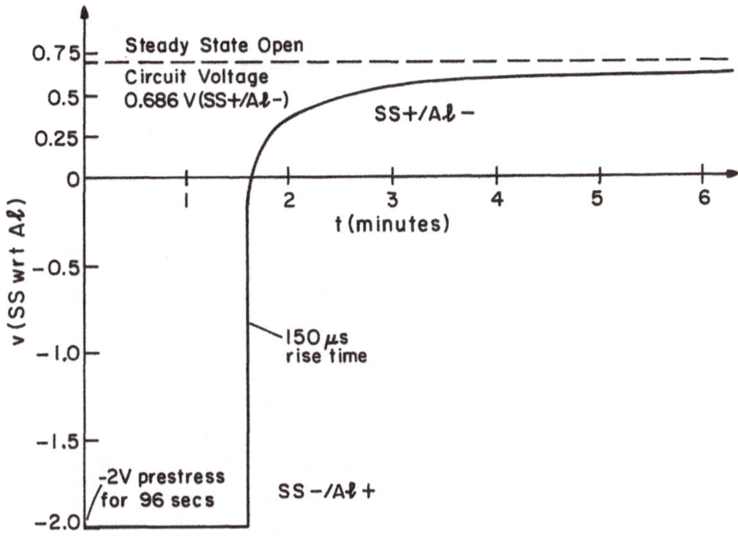

Fig. 39. Open circuit voltage of stainless steel (SS) with respect to aluminum (Al) after 96 seconds of -2V low voltage prestress to highly purified water.

because a shadow up to 45% larger than the inner electrode is projected onto the film. A bright ring of light intensity which typically surrounds the shadow implies that the beam is refracted rather than absorbed. Refraction does not occur with stainless steel electrodes with the inner electrode positive, but the refraction occurs for the inner electrode negative for both stainless steel and aluminum. This effect is not due solely to a field-dependent refractive index because it does not appear at early times when the field strength is largest. Figure 40 indicates electrode shadow growth with maximum in times of order 1 ms, comparable to ion migration times, and subsequent decay times of order 10-15 ms, comparable to the terminal voltage decay rate (LaGasse et al., 1985).

The radial variation required in the index n(r) to deflect a light ray from $r = R_i$ to $r = 1.45R_i$, the size of the largest shadow, can be estimated from the ray deflection relation derived from a differential form of Snell's law:

$$\frac{d^2r}{dz^2} = \frac{1}{n(r)}\frac{dn(r)}{dr} . \tag{22}$$

Assuming a small linear variation of the index in the shadow region of the form $n(r) \simeq n_o - \Delta n(1.45R_i - r)/0.45R_i$ where $n_o \simeq 1.33$ is the usual value for water and $\Delta n \ll n_o$, (22) can be linearized and integrated to give $\Delta n/n_o \simeq 8 \times 10^{-6}$.

Aligned Polarizers | Crossed Polarizers | Aligned Polarizers | Crossed Polarizers

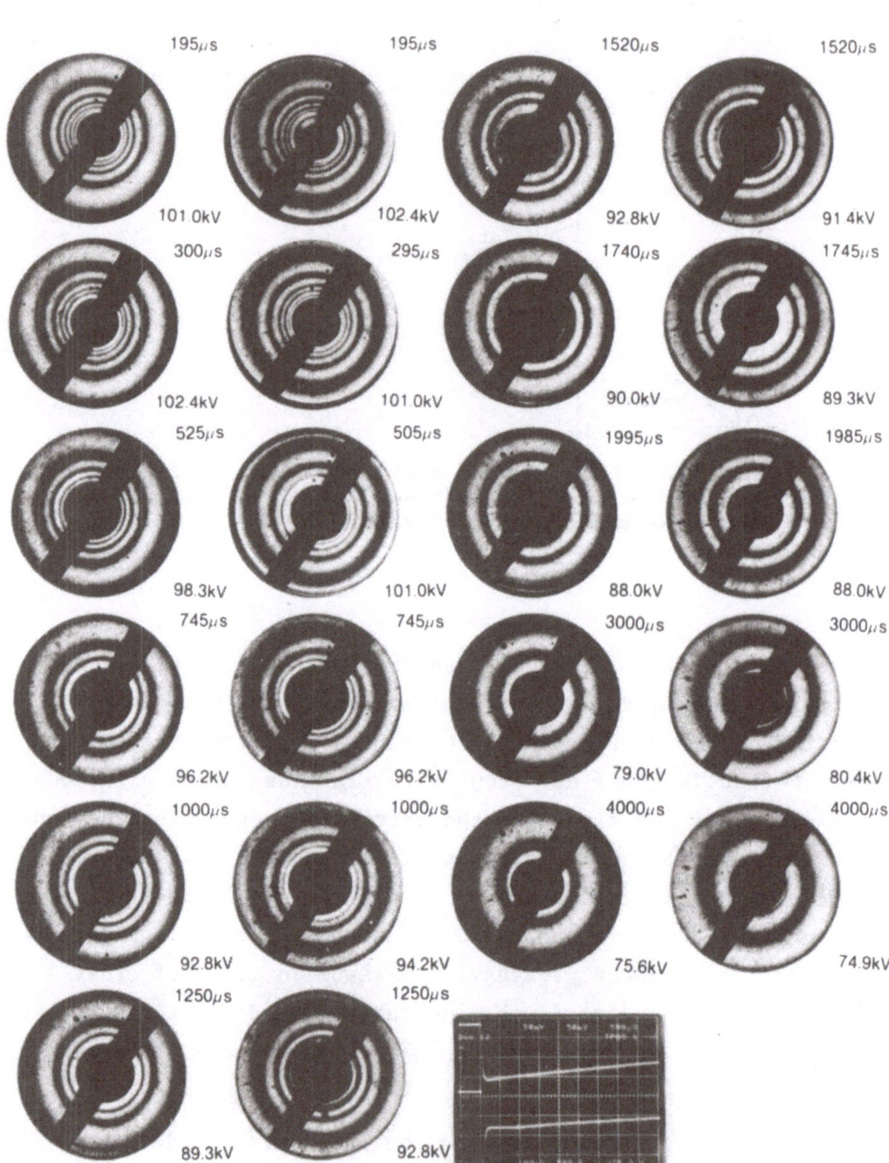

Aluminum (-, inner)/
Aluminum (+, outer)

$\rho \approx 70.0\ M\Omega$-cm T ≈ 2.0°C 42kV

Fig. 40. Kerr electro-optic field mapping measurements with coaxial cylindrical aluminum electrodes with Marx charging voltage of 42 kV/stage x 3 stages = 126 kV. Note how the diameter of the inner cylinder seems to grow with time during the voltage pulse (upper trace). The lower trace is the terminal current. $E_m \approx 35$ kV/cm.

For the case of a stainless steel inner electrode, ~70 kV can be applied without breakdown when the inner electrode is positive, while ~100 kV can be applied without breakdown with a negative inner electrode. If aluminum is the inner electrode, the effect is more extreme: only ~20 kV can be applied without breakdown when the inner electrode is positive, while ~100 kV can be applied without breakdown when the inner electrode is negative. This low positive voltage prevented any field mapping with the aluminum inner electrode positive and significantly reduces the number of data points for positive stainless steel inner electrodes.

Numerous studies in water using stainless steel electrodes have shown similar results where positive breakdown occurs at ~1/2 the field strength of negative breakdown.

The data in Fig. 40 with aluminum electrodes indicates negative space charge injection with time from the inner cylinder and electric field shielding near the inner electrode. As in Figs. 15-17, since the line integral of the electric field between electrodes is equal to the instantaneous voltage, a decrease in field near the inner electrode due to space charge shielding results in a field enhancement in the central region and near the non-injecting outer electrode. The maximum slope in electric field occurs when t~3000 μs and corresponds to a space charge density $q \simeq 0.3\ C/m^3$ at $r/R_o \simeq 0.385$.

CONCLUDING REMARKS

Design criteria for high voltage apparatus depend strongly on the electric breakdown strength and typically model the dielectric as a slightly lossy dielectric, simply described by its permittivity and resistivity. For insulating dielectrics there can be significant space charge injection when high voltage is applied, causing anomalous voltage-current characteristics and distortions in the electric field easily measured using the Kerr electro-optic effect. The sign and magnitude of the space charge depends strongly on the electrode material and voltage polarity. Although injected space charge can increase the attainable system voltage and therefore the stored energy, the usable energy delivered to a load is less than the space-charge-free capacitive energy storage value of $(1/2)CV^2$ due to energy dissipated as the injected charge migrates to the electrodes. The most interesting case is that of bipolar homocharge injection, which decreases the electric fields at both electrodes and for which in small-scale laboratory experiments in highly purified water have allowed up to 40% higher voltage without breakdown, thereby a doubling of stored energy. This increase in stored energy due to higher voltage operation greatly offsets the slight extra dissipation due to conduction, which we believe can be designed to be negligibly small.

It is best to inhibit unipolar charge injection that increases the electric field at the non-charge-injecting electrode leading to early electrical breakdown and to encourage bipolar homocharge injection at both electrodes where injected charge shields the electrodes, causing lower electrode fields. Even though this causes the electric field to be larger in the dielectric volume, the intrinsic strength of the dielectric is larger than at interfaces. To increase the magnitude of bipolar injection, special charge injecting coatings for the anode and cathode or low voltage prestressing should be investigated that allow higher voltage without breakdown.

ACKNOWLEDGMENTS

This work was supported by the Pulse Power Technology Program of the Naval Surface Weapons Center, Dahlgren, VA; by a consortium of electric utility organizations as part of the MIT Electric Utilities Program; by the National Science Foundation under Grant No. ECS-8517075; by the Electric Power Research Institute under Contracts RP-1536-7, RP-1499-8, and RP-1536-14; and by Computer Services Corp. Much of the work reported here comes from the thesis research of MIT graduate students K.G. Rhoads, M. LaGasse, R. Carreras, S.M. Gasworth, D.J. Lyon, T.D. Wang, S.M. Kardon, D.W. Ericson, S. Shepard, and S.H. Voldman; from University of Florida thesis research of R.J. Sojka and T.J. McGuire; and from the assistance of MIT Visiting Scientists from Japan, T. Takada, Y. Ohki, H. Matsuzawa, M. Ishii, M. Yoda, M. Masui, and M. Hikita and from China, G. Sun. Appreciation is also extended to MIT secretary Barbara Lakeberg for her quick and accurate word processing which allowed this paper to be finished on time with no sacrifice to the usual busy office she runs.

REFERENCES

Bockris, J.O'M., and Reddy, A.K.N., 1970, "Modern Electrochemistry," Plenum/Rosetta, New York, N.Y.

Buckingham, A.D., and Dunmur, D.A., 1968, Kerr effect in inert gases and sulphur hexafluoride, Trans. Faraday Soc., 64:1776-1783.

Buckingham, A.D., Bogaard, M.P., Dunmur, D.A., Hobbs, C.P., and Orr, B.J., 1970, Kerr effect in some simple non-dipolar gases, Trans. Faraday Soc., 66:1548-1553.

Carreras, R.F., and Zahn, M., 1986, Kerr electro-optic field mapping measurements in high voltage stressed liquid and gaseous sulfur hexafluoride, in: "1986 Annual Report for the Conference on Electrical Insulation and Dielectric Phenomena", pp. 287-292.

Chandrasekhar, S., 1961, in: "Hydrodynamic and Hydromagnetic Stability," Oxford Press, London, UK, Chapter 7.

Coelho, R., 1979, "Physics of Dielectrics for the Engineer," Elsevier, Amsterdam, The Netherlands.

Crofts, D.W., 1986, Static electrification phenomena in power transformers, in: "1986 Annual Report for the Conference on electrical Insulation and Dielectric Phenomena", pp. 222-236.

Dow Chemical Company, U.S.A., "Experimental ethylene carbonate XAS-1666.00L," Organic Chemicals Department, Midland, MI 48640.

Felici, N.J., 1972, DC conduction in liquid dielectrics (Part II): electrohydrodynamic phenomena, Direct Current, 2(4):147-165.

Fenneman, D.B., 1982, Pulsed high voltage dielectric properties of ethylene glycol/water mixtures, J. Appl. Phys., 53(12): 8961-8968.

Gasworth, S.M., 1985, "Electrification by Liquid Dielectric Flow," Ph.D. Thesis, Dept. of Electrical Engineering, M.I.T., Cambridge, MA.

Gasworth, S.M., Melcher, J.R., and Zahn, M., 1986, "Electrification Problems Resulting from Liquid Dielectric Flow," Tech. report El-4501, EPRI.

Gasworth, S.M., 1986, Electrification by Liquid Dielectric Flow, in: "1986 Annual Report for the Conference on Electrical Insulation and Dielectric Phenomena", pp. 192-199.

Gehman, V.H., Jr., 1987, Theoretical considerations of water-dielectric breakdown initiation for long charging times, 1987 Pulsed Power Conference.

Hebner, R.E., Jr., Sojka, R.J., and Cassidy, E.C., 1974, "Kerr Coefficients of Nitrobenzene and Water," NBS Report No. 74-544.

LaGasse, M.J., Zahn, M., and Huang, J., 1985, Kerr electro-optic field mapping measurements in highly purified water between coaxial electrodes, in: "1985 Annual Report for the Conference on Electrical Insulation and Dielectric Phenomena", pp. 89-95.

Lee, W.H., 1976, Cyclic carbonates, Chem. Non-Aq. Solv., 4:167-245.

Lyon, D.J., 1987, "Couette Flow Measurement of Equilibrium and Energization Charging in Transformer Insulation," M.S. Thesis, Dept. of Electrical Engineering, M.I.T., Cambridge, MA.

McLeod, A.R., 1987, Water breakdown measurements of stainless steel and aluminum alloys for long charging times, 1987 Pulsed Power Conference.

Melcher, J., Lyon, D., and Zahn, M., 1986, Flow electrification in transformer oil/cellulosic systems, in: "1986 Annual Report for the Conference on Electrical Insulation and Dielectric Phenomena", pp. 257-265.

Oommen, T.V., 1986, Static electrification properties of transformer oil, in: "1986 Annual Report for the Conference on Electrical Insulation and Dielectric Phenomena", pp. 206-213.

Rhoads, K.G., and Zahn, M., 1987, "Kerr Electro-Optic field and Charge Mapping Measurements Focussing on the Metal Electrode/Water Interface," 1987 Pulsed Power Conference.

Ryutov, D.D., 1974, Diffusion Electrodes for Investigation of the Breakdown of Liquids Dielectrics, translated from Zhurnal Prikladnoi Mekhaniki i Tekhnicheskoi Fiziki, No. 4, pp. 186-187, Jul-Aug 1972, by Consultants Bureau pp. 596-597.

Taylor, G.I., 1923, Stability of a viscous liquid contained between two rotating cylinders, Phil. Trans. of the Royal Society A, 223:289-343.

Vorob'ev, V.V., Kapitonov, V.A., and Kruglyakov, E.P., 1974, Increase of Dielectric Strength of Water in a System with 'Diffusion Electrodes,' JETP lett., Vol. 19, pp. 58-59.

Vorob'ev, V.V., Kapitonov, V.A., Kruglyakov, E.P., and Tsidulko, Yu.A., 1980, Breakdown of water in a system with diffusion electrodes, Sov. Phys.-Tech. Phys., Vol. 25, No. 5, pp. 598-602.

Zahn, M., and McGuire, T.J., 1980, Polarity effect measurements using the Kerr electro-optic effect with coaxial cylindrical electrodes, IEEE Trans. on Elec. Insul., EI15(3):287-293.

Zahn, M., Fenneman, D.B., Voldman, S., and Takada, T., 1983, Charge migration and transport in high voltage water/glycol capacitors, J. Appl. Phys., 54(1):315-325.

Zahn, M., Takada, T., and Voldman S., 1983, Kerr electro-optic field mapping measurements in water using parallel cylindrical electrodes, J. Appl. Phys., 54(9):4749-4761.

Zahn, M., and Takada, T., 1983, High voltage electric field and space charge distributions in highly purified water, J. Appl. Phys., 54(9):4762-4775.

Zahn, M., and Ericson, D.W., 1984, Electro-optic field mapping measurements in ethylene carbonate, in: "1984 Annual Report for the Conference on Electrical Insulation and Dielectric Phenomena", pp. 327-333.

Zahn, M., Ohki, Y., Rhoads, K., LaGasse, M., and Matsuzawa, H., 1985, Electro-optic charge injection and transport measurements in highly purified water and water/ethylene glycol mixtures, IEEE Trans. on Elec. Insul., EI-20(2):199-211.

Zahn, M., Ohki, Y., Fenneman, D.B., Gripshover, R.J., and Gehman, V.H., Jr., 1986, Dielectric properties of water and water/ethylene glycol mixtures for use in pulsed power system design, Proc. of IEEE, 74(9):1182-1221.

Zahn, M., Hikita, M., Wright, K.A., Cooke, C.M., and Brennan, J., 1987, Kerr electro-optic field mapping measurements in electron beam irradiated polymethylmethacrylate, IEEE Trans. on Elec. Insul., EI-22(2):181-185.

BREAKDOWN AND CONDUCTION

AN OVERVIEW OF ELECTRICAL PROCESSES LEADING TO DIELECTRIC BREAKDOWN OF LIQUIDS

T. John Lewis

Institute of Molecular and Biomolecular Electronics
University College of North Wales
Bangor, Gwynedd, LL57 1UT, U.K.

INTRODUCTION

The electrical breakdown of a dielectric liquid under high stress requires that a number of interdependent, parallel as well as sequential electronic processes occur in the liquid and at the electrodes. In addition, there is the intervention of electrically-induced but non-electronic processes such as heating and the generation of a microscopic gas phase and changes in the chemical structure of the liquid molecules. The precise nature and sequence of these processes is extraordinarily difficult to establish largely because breakdown is an instability in which a liquid-gas, insulator-conductor transition occurs very rapidly. For example, breakdown of a n-hexane sample in a 2 mm gap between electrodes establishing a field of 10^6V cm^{-1} can be complete in less than 500 ns (Wong and Forster, 1977).

In early studies, it was natural to seek explanations of the processes in terms of the already well-confirmed mechanisms of gas breakdown such as collision ionization and streamer propagation. Since that time, the very considerable advance in understanding the electronic properties of the amorphous solid state offers opportunity for a much wider appraisal of the breakdown mechanisms of liquids. They are, as condensed phases, in many ways closer to the solid than to the gaseous state, at least through the initiatory stages of breakdown if not at the onset of final dielectric collapse. An important feature of the improved understanding is the possibility to consider in detail the electronic processes of an electro-chemical nature which are likely to occur at metal electrode-dielectric liquid interfaces. As will be discussed below, the processes at these interfaces play a vital role in breakdown initiation.

Most experimental studies of high-field breakdown phenomena have centered on hydrocarbon liquids of short-chain length, n-hexane being representative of these, but there has also been a significant amount of work on liquefied gases. In the present treatment, the focus will be on the former although many of the mechanisms discussed will be applicable without serious modification to liquefied gases.

The intention is to discuss first the elementary electronic processes which can be expected to occur in a liquid under electrical stress and then to consider how these together can lead to instabilities which mark the onset of the breakdown process.

ELECTRONIC ENERGY STATES

Charge transport between metal electrodes in a dielectric liquid can be expected under sufficiently high electrical stress even when the liquid is very pure. The charge may arise from ionization of trace impurity or, in the presence of background radiation, of the liquid itself. It may also arise by charge injection from the electrodes across the electrode-liquid interface.

The charged species, while being transported, interact with the liquid. The interactions will consist of short-range attractive and repulsive forces involving the electron orbitals of those molecules in close proximity to the charge and longer-range essential attractive forces arising from a polarization response of the other molecules surrounding the charge. Polarization can be separated into a fast electronic response and a slower dipolar one in which there is some degree of ordering of the liquid structure.

We consider the likely electronic energy states which these charges might occupy in the liquid. The weakness of the intermolecular binding forces of typical non-polar dielectric liquids will mean that the electronic states of the liquid will derive directly from those of individual isolated molecules. There are two charged or ionized molecular states to consider. One is that of the negative ion in which a molecule of electron affinity A_g acquires an excess electron and the other is that of a positive ion in which a molecule of ionization energy I_g loses an electron. In the liquid phase, the interactions mentioned above lead to a polarization energy P to be associated with the ion which will have fast electronic and, in a polar liquid, slower dipolar reorganization components, P_e and P_d respectively.

Electron and negative ion states

The lowest energy states of an excess electron in a non-polar liquid in which the electron is quasi-free and delocalized so that it is not bound to a particular molecule defines a conduction band edge which has an energy V_o with respect to the vacuum level. This idea of a band edge and quasi-free conditions is most applicable to the liquefied rare gases where energy fluctuations from site to site will be small. It is still meaningful for more complex molecular liquids but temperature-induced fluctuations in local configurations will lead to Anderson-like localization as in non-crystalline solids with the excess electron wavefunction decreasing over several molecular distances. These localized E_t states will extend below V_o as indicated in Fig. 1 (Jortner, 1982).

Once an electron becomes weakly localized in this way, it is always possible, depending on thermal fluctuations, for the slower reorganization component of P_d to increasingly solvate the electron on a time scale corresponding to dielectric relaxation to give a true negative ion state.

Thus, we have $V_o = A_g + P_e$ and in the case of n-hexane the reported values of V_o lie between +0.1 and +0.04 eV above the vacuum level. The localized states E_t below V_o have a most probable value $\bar{E}_t$ reported as -0.2 eV (Noda et al., 1975) and -0.26 eV (Schiller et al., 1973) with a likely Gaussian distribution (Dodelet and Freeman, 1977) with a standard deviation of 0.12 eV (Fig. 1).

The complete solvation of the electron to form a stable negative ion will involve the polarization component P_d but this will be small in non-polar liquids such as the hydrocarbons. The situation will be quite different if polar impurities are present as will be discussed below.

Positive ion states

Positive ions also arise in pure liquids under high electrical stress as will be described in more detail below. The positive ion of the pure liquid can be considered as a positive hole state within a band of occupied hole states consisting of the neutral molecules of the liquid. The occupied 'donor' states will have energies E_d within the band, whose width will be determined by liquid fluctuations and described by a Gaussian distribution with mean energy $\bar{E}_d$ given by

$$\bar{E}_d = -I_g - P , \qquad (1)$$

where here again the polarization will have a fast electronic and a slower dipolar component. Since the polarization energy P_e is likely to be of ~2.5 eV (see Sanche, 1979, for example) and I_g is ~9 eV for n-hexane, $\bar{E}_d$ will be -6.5 eV. Thus an effective band gap $\bar{E}_t - \bar{E}_d$ exists (Fig. 2) which, for n-hexane, would be rather more than 6 eV.

In addition to ionized positive hole states, it is also possible to have excited states of an excitonic nature in which the excited electron

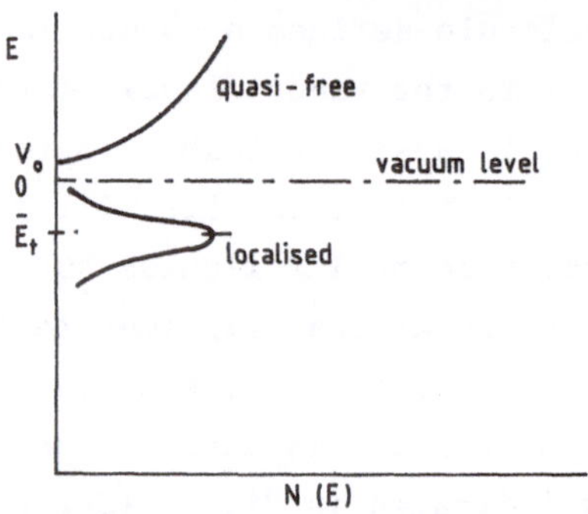

Fig. 1. Energy distribution function N(E) for excess electrons in quasi-free states $E \geq V_o$ and in localized structural states $E_t < V_o$ with a Gaussian distribution about $\bar{E}_t$.

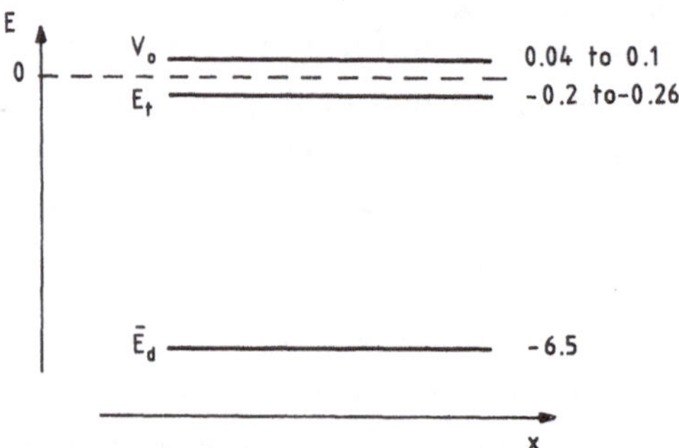

Fig. 2. Energy band scheme for quasi-free electron (E_t) and positive hole (E_d) states of a pure liquid. The E_t states extend into the quasi-free region > V_o. There is an effective band gap $\bar{E}_t - \bar{E}_d$. Estimated energies for n-hexane shown in eV.

remains bound to the molecular center. These states, although electrically neutral, are energetic and could be important in consideration of energy associated with the electrical breakdown processes.

Impurities

Even the most purified liquids will contain a small concentration of impurity molecules and these could modify the electronic states described above. If the impurity is polar the modification could be out of all proportion to the concentration. The reason is that the non-uniform radial field produced in the neighborhood of a charged center will produce dielectrophoretic forces which will cause the impurity molecules to cluster about the center. Thus, for example, water molecules could be swept up forming solvation shells of high polarizability about electrons

(Jortner and Gaathon, 1977) so trapping the electrons to become much less mobile ions (Fig. 3).

The polarization shell also alters the polarization contribution to the total energy of the ion state. We may get a crude estimate of this by using the Born approximation for the polarization energy for an ion of radius a which is

$$P = \left(-e^2/\ 8\pi\ \varepsilon_o a\right) \left(1 - \varepsilon_1^{-1}\right), \tag{2}$$

where ε_1 is the relative permittivity of the pure liquid. Then the additional polarization energy caused by the impurity shell of radius r and relative permittivity ε_p is

$$\Delta P = \left(-\ e^2/\ 8\pi\ \varepsilon_o a\right) \left(1 - a/r\right) \left(\varepsilon_1^{-1} - \varepsilon_p^{-1}\right) . \tag{3}$$

If say a = 0.2 nm, r = 0.3 nm and $\varepsilon_p >> \varepsilon_1$ then $\Delta P \simeq -0.6$ eV which is a significant increase in the magnitude of P, sufficient to stabilize the electron as a negative ion. Of course, at such 'inner shell' distances, the Born approximation is very crude but the detailed treatment given by Kebarle et al. (1977) for the solvation of negative ions shows how important the effect might be.

CHARGE TRANSPORT

Positive and negative ions which have become well solvated will move with their polarization clouds as small polarons with mobilities in the range 10^{-7} to 10^{-8} $m^2V^{-1}s^{-1}$ which correspond to ion mobilities in electrolytes (Adamczewski, 1969).

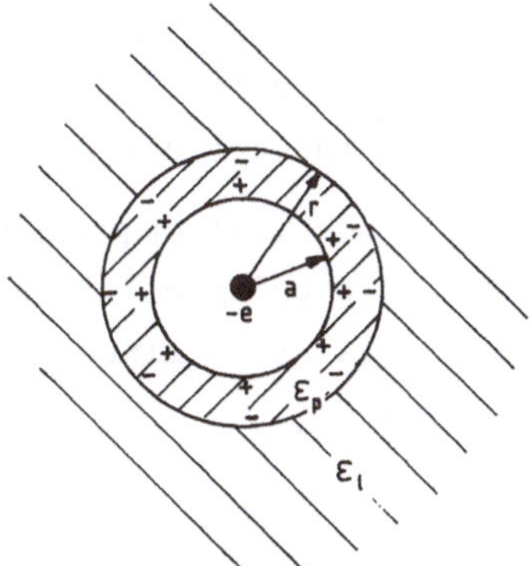

Fig. 3. Polarization shell about an electron in a non-polar liquid (ε_1) caused by a polar impurity (ε_p) collected by dielectrophoretic forces. The effective 'inner shell' radius of the electron is 'a' and that of the polarization shell is 'r'.

The positive ion considered as a positive hole, however, is able to move in a mode analogous to hole transport in semiconductors namely by counter transport of electrons by quantum-mechanical resonance tunnelling transfer from an occupied to an empty state E_d of equivalent energy on an adjacent molecular site. The principle is illustrated in Fig. 4a. If tunnelling occurs at an energy E in the E_d band through an inter-molecular potential energy barrier V(x), assumed here to be one-dimensional, then the probability of the transition depends on the factor

$$\exp\left\{(-4\pi/h)\int_{x_1}^{x_2} dx\,[2m\,(V(x)-E)\,]^{1/2}\right\}, \tag{4}$$

where m is the effective mass of the electron and x_1, x_2 are the spatial limits for V(x) = E (Schiff, 1955). The rate of hole transfer thus depends strongly on V(x) and on the closeness of approach of the two molecules. It also depends on the two states having the common energy E and this will be given by the product of the Gaussian probabilities.

Transitions can occur either way but in an applied field ξ, the barrier V(x) becomes modified to $V(x)-e\xi x$ (Fig. 4b) and a nett electron drift in the field direction will occur from which an equivalent hole mobility may be deduced. Pulse radiolysis studies (Mehnert et al., 1982) have indicated a positive hole mobility of $2 \times 10^{-6}\ m^2V^{-1}s^{-1}$ for n-heptane, at least an order of magnitude greater than the positive ion mobility. Similar studies on cyclohexane (Hummel and Luthjens, 1973) have yielded mobilities of the same order. Under conditions of high stress corresponding to breakdown, a higher mobility might be expected since the transition probability is a strong function of the barrier size and this could be much reduced by the field.

The most important category of mobile charge for high field conduction and breakdown is the excess electron generated either by electrode injection or by ionization and decoupling from the counter positive charge and moving quasi-free in the conduction band.

In an applied field electrons will acquire energy and momentum which will be lost in scattering and trapping collisions. If the liquid is pure, the structural traps E_t will be dominant and it is possible to express the electron mobility as

$$\mu_e = \mu_e^o\,\tau_f\,/\,(\tau_f + \tau_t)\,, \tag{5}$$

where μ_e^o is the mobility of a delocalized free electron, τ_f is the mean free time between traps and τ_t the time in a trap. If necessary μ_e^o may be further expressed in a relaxation time approximation in terms of the thermal velocity of the electron. The trapping time τ_t in a state E_t is

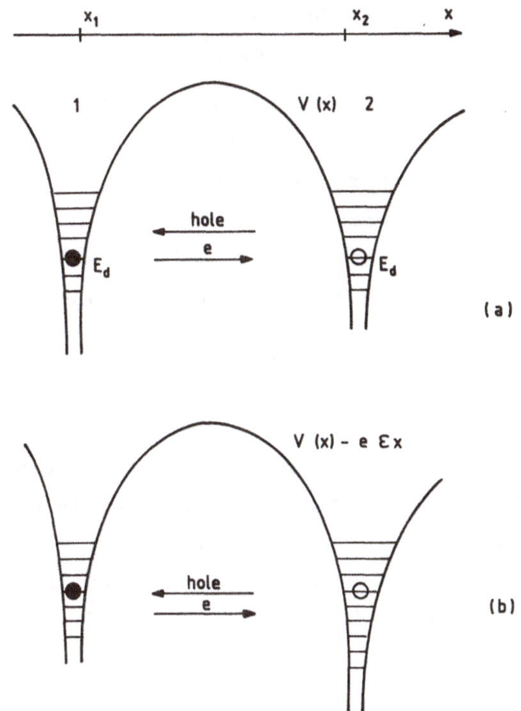

Fig. 4. (a) Transfer of positive hole by electron tunnelling from occupied E_d state on molecule 1 to empty E_d state on molecule 2.
(b) Application of field ξ increases tunnelling probability by weakening barrier $V(x)$ to $V(x) - e\xi x$ and causing a net electron drift in field.

expected to result from a thermally activated process and can be written as $\tau_o \exp(E_t/kT)$ where τ_o is the reciprocal of an attempt to escape frequency. For high electric fields such that $\tau_f \ll \tau_t$

$$\mu_e \simeq \mu_e^o \, \tau_f/\tau_t \; . \qquad (6)$$

Freeman and co-workers have made an extensive study of electron mobilities in a variety of liquids (Gee and Freeman, 1983) while Schmidt (1977, 1984) has provided a summary of values. It is found that μ_e decreases through the alkane series C_4 to C_{12} from 40 to 3 x 10^{-6} m^2 V^{-1} s^{-1}, consistent with the idea that increasing departure from sphericity will increase structural imperfection and traps. For n-hexane we find μ_e = 9 x 10^{-6} m^2 V^{-1} s^{-1} and E_t = 0.19 eV. Schiller et al. (1973) have suggested that μ_e^o is between 8 and 9 x 10^{-3} m^2 V^{-1} s^{-1}. Fueki et al. (1972), using a semi-empirical approach, find μ_e^o = 17 x 10^{-3} m^2 V^{-1} s^{-1} and E_t = 0.24 eV. Interestingly, methane liquid at 110K with a close packing spherical structure yields μ_e = 400 x 10^{-6} m^2 V^{-1} s^{-1}, almost identical with the electron mobility in liquid argon. In these close packed liquids structural traps should be shallow and thus these high mobilities probably represent values of μ_e^o for quasi-free electrons.

Equation (5) predicts that the mobility will be temperature dependent since τ_t results from a thermally-activated process. It is interesting that the electron mobility for a range of hydrocarbons including n-hexane is in agreement with this and that, moreover, the activation energy is in the range 0.1 to 0.2 eV (Schmidt, 1977) which is close to the value given above for $\bar{E}_t$ characterizing the tail states of the conduction band.

Increase in the applied electric field should reduce the potential barriers of the structural traps in the field direction, so decreasing τ_t and causing μ_e to tend towards the limit μ_e^o. If indeed a limiting mobility of $\mu_e^o \sim 10 \times 10^{-3}\ m^2\ V^{-1}\ s^{-1}$ were to be reached in a near-breakdown field of $10^8 Vm^{-1}$, then the electron kinetic energy associated with the corresponding drift velocity would exceed 3 eV and the mean rate of loss of energy to the liquid would be about 10^{14} eV s^{-1}. Magee (1977) has estimated a mean loss of energy of 7 x 10^{13} eV s^{-1} for electrons of less than 6 eV in hydrocarbon liquids.

Under these infra-breakdown conditions the electron undergoes a variety of scattering and energy-losing collisions with the liquid. In the electron energy range below about 1.25 eV [the collective regime described by Sanche (1979)] stimulation of various vibrational modes of the liquid molecules seems highly likely. These modes in the hydrocarbons are to be associated with vibrations of the CH, CH_2 and CH_3 groups at wave numbers in the infra-red and with energy quanta of 0.12, 0.18 and 0.37 eV respectively (Lewis, 1956; Hiraoka and Hamill, 1973). To acquire a kinetic energy of 0.37 eV in a field of 10^8 V m^{-1}, an electron would require a free path in the field direction in excess of 37Å and it is to be noted that Holroyd et al., (1972) have suggested a mean range of 42Å between strong scattering collisions in n-hexane.

Stimulation of vibrational modes of the molecules is probably the most important energy transfer mechanism in the pre-breakdown stage for hydrocarbon liquids. A likely form for a scattering collision cross section as a function of kinetic energy K of the electron for one of these modes is sketched in Fig. 5a. It has an onset energy K_i corresponding to the energy quantum $h\nu_i$ of the vibrational mode, a peak Q_i^m at $K_i^m > K_i$ and, thereafter, a steady fall. The mean free path for this collision is then $[NQ_i(K)]^{-1}$ where N is the number density of the scattering centers (concentration of hydrocarbon groups). The energy loss per unit path length will be $NQ_i h\nu_i$ while the energy gain from the electric field ξ will be $e\xi$ per unit path. Thus, as shown in Fig. 5b there would be, for

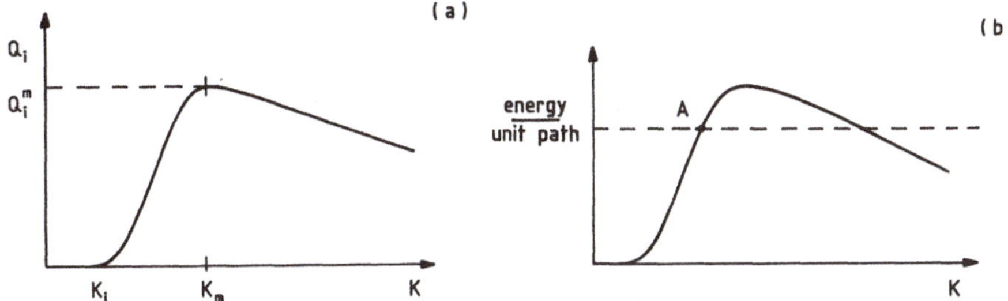

Fig. 5. (a) Idealized form of the collision cross section Q_i for energy transfer to an ith vibrational mode of the system as a function of the kinetic energy K of an electron.
(b) Energy per unit length in field direction as a function of K: broken line, gain from field; solid line, loss to liquid. Steady state condition reached at A.

any electric field in a given range, a steady state point A where the gain is balanced by the loss in scattering collisions.

In the case of hydrocarbons with the possibility of CH, CH_2 and CH_3 vibrational modes, a composite collision cross section $Q = Q_1 + Q_2 + Q_3$ representing the sum of the cross sections of the 3 modes can be expected and a corresponding energy loss per unit path length will be $N(Q_1h\nu_1 + Q_2h\nu_2 + Q_3h\nu_3)$. The exact forms of the cross sections are not known but a possible composite cross section is shown in Fig. 6 with a succession of steady-state intersection points such as A_n corresponding to equal increments in the applied field ξ. From these, since the kinetic energy is 1/2 mv^2, it is then possible to obtain a plot of mean electron velocity v as a function of field ξ as shown in Fig. 7. Assuming that v will be a reasonable approximation to the drift velocity then, depending on the relationship between the Q_i and energy, both super and sub-linear

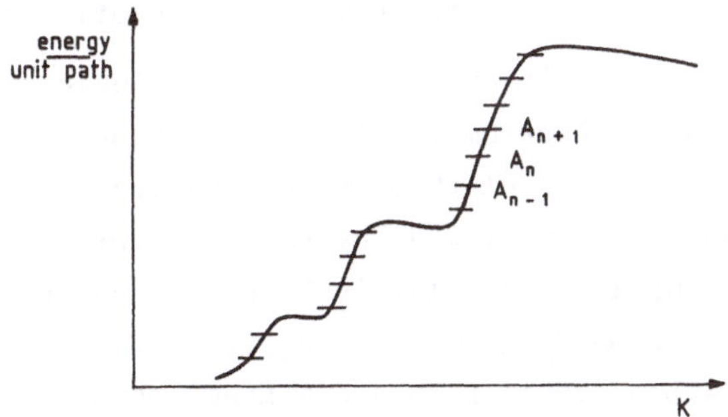

Fig. 6. Energy gain per unit length based on a composite collision cross section Q for a hydrocarbon liquid. The intersection points A_n etc. represent steady states for equal increments of applied field.

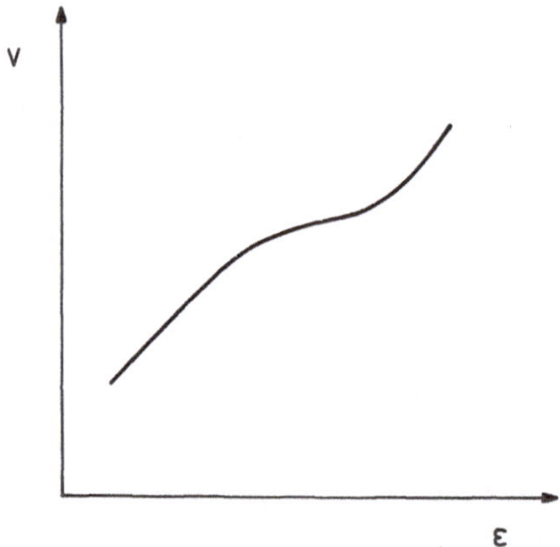

Fig. 7. Electron drift velocity in field direction v ($\propto K^{1/2}$) as a function of applied field ξ deduced from intersection points A_n in Fig. 6.

dependence of the drift velocity on electric field strength can be obtained. Dependence of this sort has been observed for a number of hydrocarbon liquids (Schmidt, 1984) and may well be good evidence for the vibrational scattering processes.

ELECTRODE PROCESSES

We now turn attention to conditions at the electrodes. These play vital roles in establishing the pre-breakdown conditions in the liquid under high electric stress and in triggering the breakdown itself. It has been natural to invoke electron injection at the cathode as an important component since high fields will lower the potential barrier to electron transfer across the interface whether it occurs by a thermally activated or tunnelling process. However, employment of the Schottky formula for field-assisted thermionic emission or the Fowler-Nordheim one for tunnel emission which are appropriately applicable only for electron transfer to a vacuum is a much too simplified solution to the problem.

Under zero field conditions the metal and liquid in contact will attempt to reach an equilibrium by electron exchange. The energy states involved will be for the metal those both occupied and empty near the Fermi energy E_f and for the liquid the band of empty electron states with a band edge at V_o and localized levels E_t below and the band of occupied donor states centered on $\bar{E}_d$ as discussed earlier. We have already noted that for a hydrocarbon liquid such a n-hexane, we expect V_o to be near the vacuum level and $\bar{E}_t$ a few tenths of an eV below while $\bar{E}_d$ would be about -6.5 eV. The contact situation would then be as in Fig. 8 where, if we assume E_f for the metal electrode to be about -4.5 eV and that a Fermi energy could be defined for the liquid as midway between $\bar{E}_t$ and $\bar{E}_d$

(i.e. at about -3.3 eV, Fig. 2), equilibrium would require electron transfer from the liquid to the metal. This would leave a space charge of positive holes (ions) within the liquid in the neighborhood of the electrodes.

Conditions will change when a bias potential is applied between the electrodes. At the cathode the additional potential difference across the interface region will encourage a reverse transfer of electrons back into the liquid while at the anode more positive ions will be injected. Carrier injection of this sort will make the liquid more conductive but it is also possible for an applied potential difference to have the opposite effect. If residual ions already exist in the liquid as a result of impurity dissociation, then they can be transported to the electrodes and neutralized there provided the over-potentials are sufficient as shown in Fig. 9 (Barret et al., 1975).

Efficient electron transfer into the liquid from the cathode will require a high field at that electrode. This can be obtained if any positive ions or holes in E_d states which drift to that electrode are not readily neutralized and form a space charge. Such a situation is likely to be common since, in practice, most electrodes will be covered with natural oxide or other semi-insulating layers. Hydrocarbon deposits, perhaps arising from electropolymerized liquid may also be adsorbed to the electrodes. On first consideration it might be concluded that such layers should inhibit charge injection but, in fact, they could enhance it considerably.

Cathode Processes

A number of authors have determined the emission characteristics of similarly-covered metal electrodes placed in a vacuum rather than a liquid and Latham and co-workers (1982, 1984, for example) have provided a very plausible model to explain them. The essential ideas are expressed in Fig. 10. It is suggested that micro-regions of the electrode surface of perhaps 10^{-8} m radius contain oxide or other semi-insulating inclusions to a thickness of about 10^{-7} m. The layer is considered as a wide band-gap (2 to 3 eV) insulator but with many imperfections which will act as carrier traps as in an amorphous solid. It is important that a relatively narrow (Schottky) barrier can be expected at the metal-insulator interface. When a field is applied, it penetrates the layer and encourages electrons to tunnel from the metal into the conduction band of the insulator filling electron traps and causing electron accumulation at the insulator-vacuum interface (Fig. 10).

When the applied field is sufficiently strong and the traps become filled, free electrons of high energy will be produced which are swept to

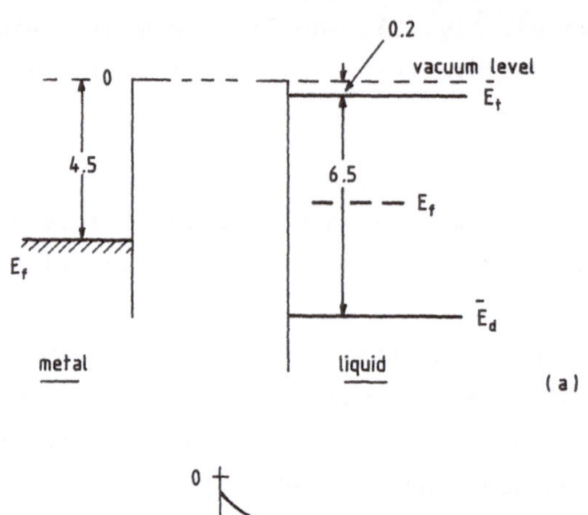

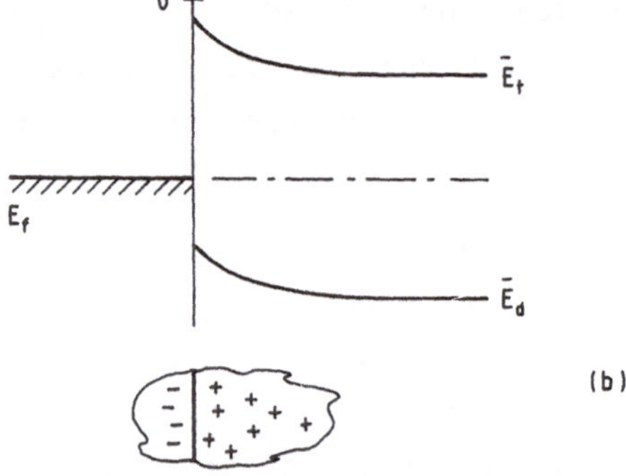

Fig. 8. Metal electrode and liquid interface (a) before contact is made and (b) after contact equilibrium is reached with positive hole (ion) space charge (Helmholtz and Gouy) layer in liquid.

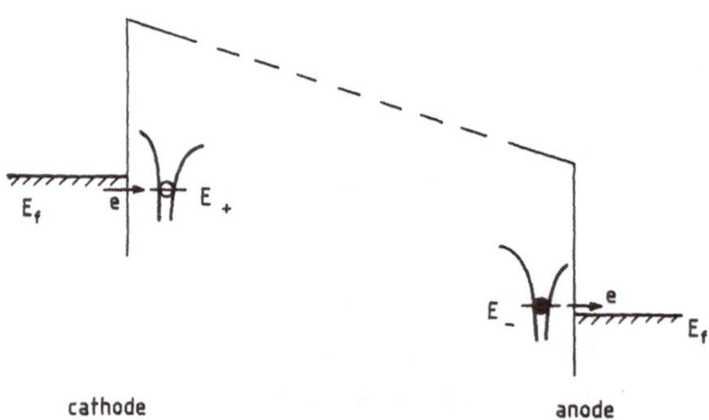

Fig. 9. Neutralization of ionic impurities (E_+,E_-) at electrodes under sufficient applied voltage (overpotential) to give $E_+ \leq E_f$ and, or $E_- \geq E_f$.

the outer face of the oxide to give a pool of electrons with high effective temperature T_e. Some holes may also be produced by collision ionisation and these will drift back to the metal face to enhance electron tunnelling.

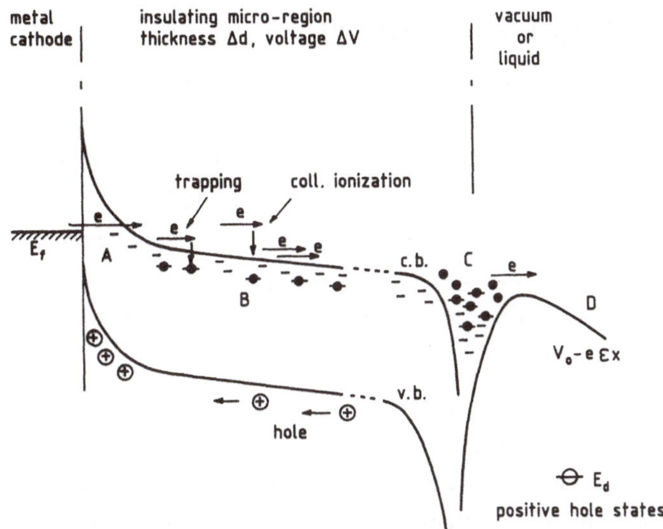

Fig. 10. Energy states for an insulating micro-region on a metal cathode under an applied field (after Latham, 1982). Electrons tunnel from cathode to conduction band of insulator through Schottky barrier, A. Electron traps become filled, B. Electrons accumulate at electron-affinity barrier at insulation-vacuum interface, C. Holes produced by collision ionization drift to A and enhance electron tunnelling. Electrons with enhanced kinetic energy emitted over barrier C into liquid conduction band, D. Positive hole states of liquid E_d, drifting to interface, readily transfer holes to insulator valence band and, thence, to barrier A.

The condition for electron emission into the vacuum would be that the applied field should generate a potential drop to compensate for the Schottky barrier, any resistive drop in the layer and the electron affinity at the vacuum interface. Latham estimated that applied fields of ~2 x 10^7 V m^{-1} would be sufficient for this. The emission would follow the Richardson-Dushman law for thermionic emission, namely the current density would be

$$J = AT_e^2 \exp(-e\chi/kT_e) , \qquad (7)$$

where A is the usual emission constant, T_e the elevated electron temperature and χ the electron affinity barrier. The temperature T_e is determined by assuming that all the potential energy drop across the layer ΔV is turned into kinetic energy i.e. $e\Delta V = \frac{3}{2} kT_e$ and that $\Delta V = \Delta d/d\varepsilon\, V$ where Δd/d is the ratio of oxide layer thickness to electrode spacing, ε the relative permittivity of the oxide and V is the applied voltage. On substitution, the total current from a micro-region of area 'a' is then

$$I = Ja = K\,V^2 \exp\,[-\chi/(2\Delta d/3d\varepsilon)\,V]\ . \qquad (8)$$

The importance of this result is that it yields a straight line on a Fowler-Nordheim, log (I/V^2) versus V^{-1} plot with a slope $-\,3\,\chi d\varepsilon/2\Delta d$ which should be compared with the slope of a similar plot for field-emission from a clean metal which is $-\,2.84 \times 10^9\, d\phi^{3/2}\,\beta^{-1}$ where ϕ is the work function of the metal and β is a field enhancement factor due, for example, to a surface asperity. Thus, the physical interpretation of the slope of the F-N plot and particularly the field-enhancement factor β will be very different in the Latham model.

Latham's model (a similar one has been proposed by Kao, 1984) requires little, if any, modification if it is applied to electron emission into a pure hydrocarbon liquid. Since the conduction band states at an energy V_0 are close to the vacuum level, the oxide outer barrier χ will be hardly altered. Moreover, the electric fields at breakdown in a liquid are generally much greater than those suggested by Latham as necessary for barrier lowering and the relative permittivities of oxide and liquid will be more favorable than oxide and vacuum. There is a further factor of considerable significance, namely that in the liquid case it will be possible for positive ions or holes from the liquid to be fed back to the oxide surface. Moreover, the energy level E_d of this band of states will be about the same as that of the valence band of the oxide. As a consequence, the hole states could be transferred readily from the liquid through the oxide to the metal interface where they will enhance electron tunnelling.

The features of the Latham model which make it appropriate for adoption as the mechanism for electron emission into liquid hydrocarbons can be summarized as follows.

a. It describes a highly localized process dependent on electrode surface conditions.

b. The emission consists of energetic 'hot' electrons, an important factor for breakdown initiation as will be discussed later.

c. Emission is expected to become efficient in fields of 10^7 V m^{-1} or greater which correspond to incipient breakdown fields for hydrocarbon liquids.

d. Fowler-Nordheim-like current voltage plots are predicted as found for liquids (see Schmidt, 1984) but the interpretation of the parameters for the plots are quite different and more realistic.

Anode processes

At the anode neutralization of negative ions and quasi-free electrons acquired from the liquid and the possible injection of positive holes or ions need to be considered.

Since the excess-electron conduction band for hydrocarbon liquids will be in the neighborhood of the vacuum level, neutralization of electrons at a clean metal anode should be facile (Fig. 11). Positive hole injection would be difficult, however, because the E_d band of positive hole states of the liquid will be at about -6.5 eV and, thus, well below the Fermi level of bare anode metal. Consequently, a barrier of about 2 eV or so to electron transfer and, thus, to hole production at a bare metal anode will exist.

The situation in the neighborhood of a micro-region of oxide such as was considered for the cathode will be quite different. Here any quasi-free electrons arriving from the liquid are likely to be trapped in surface and bulk states of the oxide establishing a strong field across it as shown in Fig. 11. This field would be reinforced by an applied field and could produce the situation shown where the metal-oxide barrier is reduced to permit electron tunnelling from the valence band of the oxide. This will generate holes in the oxide and, as already suggested, these should move readily into the liquid at the oxide-liquid boundary.

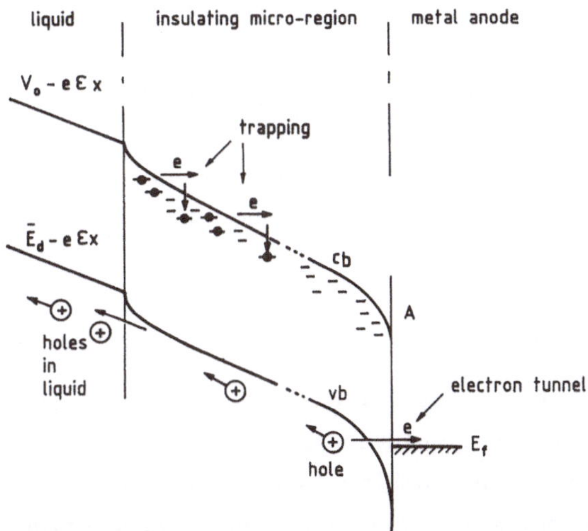

Fig. 11. Situation corresponding to Fig. 10 but at an anode. Electrons from liquid states ~V_o are held in bulk and interface trap states of insulator micro-region, establishing a strong field across region. This field, enhanced by applied field, reduces barrier at insulator-metal interface A so that electrons tunnel from valence band of insulator to metal states near E_f. Holes drift to outer liquid-insulator boundary and readily move into liquid as positive hole (ion) states.

Thus hole injection at the anode has features similar to electron injection at the cathode. In particular, it is localized at the oxide micro-regions and depends for its efficiency on the arrival of heterocharge from the liquid. Such heterocharge might be generated at the cathode.

Electrochemistry

The mechanisms proposed above for electrode processes of charge injection as well as the modelling of the ion states within the liquid are special cases of redox electrochemistry in which the liquid, rather than being an electrolyte, is initially hardly ionized at all and the redox processes are induced only by the application of very considerable overpotentials. The role of undischarged ions, which in normal electrolytes are present as Helmholtz and Gouy layers confined to regions adjacent to the electrodes, are, in dielectric liquids, less well defined. The situation becomes even more uncertain in electrochemical terms when such things as oxide phases intervene between metal and liquid. Felici and co-workers (see for example Felici, 1982) has drawn attention to the usefulness of electrochemical concepts in understanding high-field conduction in dielectric liquids.

There is another aspect of electrochemistry which could be important, particularly in a practical sense, which is a consequence of the likely injection of 'hot' electrons of several eV energy at the cathode. Such electrons could cause molecular dissociation rather than ionization, as Kao (1984) has pointed out, since the energy for bond breaking is between 3.5 and 4.5 eV for hydrocarbons. The free-radicals produced would, by initiating polymerization, encourage the formation of insulating layers at the electrodes. This may be an important feature of the high-field conduction and breakdown processes to which insufficient attention has been paid.

BREAKDOWN INITIATION

Having reviewed the basic electronic processes that are likely to operate in a dielectric liquid under electrical stress, we now consider how these processes might combine to produce a breakdown instability. The appropriate combination should result in predictions about the instability which are in agreement with experimental observation.

Study of liquid breakdown has been extensive and has continued for many years. To some extent the essential features are often lost in a wealth of detail which probably pertains to special conditions of particular experiments. However, the following appear to be the main features of the breakdown of 'pure' liquids obtained from experiment which any satisfactory model of the process should be capable of predicting. The purely electrical evidence is as follows:

- The breakdown strength is influenced by hydrostatic pressure and by the presence of dissolved impurities. In particular, dissolved oxygen, which is an electronegative gas, raises the strength.
- Repeated measurements of the breakdown strength with the same pair of electrodes can often lead to an increased strength even though the electrodes become marked by discharge craters and asperities.
- The application of pulse voltages and the measurement of time-lags to breakdown reveals a statistical feature of the breakdown process which is influenced by conditions at the cathode. While the statistical time-lag can be as long as several microseconds, there is a formative time for the breakdown process which depends on overvoltage, electrode spacing and uniformity of field and can be very short. For example, breakdown of a 2 mm uniform-field gap in n-hexane at a field of 10^6 V cm^{-1} can be completed in less than 500 ns (Wong and Forster, 1977; Beddow and Brignell, 1966).
- In general, craters and asperities caused by a preceding breakdown do not become sites for subsequent breakdown. Under certain circumstances, the condition of the anode surface influences the breakdown strength.
- When stressed to just below the level for the onset of breakdown a fluctuating current flows and light pulses can be emitted.

Turning to optical evidence of the breakdown process, laser-Schlieren techniques, which are capable of detecting microscopic local changes in liquid density on a nanosecond time scale, have been used. They have revealed that tree-like structures of low density can extend about 0.5 mm out from the cathode as little as 50 ns after the application of a breakdown field. Several 'trees' can appear simultaneously from the cathode surface and breakdown will ensue from one of them. An alternative sequence leading to breakdown which appears to be initiated by an outward growth of low density regions from the anode which is extremely fast has also been observed. The cathode and anode initiated structures are different in appearance, with the latter being much more filamentary and diffuse.

Several of the features of the breakdown process listed above can be accounted for in terms of the mechanisms described earlier.

The observation of highly localized and separated sites for breakdown on both electrodes and particularly the cathode is supportive of the idea of patches of oxide or other insulating film being essential for the breakdown process. Such layers would be destroyed by a total breakdown discharge and could not rapidly reform in a pure liquid environment which excluded oxygen. Statistical effects could also be expected if sites have to be triggered into activity, for example, by the arrival of field-enhancing heterocharge from the other electrode. Emission from these sites of electrons or holes might be expected to be fluctuating and erratic, therefore.

Breakdown is a very fast process, at least in uniform fields, but the reported values of electron and hole mobility and the likely field-dependence of these will mean that charge carriers could make a transit of the electrode gap within the formative time of breakdown. For example, under breakdown fields when hot electrons can be expected (see below), the limiting value of mobility μ_e^o ($\sim 9 \times 10^{-3}$ m^2 V^{-1} s^{-1}) would yield a transit time across a 1 mm gap under a stress of 10^8V m^{-1} of only 1 ns. This is obviously an extreme situation but it does illustrate the possibility.

The reported positive hole mobilities are very much less but the quantum-mechanical resonance nature of the transport will make it extremely dependent on field-strength and at breakdown fields much higher mobilities can be expected.

Under pre-breakdown conditions with a field of $\sim 10^8$ V m^{-1} impressed, there would be plenty of time for charge carriers to cross the electrode gap and, thus, for cathode and anode events to influence each other to a marked degree as breakdown begins.

Cathode initiation

An important aspect of the Latham model is that the electrons entering the liquid at a cathode site will not only be energetic but will have a correspondingly large directed momentum normal to the electrode surface. Thus, they will be injected into the liquid as a column or jet with a cross-sectional area corresponding in size to the insulating micro-regions of the cathode, Fig. 12. Moreover, if their energy is in excess of that for the maximum in the composite vibrational collision cross section (Fig. 6), which is likely to be less than 1 eV, then, gaining more energy from the field than is lost in vibrational collisions, they could accelerate to higher energies. At the same time, the energy that is lost will increase the thermal energy of the liquid in the column and cause it to begin to expand. Electrons continuing to be emitted into the column will, therefore, find conditions increasingly favorable for acceleration in the electric field. The so-called molecular regime (Sanche, 1979) will begin to apply in which the collisions become molecular in character. The relevant processes are then molecular

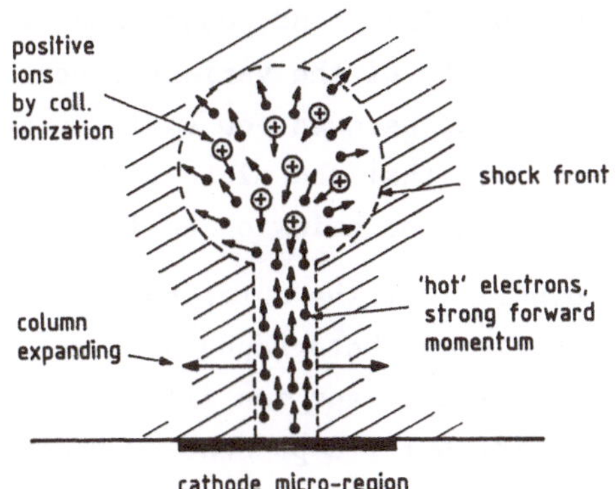

Fig. 12. Initiation of cathode 'tree' (column and expanding head). Hot electrons with strong forward momentum establish radially expanding column. Molecular collision processes at head of column produce lower energy electrons. Head expands spherically under Coulomb forces with shock front. Positive ions feed back down column to reinforce emission. Low density 'tree' forms an improved path for electron-molecule exciting and ionizing collisions.

dissociation, excitation and ionization with characteristic energies for hydrocarbons of about 4,6 and 9 eV respectively.

The occurrence of one the above types of electron-molecule collision at the end of a free path will result in destruction of the forward momentum and loss of energy by the electron so that it then becomes a low energy low velocity particle. If ionization occurs then two such electrons will be produced with an attendant positive ion which will move back down the column to the cathode patch to maintain emission.

The conversion of 'hot' mobile electrons into low-energy ones together with the production of additional low-energy electrons by liquid ionization at the head of the column will mark the onset of a spherical expansion under the Coulombic forces of the consequent cloud of negative charge, Fig. 12. The abrupt conversion of forward momentum of the column together with considerable electron energy release to the liquid will create an explosive situation in the head and set up conditions for shock-wave propagation outwards from it.

Of the other processes, free-radical components from molecular dissociation are likely to lead to deposits on the cathode surface while molecular excitation and subsequent return to the ground state will cause light output. Excited molecules might also be involved in further production of free electrons (and positive ions) via the Auger process.

Following the initiatory phase, electrons will continue to move out from the cathode along the column into an expanding head where the density is decreasing and where ionizing and exciting collisions will become increasingly likely. On the other hand, continuing radial expansion might reduce the volume density of the various processes to the point where feed-back to the cathode becomes too small to maintain the system. There is also the possibility that the micro-region of insulating oxide on the cathode might fail and the emission collapse.

It is important to note that the outer boundary of the head of the column as observed by Schlieren photography does not indicate the limit of electrical activity. Electron and ion flow could continue out into the liquid beyond that point but its presence would not be made visible by sufficient liquid expansion. If the emission is maintained the possibility of continued growth of ionization outwards from the head of the column to bridge the electrode gap and complete breakdown is possible.

Column expansion may be expected to be inhibited by increased hydrostatic pressure in the liquid and a consequence would be less efficient ionization and a higher breakdown strength. On the other hand, the ability to apply a higher electrical field would improve cathode emission and lead to more energetic electrons being emitted into the liquid (Hebner et al., 1987). Electron-attaching impurity, such as dissolved oxygen, may also be expected to increase the strength since it will reduce the overall ionization efficiency in both the column and the head and in the region beyond.

Anode initiation

Turning now to consider breakdown initiation by an anode event, the question is whether positive hole injection at an insulating micro-region could lead to the development of the low-density rapidly propagating filamentary structures which have been observed.

The only possibility for very rapid positive charge development from the anode is hole propagation by the resonance tunnelling mechanism described earlier (Fig. 4). This depends sensitively on inter-molecular spacing which must be as small as possible if a high tunnelling probability is to be achieved. Hole propagation pathways will be established, therefore, only in high density regions of the liquid. At each tunnelling step in the field direction, energy gained will be transferred to the liquid. In a field of 10^8 V m^{-1} about 75 meV would be available as a consequence of each resonance transfer and this would be converted via vibrational modes to thermal energy. Since only 100 meV per molecule is required to raise the temperature of n-hexane from normal temperature to the boiling point this transfer is significant. It will lower the local liquid density and, thus, reduce the tunnelling efficiency for any subsequent hole transfer along the same path. This is quite unlike the condition for cathode initiated events where electron transport is enhanced, not reduced, by lower liquid density.

Anode-initiated discharges will, therefore, be characterized by filamentary low-density pathways marking the one-time passage of a hole. Each hole will seek a path through high density regions in which resonance transfer is facile and, thus, the paths will be distinct and separate.

Hole propagation may be expected to be pressure-dependent. Increased density will encourage efficient tunnelling and increase the transfer probability integral (see earlier) which is very sensitive to the intermolecular spacing. Thus, the very marked increase in propagation velocity with pressure for anode initiated events in n-hexane observed by Hebner et al., (1987) are to be expected. Since hole transport requires electron transfer, mobilities comparable with electron mobilities are conceivable.

Electron and hole pathways have a degree of complementarity. A hole pathway along which the density had been lowered will be a preferred pathway for electrons travelling in the opposite direction and electron-molecule collision excitation and ionization will be enhanced. The converse will not be true, however, and holes would not propagate along electron pathways of reduced density. Positive ions, as distinct from holes and as molecular entities, will, however, become more mobile in regions of low density and can be expected to propagate back to the cathode along previous electron pathways. Thus, there is the possibility of rapid positive hole propagation in dense liquid until an already developed, low-density electron pathway is encountered, whereupon subsequent propagation of the positive charge to the cathode would be as a less-mobile ion. At the same time electrons in the electron pathway would have a low-density path to the anode immediately available to them.

CONCLUSIONS

The fundamental processes of charge carrier generation and motion in hydrocarbon liquids reviewed above appear to be sufficient to explain the onset of electrical breakdown under a variety of conditions.

The overall description of the breakdown instability has two parts. In the first pre-breakdown part, the conditions in the condensed liquid and at the electrodes are best described in the terms of the electronic solid state and the electronic states of the liquid and of the liquid-electrode interfaces are important. In the second part, at the onset of breakdown as energy is transferred from the electric field to the liquid via charge carriers, this description gives way to one of the individual molecules and gas-phase collision processes.

Thus, features of both disciplines contribute to the overall description and the crucial step which irrevocably commits an electrode gap in a liquid to breakdown may, depending on experimental conditions, be in either the condensed or gaseous phases.

An essential first step, however, does appear to be the establishment of an adequate localized charge carrier injection at an electrode, most probably of electrons at the cathode. Without a sufficient density of charge carriers it is probable that energy transfer to the liquid is insufficient to induce a final gaseous phase. The need for an adequate density of carrier injection leads to strong dependence on electrode conditions and to the statistical aspects of the breakdown process.

REFERENCES

Adamcjewski, I., 1969, "Ionization, Conductivity and Breakdown in Dielectric Liquids", Taylor and Francis, London.

Barret, S., Gaspard, F., and Mondon, F., 1975, Monitoring of injection processes in dielectric liquids in "Conduction and Breakdown in Dielectric Liquids", J.M. Goldschvartz, ed., Delft University Press, Delft.

Beddow, A.V., and Brignell, J.E., 1966, Nanosecond breakdown time lags in a dielectric liquid, Electronics Lett., 2:142.

Dodelet, J.P., and Freeman, G.R., 1977, Electron mobilities in alkanes through the liquid and critical regions, Canadian J. Chem., 55:2264.

Felici, N.J., 1982, A tentative explanation of the voltage-current characteristic of dielectric liquids, J. Electrostatics, 12:165.

Fueki, K., Feng, D-F, and Kevan, L., 1972, A semi-empirical estimate of the scattering cross section and mobility of excess electrons in liquid hydrocarbons, Chem. Phys. Lett., 13:616.

Gee, N., and Freeman, G.R., 1983, Effects of molecular properties on electron transport in hydrocarbon fluids, J. Chem. Phys., 78:1951.

Hebner, R.E., Kelly, E.F., Fitzpatrick, G.J., and Forster, E.O., 1987, The effect of pressure on streamer inception and propagation in liquid hydrocarbons, "Conference Record, 9th International Conference on Conduction and Breakdown in Dielectric Liquids", Salford, U.K.

Hiraoka, K., and Hamill, W.H., 1973, Characteristic energy losses by slow electron impact on thin film alkanes at 77°K, J. Chem. Phys., 59:5749.

Holroyd, R.A., Dietrich, B.K., and Schwarz, H.A., 1972, Ranges of photoinjected electrons in dielectric liquids, J. Chem. Phys., 76:3794.

Hummel, A., and Luthjens, L.H., 1973, Ionization in the track of a high energy electron in liquid cyclohexane; pulse radiolysis of solutions of biphenyl in cyclohexane, J. Chem. Phys., 59:654.

Jortner, J., 1982, Discussion in electron and proton transfer, Farad. Disc. Chem. Soc. No. 74:193, Roy. Soc. Chem., London.

Jortner, J., and Gaathon, A., 1977, Effects of phase density on ionization processes and electron localization in fluids, Canadian J. Chem., 55:1801.

Kao, K.C., 1984, New theory of electrical discharge and breakdown in low mobility condensed insulators, J. Appl. Phys., 55:752.

Kebarle, P., Davidson, W.R., French, M., Cumming, J.B., and McMahon, T.B., 1977, Solvation of negative ions by protic and aprotic solvents: Information from gas phase ion equilibria measurements, Farad. Dis. Chem. Soc., No. 64:220, Roy. Soc. Chem., London.
Latham, R.V., 1982, The origin of pre-breakdown electron emission from vacuum insulated high voltage electrodes, Vacuum, 32:137.
Latham. R.V., and Athwal, C.S., 1984, Switching and other non-linear phenomena associated with pre-breakdown electron emission currents, J. Phys. D. Appl. Phys., 17:1029.
Lewis, T.J., 1956, Mechanism of electrical breakdown in saturated hydrocarbon liquids, J. Appl. Phys., 27:645.
Magee, J.L., 1977, Electron energy loss processes at subelectronic excitation energies in liquids, Canadian J. Chem., 55:1847.
Mehnert, R., Brede, O., Bos, J., and Naumann, W., 1982, Transfer of the positive charge in non-polar liquids studied by pulse radiolysis, J. Electrostatics, 12:107.
Noda, S., Kevan, L., and Fueki, K., 1975, Conduction state energy of excess electrons in condensed media, liquid methane, ethane and argon and glassy materials, J. Phys. Chem., 79:2866.
Sanche, L., 1979, Transmission of 0-15 eV monoenergetic electrons through thin-film molecular solids, J. Phys. Chem., 71:4860.
Schiff, L.I., 1955, "Quantum Mechanics, 2nd Ed.", McGraw-Hill, New York.
Schiller, R., Vass, Sz., and Mandics, J., 1973, Energy of the quasi-free electrons and the probability of electron localization in liquid hydrocarbons, Int. J. Radiat. Chem., 5:491.
Schmidt, W., 1977, Electron mobility in non-polar liquids: The effect of molecular structure, temperature and electric field, Canadian J. Chem., 55:2197.
Schmidt, W., 1984, Electronic conduction processes in dielectric liquids, IEEE Trans. Elec. Insul., EI-19:389.
Wong, P., and Forster, E.O., 1977, High speed Schlieren studies of electrical breakdown in liquid hydrocarbons, Canadian J. Chem., 55:1890. See also IEEE Trans. Elec. Insul., EI-12:435.

ELECTRON SCATTERING AND DIELECTRIC BREAKDOWN IN LIQUID AND SOLID DIELECTRICS

H.R. Zeller and E. Cartier

Brown Boveri Research Center
5405 Baden Switzerland

We describe an experiment which yields data on the momentum loss and energy loss rate of hot electrons. Experiments on solid and liquid long-chain paraffins show a mobility edge a few tenths of an eV wide in the liquid, and no significant differences between solid and liquid for $E_{kin} \geqq 0.5eV$.

Polar-mode scattering leads to a maximum in the energy loss rate around $E_{kin} = \hbar\omega^{LO} \simeq 0.36eV$ (C-H stretch mode). Above $E_{kin} \simeq 2eV$ the momentum loss is controlled by deformation potential scattering (acoustic and nonpolar phonons).

The measured scattering functions correctly account for the observed breakdown field in solid paraffin.

INTRODUCTION

A solid or liquid dielectric inserted between two electrodes can support only a limited voltage. Several physical mechanisms can lead to a current instability and to breakdown, e.g., thermal instabilities in materials with thermally activated conductivities, transitions from trap-controlled transport to band transport, impact ionization, etc. (Zeller, 1987). Which one of the different mechanisms ultimately determines the dielectric strength depends on materials parameters, geometry, voltage pulse forms (including history), temperature, etc. (O'Dwyer, 1973).

Breakdown in general implies a current-controlled (S-shaped) negative differential resistance region. As a consequence, the current distribution becomes filamentary in nature.

In this lecture we will concentrate on a single breakdown mechanism, i.e., impact ionization. Impact ionization as a mechanism for breakdown in condensed mater was first investigated by Fröhlich (1937). He considered optical phonon scattering in alkali halides. There the phonon energy $\hbar\omega^{LO} \approx kT$, and hence efficient scattering occurs at thermal electron kinetic energies E_{kin}. As soon as the electrical field F is such that $E_{kin} > \hbar\omega^{LO}$, then the scattering becomes less efficient. This leads to an unstable situation in which the energy gain from the field can no longer be balanced by scattering. Once an electron has exceeded $\hbar\omega^{LO}$ it will rapidly gain energy until impact ionization occurs.

This simple picture is unphysical for several reasons. Two of them are minor and can easily be corrected. The first is that the average electron model is not suitable to describe impact ionization, and the second that for practical dielectrics $\hbar\omega^{LO} > kT$. However, there are more serious objections. Polar-mode scattering is not the only scattering channel, and, particularly in wide band-gap and non- or weakly-polar materials, other scattering channels are expected to be of comparable importance. Polar-mode scattering is completely absent in rare gas liquids. Also, Fröhlich's model is basically a band model, and it has been argued that the extremely small mobilities observed in many solid dielectrics exclude a band structure model.

An instantaneous thermal destruction of a solid requires an energy deposition of the order of 10^9 J/m^3. With a breakdown field of the order of 10^9 V/m this results in a charge flow of the order of 1 Asec/m^2. Assuming a homogeneous flow and essentially only displacive currents (as would be the case in an avalanche) we would obtain a self-field of the order of a few 10^{10} V/m, depending on the dielectric constant ε (O'Dwyer, 1973). Often it has been concluded that this completely excludes avalanche breakdown. In fact, it only excludes a homogeneous avalanche front and implies that in filamentary avalanches the self-fields are a controlling factor very much as in the case of leaders in gaseous discharges.

PHENOMENOLOGICAL MODELS

In most practical cases breakdown starts from a point of strong local field enhancement, e.g., at an irregularity of an electrode. If the current-voltage characteristic is of threshold type with a threshold field F_{th}, then any field enhancement above F_{th} will be rapidly counteracted by the self field of the injected space charge. As long as the applied voltage V is smaller than $F_{th} \cdot d$, no breakdown will occur (d is the electrode spacing). In steady state, $|\vec{F}| < F_c$ outside (Hibma and

Zeller, 1986). Space-charge injection thus smoothes and limits local field enhancements, and, unless the overall breakdown voltage is reached, it is self-limited.

The picture completely changes if we reach a current-controlled negative-differential-resistance (NDR) regime. As a consequence of NDR the current distribution decays into filaments. In the simplest model two parameters are needed. One is the breakdown field, or threshold field, for growth of a filament F_b, and the other is the channel field F_{ch} (Zeller, 1987; Wiesmann and Zeller, 1986). The contraction into filaments leads to the situation that a local field enhancement becomes self-enhanced and propagating at the tip of the filament. If $V > F_{ch} \cdot d$, then the filament will reach the counter-electrode and breakdown will occur.

For thin and long filaments (radius r << length l) in a homogeneous field F ($F > F_{ch}$), an approximate analytical solution can be given. Assuming that the filament originates at a planar electrode the line charge distribution $\tau(z)$ in a filament becomes (Landau and Lifshitz, 1960)

$$\tau(z) \sim \frac{4\pi\varepsilon\varepsilon_o z(F-F_{ch})}{\log\left[\frac{4}{r^2}(l^2-z^2)\right]-2} \qquad (l \gg r) . \tag{1}$$

The above analytical expression breaks down at the end of the filament ($z = l$). Nevertheless, we clearly note a strong peak in the line charge at $z \sim l$, and, as a result, a strong field enhancement in front of the filament tip. This field enhancement may lead to impact ionization even in cases in which the original instability was due to other reasons.

Next we discuss impact ionization. We assume that a seed electron starts an avalanche. The probability that, provided an electron is at position x^1, the next ionization event occurs at $x(x > x^1)$ is $P(x) = \gamma_i \exp[-\gamma_i(x-x^1)]$. γ_i is the ionization rate or inverse ionization length. If the hole mobility is small compared to the electron mobility then we may treat the holes as immobile on the timescale of avalanche formation and obtain, for the line charge along the filament, after N generations

$$\tau(x) = e \cdot 2\gamma_i \exp(-\gamma_i x)\left[\frac{(2\gamma_i x)^{N-1}}{(N-1)!} - \sum_{k=1}^{N-1}\frac{(2\gamma_i x)^{k-1}}{(k-1)!}\right] . \tag{2}$$

Now we are going to show that for all practical purposes impact ionization at the destructive level is self-field controlled even in thin filaments. The following will be order-of-magnitude estimates and we neglect factors of order one.

The radial field at the surface of a filament with radius r and line charge τ is

$$F_s \sim \frac{\tau}{2\pi\ \varepsilon\varepsilon_o\ r}\ . \tag{3}$$

The energy density within a filament is of the order $\frac{\tau \cdot F}{\pi r^2 \gamma_i}$. Assuming that $F \sim F_b$ (breakdown field) and $F_s \ll F_b$ (otherwise defocussation would occur), we obtain, with U_{crit} = critical energy density for destruction:

$$U_{crit} \cdot r \ll \frac{2\ \varepsilon\varepsilon_o\ F_b^2}{\gamma_i}\ . \tag{4}$$

For example, with $F_b = 5 \cdot 10^8$ V/m, $\varepsilon = 3$, $(\gamma_i)^{-1} = 50\ \mu$, and $U_{crit} = 10^9$ J/m^3, self-field effects are important if the filament radius is not much smaller than 0.7 µm. A typical momentum scattering length is lp = 20Å. Within two ionization events there will be in this example 25,000 scattering events. This leads to a radial spread of the order of $l_p \sqrt{\frac{1}{\gamma_i lp}} \sim 0.3$ µm. In other words, within a few generations the statistical spread of the avalanche becomes larger than 0.7 µm. Self-field effects thus will dominate before N_{crit} is reached.

The electrons and holes generated by impact ionization drift in opposite directions. This generates a dipole field which counteracts the external field. Since γ is a very steep function of F the self-fields have a regulation effect on the field distribution. Whenever $F > F_c$, then rapid impact ionization sets in which reduces the field. Since an avalanche never represents a stationary solution, the net result is not $F = F_c$, as in the threshold mobility situation (Hibma and Zeller, 1986). In practice, numerical simulation models have to be used to describe the evolution of an avalanche until destruction of a solid or liquid dielectric.

SCATTERING OF HOT ELECTRONS

The scattering of hot electrons is controlled by two scattering functions, both of which depend on energy. The first is the momentum loss rate, $\gamma_p\ (E_{kin})$. Within the average electron model the drift velocity v_d of a charge carrier is given by

$$v_d = \frac{e \cdot F}{m^* \cdot \gamma_p\ (E_{kin})}\ . \tag{5}$$

The kinetic energy, on the other hand, depends on the energy loss rate $\gamma_u(E_{kin})$ as:

$$E_{kin} = \frac{e \cdot F \cdot v_d}{\gamma_u (E_{kin})} = \frac{e^2 F^2}{m^* \gamma_p (E_{kin}) \cdot \gamma_u (E_{kin})} . \tag{6}$$

Instead of the scattering rates it is also possible to use scattering or loss lengths l_p, l_u given by $l = \sqrt{\frac{2E}{m^*}} \frac{1}{\gamma}$. This results in

$$E_{kin} = eF \sqrt{\frac{l_u l_p}{2}} . \tag{7}$$

Of course, the average electron model is inadequate to discuss impact ionization. A high energy tail in the kinetic energy distribution can cause avalanche breakdown even if the average electron energy is still small. Better approximations of the Boltzmann transport equation show that the energy distribution $n(E_{kin})$ depends on the integral (Sparks et al., 1981)

$$\int_0^{E_{kin}} m^* \gamma_u (E') \gamma_p (E') dE' , \tag{8}$$

which is equivalent to

$$\int_0^{E_{kin}} \frac{2E' dE'}{l_u(E') l_p(E')} . \tag{9}$$

We have designed a method to experimentally determine the scattering functions (Pfluger et al., 1984a, 1984b). A metal surface is covered with a thin film of the dielectric. The film thickness d is of the order of a mean free path, i.e., of the order of 100Å. Photoelectrons are then excited from the metal into the dielectric by UV irradiation. The electrons escaping from the dielectric into the vacuum are collected and their kinetic energy distribution is determined. For various reasons we have performed our experiments on long-chain linear alcanes $C_n H_{2n+2} (n >> 1)$. First, the alcanes are electronically very simple model compounds for polyethylene ($n = \infty$). Second, they have a negative electron affinity which means that electrons do not encounter a barrier when escaping into the vacuum. Also, the films can be prepared sufficiently clean to avoid trapping of electrons and charging of the films. It turns out that films grow very well on Pt substrates, probably because Pt as a well-known hydrocarbon catalyst strongly interacts with the C-H bond. The Fermi level of Pt is very well matched to the Fermi level of the alcane. This is important because otherwise substantial band bending would occur (Cartier et al., 1987). The energy-level diagram of the experimental set up is shown in Fig. 1.

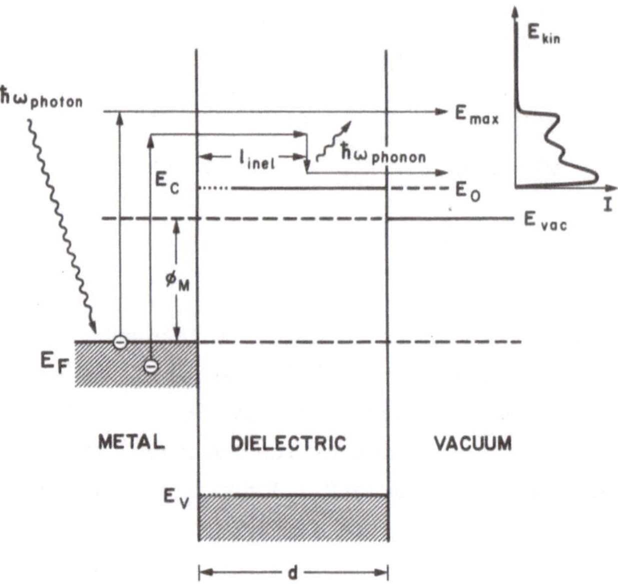

Fig. 1. Energy-level diagram of the substrate/overlayer experiment showing the case of negative (or zero) electron affinity and sub-band-gap photon excitation.

Injected photoelectrons may either ballistically cross the dielectric and escape into the vacuum or undergo scattering processes. Purely elastic processes lead to momentum randomization and will cause a fraction of the electrons to travel back into the metal instead of being emitted into the vacuum. The decrease in the number of emitted electrons as a function of d is thus a measure of γ_p. The downshift in kinetic energy of the emitted electrons, on the other hand, is a measure of the energy loss rate γ_u. Both processes can be easily seen from Fig. 2. With increasing film thickness the escaping electrons rapidly loose kinetic energy and decrease in number.

As will be discussed later the major energy loss process in the energy range of interest is the emission of a LO phonon which corresponds to the C-H stretch mode, and has an energy of $\hbar\omega^{LO} = 0.36$ eV. We make active use of this in the data analysis. If we consider the top-most interval of width $\hbar\omega^{LO}$ in the energy distribution then each momentum scattering event will leave the electron in this interval, but each loss process will remove the electron from the interval. Since it is the top-most interval no new electrons are supplied from higher intervals. If the elastic scattering is isotropic one finds that for $d^2 \gg l_u \cdot l_p$ the electron intensity in the interval varies as (Bernasconi et al., TBP)

$$I(d) = Io \exp(-d/l_{eff}) , \tag{10}$$

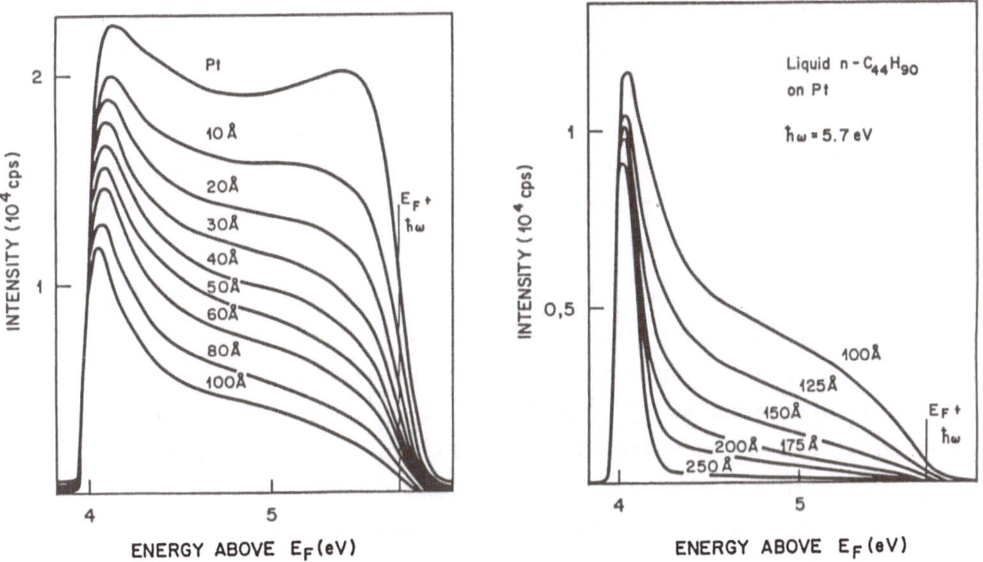

Fig. 2. Kinetic energy distribution of electrons emitted into the vacuum as a function of overlayer thickness. Zero kinetic energy corresponds to about 4 eV. Note the downshift in energy due to inelastic processes and the pile up of electrons with E_{kin} < phonon energy at large d.

where $l_{eff} = 1/2 \sqrt{l_u \cdot l_p}$. A simple analysis of the data thus directly gives the product lu · lp needed to compute the kinetic energy distribution as a function of electrical field and the impact ionization rate. An example of such an analysis is given in Fig. 2. The strong energy dependence of l_{eff} is evident from the raw data.

A determination of the individual values of l_u, l_p requires data analysis at d $\lesssim l_u, l_p$. We found that this does not lead to reliable values for solid films, probably because of changes in film morphology and structure as a function of d. Much better data could be obtained from liquid films (Cartier et al., 1987). To be compatible with UHV requirements the experiments were performed on $C_{44}H_{90}$ a few degrees above the melting point at 86C. Figure 3 shows the energy dependence of l_u, l_p for this material. The values for l_{eff} for solid ($C_{36}H_{74}$) and liquid ($C_{44}H_{90}$) films are virtually identical for $E_{kin} \gtrsim 0.5$ eV.

Next, we are going to discuss theoretical models for scattering processes. In the first model, we treat the electron as a point charge which interacts with the solid by virtue of its Coulomb field. The solid is characterized by its dielectric function ε(q, w). Since it is the longitudinal field of the electron (as opposed to the transverse field of the photon) which couples to the solid (or liquid), the scattering rate scales with Im[1/ε(q, w)] (Landau and Lifshitz, 1960; Sparks et al., 1981). This expression becomes large whenever ε is small, which means phenomenologically that the electron couples to the plasmon, or longitudinal, modes of the system. For small energies this will be the longitudinal polar phonon modes, and for larger energies, electronic excitations such as plasmons. In particular, for alcanes at small energies the dominant contribution to $\varepsilon(q, w)$ comes from the C-H stretch mode at 0.36 eV. Since it is a molecular vibration there is no q dependence and the scattering is nearly isotropic.

In his classical paper, Fröhlich (1937) has studied electron scattering by polar modes with regard to dielectric breakdown. If $kT \ll \hbar\omega^{LO}$, then the scattering rate has a maximum at $E_{kin} \sim \hbar\omega^{LO}$. At high energies γ_u decreases as $E_{kin}^{-3/2}$ and γ_p as $E_{kin}^{-1/2}$. This is a purely phenomenological consequence of any oscillator type $\varepsilon(w)$ and is qualitatively true also for electronic transitions. In particular, for $E_{kin} \lesssim$ 100 eV, the dielectric function $\varepsilon(w)$ is essentially given by sum-rule arguments, i.e., by the volume density of valence electrons. Since the valence electron density is approximately constant for condensed matter, this results in a quasi-universal scattering function (Sea and Dench, 1979) in energy range of about 100 eV to a few keV.

Polar-mode scattering is not the only scattering mechanism. An important scattering channel is deformation potential scattering (Sparks et al, 1981). At low energies it involves scattering with acoustic phonons. In simple solids the acoustic phonons have a small energy, and deformation potential scattering may be treated as quasi-elastic.

More complex compounds, in particular molecular solids or liquids, may exhibit nonpolar modes with energies in the LO phonon range, and thus deformation potential scattering may significantly contribute also to the energy loss rate. Essentially based on phase-space arguments, it is expected that the scattering rate should continuously increase and dominate polar-mode scattering at $E_{kin} \gtrsim$ a few eV. Liquids do not possess a well defined dispersion relation for phonons and thus the phonon dispersion has to be replaced by the structure factor S(q,w) (Cohen and Lekner, 1967).

A special situation is found in materials with no LO modes, such as liquefied rare gases. Here deformation potential scattering is the dominant vibronic scattering channel at small energies.

In disordered media such as liquids, disorder scattering becomes important. Depending on conditions disorder scattering leads to a divergent scattering rate at small energies which is identical to localization. Disorder scattering is mostly elastic and rapidly decreases with increasing energy.

We stress that the list of scattering channels given here is by no means exhaustive. The list appears to be sufficient, however, to discuss the experimental data on liquid and solid alcanes. Figure 3 shows the elastic and inelastic scattering length for liquid $C_{44}H_{90}$. The data analysis is based on the assumption that scattering is either quasi-elastic (acoustic phonons, etc.) or inelastic with LO phonon emission. Since LO phonon emission also randomizes momentum we have $l_u = l_{inelastic}$, $l_p^{-1} = l_{elastic}^{-1} + l_{inelastic}^{-1}$. From the data of Fig. 3 for liquid $C_{44}H_{90}$ we clearly see the disorder contribution to the elastic scattering at $E_{kin} \lesssim 0.3$ eV. It is remarkable that the effect of disorder is limited to fairly small energies. This means that thermal exitations of localized electrons into extended states are frequent. The width of the disorder-induced mobility edge is of the same order as the activation energy for the mobility of thermalized conduction electrons (Schmidt, 1984) in short-chain liquid alcanes.

The inelastic scattering length shows qualitatively the behavior predicted by polar-mode scattering. In fact, $l_{inelast}$ can be semiquantitatively predicted from the experimentally determined $\varepsilon(w)$ (Pfluger et al., 1984b). However, it is evident from Fig. 3 that quasi-elastic

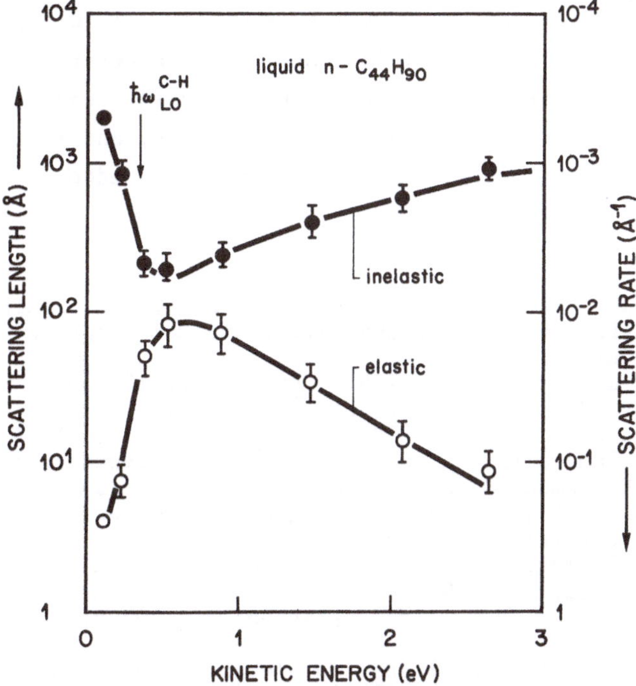

Fig. 3. Elastic and inelastic (LO phonon excitation) scattering processes in liquid n-$C_{44}H_{90}$. LO phonon emission has a maximum near the phonon energy compatible with the Fröhlich model. At small energies the elastic scattering diverges (mobility edge); at high energies it is dominated by acoustic phonon scattering.

processes are of comparable importance at $E_{kin} \approx \hbar\omega^{LO}$, and clearly dominate at higher kinetic energies.

Experiments on solid alcane films give basically the same results at $E_{kin} \gtrsim 0.5$ eV. At smaller energies no data analysis is possible. From the fairly high drift mobilities in crystalline alcanes it can be concluded that the only difference consists in the absence of the disorder-induced mobility edge in the solid.

From the experimental data on $l_{eff}(E_{kin})$ it is possible to calculate a breakdown field (Sparks et al, 1981). For solid $C_{36}H_{74}$ we find $F_b \sim 1$ MV/cm. This compares well with the experimental value of 0.9 - 1.4 MV/cm (Yoshino et al., 1979).

Since in liquid alcanes a localized electron in the conduction band can be relatively easily thermally excited into an extended state, we would expect F_b to be only slightly higher in the liquid than in the solid. We could go one step further and argue that polyethylene, since it is partially amorphous, resembles a liquid alcane and also is expected to exhibit $F_b \sim 1$ MV/cm. This is definitely not confirmed by experiment. Polyethylene can withstand fields up to ~ 5 MV/cm (O'Dwyer, 1973).

The reasons for this discrepancy are not clear. It is unlikely that at $E_{kin} \gtrsim \hbar\omega^{LO}$ the scattering will be significantly different in polyethylene than in alcanes. It is conceivable, however, that the fairly high trap concentration in technical polyethylene leads to a significant increase in scattering at quasithermal energies, and to a very efficient electron capture at energies within the mobility edge.

REFERENCES

Bernasconi, J., Cartier, E., and Pfluger, P., to be published.

Cartier, E., Pfluger, P., Pinaux, J.J., and Rei Vilar, M., Appl. Phys, A. (to appear 1987).

Cohen, M.H., and Lekner, J., 1967, Phys. Rev., 158:305.

Fröhlich, H., 1937, Proc. Roy. Soc., A160:230.

Hibma, T. and Zeller, H.R., 1986, J. Appl. Phys., 59:1614.

Landau, L.D., and Lifshitz, E.M., 1960, Electrodynamics of Continuous Media, Pergamon Press, London, New York, Paris.

O'Dwyer, J.J., 1973, The Theory of Electrical Conduction and Breakdown in Solid Dielectrics, Oxford University Press, London.

Pfluger, P., Zeller, H.R., and Bernasconi J., 1984a, IEEE Trans. on Electr. Insul., EI-19:419.

Pfluger, P., Zeller, H.R., and Bernasconi J., 1984b, Phys. Rev. Lett., 53:94.

Schmidt, W.F., 1984, IEEE Trans. on Electr. Insul., EI-19:389.

Sea, M.P., and Dench, W.A., 1979, Surf. Interface Anal., 1:2.

Sparks, M., Mills, D.L., Warren, R., Holstein, T., Maradudin, A.A., Sham, L.J., and King, D.F., 1984, Phys. Rev., 24:3519.

Wiesmann, H.J., and Zeller, H.R., 1986, J. Appl. Phys., 60:1770.

Yoshino, K., Harada, S., Kyokane, J., and Inuishi, Y., 1979, J. Phys. D., 12:1535.

Zeller, H.R., 1987, IEEE Trans. on Electr. Insul., EI-22:115.

STREAMERS IN LIQUIDS

Robert E. Tobazéon

Laboratoire d'Electrostatique et de Matériaux Diélectriques*
Centre National de la Recherche Scientifique, 166 X
38042 Grenoble Cedex France

INTRODUCTION

Faults in devices subjected to an electric field are mostly due to electric breakdown of the insulations. Such a breakdown is the ultimate stage of irreversible processes which make any material switch from the insulating state to a conducting one (via an arc between two conductors). The situation is more or less catastrophic according to the nature of the insulator: solids, afterwards, are generally unable to sustain a further application of the voltage, whereas fluids (gases, liquids) can often do so, since they are easily renewed in the gap (by natural or forced convection).

Several features of breakdown are common to all kinds of insulating materials:

- Breakdown is localized, the main part of the insulation keeping its original properties;
- Scale effects are important: the lower the gap or the area or the volume subjected to the field, the higher the mean breakdown field necessary;
- The time duration and the slope of the voltage wave (impulse, AC, DC) play a dominant role: breakdown can take place after very short delay times (<1 ns), or (fortunately) after rather long ones (years);
- Electronic processes are always involved in the early stages of the successive steps leading to breakdown, though these, however,

---

* Laboratoire Associé à l'Université Scientifique, Technologique et Médicale de Grenoble

can take place as a result of cooperative effects of a wide variety of more or less intercorrelated phenomena (thermal, mechanical, chemical, electrohydrodynamical,...);

- Tree-like patterns, streaming out from highly stressed regions, can be visualized, and were primarily called streamers in gases, a term later used for similar forms in liquids and solids;
- The breakdown strength is not an intrinsic property, but depends on the shape and the duration of the applied voltage, the electrode geometry, and the treatment and handling of the tested samples.

For certain classes of solids, theoretical models allow us to predict breakdown strengths via electronic mechanisms (crystals), electromechanical quenching (soft polymers), thermal processes, etc... In gases, the Townsend's criterion has been known from the beginning of the century and the behavior of streamers began to be widely interpreted in the early forties (Raether, 1964). A liquid's main role in electrical insulation is to fill air layers (or pockets) in order to eliminate partial discharges. Research on breakdown drew little attention, and knowledge on this subject was restricted, often contradictory, and widely empirical until the fifties. Then, experimental efforts were focused on simple, pure liquids, and new techniques (mainly optical) were developed. However, because of the lack of knowledge on the liquid state, people applied to liquids the theories of gases or solids. So, at the end of the last decade, there were two main schools of thought (Sharbaugh et al., 1978):

- Breakdown results from electron multiplication, a gaseous phase being possibly generated (as a consequence of ionizing collisions);
- Gas bubbles are formed first (several mechanisms are possible), grow, ionize, and lead to breakdown.

Over the past ten years progress in the field of electronic conduction in liquids (Schmidt, 1984) and of streamer mechanisms (Yoshino, 1980; Devins et al., 1981) has been accomplished. This is due more to the excellent work of a number of small groups during that time, rather than to any sudden burst of interest dragging both experimentalists and theoreticians into the field. However, it can unfortunately be said that, as a general rule, throughout the world, research on liquid breakdown has been progressively reduced (at present, the number of people in this field is only a small fraction of the number of those working on gases or solids).

This paper is an attempt to present and classify the main features of streamer generation and propagation in liquids, to discuss the various mechanisms involved, and to establish correlations between streamers and breakdown voltage.

GENERAL FEATURES OF STREAMERS IN LIQUIDS

Breakdown time-lag and successive steps leading to breakdown.

For a fixed voltage, a certain time delay t_d is elapsed before breakdown occurs. The voltage-time curves (V-t curves or "life-curves") always decrease, whatever the shape of the voltage wave (impulse, AC, DC), the geometry of the voltage gap (uniform field, divergent field), and the liquid. The long-term breakdown voltage is often more than one order-of-magnitude lower than the short-term one, especially in the case of a fast rise-time step voltage, as illustrated in Fig. 1. Several points are worthy of note:

- Time-lags of a few nanoseconds usually occur with uniform fields and short gaps (< 0.1 mm);
- Short-time breakdown fields are of the order of MV/cm (uniform field) or several tens of MV/cm (the harmonic field is calculated at the tip of needles currently used in point-plane geometry);
- The field at the tip of very sharp points (~0.1 μm in radius) has been reported to have attained the exceptional value of 100 MV/cm within minutes (no detectable current flowed under DC voltage) in liquid nitrogen (Sibillot and Coelho, 1974); at the other extreme, a one-meter gap of transformer oil in a gap of one meter was unable to sustain a 1-MV "lightning surge", breakdown having occurred below 10 kV/cm (Kamata and Kako, 1980).

The breakdown time-lag t_d corresponds to the following successive steps, in widely used and well-defined situations - divergent fields,

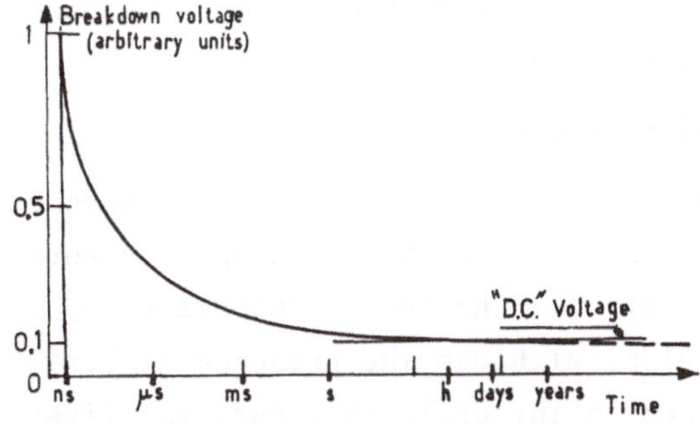

Fig. 1. Breakdown voltage versus breakdown time-lag under voltage step (schematic).

rectangular voltage (or lightning surge) - for various kinds of liquids and a very wide range of gaps (0.1 mm to 1 m):

- An "initiation phase" lasting a time t_i ("initiation time", from less than 1 ns to say μs), until the onset and development of a precursor event able to give rise to the next phase;
- A "propagation phase" of an event identified as a more or less ramified, luminous and conducting tree-like pattern, called a "streamer", streaming out from the high-field region, and taking time t_p ("propagation time", ns to tens of μs) to cross most of the gap;
- An "arcing" phase, where the "main-stroke" is established, the "arcing time" t_a being less than 1 ns; this final stage, a very destructive one if the current is not limited, has been by far the least studied.

This separation into distinct phases is questionable especially when breakdown takes place at very short times (ns), e.g., for short gaps, high over-voltages, and quasi-uniform fields. Breakdown could then take place via means somewhat different from the so-called streamer mechanism in liquids. It generally remains valid with uniform fields at moderate over-voltage, streamers in most cases issuing from one electrode, but sometimes from a solid particle or a gaseous bubble. It seems that the development of streamers across the gap is much more limited than in divergent fields before the establishment of the arc. When the breakdown time-lag under constant voltage is much longer than that expected from the above-described streamer mechanism (e.g., > or >> 100 μs), this waiting time (expected to be several years for DC or AC insulations) is the time during which there is an evolution of something either in the liquid (particles, bubbles,...) or at the electrodes (protrusions, deposits,...) able to trigger a fast series of events, the random nature of the appearance of which makes their study difficult. No visualization of the phenomena has been undertaken (under both DC and AC) until recently, where the generality of the streamer mechanism has been seen (Lesaint and Tobazéon, 1986; Lesaint, 1987).

Streamer shapes and characteristics

In gases, the prebreakdown mechanisms have been extensively studied, and a terminology well established including such terms as pre-streamers, streamers, leaders, etc..., the leader channel being specific to long distances. In liquids, although the presence of luminous and very fast events was recognized in the early thirties, the first detailed studies appeared much later. They were devoted to impulse voltage applied to large liquid gaps (5-20 cm), and drew attention to the phenomenological similarity between discharges across wide gas gaps and discharges in liquids. The luminous channels were called "initial streamers" and "pilot streamers" (Liao and Anderson, 1953), whereas others claimed that "leaders" developed more or less stepwise (Komel'kov, 1962; Stekol'nikov and Ushakov, 1966). In these papers, the patterns were visualized using their own emitted light. Later on, Schlieren or shadowgraph techniques were preferred and mainly short gaps were investigated (<2.5 cm). Many people called a "streamer" almost any detectable tree-like event. From the analysis of the literature, we have selected some general characteristics of streamers.

Whatever the liquid (nonpolar, polar, liquefied gas):

- The streamer possesses an optical index different from that of the liquid;
- The tree-like patterns continue to develop (see Fig. 2), the mean velocity of the fastest branch being vastly different from one liquid to another;
- "Slow" streamers (velocity < 1 km/s) are "bushlike", while "fast" streamers are filamentary (Fig. 3);
- For a given liquid, positive streamers (originating from the anode) are always faster than the negative;
- The streamer produces a current possessing more or less discrete peaks and emits light (UV to the visible); the current and the emitted light intensity have similar shapes;

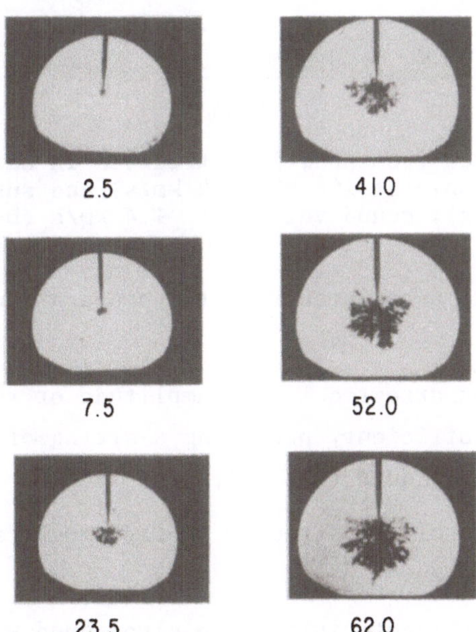

Fig. 2. Negative streamer growth in a white naphtenic mineral oil (Marcol 70). Gap: 1.27 cm; voltage step: 185 kV; time in microseconds (Devins et al., 1981).

Fig. 3. Positive filamentary streamer in silicone oil (Chadband, 1980).

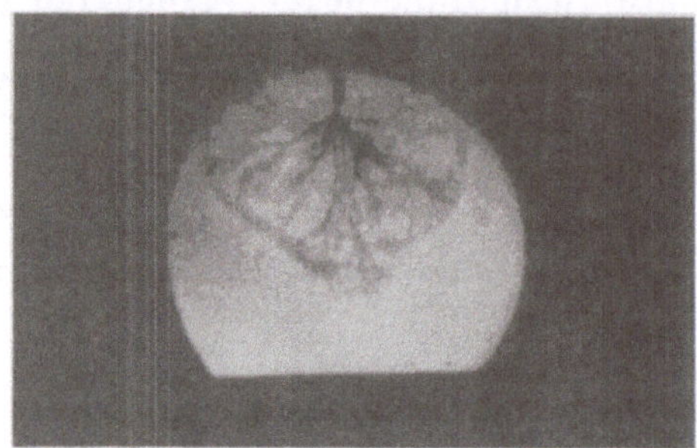

106 kV

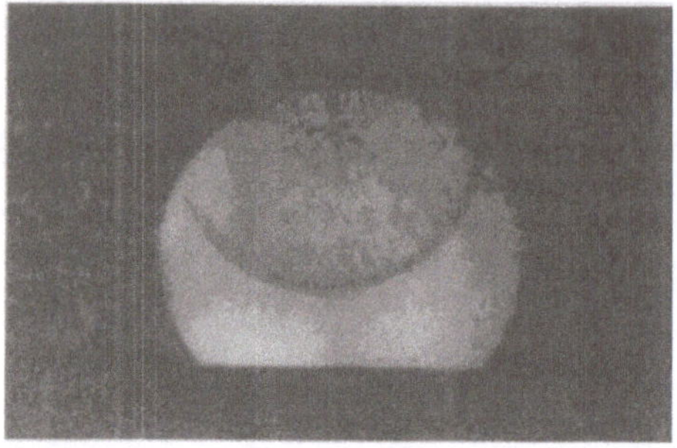

63 kV

Fig. 4. Shock waves from positive streamers in Marcol 70. Streamer is supersonic at 106 kV (1.8 km/s) and subsonic at 63 kV (1.07 km/s); sound velocity: 1.4 km/s (Devins et al., 1981)

- Shock waves are associated with the propagation of the streamer (Fig. 4);
- The streamer is arrested if the amplitude or the duration of the voltage is insufficient, producing a string of microbubbles which dissolve in the liquid (Fig. 5);
- The progression of the streamer is impeded by an increase of the hydrostatic pressure;
- Most of these characteristics are also found with uniform fields, no matter what shape the voltage wave (Fig. 6 illustrates negative streamers).

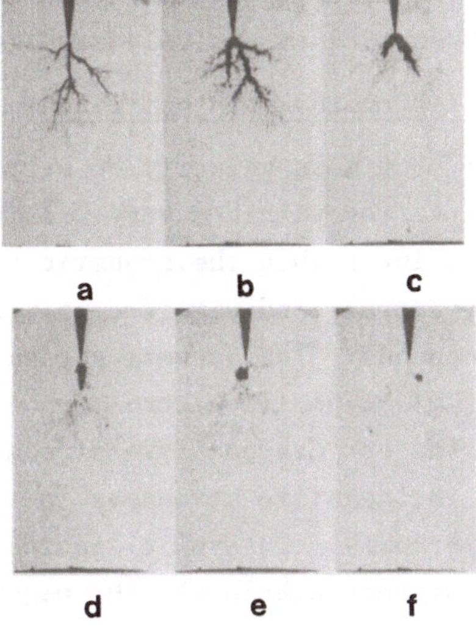

Fig. 5. Illustrating the decay of a positive streamer in a string of bubbles (Chadband, 1980).

Fig. 6. Negative streamers and spark channel in n-hexane. Electrode separation: 0.55 cm. Crest voltage: 225 kV (Forster and Wong, 1977).

Several facts support the view that the streamer is a gaseous phase: optical index, shock waves indicating a phase change, appearance of microbubbles, influence of the pressure. So, both mechanical and electronic processes (presence of current and emitted light) are operating during the streamer propagation. The propagation phase is the one most considered and also the one for which relevant experimental results have been obtained. Therefore, we shall begin with this topic.

STREAMER PROPAGATION

Most of the experimental data have been obtained with point-plane and rod-plane geometries and an impulse voltage: rectangular voltage or "lightning surge" (1-μs rise time, one-half of the maximum voltage at

50 μs). As in the cases of gases, positive streamers, regarded as the most "dangerous", have been more extensively studied.

<u>Influence of the chemical structure of the liquid</u>

As shown in Table 1, the mean velocity of streamers (evaluated via the Schlieren technique) varies within a very wide range, over 3 orders-of-magnitude. It is much lower than the acoustic velocity in cyclohexane (50 m/s for negative streamers), and significantly supersonic (up to 80 km/s) in halogenated compounds. The slowest streamers are negative in pure hydrocarbons (<0.5 km/s), positive streamers in these liquids being faster (1-3 km/s). The fastest are positive streamers in halogenated compounds (10-80 km/s), the negative streamers in these liquids being also very rapid (up to 25 km/s). In cyclic liquids of different structure (cis- or trans-decahydronaphthalene), the negative streamer velocity is comparable to that of cyclohexane, but the presence of a single atom of chlorine leads to a large increase in this velocity, by a factor of 10 in chlorocyclohexane (Béroual and Tobazéon, 1986). Table 2 gives, for comparison, some values of streamer velocities in air and in an amorphous solid polymer.

Table 1. Mean value of streamer and shock wave velocities in dielectric liquids (point-plane geometry). Applied voltage: 42.8 kV; gap spacing: 4.0 mm; tip radius: 10μm (Sakamoto and Yamada, 1980).

Liquids	Streamer velocity[a]		Shock wave velocity[a]
	+	-	
n-Pentane	3.6	0.11	0.78
n-Hexane	3.2	0.19	1.1
n-Heptane	3.4	0.22	0.99
n-Octane	3.3	0.13	0.94
Cyclohexane	2.8	0.056	1.0
Benzene	1.2	0.12	1.2
Toluene	1.2	0.60	1.2
Dodecylbenzene	2.3	0.081	1.5
Transformer oil	2.2	1.1	1.4
Carbon tetrachloride	30	8.4	0.90
Monochlorobenzene	76	0.45	1.2
Trichlorobenzene	62	25	1.2
Iodobenzene	15	9.1	-

[a] in units of 10^5 cm/s

Table 2. Mean value of streamer and shock wave velocities in air (Raether, 1964) and in an amorphous polymer: the polymethylmethacrylate (PMMA) in point-plane geometry (Kitani and Arii, 1980) and in a uniform field (Budenstein, 1980).

Insulator		Streamer velocity[a] +	Streamer velocity[a] -	Shock wave velocity[a]
Air	Point-plane d = 2 cm	300	15	0.3
	Uniform field d = 3.6 cm	> 1,000	700 - 900	
PMMA	Point-plane d = 0.03 cm	150	25	1
	Uniform field d = 1.91 cm	> 200	-	

[a] in units of 10^5 cm/s

Influence of additives

In the pioneering work of Devins et al. (1981), it has been strikingly demonstrated that small concentrations of compounds added to hydrocarbons had a remarkable effect on the streamer propagation, enlightening the importance of electronic processes:

- Electron scavengers, such as ethyl chloride and sulfur hexafluoride, considerably increase (by say one order-of-magnitude) the velocity of slow negative streamers (Fig. 7), no measurable effect being observed on the positive streamer velocities;
- Low ionization potential compounds, for instance N, N′-dimethylaniline (DMA), does not change the negative streamer velocity, but speed up (by a factor 2 to 3) the positive streamers (Fig. 8); the velocity of positive streamer in mixtures of n-hexane and carbon tetrachloride vary in the way shown in Fig. 9.

Until now, no additive has permitted a slowing down of streamers.

Correlation between shape and velocity of streamers

It was commonly accepted that only positive streamers were filamentary (with a high velocity), the negative ones being bush-like. In fact, the increase of negative streamer velocity by electron scavengers renders them more filamentary and they resemble more closely the positives

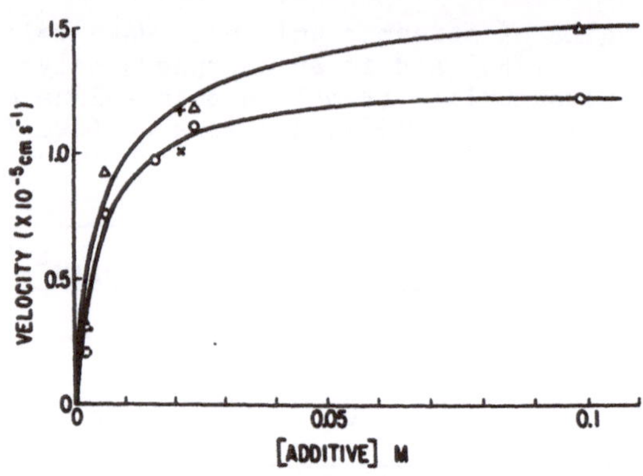

Fig. 7. Acceleration of negative streamers by C_2H_5Cl (open points) and SF_6 (O,+): gap, 2.54 cm; voltage, 180 kV. (Δ,x): gap, 1.27 cm; voltage, 150 kV (Devins et al., 1981).

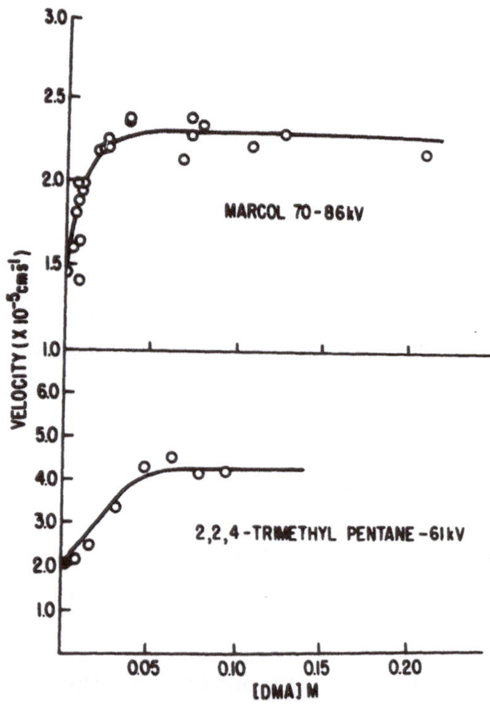

Fig. 8. Effect of DMA on the positive streamer velocity in Marcol 70 and 2,2,4-trimethylpentane. Gap: 2.54 cm (Devins et al., 1981).

(Devins et al., 1981); negative streamers can be very rapid (see Table 1). A study of the influence of the additives on velocity and shape of streamers in a wide variety of liquids has been undertaken by Béroual and Tobazéon (1986). They conclude that whatever the liquid and the polarity, there is a very close correlation between the shape and the amplitude of the velocity. Slow (subsonic) streamers are bush-like, fast

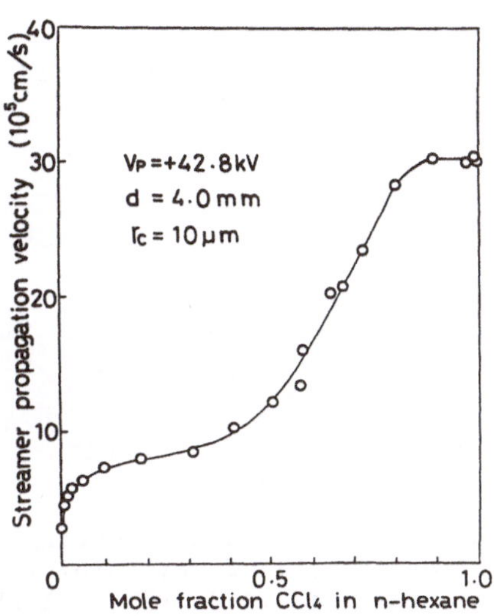

Fig. 9. Effect of CCl_4 molar fraction on the positive streamer velocity in n-hexane (Sakamoto and Yamada, 1980).

(supersonic) are filamentary, the higher the velocity, the more filamentary the streamer.

Streamer current and emitted light

The current flow during the propagation of the negative streamers in Marcol 70 consists of closely spaced pulses, the light emitted by the streamer being pulse-like, in contrast with the positive current which grows continuously (Devins et al., 1981). As a general rule, for low gaps (< 5 cm), the negative currents for slow streamers (< 0.5 km/s) comprise a series of short pulses (< 10 ns) of increasing amplitude and number throughout the propagation. Positive streamers generate currents growing in a more continuous manner, up to a maximum, when they reach the insulator covering the plane (an arrangement used to protect the experimental equipment against breakdown). The emitted light has a similar appearance to the instantaneous current for both polarities (see Fig. 10) as reported by several authors. It appears, in fact, that currents increasing in a continuous manner are produced by fast streamers (> 2 km/s) whatever the polarity (Béroual and Tobazéon, 1986).

For large gaps (5 cm < L < 1 m), experimental results are scarce; in transformer oil, spaced pulses are present for both polarities (Liao and Anderson, 1953; Stekol'nikov and Ushakov, 1966; Kamata and Kako, 1980).

With AC voltages, the positive current flows have the same appearance as under impulse voltages, and bear some similarity in various liquids and gases (Fig. 11). In transformer oil, under AC, negative and

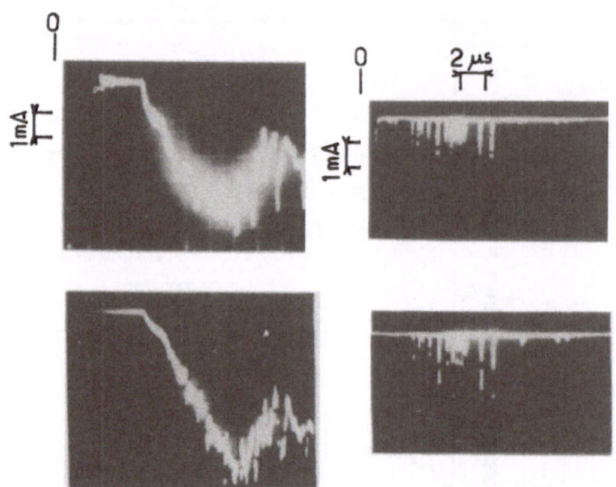

Fig. 10. Current and light emission during streamer growth in cyclohexane. Gap: 2 mm; Point radius: 3 µm. Point positive (at the left), $V_+ = 21$ kV. Point negative, $V_- = 30$ kV (Béroual and Tobazéon, 1986).

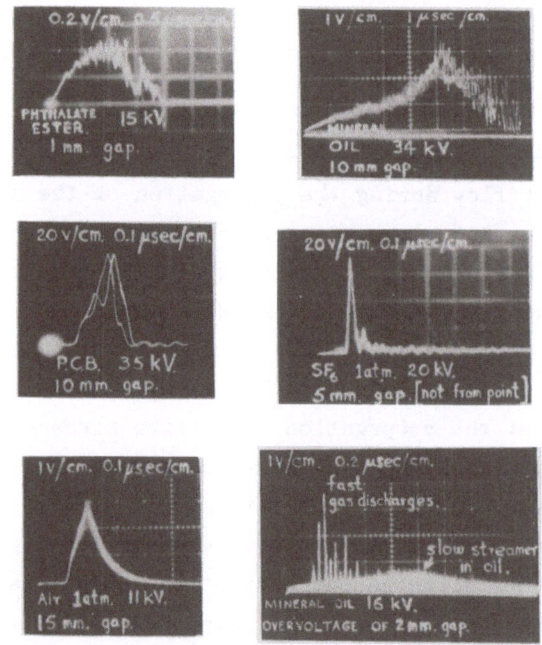

Fig. 11. Current pulse wave shapes for AC-point-to-insulated plane discharges in various insulating liquids and gases (Harrold, 1974).

positive streamers behave as under a step voltage: shapes, currents, emitted light, and velocities are comparable (Lesaint and Tobazéon, 1986).

Velocity profile in the gap

For most of the liquids, whatever the polarity and the streamer shape, the velocity distribution goes through a more or less marked

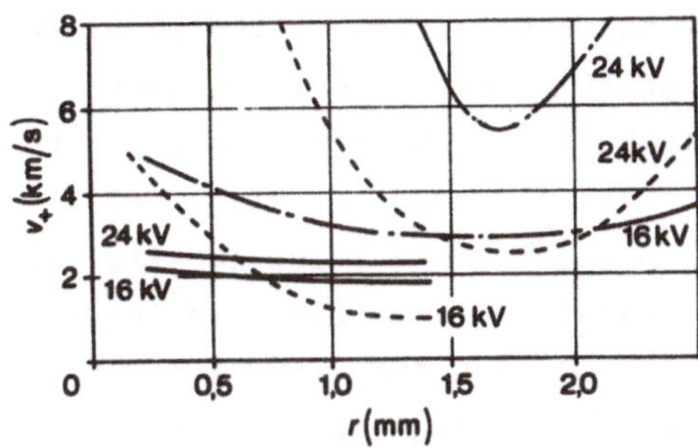

Fig. 12. A comparison of the variation of positive discharge tip velocities with growth. Full curve, transformer oil; chain curve, 10 Cst Silicone oil; broken curve, 1,000 Cst silicone oil (Chadband, 1980).

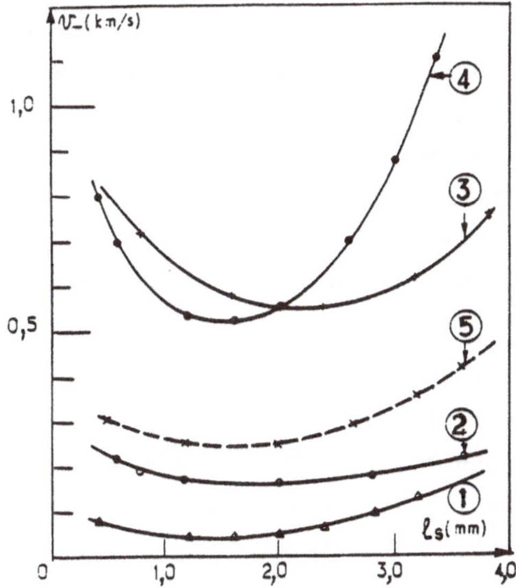

Fig. 13. Velocity distribution of streamers in cyclohexane (CH), pure or containing various additives. Gap: 4 mm; Point radius: 3 μm; Voltage: 50 kV. Point negative: (1) Pure CH, (2) CH + 0.2% anthracene, (3) CH + 0.3% CCl_4, (4) CH + 1% CCl_4 (5) Point positive: pure CH (Beroual and Tobazeon, 1986).

minimum located between one-half and two-thirds of the inter electrode distance (Figs. 12 and 13).

Two purely electrostatic models have been proposed to approximate the velocity of the leading edge of the streamer with the assumption that its velocity is proportional to the local electric field, the streamer being considered to be a conductor. Chadband and Wright (1965) at first proposed to simulate the streamer by an inflating conducting sphere that approaches the plane, while remaining in contact with the point. According to this model, the field, and therefore the velocity, possesses a minimum at about two-thirds of the way across the gap. The second model considers the streamer as an extension of the point (Chadband, 1980); in this case, the tip velocity is expected to increase as the streamer grows. The field distribution with this assumption has been calculated by a charge simulation method (Devins et al., 1981). Such representations could be useful to describe at least qualitatively the propagation phase; analytical expressions for the field in the gap have been derived (Béroual and Tobazéon, 1986) with the view of estimating the influence of the experimental parameters (voltage, gap distance, tip radius). Indeed, these experimental distributions are similar, in a wide variety of liquids, to that of the field on a growing conducting sphere, even at supersonic velocities (where filamentary streamers are propagating), a somewhat surprising fact since what was expected was a continuous increase in velocity. Interestingly enough, we obtain an estimate of the capacitive current produced either by the conducting moving sphere or by the single filament. Neither the shape, nor the amplitude of the measured currents correspond to either of these models. Moreover, currents corresponding to filamentary streamers are much higher than those of bush-like streamers (unlike those predicted by the above estimation). It may be concluded that there is an important conducting volume surrounding the filamentary streamer.

A few exceptions to the velocity distribution with a minimum were reported in transformer oil under impulse voltage (Chadband, 1980; Devins et al., 1981), where the velocity of positive streamers was essentially constant across the gap, and in n-hexane, where the velocity of supersonic positive streamers increased from the point to the plane (McKenny and McGrath, 1984).

Recent work under AC voltage (Lesaint and Tobazéon, 1986) has evidenced that both the negative and positive streamers propagate throughout the gap at a constant velocity, close to the value measured under impulse (respectively 1 km/s and 1.9 km/s). The positive streamers growth is limited to an extent proportional to the voltage, the propagation of negative ones being rather controlled by the mean field (no propagation below 40 kV/cm; always propagation over 80 kV/cm).

Influence of the hydrostatic pressure

It has been known for a long time that an increase of the hydrostatic pressure considerably improved the AC or DC breakdown voltages of liquids. As reported in the book of Nikuradse (1934), the AC breakdown voltage under uniform fields of various oils is increased by a factor 3 to 4 at 40 bars (a "saturation" taking place over 50 bars). In liquid nitrogen under impulse voltage, up to 12 bars, the point-to-plane breakdown voltage always increases with the point negative, but saturates at 10 bars (as observed in uniform field); the shorter the impulse voltage, the lower the increase of breakdown voltage (Yoshino, 1980). This is consistent with the results of Kao and McMath (1970) showing that the electric strength between spherical electrodes was much less dependent on the pressure when the rate of rise of the ramp voltage was increased.

A systematic study of the influence of pressure on streamers has been carried out under rectangular voltage steps (Béroual and Tobazéon, 1986, 1987). Figure 14 shows a moderate increase of pressure impedes the propagation of slow streamers in several liquids (similar observations were made previously in transformer oil by Hizal and Dincer, 1982); at the same time, the number and the amplitude of the current (and light) peaks are reduced. For a given pressure, the streamer completely vanishes (as do current and emitted light). The effect is similar for fast streamers, but the pressure necessary to suppress their propagation is higher (Fig. 15). Reducing the pressure below atmospheric has no appreciable effect. The velocity of streamers is not significantly

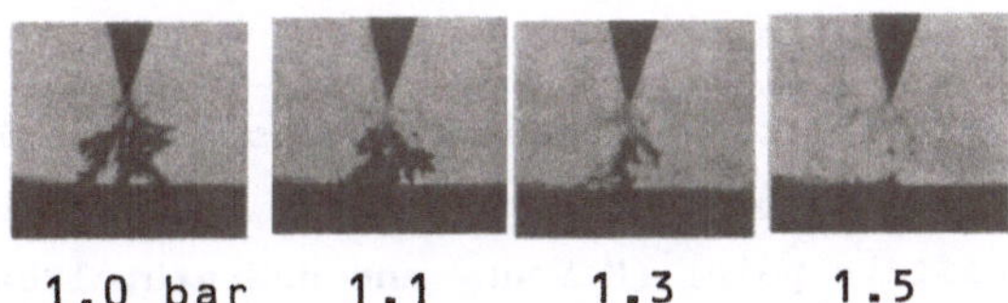

Fig. 14. Influence of the pressure on the negative streamer growth in cyclohexane. Gap: 1 mm; Point radius: 5 μm; Voltage: 28 kV; (Béroual and Tobazéon, 1987).

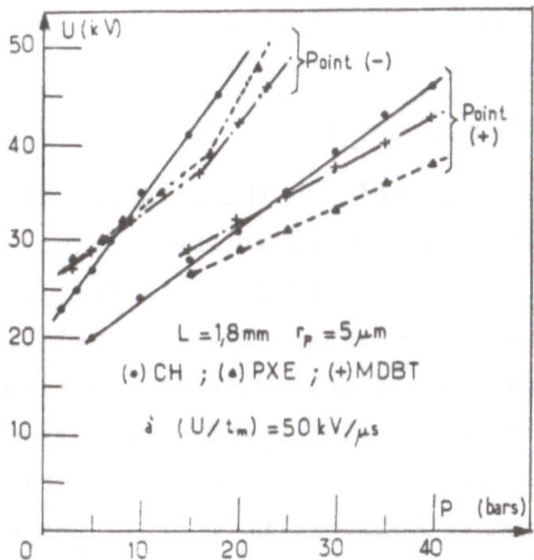

Fig. 15. Pressure necessary to suppress streamer growth under a step of voltage (duration: 50 μs); Cyclohexane (CH); Phenylxylylethane (PXE); Blend of mono- and dibenzyltoluene (MDBT) (Béroual and Tobazéon, 1987).

modified whatever the pressure (this had been reported for subatmospheric pressures by Devins et al., 1981).

Pressure plays therefore an analogous role for each polarity, suggesting that in both cases a gaseous phase is involved.

Influence of the temperature

Even near either the freezing point or the boiling point, temperature has but a minor influence upon the streamer's behavior at atmospheric pressure. The breakdown strengths (10-μs voltage pulses, plane electrodes at a spacing of 127 μm) of hexane liquid are somewhat lower than those of the vapor having the same density, the values converging as one approaches the critical point; it was suggested that the breakdown mechanisms were similar in this region (Sharbaugh and Watson, 1977). On the other hand, below the melting point of cyclohexane, much higher voltages were required to detect prebreakdown events (which could not be determined reliably with our equipment), indicating that prebreakdown mechanisms were drastically changed (Béroual, 1987).

Various modes of propagation

For a fixed voltage, two propagation modes are possible, the velocity being either subsonic or supersonic. This has been observed in n-hexane with a positive point (McKenny and McGrath, 1984) and in transformer oil with a negative point (Yamashita and Amano, 1985). In cyclohexane (and other liquids), the two modes can be produced whatever the polarity, the probability to get the supersonic regime being higher when the pressure is increased, or the temperature lowered, or the voltage raised (Béroual and Tobazéon, 1987). We may notice that, at elevated pressures, much higher voltages are needed for the streamers to cross the gap; their velocity is therefore increased, and they become more filamentary.

Influence of the physical properties

They only play a minor role on the streamer behavior. Most of the streamers are nearly the same whatever the liquid: thermal properties (specific heat, heat of vaporization), mechanical properties (superficial tension, compressibility); consequently, it is almost impossible to appreciate their influence. Liquid mass per unit volume is nearly independent of pressure and temperature, and also not very different from one liquid to another (600 to 1,500 kg/m^3, except liquid xenon; 3,520 kg/m^3 at 164°K). Contrasting with gases, no correlation exists between this property and either breakdown or streamer characteristics.

Although they can differ by orders-of-magnitude according to the nature of the liquid, viscosity or vapor pressure do not play a significant role:

- The viscosity of silicone fluids of identical chemical nature has been varied by 3 orders-of-magnitude with no significant change in streamer velocity and shape (Arii et al., 1975; Chadband, 1980; Béroual and Tobazéon, 1987).
- Vapor pressure is nearly the same in cyclohexane and in CCl_4, but the streamer velocity can differ by 2 orders-of-magnitude, whereas this velocity is comparable in liquids the vapor pressure of which differs by 4 orders-of-magnitude.

At the present time, no satisfactory correlation has been established between streamer propagation and either the electronic properties (mobility, lifetime of electrons or holes) or the optical properties (interaction with radiation, luminescence) of liquids.

STREAMER INITIATION

The statistical study of breakdown time-lags in liquids has been the subject of numerous studies; Lewis (1985) and his coworkers have proved the existence of a field-dependent "statistical time-lag", and showed how curves like the one of Fig. 1 could be predicted. The influence of the conditioning process of the electrodes and of the test method on the time-lag distribution density functions are extensively discussed in the book of Gallagher (1975). From breakdown time-lag measurements under rectangular voltage, it is possible to separate the initiation time (its statistical nature being questionable if it is field-dependent) from the so-called "formative time-lag", which is in fact the streamer propagation time (at least, in most of the experimental situations). With AC or DC, another aspect of initiation phenomena will be presented.

Initiation under rectangular voltage waves

Figure 16 is an example of linear time-lag distribution obtained in a silicone oil. From such a so-called "von Laue curve" it is possible to get t_i and t_p via the following relationship (Grey Morgan, 1978):

$$N_t/N_o = \exp\left[-(t-t_p)/t_i\right] \text{ , where}$$

N_o = total number of shots and N_t = number of shots for which no breakdown has occurred until time t. From similar curves obtained with uniform or divergent fields in silicone oil (Arii et al., 1975) or in liquid argon (Yoshino, 1980), we can conclude that t_p is in reasonable agreement with known values of streamer velocities, and that t_i varies very rapidly with the applied voltage (the more so when the point is

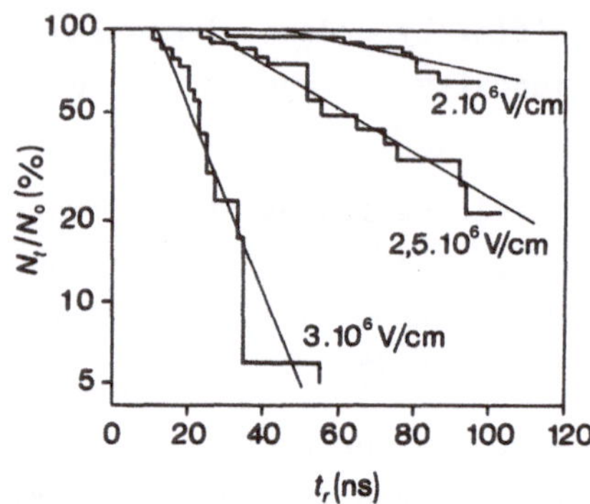

Fig. 16. Von Laue plots in a 10^3 Cst silicone oil uniform field; gap: 80 μm (Arii et al., 1975).

positive). However, the scarcity of experimental results does not permit us to obtain useful information about initiation phenomena from this method.

Rzad et al. (1979) have shown that breakdown in a rod-to-plane geometry was controlled either by initiation phenomena or by propagation of streamers. Under rectangular voltage waves, with a fixed gap of distance L, by varying the radius of curvature r_o of the rod, they measured breakdown time-lags t_d as shown by Fig. 17. At high voltages (with the rod negative), t_d was nearly independent of r_o, the initiation time being negligible, and $t_d \sim t_p$. As the voltage is decreased and r_o increased, the initiation time increases, and $t_d \to t_i$. From these results we can conclude that t_p corresponds to the velocity of negative streamers; a rough estimation of t_i indicates that it decreases very rapidly when V is

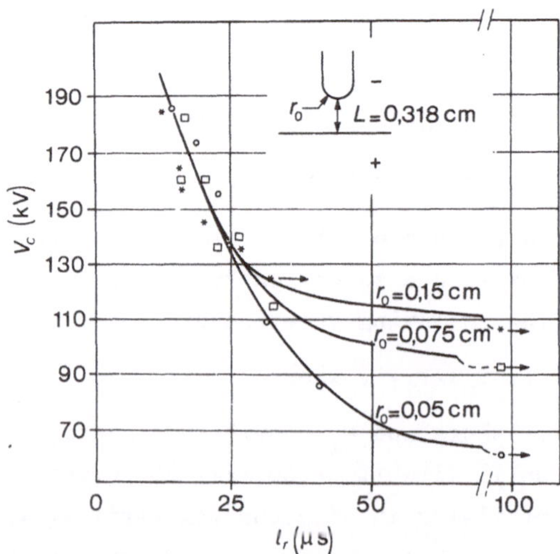

Fig. 17. Time-to-breakdown vs breakdown voltage in Marcol 70 for 100-μs rectangular voltage waves with rod-plane geometries. Radius of curvature of the rod; (o): 0.05 cm; ( ): 0.075 cm; (*): 0.15 cm (Rzad et al., 1979).

increased and r_o decreased. Breakdown time-lags with a positive point were much shorter but no detailed study was performed.

It is possible, however, to get more useful information on streamer initiation: the method consists of, under a rectangular voltage, measuring t_i, defined as the time at which a small optical perturbation (e.g., < 10 μm in diameter) is detectable by the Schlieren technique, or by light emission, or also by current measurements characteristic of the early streamer propagation. It was shown that when the rise time of the voltage t_r was much lower than t_i, for several liquids, initiation times were shorter with the positive point than with the negative point, but from the V-t curves it was only deduced that one mechanism appeared to be rejected-cavitation, as will be discussed later - and it was not feasible to conclude much more (Béroual and Tobazéon, 1987); for elevated voltages, streamer initiation was produced during the rise time of the voltage wave.

In conclusion, the measurement of the initiation times under impulse voltage has not been very fruitful in the search for physicochemical mechanisms involved in streamer generation. The use of fast sensitive methods of light and current detection could be developed to analyze the early phase in the production of the streamer. This is, to a certain extent, easier for DC voltages and for a slowly varying wave (AC, 50 Hz).

Prebreakdown events under DC voltage

With a view to studying high field conduction in liquids, the point-to-plane geometry has been widely employed. The pioneering work on the subject by Halpern and Gomer (1969) has highlighted the importance of electronic-field emission at the cathode, and liquid-field ionization at the anode. Some common features characterize high field DC conduction:

- Current fluctuations in the form of pulses (Halpern and Gomer, 1969; Sibillot and Coelho, 1974; Arii and Schmidt, 1984; Denat et al., 1987), both regularly (point-cathode) and randomly (point-anode) distributed (Denat et al., 1987);
- Light emission is connected with current burst (Sibillot and Coelho, 1974; Arii and Schmidt, 1984; Denat et al., 1987);
- Gas bubbles can be generated near the point to be subsequently ejected into the bulk of the liquid, without producing breakdown, in hydrocarbons (Singh et al., 1972; Denat et al., 1987) or in cryogenic liquids (Halpern and Gomer, 1969; Sibillot and Coelho, 1974);
- Streamers have recently been visualized with a positive point in cyclohexane (Denat et al., 1987) and in hexane with a negative point (MacGrath and Marsden, 1986).

Random aspect of initiation under AC voltage (50 Hz)

Figure 18 shows the variation of the number of prebreakdown events per unit time in transformer oil as a function of the applied voltage for a point-to-insulated-plane geometry; these prebreakdown events include all detectable peaks of light (or of current), whatever the polarity. The general features of these events were (Lesaint, 1987):

- The first detected events were produced when the point was negative; the positives, produced at higher voltages, were always less numerous for a given time duration;
- Above a voltage threshold (the initiation voltage), the frequency of events tended to increase exponentially with voltage;
- Whatever the radius of curvature of the point, the slopes of the curve were comparable, which suggests that initiation phenomena are of the same character;

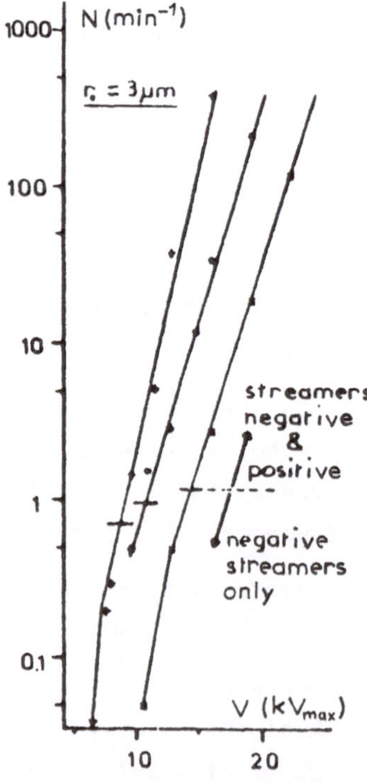

Fig. 18. Number of prebreakdown discharges versus the maximum applied voltage. Transformer oil: frequency, 50 Hz; point radius, 3 µm (Lesaint, 1987).

- The random aspect of initiation of these events is typical of long-time voltage application.

<u>The succession of prebreakdown events</u>

Although there were several arguments favoring the idea that a gaseous phase was formed prior to streamer appearance, experimental evidence of bubble generation has only recently been produced, due to the development of a fast sensitive method of visualization coupled with the measurement of current and emitted light (Lesaint and Tobazéon, 1986). In a point-plane geometry with a slowly increasing AC voltage, the first detectable event in transformer oil appeared when the former is negative (Fig. 19): A single burst of current (< 10 ns), associated with a light burst, gives rise to a microbubble (10 µm in diameter). Thereafter, slight expansion of its volume is observed, followed by detachment from the point, and motion toward the plane at a velocity of several m/s; neither current nor light are recorded during this phase. Blurring occurs rapidly and the bubble disappears a few µs later (Fig. 20). A very similar sequence of bubble generation has been recorded in cyclohexane under AC voltage (Denat et al., 1987). A slight increase of the voltage produces other current and light bursts of increasing amplitude, following the initial burst, the observed pattern having at first a bush-like aspect and dying out in the bulk of the liquid (partial propagation); a further increase of the voltage produces a larger number of current peaks, branching of the bush-like figure which is similar to the slow streamers observed under impulse voltage.

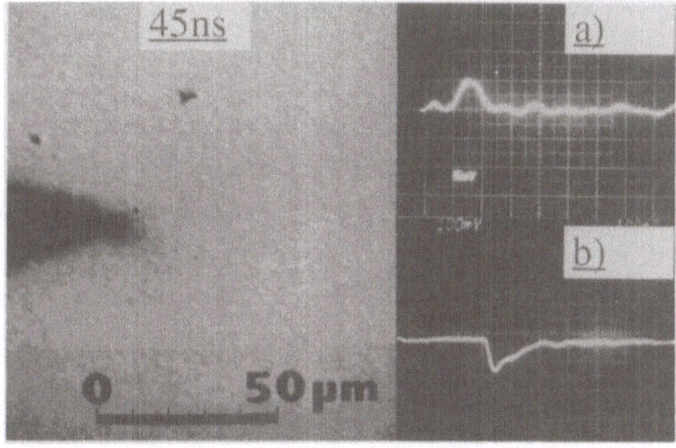

Fig. 19. Generation of a bubble in transformer oil. Negative point radius, 3 µm; AC voltage, 11.2 kV_{rms}; (a) Current: 10 ns/div, 10 µA/div; (b) Emitted light: 100 ns/div; Delay time-to-visualization: 45 ns (Lesaint and Tobazéon, 1986).

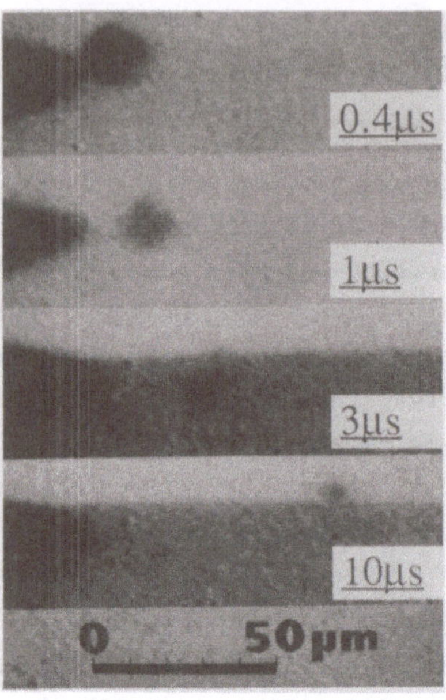

Fig. 20. The bubble growth, motion and disappearance into the liquid as a function of time; no current is detectable (Lesaint and Tobazéon, 1986).

When the point was positive, the initiation voltage was higher, but we observed only occasionally the generation of a bubble. As a general rule, in transformer oil, the streamer regime is immediately established and similar to the one observed under impulse voltage: The current increases in a regular manner (it contains also irregular peaks), and the streamer comprises thin filaments (a few μm) embedded in a sphere centered at the point, the number of filaments increasing with the voltage.

The influence of pressure

As previously stated, an increase in the hydrostatic pressure impedes the development of streamers and diminishes the number of current peaks. In cyclohexane, with a DC voltage and a negative point, the first current pulse is not suppressed even at high pressures (50 bars). Its duration and amplitude are independent of the pressure, and the initiation voltage increases only slightly; over 10 bars, no bubble is detectable (Denat et al., 1987). Similar observations have been made with AC voltage in transformer oil (Lesaint, 1987). With the point positive, in cyclohexane under DC voltage the number of pulses per unit time is independent of the pressure, but their shapes are significantly different as the pressure is raised; over 10 bars, no bubble has been detected (Denat et al., 1987).

The influence of the nature and the treatment of the electrodes

No significant correlation has been obtained between the work function of the cathode and either the streamer initiation or the breakdown voltage except perhaps in liquid helium (Yoshino, 1980).

Undoubtedly, the quality of the electrode surface plays an important role as demonstrated (especially with quasi-uniform fields) by conditioning phenomena--cleaning and polishing of the electrodes, formation of oxide layers, flow of current modifying the initiation voltage. It has been reported often that the radius of curvature of sharp points could be appreciably modified, the observed erosion being caused by mechanisms not clearly understood (sputtering, cavitation). Similar phenomena are likely to occur on local protrusions at the surfaces of plane or spherical electrodes.

The influence of additives

There are few readily available results showing a significant influence of additives upon the initiation voltage of streamers (either under impulse or under AC voltage). With DC, adding aromatic compounds to transformer oil or liquid paraffin significantly increased the breakdown voltage, both in a divergent field with a negative point, and in the case of a uniform field (Angerer, 1965). With DC voltages, we expect that space-charge phenomena play an important role; homocharges could reduce the field near the point (or near asperities of planar surfaces) lowering consequently the initiation voltage. Ionic additives have only a negligible effect under rectangular voltage (Devins et al., 1981; Béroual and Tobazéon, 1986). Much more work is needed in this fundamental area.

STREAMER PROPERTY AND BREAKDOWN VOLTAGE CORRELATION

We shall take transformer oil as an example since it is a liquid of major technological importance in high voltage insulation. Moreover, there is a multitude of data on breakdown tests in the literature than we can tentatively unify in the light of our own results.

AC voltage

With divergent fields, there are two extreme situations:

- High mean fields (> 80 kV/cm) and low divergence: The propagation of either positive or negative streamers is possible, and breakdown will be controlled by streamer initiation; since the negative initiation voltage is the lowest, breakdown will be produced when the point is negative;

- Low mean fields (< 30 kV/cm) and high divergence: All the generated streamers (positive or negative) cannot propagate across the gap; therefore, breakdown will take place more often if the point is positive (since positive streamers are much more easily propagated).

Figure 21 summarizes the different possibilities of breakdown observed for various liquid gaps. For sufficiently high liquid gaps the slope approaches 10 kV/cm; a constant value close to this has been obtained for large gaps (2 to 5 cm), the maximum breakdown voltage increasing linearly with the gap (Epri Report, 1978; Kamata and Kako, 1980). Over 50 cm (breakdown voltage: 500 kV), the increase of voltage then ceasing to be linear (500 kV for 1 m).

With uniform fields and conveniently filtered liquids, the mean applied field will always be much higher than the minimum field of propagation for a negative polarity (40 kV/cm), as well as the voltage propagation length ratio, characteristic of the streamer length for a positive polarity (35 kV/cm). Therefore, breakdown will always be controlled by streamer initiation, and in most cases will be due to negative streamers (lower initiation voltage).

Impulse voltage

With divergent fields, breakdown is lower for a positive point, sometimes much lower than for a negative one by a factor of at least 3 in certain types of oils (EPRI Report, 1978, 1979), which exemplifies the "danger" of (easily propagated) positive streamers. In transformer oil, the shape of the voltage-propagation-length curves with the point positive is similar to those obtained using AC (the breakdown voltages are

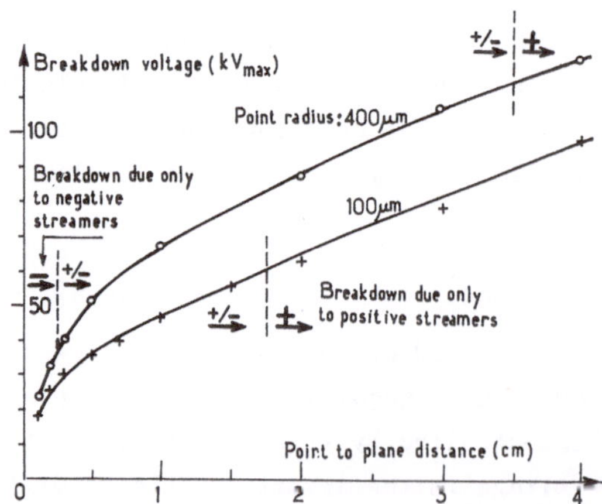

Fig. 21. Breakdown voltage of mineral oil versus gap distance. Point-to-plane geometry; voltage linearly increased at a rate of 1 kV/s (Lesaint, 1987).

somewhat higher than the maximum amplitude of the AC wave). With the point negative, the breakdown voltage increases approximately linearly with the gap, the slope being higher than in the AC case.

With uniform fields, in narrow gaps (< 1 mm) very high field values (> 1MV/cm) can be reached; there is a progressive reduction (down to a factor of 10) of the breakdown field when the gap, the electrode area and the liquid volume are increased, showing once more that initiation controls breakdown. However, most of the results obtained under lightning surge are difficult to interpret, streamers being generated and/or propagated either on the front or on the tail of the voltage wave.

Finally, there are semi-quantitative correlations between tests performed in divergent and uniform fields contrary to that stated by Hosticka, 1979. The addition of electron scavengers or aromatic compounds lowers the negative impulse breakdown in a divergent field (due to the increase of the streamer velocity), the positive impulse breakdown being unaffected (there being no influence on positive streamer velocity). For both AC and DC, there is a tendency for the breakdown voltage to increase (higher initiation voltage). These additives also produce an improvement in uniform fields no matter what the voltage, probably via the increase of negative streamer initiation voltage. We know that water can reduce the AC and DC voltage breakdown under uniform fields, usually when the water content reaches saturation (EPRI Report, 1979), but not the impulse voltage; the influence of water on streamer behavior remains as yet unknown.

DISCUSSION OF POSSIBLE MECHANISMS INVOLVED IN STREAMER INITIATION AND PROPAGATION

Initiation field

The initiation mechanisms with divergent fields are controlled by phenomena at the liquid-metal interface. The field at the surface of the point or of the rod electrode very likely plays a fundamental role. The harmonic field E_0 can be calculated either by the spherical or hyperboloidal approximation; in this case

$$E_0 = 2\ V/r_0 \ln (4L/r_0)$$

(V = applied voltage; r_0 = tip or rod radius; L = liquid gap). The hypothesis of a constant streamer initiation field does not hold in most cases, since with the measured streamer initiation voltages, the harmonic field values obtained via the above approximations are much higher with sharp tips (> 10 MV/cm, for r_0 < 3 µm; < 1 MV/cm, for r_0 > 100 µm). This has been observed for both impulse (Rzad et al., 1979; Chadband, 1980; Béroual and Tobazéon, 1987) and AC voltages (Lesaint, 1987). The actual field can be either reduced by injection of homocharges by the point, or enhanced by protrusions or particles on electrodes with large radius of curvature.

The injection of homocharges by the point is readily evidenced by measuring the current I under a DC voltage. The space-charge-limited current law, obtained either with the spherical or with the hyperboloidal approximation (the charge carrier mobility being assumed independent of the field), takes the form:

$$I^{1/2} = p\ (V-V_t)\ .$$

Such a variation has been found in a wide variety of liquids; for example, simple pure liquids (Halpern and Gomer, 1969; Schmidt, 1984; Yamashita and Amano, 1985; Denat et al., 1987), or in transformer oil (Lesaint, 1987). The field E_t corresponding to the threshold voltage V_t is very nearly independent of r_o (in the range 1 - 10 μm). To give some examples of this in highly purified cyclohexane, one obtains with a negative point, E_{t-} = 7 MV/cm, and E_{t+} = 5 MV/cm with the point positive (Denat et al. 1987); in transformer oil, E_{t-} = 3 MV/cm and E_{t+} = 5 MV/cm (Lesaint, 1987). The amplitude of the measured currents cannot be explained either by field emission or by field ionization, which would produce negligible currents for these values of threshold fields.

We know that particles can considerably reduce the AC or DC voltage breakdown under quasi-uniform field. A single, one-micron diameter particle would lead to a lowering of pulse strength by a factor of 2 to 3 (Sharbaugh, 1972). Furthermore, it has been shown that a streamer could initiate at the surface of a particle in the bulk of the liquid, between parallel electrodes (Thomas and Forster, 1975). The enhancement of the field either by particles in the bulk, or by asperities on the electrode surfaces, is likely to explain at least partly the low values of E_o for large r_o. The field enhancement by a conducting body is highly dependent on its shape: three for a conducting sphere, much more for elongated bodies (Félici, 1979).

High field conduction in liquids

Conduction mechanisms in dielectric liquids at very high fields (say, over 1 MV/cm) are not yet fully elucidated. The origin and nature of charge carriers can widely vary according to the liquid and to the experimental situations. The study of high fields has mostly been carried out in the point-to-plane geometries (breakdown occurs too easily with parallel electrodes). Only two mechanisms of charge generation have been considered--electron injection at the cathode by field emission, and field ionization of the liquid at the anode. These mechanisms have been well verified in certain liquified gases (Halpern and Gomer, 1969; Schmidt, 1984) and in a few organic liquids (Dotoku et al., 1978, Schmidt 1984). A recent work by Denat et al. (1987) has shown that in highly purified cyclohexane both field emission and field ionization could be observed, but only for tips of very small radius of curvature (r_o < 0.5 μm). For larger tip radii (r_o > 1 μm), the current-voltage curves, and the time variation of the current at constant voltage, were similar to those of corona discharge in gases. For negative points, above the threshold voltage corresponding to a field of 7 MV/cm (as we mentioned in the preceding paragraph), the current is constituted of regular pulses (analogous to Trichel pulses). A similar regime has also been observed in liquid argon and liquid xenon (Arii and Schmidt, 1984). Each individual current pulse contains an initial burst of say 200 μA, of 10 ns duration, independent of pressure and temperature, which leads to bubble formation at low enough pressure (see above). For positive point with r_o > 1 μm, a current pulse regime was also observed.

The origin of the positive space charge necessary to obtain these Trichel-like pulses is not, according to Denat et al., to be found in the discharge inside a bubble created at the point, as proposed by Arii and Schmidt (1984) in LAr and LXe, but could be due to electron avalanches in the liquid. An electronic multiplication has effectively been experimentally measured in LAr and LXe (Derenzo et al., 1974), but is presently questionable in hydrocarbons. Since electronic mobilities have only been measured at fields lower than 300 kV/cm (Schmidt, 1984), it is open to question if an electron can obtain sufficient energy to ionize the liquid. The subject of electron avalanches in liquids has been an old and controversial one; Nikuradse (1934) found that current varied in liquids exponentially with the gap and put forward a theory of conduction in liquids analogous to the one of gases. Later on, it was strikingly demonstrated that such current variations resulted from ion injection into the liquid at the cathode or at the anode, or both. Sharbaugh and Watson (1962) found that the reciprocal of electric strength against the logarithm of the gap was linear, as expected from an electron avalanche build-up. However, a decrease in the diameter of the spherical electrodes used produced a somewhat higher electrical strength so that the avalanche "evidence" turned out to be an area effect. These authors also measured currents with flat electrodes as a function of electrode spacing under constant field conditions. They could not detect any α-process up to breakdown at around 1MV/cm; if it existed at all, the product of αL was less than 0.1 (Sharbaugh, 1972).

Ionic processes could also play a significant role, especially for nonpolar liquids not having exceptional levels of purity, and for polar liquids. At very high fields, the following phenomena are able to produce large current density and noticeable space charges:

- Ion injection at the anode or at the cathode;
- Field-enhanced dissociation of tiny amounts of ionizable impurities, or of the liquid itself;
- Increase in the ionic mobility, its possible variation with the field being identical to the electronic one (Bockris and Reddy, 1970), e.g., an enlargement of more than an order-of-magnitude for $E = 5$ MV/cm.

Electroconvection, e.g., convective transport of charge carriers (whatever their nature and polarity) via the coulombic force, is always at work; the so-called electrohydrodynamic mobility, $K_h = (\varepsilon/\rho)^{1/2}$ (ε = permitivity, ρ = mass per unit volume), is an upper limit for the apparent mobility corresponding to the convective transport of charges. Its value is typically 1.5×10^{-7} m^2/Vs in nonpolar liquids, i.e., much higher than the ionic mobility (< 10^{-8} m^2/Vs), and lower, or much lower, than electron or hole mobilities. Above the threshold field, the current is space-charged limited, the mobility being close to the electrohydrodynamic one. This is a universal phenomenon encountered in ultra-pure liquids (Denat et al., 1987) or in technical liquids (Lesaint, 1987). This means that, as a general rule, (i) electrons are trapped very near the point and converted into ionic carriers; (ii) liquid motion is always present; an estimate of the velocity in the vicinity of the point gives several m/s (a value confirmed by experimental observations). Consequently, either space-charge distribution or thermal processes could be drastically modified, even within very short periods of time following the voltage application. Indeed, the time delay to put the liquid in motion in the unipolar injection regime is given as $t_d \propto \eta / \varepsilon E^2$ (η = viscosity). An estimate shows that t_d could be less than 1 μs for a few MV/cm (see Tobazéon, 1986, and references therein).

Generation of the first bubble

When a liquid is heated under constant pressure, or when the pressure is reduced at constant temperature, vapor or gas and vapor-filled bubbles are generated and grow. The former process of cavity production is known as boiling, the latter as cavitation. The hypothesis of boiling was checked by Sharbaugh and Watson (1962) under pulses of a few μs in a uniform field. The current was assumed to be space-charge limited, and the following relationship derived from the heat balance:

$$AE_b^{3/2}\ t = m\ [C_p(T_b - T_a) + l_v]\ ,\ \text{where}$$

A = constant, E_b = breakdown field, t = pulse length, m = mass of liquid, C_p = specific heat, T_b = boiling point, T_a = ambient temperature, and l_v = heat of vaporization. The dependence on pressure and molecular structure (which change the boiling point of liquids) fitted well with this theory. These results were, however, limited to a series of alkanes, and there is no experimental proof that bubble generation is the effective mechanism.

If all the electrical energy dissipated corresponding to the first current pulse in transformer oil under AC (Lesaint and Tobazéon, 1986), or in cyclohexane under DC (Denat et al., 1987), is converted into heat, this would lead to a gas volume which compares fairly well with the actual observed volume of the first bubble. We must point out that the electrostrictive pressure (25 bars at 5 MV/cm) seems to play a minor role, as do electroconvective phenomena.

Cavitation may appear if the pressure is locally reduced to below the vapor pressure of the liquid. Pressure fluctuations could be produced by the motion of the liquid induced by electrohydrodynamic phenomena, or in the wake of moving microparticles. The fluid velocity could locally exceed 10 m/s, producing a lowering in the pressure of say 1 bar (according to Bernouilli's theorem), leading to bubble generation by cavitation. However, the expected lengthening of the time delay of bubble appearance with an increase in viscosity has not been observed (in viscous liquid silicones and polybutenes as previously mentioned). Nevertheless, this mechanism is not to be totally rejected, especially for long-term voltage application.

Streamer propagation

It is possible to obtain an evaluation of the dominant mechanisms involved during streamer propagation via the following energy balance argument (Félici, 1987): The total energy available (the electrical energy input) must be balanced by the sum of the energy densities needed to generate a unit volume of the expanding streamer. Following this argument, the possible mechanisms are:

- Vaporization of the liquid; this requires an energy density of the order of 1 J/cm^3 (for most liquids).
- Dissociation and/or ionization of molecules; this necessitates at least 10 to 100 times more energy than that needed for vaporization.

- Displacement of the liquid repelled by the streamer, e.g., $(1/2)\rho v^2$ (ρ = mass of the liquid per unit volume, v = velocity of the streamer).

- Capillary energy, which is very small compared to the previous mechanisms.

Table 3 gives an example of such an energy balance. Energy densities are calculated or measured (using charge or current measurements) and the streamer volume is estimated from photographs; the mean velocities measured were 50 m/s for negative streamers and 2 km/s for the positives. For negative streamers the electrical energy is comparable to both the energy needed for vaporization and for displacement. As will be seen later, chromatographic analysis of the gases generated by the streamer indicates that vaporization would require the major part of the energy (Table 4). For positive streamers, both dissociation and ionization are dominant, vaporization being also possible. This was obtained under a rectangular wave voltage, but the same conclusions were arrived at in transformer oil with AC voltages (Lesaint, 1987).

To interpret the mechanisms responsible for the propagation of the positive streamer, Devins et al. (1981) assume that field ionization occurs in the liquid, and they use the Zener theory of tunneling in solids to calculate the concentration of positive and negative carriers contained in a cylindrical channel of radius r_s. Assuming a spherical field distribution (the field at the tip of the channel is then $E_o = V/r_s$), they derived the following expression for the streamer velocity:

Table 3. Measurement or evaluation of the energy densities involved during streamer propagation in a blend of mono- and dibenzyltoluene (MDBT) (Béroual, 1987).

Energy densities (J/cm^3)	Polarity	
	+	-
Electrical energy	74	1.2
Vaporization	1.3	1.3
Displacement	2.000	1.3
Dissociation	< 43	
Ionization	< 43	
Capillary energy	1.5×10^{-2}	

Table 4. Comparison of charges and gas produced in MDBT by a series of streamers of positive or negative polarity. Applied voltage = 30 kV; gap spacing = 2 mm; tip radius = 3 μm (Béroual, 1987).

	+	-
Total number of streamers	717	9,994
Number of fast streamers	690	0
Number of slow streamers	27	9,994
Total charge of fast streamers (nC)	2,719	0
Total charge of slow streamers (nC)	6	6,000
Total charge (nC)	2,725	6,000
Total volume of dissolved gases (cm^3)	1.1×10^{-3}	6.6×10^{-5}

$$v = \left(\frac{a\, e^3\, E_o^{\,3}}{\pi\, m\, V_i^{\,2}}\right)^{1/2} \frac{r_s}{\gamma}\, \mathrm{erfc}\left(\frac{\pi^2\, m\, a\, V_i^{\,2}}{e\, h^2\, E_o}\right)^{1/2},$$

where e and m are the electron charge and mass, respectively; h, Planck's constant; V_i, the liquid-phase ionization potential; a, the molecular separation; γ, the Zener coefficient. As pointed out by the authors, v increases very sharply when E_o is increased and V_i is decreased, and a relationship between E_o, r_s and v could explain the regulatory mechanism responsible for the constancy in positive streamer velocity. They suggest that E_o remains constant, which could be due to a variation in r_s and/or γ.

Concerning negative streamer propagation, Devins et al. (1981) considered a two step model--electron injection and trapping followed by ionization within the liquid (there results a plasma similar to that produced under positive polarity). The negative velocity is determined by the time spent in one or the other step (injection or trapping). Their model is supported by the facts that the addition of electron scavengers reduces the trapping distance and the time t_1 spent in the first step, thus increasing the negative velocity, and that further addition of low ionization potential additives increases the rate of ionization, and reduces the time t_2 spent in the second step, thus again increasing the velocity.

Indeed, Devins et al.'s model gives a valuable description of fast streamer behavior (action of the additives, presence of a regulatory mechanism for the constancy of the velocity) but it fails to explain several experimental results. For instance, the expected variation in the velocity with the density of the liquid is not found, nor does it explain the experimentally observed minimum in the velocity distribution. The streamer model proposed by Chadband and Sufian (1985) was of a discharge propagating under the combined action of the local Coulomb field and the Laplacian field. The tip propagates by attracting negative charge from the liquid ahead of the tip into the tip itself, the charge in the original tip is counterbalanced by an equal negative charge, and a new positive tip is established a certain distance ahead. The electric field just ahead of the tip is calculated using the method of images. This model of streamer growth predicts a velocity minimum. The observed effects of varying the gap length or of adding low-ionization additives to the liquids are consistent with this model also. These authors considered that their model was one of tip propagation by positive hole movement under the applied field. The role of the holes was also put forward by Lewis (1985) who suggested that the hole transport would be facilitated along the walls of the plasma column (the positive streamer). This could explain why streamers remain filamentary and why streamer propagation is very rapid (the hole mobility is high).

As regards the initial phase of streamer expansion, Watson (1981, 1985) has proposed two models considering the streamer as a growing cavity. Assuming that the low-density cavity is roughly spherical in shape and equipotential with the cathode, he found that the wall velocity (of the cavity) was

$$v = (\varepsilon/3\rho)^{1/2}\ (V/R)$$

by equating the work of the electrostatic pressure to the kinetic energy of a fluid volume; R, the radius of the cavity, is a function of time and is characterized by the expression

$$R = [Vt\ (2\varepsilon/3\rho)^{1/2}]^{1/2}\ .$$

In his second model, he considered a hypothetical point source emitting a volume of fluid (ionized gas or plasma) per unit time generating a surface separation similar to a streamer, which experiences a force due to the streamer velocity. He found as velocity of the cavity

$$v = (\varepsilon/3\rho)^{1/2}\ (V/R_s)\ ,$$

R_s being the radius of the extremity of the channel left behind the moving sphere. Comparing with some experimental results, Watson found reasonable agreement, the velocity calculated being 110 m/s with this formula, compared to the data of Chadband and Wright (1965) in hexane (50 to 200 m/s). Even with the thinner observed radii, the maximum velocity cannot be over 1 km/s, whereas experimental values can exceed 50 km/s. Watson, with the view of explaining the propagation of fast streamers with a steady velocity, considered the role of the electrostatic force in the growth of the wave-like instabilities of the surface of the vapor cavity, and calculated the growth rate using a two-dimensional electrohydrodynamic model. The fit with some experimental data is good, but limited to a small number of results.

Nature and conductivity of the streamer

Spectroscopic studies of the light emitted during the initiation of propagation of streamers present a promising method. Unfortunately, we have only partial results. It has been known for a long time that the spectrum of the emitted light extends from the UV to the visible range. During the prebreakdown phase corresponding to the propagation of negative streamers in iso-octane and cyclohexane between parallel electrodes, Forster and Wong (1980) carried out a spectral analysis of the emitted light and concluded the presence of atomic and molecular hydrogen, carbon (C_2 and C_3), and tiny amounts of metal (coming from the electrodes). Sakamoto and Yamada (1982) observed in a point-to-plane geometry, superposed on the continuous spectrum, several peaks attributed to the H-Balmer series and to the Swan bands of C_2. The polarity of the point being unspecified in the paper, it is rather difficult to identify the physico-chemical processes involved for one or other polarity. The spectral analysis associated with the well-identified propagation of either positive or negative streamers in a point-plane geometry has been recently undertaken in pure cyclohexane; it reveals the presence of a H-line, some C_2 Swan bands and an important continuum which would correspond to dissociation and fragmentation of liquid molecules, and to recombination of molecular fragments. Spectra of the emitted light are different according to polarity and streamer velocity; the more energetic the streamer (that is the faster), the higher are the characteristic peaks emerging from the continuum (Béroual and Denat, 1987).

Gas chromatography can also be a useful tool. We have tried to correlate the intensity of the charge involved during the propagation to the volume and the nature of gases dissolved afterwards in the liquid. The principle of the method was to subject liquid samples initially degassed and then saturated with nitrogen, to a very great number (hundreds or thousands) of rectangular pulses with the same polarity of the point, and to record the individual and the total charge of the streamers. Since we know the negative and positive streamer velocities and, therefore, the time needed for them to cross the gap, we can select the integration time. The gases dissolved in MDBT (a blend of mono- and dibenzyltoluene) were mainly C_2H_2 and H_2, with some CH_4, C_2H_4 and C_2H_6. The dissolved gas volumes were much higher with the point positive. On the other hand, the total volume of dissolved gases with the point positive was comparable with the apparent volume estimated from photographs of streamers, whereas the apparent volume of negative streamers was much higher (by a factor of 10^3). Table 4 gives an example of the results, from which it was concluded that for negative streamers vaporization was the dominant mechanism since the volume of decomposition products was very small, whereas dissociation and ionization were predominant for positive streamers (Béroual, 1987).

The Kerr technique makes it possible to study the electric field distribution in the bulk of the liquid and to follow its time evolution. This technique has been successfully employed by Kelley and Hebner (1981) in a point-plane geometry; the liquid used was nitrobenzene, a strongly polar liquid with a very high Kerr constant. For both polarities, streamers were bush-like, and these authors concluded that: (i) streamers behave as conductors, equivalent to inflating spheres growing toward the plane and remaining in contact with the tip; (ii) no space charge between the leading edge of the streamer and the plane was detectable. This method is not very sensitive (especially in hydrocarbons which possess very low Kerr constants). It could be used in an attempt to evaluate the field (expected to be very high) at the tips of filamentary streamers, with the view to reconcile the discrepancies between the measured currents and those that would produce conducting filaments.

Indeed, it would be desirable to make more accurate measurements of the fast current peaks (time-resolved) coupled with a more detailed visualization (in space and time) of the streamer propagation and of the light which is emitted by it. It is likely that the use of large distances could facilitate this task.

CONCLUSION

This overview of the streamer behavior in liquids does not pretend to be exhaustive. I would like to express my apologies to those whose work has not been referred to; it was rather due to lack of time and space than for other reasons. I would like also to acknowledge all those who have helped me in the course of this work. I have not avoided the temptation of presenting, perhaps too often, the work done in my own laboratory. This can be understood.

Obviously a better understanding of streamer mechanisms will necessitate an improvement of our knowledge on high-field conduction under DC and fast-rising step voltages, both with divergent and uniform fields. With the view of specifying the role of electrons and ions, efforts might be directed toward the choice of "model" liquids (and selected additives) possessing specific properties, such as tetramethylsilane (high electron mobility), liquid xenon (in which an α-process has been observed), and carbon tetrachloride (in which streamers are very rapid). Another interesting approach would be to make comparative studies in the same compound, liquid or gaseous, by varying the pressure up to the critical point, and also to study the liquid/gas interface.

As concerns the experimental studies, the development of simultaneous measurements is recommended, using both optical (spectroscopy, visualization, Kerr plots) and electrical (such as time-resolved current pulse measurements) methods.

As a final remark, the reader may be informed that the Ninth International Conference on Conduction and Breakdown in Dielectric Liquids was held in Salford (July, 1987), and that several papers will be published in a special issue of the IEEE Trans. on Elec. Ins. in early 1988.

REFERENCES

Adamczewski, I., 1969, "Ionization, conductivity and breakdown in dielectric liquids", Taylor and Francis, London.

Angerer, L., 1965, Effect of organic additives on electrical breakdown in transformer oil and liquid paraffin, Proc. IEE, 112:1025.

Arii, K., Hayashi, K., Kitani, I., and Inuishi, Y., 1975, Breakdown time-lag and time of flight measurement in liquid dielectrics, Proc. of the 5th Conf. on Cond. and Breakd. in Diel. Liquids, J.M. Goldschwartz, ed., Delft, University Press, Delft., 163.

Arii, K., and Kitani, I., 1984, Bubble generation in liquid hydrocarbon in divergent fields, Jap. J. Appl. Phys., 23:49.

Arii, K., and Schmidt, W.F., 1984, Current injection and light emission in liquid argon and xenon in a divergent electric field, IEEE Trans. Elec. Ins., EI-19, 1:16.

Béroual, A., 1987, Phénomènes de propagation et de génération des streamers dans les diélectriques liquides en géométrie pointe-plan sous créneau de tension, Doctoral Thesis, Grenoble.

Béroual, A., and Denat, A., 1987, Analyse spectrale de la lumière émise par les streamers dans les diélectriques liquides, C.R. Acad. Sc. Paris (in press).

Béroual, A., and Tobazéon, R., 1985, Effects of hydrostatic pressure on the prebreakdown phenomena in dielectric liquids, Ann. Report. Conf. on Electr. Insul. and Dielectr. Phenomena, NAS-NRC, 44.

Béroual, A., and Tobazéon, R., 1986, Prebreakdown phenomena in liquid dielectrics, IEEE Trans. Elec. Ins., EI-21, 4:613.

Béroual, A., and Tobazéon, R., 1987, Propagation et génération des streamers dans les diélectriques liquides, Revue Phys. Appl., 22:189.

Bockris, J. O'M., and Reddy, A.K.N., 1970, "Modern electrochemistry", Plenum Press, New York.

Budenstein, P.P., 1980, On the mechanism of breakdown in solids, IEEE Trans. Elec. Ins., EI-15, 3:225.

Chadband, W.G., 1980, On variations in the propagation of positive discharges between transformer oil and silicone fluids, J. Phys. D:Appl. Phys., 13:1299.

Chadband, W.G., and Sufian, T.M., 1985, Experimental support for a model of positive streamers propagation in liquid insulation, IEEE Trans. Elec. Ins., EI 29, 2:239.

Chadband, W.G., and Wright, G.T., 1965, Prebreakdown phenomena in the liquid dielectric hexane, Brit. J. Appl. Phys., 16:305.

Denat, A., Gosse, J.P., and Gosse, B., 1987, Conduction du cyclohexane très pur en géométrie pointe-plan, Revue Phys. Appl., 22:162.

Derenzo, S.E., Mast, T.S., Zaklad, H. and Muller, R.A., 1974, Electron avalanches in liquid xenon, Phys. Rev., A9:2582.

Devins, J.D., Rzad, S.J. and Schwabe, R.J., 1981, Breakdown and prebreakdown phenomena in liquids, J. Appl. Phys., 52:4531.

Dotoku, K., Yamada, H., Sakamoto, S., Noda, S., and Yoshida, H., 1978, Field emission into nonpolar liquids, J. Chem. Phys., 69:1121.

EPRI Report, 1978, Uniform and non-uniform field electrical breakdown of naphthenic and paraffinic transformer oils, RP 562-1, Contract Nr CCR-78-07 (prepared by the General Electric Co.).

EPRI Report, 1979, Study to determine the potential use of silicone fluids in transformers, HCP/T 2115, Contract Nr EX.76.C.01.2115 (prepared by the General Electric Co.).

Félici, N.J., 1979, Bubbles, partial discharges and liquid breakdown, The Institute of Physics (GB), Inst. Phys. Conf. Ser., 48:181.

Félici, N.J., 1987, Liquides et gaz: les mécanismes de claquage sont-ils comparables? Revue Phys. Appl., 22:191.

Forster, E.O., and Wong, P., 1977, High speed laser schlieren studies of electrical breakdown in liquid hydrocarbons, IEEE Trans. Elec. Ins., EI-12, 6:435.

Forster, E.G., and Wong, P.P., 1980, The dynamics of electrical breakdown in liquid hydrocarbons. V: light emission processes, Conf. Rec. of the IEEE Intern. Symp. on Elec. Insul., Boston :222.

Gallagher, T.H., 1975, "Simple Dielectric Liquids", Oxford University Press, London.

Grey Morgan, C., 1978, Irradiation and time lags, in: "Electrical breakdown of gases", Ch. 7, J.M. Meed and J.D. Craggs, eds., John-Wiley and Sons, New York.

Halpern, B., and Gomer, R., 1969, Field emission in liquids, J. Chem. Phys., 51:1031 and 1048.

Harrold, R.T., 1974, A simple electronic technique for measuring streamer velocities in insulating liquid and gases, Ann. Rep. Conf. on Electr. Insul. and Dielectr. Phenomena, NAS-NRC, 123.

Hizal, E.M., and Dincer, S., 1982, Breakdown time lags and prebreakdown phenomena in transformer oil, effects of hydrostatic pressure, J. Electrostatics, 12:333.

Hosticka, C., 1979, Dependence of uniform/non-uniform field transformer oil breakdown on oil composition, IEEE Trans. Elec. Ins., EI-14, 1:43.

Kamata, Y., and Kako, Y., 1980, Flashover characteristics of extremely long gaps in transformer oil under non-uniform field conditions, IEEE Trans. Elec. Ins., EI-15, 1:18.

Kao, K.C., and McMath, J.P.C., 1970, Time-dependent pressure effect in liquid dielectrics, IEEE Trans. Elec. Ins., EI-15, 3:64.

Kelley, E.F., and Hebner, R.E., 1981, The electric field distribution associated with prebreakdown phenomena in nitrobenzene, J. Appl. Phys., 52:191.

Kitani, I., and Arii, K., 1980, Impulse breakdown of polymer dielectrics in the ns range in divergent fields, IEEE Trans. Elec. Ins., EI-15, 2:134.

Komel'kov, V.S., 1962, Development of a pulse discharge in liquids, Soviet Physics-Tech. Phys., 6:691.

Lesaint, O., 1987, Claquage et preclaquage dans I'huile minerale sous tension alternatiye, Doctoral Thesis, Grenoble.

Lesaint, O., and Tobazeon, R., 1986, Prebreakdown phenomena in transformer oil under AC voltage, CIGRE Report, Working Group, 15-02, p. 1-9.

Lewis, T.J., 1959, The electric strength and high field conductivity of dielectric liquids, in: "Progress in Dielectrics", Vol.1, J.B. Birks, ed., Heywood, London.

Lewis, T.J., 1985, Electronic processes in dielectric liquids under incipient breakdown stress, IEEE Trans. Elec. Ins., EI-20, 2:123.

Liao, T.W., and Anderson, J.G., 1953, Propagation mechanism of impulse corona and breakdown in oil, Trans. Amer. Inst. Engrs., 72, Pt1:641.

McGrath, P.B., and Marsden, H.I., 1986, DC-induced prebreakdown events in n-hexane, IEEE Trans. Elec. Ins., EI-21, 4:669.

McKenny, P.J., and McGrath, P.B., 1984, Anomalous positive point prebreakdown behavior in dielectric liquids, IEEE Trans. Elec. Ins., EI-19, 2:93.

Nikuradse, A., 1934, "Das Flüssige Dielektrikum", Springer Verlag, Berlin.

Raether, H., 1964, "Electron avalanches and breakdown in gases", Butterworths, London.

Rzad, S.J., Devins, J.C., and Schwabe, R.J., 1979, Transient behavior in transformer oils: prebreakdown and breakdown phenomena IEEE Trans. Elec. Ins., EI-14, 6:289.

Sakamoto, S., and Yamada, H., 1980, Optical study of conduction and breakdown in dielectric liquids, IEEE Trans. Elec. Ins., EI-15, 3:171.

Schmidt, W.F., 1984, Electronic conduction processes in dielectric liquids, IEEE Trans. Elec. Ins., EI-19, 5:389.

Sharbaugh, A.H., 1972, Mechanisms of electrical breakdown in dielectric liquids. Past, Present, and Future, Ann. Rep. Conf. on Elec. Insul. and Dielec. Phenomena, NAS-NRC, 427.

Sharbaugh, A.H., Devins, J.C., and Rzad, S.J., 1978, Progress in the field of electric breakdown in dielectric liquids, IEEE Trans. Elec. Ins., EI-13, 4:249.

Sharbaugh, A.H., and Watson, P.K., 1962, Conduction and breakdown in liquid dielectrics, in: "Progress in Dielectrics", Vol. 4, J.B. Birks, ed., Heywook, London.

Sharbaugh, A.H., and Watson, P.K., 1977, The electric strength of hexane vapor and liquid in the critical region, J Appl. Phys., 48:943.

Sibillot, P., and Coelho, R., 1974, Prebreakdown events in liquid nitrogen, J. Physique, 35:141.

Singh, B., Chadband, W.G., Smith, C.W., and Calderwood, J.H., 1972, Prebreakdown processes in electrically stressed insulating liquids, J. Phys. D: Appl. Phys., 5:1457.

Stekol'nikov, I.S., and Ushakov, V.Ya., 1966, Discharge phenomena in liquids, Soviet Physics-Tech. Phys., 10:1307.

Thomas, W.R.L., and Forster, E.O., 1975, Electrical conductance and breakdown in liquid hydrocarbons, Proc. of the 5th Conf. on Cond. and Breakd. in Diel. Liquids, J.M. Goldschwarz, ed., Delft University Press, Delft, 49.

Tobazéon, R., 1986, Liquides diélectriques. Préclaquage et claquage, Techniques de I'Ingénieur, D 226, D227.

Watson, P.K., 1981, Electrohydrodynamic instabilities in the breakdown of point-plane gaps in insulating liquids, Ann. Rep. on Conf. on Elec. Ins. and Dielec. Phen., NAS-NRC, p. 370.

Watson, P.K., 1985, Electrostatic and hydrodynamic effects in the electrical breakdown of liquid dielectrics, IEEE Trans. Elec. Ins., EI-20, 2:395.

Yamashita, H., and Amano, H., 1985, Prebreakdown current and light emission in transformer oil, IEEE Trans. Elec. Ins. EI-20, 2:247.

Yoshino, K., 1980, Dependance of dielectric breakdown of liquids on molecular structure, IEEE Trans. Elec. Ins., EI-15, 3:186.

ELECTRIC CONDUCTION IN DIELECTRIC LIQUIDS

J. P. Gosse

* Laboratoire d'Electrostatique et de Matériaux
Dielectriques, 166 X
38042 Grenoble Cedex France

INTRODUCTION

Conduction in dielectric liquids has been reviewed many times (Lewis, 1959; Sharbaugh and Watson, 1962; Félici, 1971; Gallagher, 1975). Since then, few major improvements of the understanding of conduction mechanisms have been achieved: all the possible mechanisms had been previously put forward. The main contribution of the latest works has been the experimental verification of some of the above models and, more especially, the definition of their domains of validity. There are many reasons for such a slow progress. It is impossible to control the nature and concentration of impurities acting on the electric conduction of very resistive liquids, conductivity remaining more sensitive to impurities than physicochemical techniques. Industry uses liquid insulants only for particular applications; they were essentially mineral oil and chlorobiphenyl liquids which gave full satisfaction, and it needed the prohibition of the PCBs to bring about a new interest in conduction studies.

The scope of this review is restricted to ionic conduction in organic liquids, from hydrocarbons to polar liquids. Indeed, electrons which are discussed in other papers remain free over very short times (< μs) in organic liquids (Schmidt, 1984). And for the maximum electric field values considered here ($\sim 10^8$ Vm^{-1}), field emission cannot account for the large values of the current density.

We first consider briefly the transport of ions by conduction or convection. Then, we discuss the mechanisms of ion creation either in

* Laboratoire Associé à l'Université, J. FOURIER, Greenoble I.

the bulk of the liquid or at the electrodes, and present the related characteristics j(t), j(E) and tan(E) between parallel electrodes.

CHARGE TRANSPORT

In a liquid, ion diffusion results from the random walk of ions, and their drift in an electric field is only superimposed on their random walk. Hence, the drift and random walk are linked. This is expressed by the Einstein relation between the mobility K and the diffusion coefficient $D = KU$, with $U = kT/e$.

In polar liquids, ionic mobilities have been extensively studied, for instance by conductometry. They are known for many solvents (Robinson and Stokes, 1959; Janz and Tomkins, 1972). In low polar liquids, K has been deduced most often from transit time measurements, after irradiation of a thin layer of liquid or photoexcitation of the cathode, or after applying a voltage step. K values are now available in different liquids (Tables 1 and 2) but the nature of the corresponding ion is often unknown. So, to estimate the value of the mobility of a given ion in a liquid, the relation, $K\eta = e/6\pi r$, between K and the liquid viscosity η (Stokes-Einstein relation) is often used, where r is the hydrodynamic radius of the ion. In fact, this relation is only a crude approximation even for rather large ions, since the viscous forces are not the only retarding force acting on the ion, and the solvent is

Table 1. Mobility of Cl^- ion in different liquids at 20C (Lacroix et al., 1975).

Liquid	Propylene carbonate	Nitro-benzene	Chlorobiphenyl			Chloro-benzene
Dielectric constant	65	36	5.9	6.4	5.6	5.7
Viscosity η (cp)	2.48	1.8	10	50	90	0.8
Mobility $\times 10^8$ ($m^2V^{-1}s^{-1}$)	1.4	2.3	0.34	0.07	0.04	4.5
$K\eta \times 10^{11}$ (SI)	3.47	5.4	3.4	3.5	3.6	3.6

Table 2. Mobility of Pi^- in different non-polar liquids at 20C (Denat et al., 1979).

Liquid	Benzene	Cyclohexane	Hexane	n-decane
Viscosity (cp)	0.65	1.02	0.33	0.92
Mobility K $\times 10^8 (m^2\ V^{-1}\ s^{-1})$	3.3	1.4	4	1.7
$K\eta \times 10^{11}$ (SI)	2.15	1.4	1.3	1.56

represented as a continuum. Indeed, it is observed that $K\eta$ decreases when the liquid permittivity decreases (Fig. 1). In some theoretical approaches to this problem (Fuoss, 1959; Zwanzig, 1963; Hubbard and Onsager, 1977) dielectric friction has been considered. For instance, Fuoss proposed to replace the frictional factor $6\pi\eta\ r$ by $6\pi\eta\ [r+B/\varepsilon]$, where B is an empirical constant. So in Fig. 1, Walden's rule becomes $(K_+ + K_-)\eta = 4.15 \times 10^{-11}/(1 + 2.71/\varepsilon)$. Other modifications have been discussed by Franck (1966). But, since assumptions of the starting model are very crude, Walden's rule, or its expression modified by Fuoss, would be used only to get an approximate value of an ion mobility. The exact value would be experimentally measured, from transit time measurements, for instance.

About mobility measurements, two other remarks can be made. First, in many papers the variation of the product $K\eta$ with temperature is

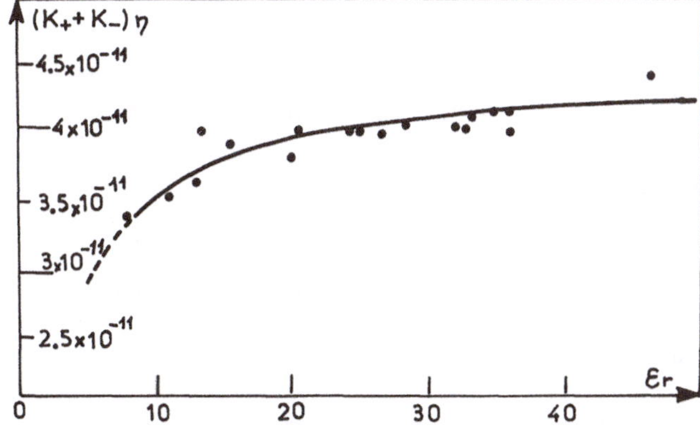

Fig. 1. Dependence of Walden product on liquid dielectric constant ε for tetrabutylammonium tetraphenylboride (Coetzee and Cunningham, 1965; Accascina et al., 1959; Fuoss and Hirsch, 1960).

studied. From the preceding discussion, the mobility should be and is found to vary nearly as η^{-1} which, contrary to ε and, thus, to the electrical forces, varies strongly with temperature (Table 3). Secondly, in some papers (Yasufuku et al., 1979) carrier mobilities are deduced from the current peak time observed after a reversal of the polarity of the voltage applied to the electrodes (Fig. 2). This is a general observation in polar and non-polar liquids. But the theoretical interpretation of the curve i(t) appears, in fact, not so clear as implicitly supposed. For instance, the mobility values deduced from the first transient and from the following ones are rather different, and T_{max} does not exactly follow the relation $T_{max} \propto d^2/V$ (Novotny and Hopper, 1979; Casanovas et al., 1985). These authors invoke electrohydrodynamical motions which increase the ion mobility and then current transport.

The voltage of convection onset has been theoretically investigated (Atten and Moreau, 1972; Schneider and Watson, 1970; Atten and Lacroix, 1979). Its value is rather well estimated when space charges are due to injected ions. Since injection is the prevailing mechanism of ion creation at high electric fields, we shall give short shrift only to this case. For strong ion unipolar injection, the theoretical critical voltage V_c ($V_c = 161\ \eta K/\varepsilon$) is very low. In Fig. 3(a), for SCL injection of Cl-ions in chlorobiphenyl (η = 50 cp at 20C), the critical voltage is 100 V. For a weak injection characterized by a finite value of the

Table 3. Temperature dependence of Walden's product for the picrate and propionate ions in n-decane (Nemamcha et al., 1987).

	T(C)	20	40	60	80
D	$\times 10^3$ (P_o)	0.92	0.68	0.54	0.43
$TIAP_i$	$K \times 10^8$ ($m^2\ V^{-1}\ s^{-1}$)	2.2	2.9	3.8	4.8
	$K\eta \times 10^{11}$ (SI)	2.02	1.97	2.05	2.06
DAP	$K \times 10^8$ ($m^2\ V^{-1}\ s^{-1}$)	1.8	2.4	2.9	4
	$K\eta \times 10^{11}$ (SI)	1.66	1.63	1.57	1.72

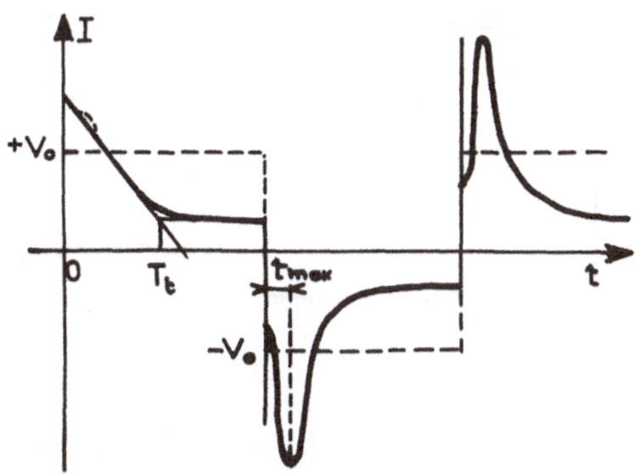

Fig. 2. Typical transient currents observed after repetitive reversals of the voltage polarity.

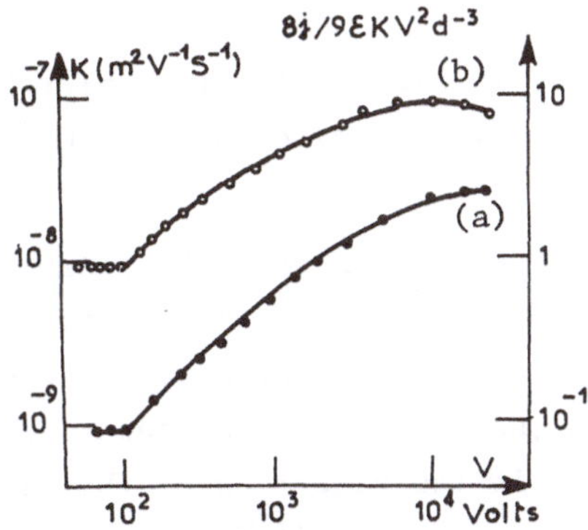

Fig. 3. (a) Dependence of the mobility of Cl^- ion in chlorobiphenyl ($\varepsilon_r = 6.4$, $\eta = 50$ cp at 20C) on the applied voltage; $d = 4.5$ mm.
(b) Dependence of the ratio j/j_{LCE} in the same liquid on the applied voltage; j is the current density and j_{LCE} the space-charge-limited current density.

charge density q_i near the injecting electrode, V_c, given by $\varepsilon V_c/\eta K = 221(q_i d^2/\varepsilon V_c)^2$, may take higher values. In Fig. 4, for weak injection in cychlohexane, we measured V_c higher than 10kV.

For strong unipolar injection and an applied voltage $V \gg V_c$, the liquid motion is turbulent (Hopfinger and Gosse, 1971). The order-of-magnitude of the liquid turbulent velocity in the direction of the field is $(\varepsilon/\rho_d)^{1/2}E = K_hE$, with ρ_d the liquid density, and K_h is the hydrodynamic mobility. Ions are carried about by this turbulent flow and have an apparent mobility $\sim 0.4K_h$ during the transient following the application of a voltage step. In stationary conditions, an apparent ionic mobility cannot be defined, but the maximum value of the current density j_c with convection is $j_{cm} = (\varepsilon K V^2/d^3) \times (K_h/3K)^{1/2}$ (Lacroix et al., 1975). j_{cm} will be greater for larger dielectric permittivity and liquid

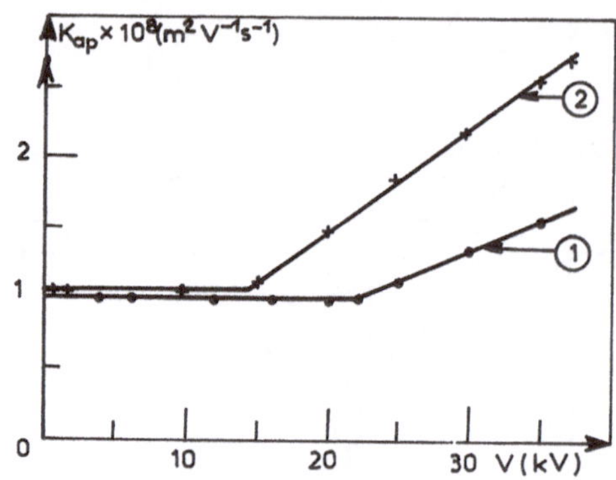

Fig. 4. Dependence of the mobility of the propionate anion in cyclohexane on the applied voltage (d = 0.8 mm); curve 1, 7.4×10^{11} Ω m; curve 2, 1.7×10^{10} Ω m.

viscosity. For instance, in Fig. 3(b), with chlorobiphenyl ($\varepsilon = 6.4$, $\eta = 50$ cp, $\rho_p = 1.5$), the hydrodynamic mobility K_h is 293 times K, and $j_{cm} = 9.9\ (\varepsilon K V^2/d^3)$.

There exists another mechanism of field-enhanced mobility, but it needs a far higher applied voltage than EHD motion, and consequently has never been observed. Indeed, a classical rate-process approach of ionic diffusion and migration (see for instance Conway, 1970) shows that the drift velocity of an ion in the direction of the field is proportional to sinh (El/2 U), with l the mean jump distance of the ion from one site of low energy to another. If $E \ll 2\ U/l$, the drift velocity is proportional to E; if $E \gg 2\ U/l$, however, it grows exponentially. l being about a few 10^{-10}m, this should occur whenever $E > 5$ to $10 \times 10^7\ V\ m^{-1}$. In common experimental conditions for d in the range 10^{-3} to some 10^{-2} m, the corresponding voltage would be more than 50 kV.

As a conclusion, if the measurement of the ion mobility K involves high applied voltage, it is necessary to verify that K remains constant in large ranges of voltage and electrode spacing.

VOLUMIC CONDUCTION

The spontaneous creation of ions in a liquid without an applied field proceeds from an equilibrium with two steps (Bjerrum model, 1926; Onsager, 1934; Eigen and De Maeyer, 1974),

$$\text{neutral molecule} \underset{k_2}{\overset{k_1}{\rightleftharpoons}} \text{ion pair} \underset{k_4}{\overset{k_3}{\rightleftharpoons}} \text{free ions}\ . \tag{1}$$

The first step is the ionization of the molecule and the formation of an ionic bond; it depends on chemical properties of the molecule and the kinetic rate constants k_1 and k_2 generally remain unknown.

The second step is the dissociation of the ion pair and results from coulombic interactions. The recombination kinetics is related to the probability for two ions of opposite sign to have their centers separated from less than the minimum distance of approach, a. Debye (1942) has established that the rate constant for diffusional encounter (k_4) of two oppositely charged particles in electrostatic interactions is

$$k_4 = e\ (D_+ + D_-)/\varepsilon\ U\ (1 - \exp\ (-e/5\ \pi\ \varepsilon\ a\ U))\ , \qquad (2)$$

where D_+ and D_- are the diffusion coefficients, $U = kT/e$, and ε is the dielectric permittivity. When the energy, $W_a = e^2/4\ \pi\ \varepsilon\ a$, is far larger than kT (as in the case of low-polar liquids), the Debye expression becomes the one given by Langevin (1903) for ionized gases if each encounter is efficient: $k_4 \# k_R = e\ (D_+ + D_-)/\varepsilon\ U = e\ (K_+ + K_-)/\varepsilon$. In both ammonium salt and ionic surfactant solutions in liquids such as cyclohexane or diphenylether, the kinetic rate constant of recombination k_R has been found to be lower than the theoretical Langevin value. The experimental value is about 0.7 or 0.8 times the theoretical one (Honda et al., 1979; Randriamalala et al., 1985). It has been suggested that the Langevin recombination rate constant overestimates the value of k_4 in liquids where the hydrodynamic interactions would decrease the mutual mobilities of ions. The expression of the kinetic rate constant of dissociation (k_3) (Eigen, 1954) is $k_3 = 3\ e\ (D_+ + D_-)/4\ \pi\ \varepsilon\ a^3\ (\exp W_a/kT - 1)$. The volumic density of anions or cations in the liquid for zero applied field is $n_o = (k_3\ k_1/k_2\ k_4)^{1/2}\ C^{1/2}$, with C the concentration of the dissociating molecule. In Table 4 are reported the values of exp W_a/kT for different liquids (at 20 C and a = 0.5 nm). For non-polar liquids, a small decrease of ε causes a very large decrease of this factor and then of the ion concentration n_a, or of the conductivity.

An electric field in the liquid increases the probability to separate the ions of an ion-pair at a distance > a. The kinetic theory of field-enhanced dissociation (Onsager, 1934) gives

Table 4. Dependence of the term exp $W_a/2\ kT$ ($W_a = -e^2_o/4\ \pi\ \varepsilon\ a$) on the liquid dielectric constant.

Liquid	Hexane	Cyclo-hexane	Benzene	Chloro-benzen	Dichloro-ethane	Nitro-benze
Dielectric constant	1.89	2.02	2.28	5.7	10.65	36
Exp $W_a/2$ kT	5.8×10^{-14}	4.1×10^{-13}	1.1×10^{-11}	4.1×10^{-5}	4.5×10^{-3}	0.2

$$k_3 (E) = k_3 (0) J_1 [(-8b)^{-1/2}]/(-2b)^{-1/2} = k_3 (0) F (E) , \quad (3)$$

with J_1 the Bessel function and $b = e|E|/8 \pi \varepsilon U$; k_3 is an increasing function of the absolute value of the electric field. This property was used by Persoons (1974) in a field-modulation technique to experimentally verify the Onsager theory. Theory and experiments were found in agreement not only in the case of an equilibrium between free ions and ion pairs (TBAP in diphenylether), but also when the solution contained more complex species such as triple-ion quadrupoles (TBAP in a mixture benzene-chlorobenzene, e.g., Honda et al., 1979).

We shall discuss the electric conduction due to creation of ions in the liquid only for the simple case of the dissociation of ion pairs $A^+ B^-$, i.e.,

$$A^+ B^- \underset{k_4}{\overset{k_3}{\rightleftarrows}} A^+ + B^- . \quad (4)$$

At zero applied field, the liquid conductivity is

$$\sigma_a = (K_+ + K_-) e (k_3 C/k_4)^{1/2} = (K_+ + K_-) q_o . \quad (5)$$

The influence of an electric field on the liquid conduction depends on the value of the dimensionless number $C_o = q_o d^2/\varepsilon V$, with d the electrode spacing and V the applied voltage. This number is one-half of the ratio of the transit time of an ion from one electrode to the opposite one to the relaxation time of the dissociation equilibrium ($\tau = \varepsilon/2\sigma_a$). When C_o>>1, the field has only a slight influence on the equilibrium, the liquid remains at thermodynamic equilibrium, and the current density is $j = \sigma(E) E = \sigma_a F(E)^{1/2} E$. When C_o<<1, the current density is limited by the kinetic rate of dissociation, and the current density is $j = k_3(E)Ced$, with C the concentration of ion pairs.

The shape of the characteristic j(E) is given in Fig. 5. With non-polar liquids, the voltage for which the current density saturates ($C_o = 1$) is very low (order of magnitude ~ 100 V). We also give in Fig. 6 (curve 1) the field dependence of the electrical losses [tanδ (E)] of solutions of the ionic surfactant Aerosol OT in cyclohexane. Transient and steady-state currents have shown that conduction is there due to ions created in the bulk of the liquid (Denat et al., 1982a). Figure 6 gives a typical example of tanδ (E) of these solutions with a small maximum followed by a minimum, and then a sharp rise with increasing electric field.

CREATION OF IONS ON A METAL ELECTRODE

As soon as the field exceeds about 100 kV cm^{-1}, in most of the cases, the field dependence of the current density is far stronger than

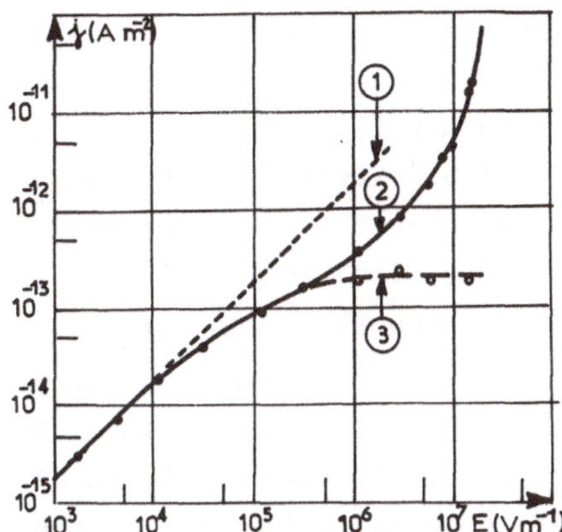

Fig. 5. Dependence of the current density in a solution of Aerosol OT in cyclohexane (d = 0.35 mm, resistivity = 5.9×10^9 Ω m) on the applied voltage. Curve 1, Ohm's law; curve 2, j(E); curve 3, j(E)/F(E).

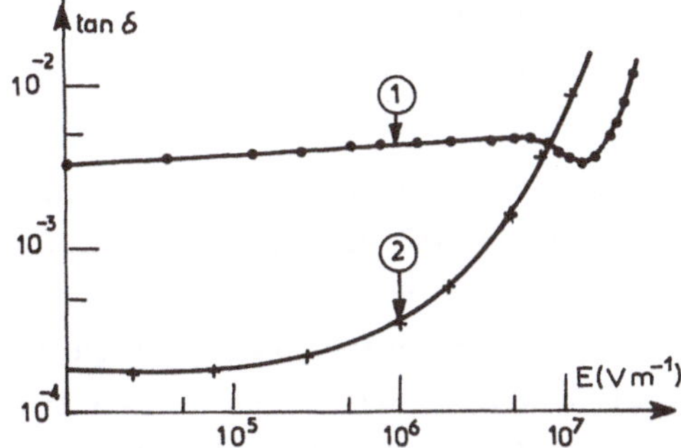

Fig. 6. Dependence of tanδ (conduction losses) on the amplitude of the electric field. Curve 1, solution of Aerosol OT in cyclohexane (d = 0.33 mm); curve 2, solution of TIAP in cyclohexane (e = 0.25 mm, resistivity = 2×10^{11} Ω m).

described by the preceding mechanism. In polar liquids, where the Kerr effect allows the measurement of the field distribution between the electrodes and its temporal variations, it is easy to show that ions are created at the electrodes.

Field emission of electrons at the cathode has been postulated by a number of workers, as well as liquid ionization. This subject has been previously discussed in other lectures. We can only add that in cyclohexane true field emission is observed for $E > 2 \times 10^9$ V m^{-1}, and liquid ionization for $E > 1.5 \times 10^9$ V m^{-1} (Denat et al., 1987). For both mechanisms j is strongly dependent on E and, for the field values considered here ($< 10^8$ V m^{-1}), j is far lower than the measured values.

It is now acknowledged that charge transfer between metal and liquid occurs in the electrified double-layer by electron tunneling between the metal and an absorbed species (see for instance Bockris and

Reddy, 1970). Its consequence is the chemical oxidation or reduction of neutral species and creation of ions. These reactions may involve either neutral impurities (oxygen and water are often present in rather large concentrations), or the liquid itself, or the metal of the electrodes. Studies of the electrical conduction of polar liquids such as nitrobenzene (Brière et al., 1969) or propylene carbonate (Félici et al., 1976) have shown: (a) when the electrode metal is easily oxidized (silver, for instance), positive ions are created at the anode according to $M \longrightarrow M^+ + e^-$; such positive injections, depending on the electrode metal, have also been observed in water (Zahn et al., 1985); (b) between unoxidizable electrodes, at high fields, the current density, which may reach 1 mA cm^{-2}, is carried by negative ions coming from the reduction of O_2 or of the liquid itself; (c) at moderate fields, the injected ions are often cations and have been thought to originate from the oxidation of impurities such as water; (d) when easily oxidizable products (such as perylene) or gas (H_2) are dissolved in the liquid, the sign of the injected ion is always in agreement with that expected from the electro chemical reaction of the impurity.

In low-polar liquids, because of their high resistivity, the chemical degradation of the liquid caused by electrode reactions is also observed, but after long-time dc voltage applications. For instance, the chemical degradation of displays using nematic liquid crystals has been attributed to the electrochemical oxidation of the liquid crystal (see Creagh, 1973). Electrochemical oxidation of oil-cooled nonvarnished conductors may limit the lifetime of high-stressed rectifier transformers (Bronger et al., 1981). In impregnated paper or "mixed" plastic film-paper capacitors, additives such as anthraquinone are used to extend dc life. The stabilizers, which are hydrogen-acceptors, are thought to react with nascent cathodic hydrogen or to produce by their own electrochemical reduction harmless products (Parkman, 1978).

Let us consider an electrode where the reaction $C + e^- \rightleftarrows C^-$ occurs. If the interface is at equilibrium, with no external generator, no net current flows across the interface, but electron transfer occurs in both directions. To the reaction $C \longrightarrow C^-$ corresponds

$$j = [C^-]\, k_f{}^\circ \exp(-W_f/kT) = j_o \,, \tag{6}$$

and in the opposite direction

$$j = [C^-]\, k_r{}^\circ \exp(-W_r/kT) = j_o \,. \tag{7}$$

If an electric field E is applied to the liquid, the potential drop in the double layer is modified by ΔV. The activation energies of formation and discharge of C^- become $W_f - (1 - \beta)\, e\Delta V$ and $W_r + \beta\, e\Delta V$, respectively, where β between 0 and 1 depends on the symmetry of the barrier. Then, the current density is $j = 2j_o sh\ [\Delta V/2\ U]$ (Butler - Volmer equation). Félici (1983) has written $\Delta V = x_A E\alpha$, with x_A the distance of closest approach, and $\alpha = \varepsilon/\varepsilon_{DL}$, the ratio of the permittivity ε of the free liquid to the permittivity in the double-layer. He notes a good agreement with the experimental characteristics j (E), but the fitting of curves with theory leads to a distance x_A that exceeds accepted values by an order-of-magnitude ($x_A \sim 30$ nm). He concludes that ions having high discharge potentials are accumulating on the electrodes and cause an additional modification of W_f and W_r. The charge density on the electrodes and its evolution with time under dc voltage cannot be foreseen at the present time. Progress is needed in this direction to explain the characteristic j (E) of a liquid.

In non-polar liquids, ions created by the reaction $C + e^- \rightleftarrows C^-$ have to surmount the image-force barrier. The top of the barrier is located at a distance $d = e/16\ \pi\varepsilon E$ from the metal (for $E = 10^7\ V\ m^{-1}$ and $\varepsilon = 2$, $d \sim 4$ nm) which is well outside the double-layer. This escape step can be written $C^- \xrightarrow{kes} C^-_{free}$, and its kinetic rate constant, kes, has been calculated from an electro-diffusion model (Blossey, 1974; Charle and Willig, 1978):

$$kes = D\left(\exp\left[-x_i/x_A - x_A\ E/U\right]\right)/(e/4\pi\varepsilon\ E)^{1/2}\ K_1\left[e_o\left(E/4\pi\varepsilon\right)^{1/2}/U\right], \quad (8)$$

where $x_i = e/16\ \pi\varepsilon U$, D is the diffusion coefficient of the injected ions, and K_1 is the modified Hankel function. We have experimentally verified the field dependence of kes in hydrocarbons, containing electrolytes such as TiAP for fields up to $4 \times 10^7\ V\ m^{-1}$ (Denat et al., 1982b). Figure 7 shows the strong field dependence of the current density in such solutions. It also shows the proportionality of j with the conductivity of the solutions. This is particular to these solutions where the injected ions are the ions of the electrolyte added to the liquid. Figure 6 (curve 2) shows the field dependence of $\tan\delta$ (E) in such solutions; the losses due to injection increase very strongly for about $E > 5 \times 10^6\ V\ m^{-1}$, and their variations are very different from those due to volumic conduction. Such measurements can be used to distinguish which mechanism of ion creation dominates in the considered liquid. The different behavior of solutions do not appear so clearly on the characteristics j(E). We give an example in Fig. 8, where cyclohexane, with some additives, presents a volumic conduction (curves 2 and 3), and with others presents a conduction due to injected ions (curve 1).

CONCLUSION

Mechanisms of liquid conduction are now fairly well understood. In the case of field strengths lower than 1 or $2 \times 10^8\ V\ m^{-1}$, electron

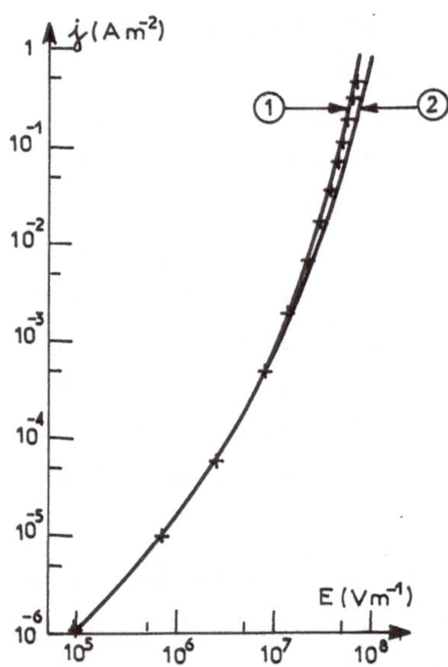

Fig. 7. Field dependence of the current density in a solution of triisoamylammonium picrate in cyclohexane (d = 0.78 mm, resistivity = 1.7 x 10^{11} Ω m). Curve 1, j α kes; curve 2, j α (kes) exp (x_A E/U); + = experimental points.

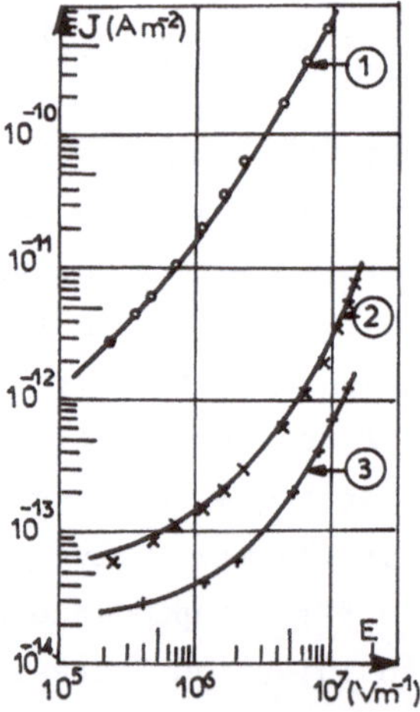

Fig. 8. Field dependence of the current density in cyclohexane with different additives. Curve 1, 10^{-4} M AOT $(C_2H_5)_4N$. The ammonium group is substituted for Na in the AOT molecule. Curve 2, 10^{-4} M AOT NH_4; curve 3, 10^{-4} M AOT.

emission or liquid ionization are ruled out, and conduction is electrochemical in nature. There are two sources of ions: (i) electrolytic dissociation of impurities or of the liquid itself; (ii) ion injection at the electrodes through electrochemical reactions. The second dominates at fields higher than 10^7 V m^{-1}. These reactions generally entail chemical degradation and limit the use of liquids in dc voltage. The kinetic rate constants involved in both mechanisms (dissociation and injection) vary with the field according to Onsager's theory and Willig's theory. Their expressions can be used to interpret the characteristic curve j(E) of liquids.

REFERENCES

Accascina, F., Petrucci, S., and Fuoss, R.M., 1959, The conductance of Bu_4 N BPh_4 in Acetonitrile-Carbon Tetrachloride mixtures at 25C, J. Amer. Chem. Soc., 81:1301.

Atten, P., and Moreau, R., 1972, Stabilité hydrodynamique des liquides isolants soumis à une injection unipolaire, J. de Mécanique, 11:471.

Atten, P., and Lacroix, J.C., 1979, Non-linear hydrodynamic stability of liquids subjected to unipolar injection, J. de Mécanique, 18:469.

Bjerrum, N., 1926, quoted in "The principles of Electrochemistry" D. A. Mac Innes, Dover Publ., New York, 1961, Kgl. Danoke Vidensk. Selskab, 9:7.

Blossey, D.F., 1974, One-dimensional Onsager theory for carrier injection in metal-insulator systems, Phys. Rev. B., 9:5183.

Bockris, J. O'M., and Reddy, A.K.N., 1970, in : "Modern Electrochemistry" Vol. 1 and 2, Plenum Press, New York.

Brière, G., Cauquis, G., Rose, B., et Serve, D., 1969, Relation entre la conduction électrique des liquides polaires et leurs propriétés électrochimiques, J. Chimie Phys., 66 (1):44.

Bronger, W., Kranz, H.G., and Moller, K., 1981, Electrolytical phenomena in liquid hydrocarbons of thermally high-stressed rectifier transformers, in : "Proc. 7th Int. Conf. on Conduction and Breakdown in Dielectric Liquids," Berlin, W. F. Schmidt, ed., I.E.E.E. El. Ins. Soc., n° 81 CH 1594-1.

Casanovas, J., Grob, R., Garbay, H., and Crine, J.P., 1985, Transient currents in silicone oils subjected to voltage steps with polarity reversal, I.E.E.E. Trans. on E.I., EI-20:183.

Charle, K.P., and Willig, F., 1978, Generalized one-dimensional Onsager model for charge carrier injection into insulators, Chem. Phys. Lett., 57 (2):253.

Coetzee, J.F., and Cunningham, G.P., 1965, Evaluation of single ion conductivities in Acetonitrile, Nitrobenzene and Nitromethane using Bu_4 N BPh_4 as reference electrolyte, J. Amer. Chem. Soc., 87 (12):2529.

Conway, B.E., 1970, Some aspects of the thermodynamic and transport behaviour of electrolytes, in : "Physical Chemistry" Vol. 9A/Electrochemistry, Eyring H., ed., Academic Press, New York.

Creagh, L.T., 1973, Nematic liquid crystal materials for displays, Proc. I.E.E.E., 61:814.

Debye, P., 1942, Reaction rates in ionic solutions, Trans. Electrochem. Socl., 82:265.

Denat, A., Gosse, B., and Gosse, J.P., 1979, Ion injections in hydrocarbons, J. Electrostatics, 7:205.

Denat, A., Gosse, B., and Gosse, J.P., 1982a, Electrical conduction of solutions of an anionic surfactant in hydrocarbons, J. Electrostatics, 12:197.

Denat, A., Gosse, B., Gosse, J.P., 1982b, High field DC and AC conductivity of electrolyte solutions in hydrocarbons, J. Electrostatics, 11:179.

Denat, A., Gosse, J.P., and Gosse, B., 1987, Conduction du cyclohexane très pur en géométrie pointe-plan, Rev. Phys. Appl., 22:1103.

Eigen, M., 1954, Uber die Kinetik sehr schnell verlaufender Ionenreaktionen in wasseriger Losung, Z. Phys. Chem. NF, 1:176.

Eigen, M., and De Maeyer, L., 1974, Theoretical Basis of Relaxation Spectroscopy, in: "Technique of Organic Chemistry", 8, 895, Friess, S.L., Lewis, E.S., Weissberger, A., eds., Interscience Publ., New York.

Félici, N.J., 1971, D.C. conduction in liquid dielectrics, A survey of recent progress, Part I, Direct Current, 2:90, Part II, Direct Current, 2:147.

Félici, N., Gosse, B., and Gosse, J.P., 1976, Aspects électrochimiques et électrohydrodynamiques de la conduction des liquides isolants, R.G.E., 85 (11):861.

Félici, N., 1983, Conduction des liquides diélectriques sous haute tension et régime de la double couche aux électrodes, Compt. Rend. Acad. Sc. Paris, 296 (II):523.

Franck, H.S., 1966, Solvent models and the interpretation of ionization and solvation phenomena, in : "Chemical Physics of Ionic Solutions," B.E. Conway and R.G. Barradas, ed., J. Wiley & sons, New York.

Fuoss, R.M., 1959, Dependence of the Walden product on dielectric constant, Proc. N.A.S., 45:807.

Fuoss, R.M., and Hirsch, E., 1960, Single ion conductances in non aqueous solvents, J. Amer. Chem. Soc., 82:1013.

Gallagher, T.J., 1975, "Simple Dielectric Liquids," Oxford Univ. Press, Oxford.

Honda, T., Evereart, J., and Persoons, A., 1979, Dependence of the field dissociation effect on electric field strength, in : "Non-linear behaviour of Molecules, Atoms and Ions in Electric, Magnetic or Electromagnetic Fields," L. Neel, ed., Elsevier, Amsterdam.

Hopfinger, E.J., and Gosse, J.P., 1971, Charge transport by self-generated turbulence in insulating liquids submitted to unipolar injection, Phys. Fluids, 14:1671.

Hubbard, J., and Onsager, L., 1977, Dielectric dispersion and dielectric friction in electrolyte solutions. I, J. Chem. Phys., 67:4850.

Janz, G.J., and Tomkins, R.P.T., 1972, in : "Non aqueous Electrolyte Handbook," Vol. 1, Academic Press, New York.

Lacroix, J.C., Atten, P., and Hopfinger, E.J., 1975, Electro-convection in a dielectric liquid layer subjected to unipolar injection, Phys. Fluid Mech., 69:539.

Langevin, M.P., 1903, Recombinaison et mobilités des ions dans les gaz, Ann. Chim. et Phys., 28:433.

Lewis, T.J., 1959, The electric strength and high-field conductivity of dielectric liquids, in : "Progress in Dielectrics 1," J.B. Birks, ed., Heywood & Co, London.

Nemamcha, M., Gosse, J.P., Denat, A., and Gosse, B., 1987, Temperature dependence of ion injection by metallic electrodes into non-polar dielectric liquids, I.E.E.E. Trans. on EI, EI-22:459.

Novotny, V., and Hopper, M.A., 1979, Transient conduction of weakly dissociating species in dielectric fluids, J. Electrochem. Soc., 126:925.

Onsager, L., 1934, Deviations from Ohm's law in weak electrolytes, J. Chem. Phys., 2:599.

Parkman, N., 1978, Some properties of solid-liquid composite dielectric systems, I.E.E.E. Trans EI, EI-13:289.

Persoons, A., 1974, Field dissociation effect and chemical relaxation in electrolyte solutions of low polarity, J. Phys. Chem., 78:1210.

Randriamalala, Z., Denat, A., Gosse, J.P., Gosse, B., 1985, Field-enhanced dissociation, the validty of Onsager's theory in surfactant solutions, I.E.E.E. Trans. EI, EI-20, 167.

Robinson, R.A., and Stokes, R.H., 1959, in : "Electrolyte solutions", Butterworths, London.

Schmidt, W.F., 1984, Electronic conduction processes in dielectric liquids, I.E.E.E. Trans. EI, EI-13:389.

Schneider, J.M., and Watson, P.K., 1970, Electrohydrodynamic stability of space-charge limited currents in dielectric liquids, Phys. Fluids, 19:1948.

Sharbaugh, A.H., and Watson, P.K., 1962, Conduction and breakdown in liquid dielectrics, in : "Progress in Dielectrics 4," J.B. Birks, ed., Heywood & Co, London.

Yasufuku, S., Umemura, T., and Tanii, T., 1979, Electric conduction phenomena and carrier mobility behavior in dielectric fluids, I.E.E.E. Trans. Electr. Insul., EI-14:28.

Zahn, M., Ohki, Y., Rhoads, K., Lagasse, M., and Matsuzawa, H., 1985, Electrooptic charge injection and transport measurements in highly purified water and water/ethylene glycol mixtures, I.E.E.E. Trans. on EI, EI-20:199.

Zwanzig, R., 1963, Dielectric friction on a moving ion, J. Chem Phys., 38:1603.

MEASUREMENT OF ELECTRICAL BREAKDOWN IN LIQUIDS

Robert E. Hebner

National Bureau of Standards
Gaithersburg, MD 20899

INTRODUCTION

The general features of electrical breakdown between two electrodes in a liquid have been known for decades. When breakdown occurs, a flash of light is emitted, an acoustic signal is emitted, the potential difference between the electrodes decreases rapidly, and the current flowing between the electrodes increases rapidly. The continuing development of light sources, high-speed cameras, and high-speed electronic measuring systems have made it possible to study the breakdown process in increasing detail. This lecture summarizes some of the results obtained using these various measurement techniques to investigate electrical breakdown in liquids under pulsed voltage.

Four basic types of measurements are discussed. The first is high-speed photography of the breakdown process. Two different applications of photography are presented. One is the measurement of the density gradients in the field which result as the breakdown develops, and the other is the measurement of the evolution of the electrical field distribution during breakdown. The second type of measurement is the recording of the voltage and current. The third is optical spectroscopy of the emitted light, and the fourth is the measurement of acoustic emission.

Having developed a number of measurement techniques, understanding of the breakdown process is gained by changing the system in known ways and determining the effect of these changes on the measured results. Parameters which have been investigated systematically by a number of investigators over the past ten or fifteen years include: types of liquids, chemical additives, particulate additives, pressure, viscosity, and the rate-of-rise of the applied voltage.

The strengths and limitations of the various diagnostic approaches are illustrated in this discussion by showing what has been learned about the various stages of the breakdown process. The second section of this lecture outlines the processes which occur during breakdown. The third section describes measurements of the phenomena which lead up to the arc. The fourth section deals with processes which occur as the arc is ignited, and the fifth section describes measurements of processes after the arc has extinguished.

THE BREAKDOWN PROCESS

High-speed, multi-frame photographs of a single breakdown have provided a clear picture of the breakdown process under pulsed voltages. The various events are summarized in Figs. 1, 2, and 3. Figure 1 shows events which preceded the initiation of an arc (Fitzpatrick et al., 1987) with a point cathode, Fig. 2 shown the related processes for an anode point, and Fig. 3 shows the arc and post-arc events (Zahn et al., 1982; Fenimore and Hebner, in press). Figure 1 shows four of the distinct modes of growth that are possible for a streamer initiating from a cathode. Descriptive names for the various modes of growth have not been standardized. In this discussion, they will be given descriptors of the form "1st cathode mode" or "2nd anode mode." The properties of the modes are given in the next section (PREBREAKDOWN PHENOMENA) and their shapes are shown in Figs. 1 and 2. The first mode (Fitzpatrick et al., 1987) is a thin pencil-like structure which can be as little as a few micrometers in diameter. This mode grows at a subsonic rate. The second mode, a

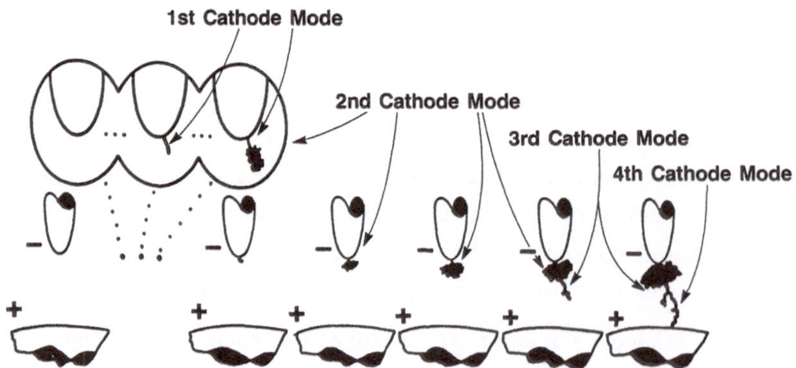

Fig. 1. Development of a streamer from a cathode point. This drawing shows four modes of growth of an idealized streamer. Each drawing is an instantaneous representation of the dynamic breakdown process. The streamer initiates between the second and third drawing from the left and completely bridges the gap in the drawing at the right. The second, third, and fourth drawings are magnified to show the initiation.

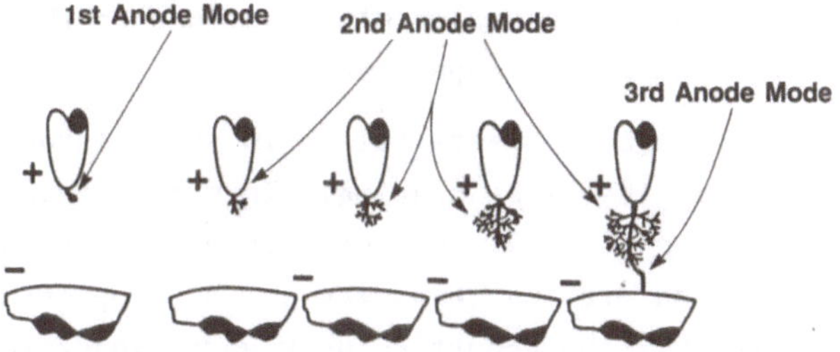

Fig. 2. Development of a streamer from an anode point. This drawing shows three modes of growth. Each drawing is an instantaneous representation of the dynamic breakdown process with the earliest time on the left, and the latest time on the right.

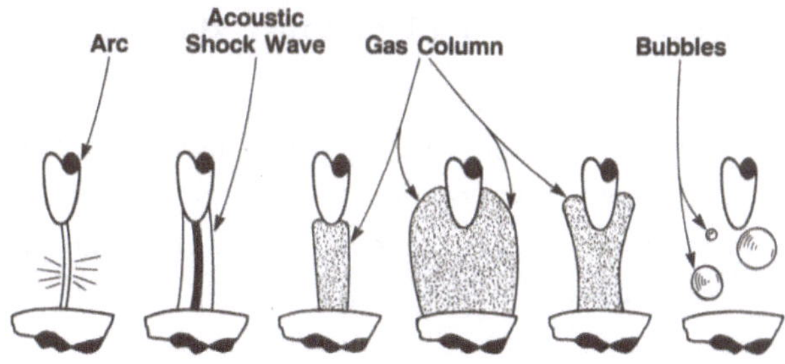

Fig. 3. Development of the postbreakdown pheonmena. These phenomena do not depend on point polarity. The arc develops at the left. An acoustic shock wave is produced and propagates away from the arc region. The arc region then expands as a gas column followed by contraction as it cools. Eventually the collapsing gas column breaks into one or more bubbles.

bushier structure, also grows at a subsonic rate (Devins et al., 1981; Kelley and Hebner, 1981a). The third mode is again a thin structure but this structure propagates at nearly sonic velocity (Kelley and Hebner, 1981a). The fourth stage has nearly the same shape as the third, but propagates an order-of-magnitude faster (Kelley and Hebner, 1981a).

Three possible stages of growth are shown for streamers which originate at the anode in Fig. 2. The first is a bushy, subsonic structure (McKenny and McGrath, 1984). The second is a more filamentary structure which propagates near the sonic velocity (Devins et al., 1981; Kelley and Hebner, 1981a; McKenny and McGrath, 1984). The third mode is an order of magnitude faster than the second (Kelley and Hebner, 1981a).

It should be noted that all of these modes of growth may not occur during each breakdown and that this may not be a complete listing of all possible modes of growth. It should also be noted that progress has been made in the attempt to model the transition from one mode to another (Watson, 1981; Watson, 1985), but that additional work is needed. The subsequent discussion provides examples of conditions in which one or more of the modes are enhanced or suppressed.

Figure 3 shows drawings of several steps in the breakdown process with the breakdown arc in the first drawing. This arc is initiated as the streamer, shown in Fig. 1 or 2, and completely bridges the interelectrode gap. The second drawing shows the arc but represents a somewhat later time so that the acoustic shock wave, generated when the arc ignited, has had time to propagate away from the arc. The third and fourth drawings show later times. The shock wave has propagated well away and the gas column, produced as the arc vaporizes the surrounding liquid, has continued to expand, even though the arc has extinguished (Fenimore and Hebner, in press). In the fifth drawing, the gas column has begun to collapse. In the final drawing, the gas column has broken into bubbles.

The processes shown in Figs. 1, 2, and 3 occur on significantly different time scales. These differences require widely different exposure times and interframe intervals to record the full range of phenomena. This situation is summarized in Table 1.

PREBREAKDOWN PHENOMENA

Introduction

This section reviews the results of typical measurements made on the various propagation modes which have been observed. In this discussion, the term "cathode streamer" refers to a streamer, in any or all of its modes of growth, which initiates at the cathode. Similarly, the term

Table 1. Typical Times of Breakdown Events under Pulsed Voltage

Event	Velocity (cm/s)	Time to Propagate 1 mm (μs)
Cathode		
1st Mode	2×10^4	5
2nd Mode	2×10^4	5
3rd Mode	2×10^5	0.5
4th Mode	4×10^6	0.02
Anode		
1st Mode	2×10^4	5
2nd Mode	4×10^5	0.2
3rd Mode	4×10^6	0.02
Shock Wave	1×10^5	1
Gas Column	5×10^3	20

"anode streamer" refers to a streamer which initiates from an anode. It should be noted that the same type processes occur both for electrodes, which produce a nearly uniform electric field (sphere-sphere or plane-plane electrodes, for example), and those which produce a highly divergent electric field (point-plane electrodes) (Hebner et al., 1985).

A summary of the observations of the various modes of growth in a number of liquids is given in Table 2. The information in Table 2 is intended to be representative but is not exhaustive. It should also be recognized that many earlier authors did not specify which mode of growth was being observed during their studies. In those cases, the assignment is made here on the basis of the information presented in the paper. The assignments are expected to be generally correct, but if a specific identification is critical in another investigation, the reader should depend on the information given in the original paper and not on the assignment in this table. The table demonstrates that a large number of investigators have detected one or more modes of streamer growth in a range of liquids.

1st Cathode Mode

The 1st cathode mode has been studied in two ways. The first has been through the use of high-speed, high-magnification photography (Fitzpatrick et al., 1987). The second has been to augment the photographic data by also recording the Kerr-effect fringe pattern (Hebner, 1986) produced as the streamer initiates. The photography records the evolution of density gradients in the fluid, while the Kerr effect permits the determination of the electric field distribution and the space-charge density.

Typical apparatus to photograph the prebreakdown is shown in Fig. 4. A variety of techniques are available for high-speed photography (Courtney-Pratt, 1986). These can be divided into single-frame photography and multiple-frame photography. As the name implies, in single-frame photography only a single photograph is taken of a breakdown event. This approach can provide very good spatial resolution but limited temporal resolution. The information obtained can be difficult to interpret because of the shot-to-shot variation of the breakdown process. Even though interpretation can be difficult, much of our understanding of the prebreakdown process has been derived from single-frame data (Devins et al., 1981; Chadband and Sufian, 1985; Forster and Wong, 1977).

More complete temporal information is gained from multiple-frame photography in which a number of sequential high-speed photographs are taken of the same breakdown event. The typical number of photographs is

Table 2. Summary of liquids investigated

Liquid	Mode	Reference
chlorocyclohexane	2nd Cathode	Beroual and Tobazeon, 1986
cyclohexane	2nd Cathode	Hebner et al., 1985 Beroual and Tobazeon, 1986a; Forster et al., 1981; Hebner et al., 1982; Beroual and Tobazeon, 1986b
	3rd Cathode	Hebner et al., 1985
	1st Anode	Beroual and Tobazeon, 1986b
	2nd Anode	Beroual and Tobazeon, 1986b; Kelley et al., 1982
	3rd Anode	Forster et al., 1981
+ carbon tetrachloride	2nd Cathode	Beroual and Tobazeon, 1986b
+ anthracene	2nd Cathode	Beroual and Tobazeon, 1986b
cyclooctane	2nd Cathode	Beroual and Tobazeon, 1986a
dibenzyl-toluene	2nd Cathode	Beroual and Tobazeon, 1986a
cis-decahydronaphalene	2nd Cathode	Beroual and Tobazeon, 1986a
trans-decahydronaphalene	2nd Cathode	Beroual and Tobazeon, 1986a
n-decane	2nd Anode	McKenny and McGrath, 1984
dodecane	2nd Anode	Chadband and Sufian, 1985
n-heptane	1st Anode	McKenny and McGrath, 1984
	2nd Anode	Devins et al., 1981; McKenny and McGrath, 1984
n-hexadecane	2nd Anode	Devins et al., 1981; McKenny and McGrath, 1984
n-hexane	1st Anode	McKenny and McGrath, 1984; Kelley et al., 1986
	2nd Anode	McKenny and McGrath, 1984; Hebner et al., 1986; Chadband and Sufian, 1985; Kelley et al., 1986
	3rd Anode	Hebner et al., 1985; Forster et al., 1981; Kelley et al., 1986
	2nd Cathode	Hebner et al., 1985; Forster et al., 1981; Kelley et al., 1986; McGrath and Marsden, 1986; Hebner et al., 1982
	3rd Cathode	Hebner et al., 1985; Kelley et al., 1986; Hebner et al., 1982
+ n,n dimethylaniline	2nd Anode	Hebner et al., 1985; Chadband and Sufian, 1985
	3rd Anode	Hebner et al., 1985
	2nd Cathode	Hebner et al., 1985
	3rd Cathode	Hebner et al., 1985
issoctane	1st Anode	Fitzpatrick et al., 1985
	2nd Anode	Hebner et al., 1985
	3rd Anode	Hebner et al., 1985; Forster et al., 1981
	1st Cathode	Fitzpatrick et al., 1985
	2nd Cathode	Hebner et al., 1985; Hebner et al., 1982
	3rd Cathode	Hebner et al., 1985; Hebner et al., 1982
nitrobenzene	2nd Anode	Kelley and Hebner, 1981
	1st Cathode	Kelley and Hebner, 1986
	2nd Cathode	Kelley and Hebner, 1981
n-nonane	2nd Anode	McKenny and McGrath, 1984

Table 2. Summary of liquids investigated (Cont.)

Liquid	Mode	Reference
n-octane	1st Anode	McKenny and McGrath, 1984
	2nd Anode	McKenny and McGrath, 1984
n-pentane	1st Anode	McKenny and McGrath, 1984
	2nd Anode	McKenny and McGrath, 1984
phenylxylylethane	2nd Cathode	Beroual and Tobazeon, 1986
polydimethysiloxane	2nd Anode	Devins et al., 1981; Kelley et al., 1982; Chadband, 1980
	3rd Anode	Kelley et al., 1982; Chadband, 1980
	2nd Cathode	Beroual and Tobazeon, 1986; Kelley et al., 1982
	3rd Cathode	Kelley et al., 1982
squalene	2nd Anode	Chadband and Sufian, 1985
toluene	1st Anode	Fitzpatrick et al., 1985
	2nd Anode	McKenny and McGrath, 1984; Hebner et al., 1985
	3rd Anode	Hebner et al., 1985; Forster et al., 1981; Fitzpatrick et al., 1982
	1st Cathode	Fitzpatrick et al., 1985
	2nd Cathode	Hebner et al., 1985; Forster et al., 1981; Fitzpatrick et al., 1982; Hebner et al., 1982
	3rd Cathode	Hebner et al., 1985
+ n,n dimethylaniline	2nd Anode	Hebner et al., 1985
	3rd Anode	Hebner et al., 1985
transformer oil	2nd Anode	Devins et al., 1981 Kelley and Hebner, 1981; Chadband, 1980
	3rd Anode	Kelley and Hebner, 1981; Chadband, 1980
	2nd Cathode	Devins et al., 1981; Kelley and Hebner, 1981; Yamashita and Amano, 1985
	3rd Cathode	Kelley and Hebner, 1981; Yamashita and Amano, 1985
	4th Cathode	Kelley and Hebner, 1981
white mineral oil	2nd Anode	Devins et al., 1981; Hebner et al., 1985; Kelley et al., 1984
	3rd Anode	Devins et al., 1981; Hebner et al., 1985; Kelley et al., 1984
	1st Cathode	Fitzpatrick et al., 1985
	2nd Cathode	Devins et al., 1981
+n,n dimethylaniline	2nd Anode	Devins et al., 1981; Hebner et al., 1985; Kelley et al., 1984
	3rd Anode	Hebner et al., 1985; Kelley et al., 1984
	2nd Cathode	Devins et al., 1981
+ sulphur hexafluoride	2nd Cathode	Devins et al., 1981
+ ethyl chloride	2nd Cathode	Devins et al., 1981
+ 2-methylnaphtalene	2nd Anode	Devins et al., 1981
2,2,4-trimethylpentane	2nd Anode	Devins et al., 1981
+ n,n dimethylaniline	2nd Anode	Devins et al., 1981

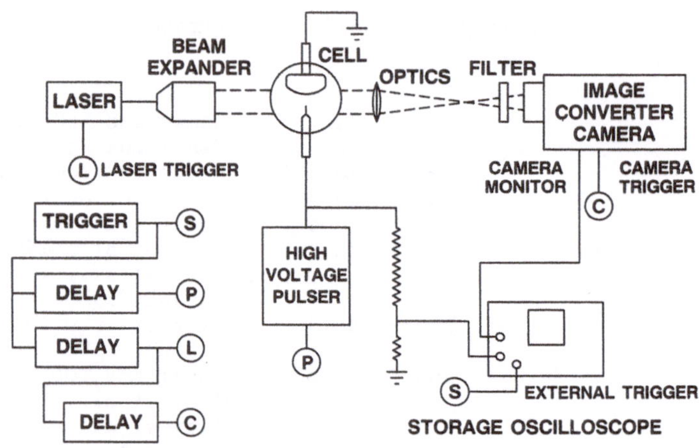

Fig. 4. Schematic of typical apparatus used to obtain high-speed photographs of prebreakdown phenomena in liquids.

between three and twelve. The critical parameters of this type of photography are the exposure time and the time between frames. Typically, the spatial resolution in a high-speed camera is not better than 10 line pairs per millimeter which means that a change smaller than 0.1 mm cannot be resolved. This change, of course, refers to a change at the film plane. In the study of prebreakdown streamers in liquids, the typical interelectrode spacing is between a few millimeters and a few centimeters. The various branches of the streamer have diameters which are less than 15 micrometers (Fitzpatrick et al., 1987; Devins et al., 1981). Because of these dimensions, the optical systems used for the study of these structures typically has a magnification between 1:1 and 100:1. From this information and the information given in Table 1, it can be seen that the required exposure time ranges between 0.5 μs for the slowest events and the lowest magnification to 0.02 ns for the fastest events. The values given in this discussion are intended to be typical values which give the general range of operation of measurement systems used to photograph prebreakdown data.

Figure 5 shows ten frames recorded during a single breakdown in toluene (Fitzpatrick et al., 1987). The magnification is 93:1. The exposure time is 10 ns and the time between successive frames is 50 ns. The voltage of the point electrode is -40 kV with respect to the plane electrode and the interelectrode gap is 1 cm. These photographs are typical of the 1st cathode mode of streamer growth. The width of the structure is 10 - 50 μm and its length is approximately 100 μm. It should be noted that the existence of this mode could be deduced from the early single-frame data (Forster and Wong, 1977), but its growth rate could not be determined.

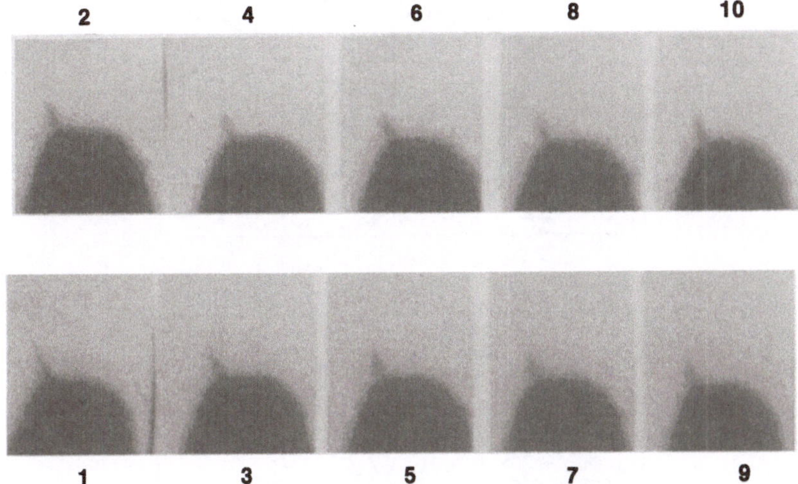

Fig. 5. Photographs of 1st cathode mode in toluene. Successive photographs were taken in the order indicated by the numbers. In these photographs, the optical magnification is nearly 100:1 and the time between successive photographs is 50 ns.

An interesting feature of this mode was discovered by combining high-speed photography with the electro-optic Kerr effect (Kelley and Hebner, 1986). Drawings of the Kerr response are shown in Fig. 6. In Kerr effect measurements, the electric field between the streamer and the electrode toward which it is propagating produces a phase shift between orthogonally polarized components of a light beam which is passed through the region. The phase shift is converted to an intensity variation by passing the light beam through a polarizer. The dark bands in Fig. 6 are regions of equal phase shift and, therefore, of equal electric field strength. In the figure, odd numbers indicate regions of maximum intensity. The variation of the electric field in going from one of these regions to another can be calculated knowing the geometry and Kerr response of the liquid. These Kerr effect measurements show that the electric field is distorted. This distortion occurs at a lower voltage than the lowest voltage at which the 1st cathode mode can be observed using the conventional shadowgraph techniques.

From the magnitude of the distortion and using Gauss's law, one can estimate the amount of charge which would be required to produce the observed field distortion. The charge required to produce the pattern shown in Fig. 6 is approximately 20 pC. From the high-speed photographs, it is possible to estimate the volume in which the charge resides and the time over which it was injected. Straightforward calculations show that the pattern in Fig. 6 would be produced when the injected charge would have deposited about 20% of the energy necessary to boil the volume of liquid in which it resides. The same calculations indicate that at the

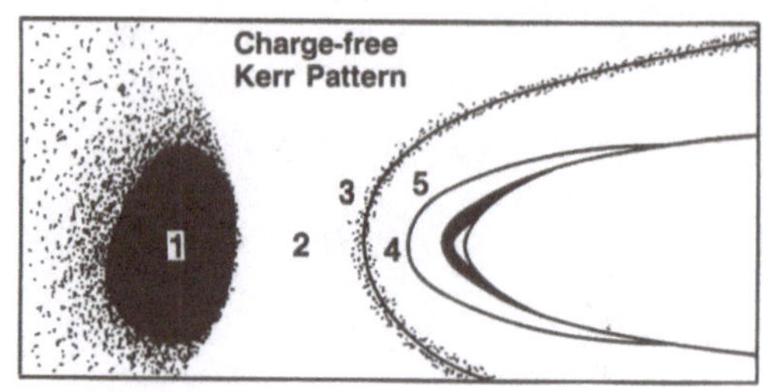

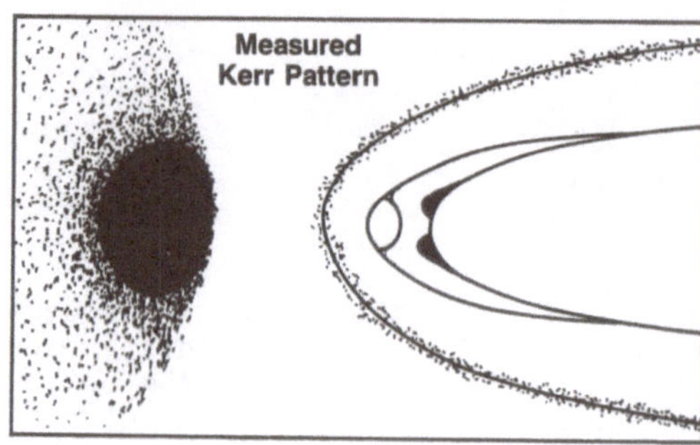

Fig. 6. Kerr pattern in the vicinity of a cathode point, which extends into the drawing from the right of the drawing, is shown in the upper sketch. The lower sketch shows the pattern taken when the injected charge distorts the electric field and, therefore, the Kerr pattern.

voltage levels at which the streamers are visible in the shadowgraphs, the deposited energy is sufficient to boil the liquid in the volume of the streamer.

2nd Cathode Mode

As discussed in the second section (THE BREAKDOWN PROCESS), the 2nd cathode mode is a bushy structure which propagates at a subsonic velocity. Among the interesting features of this mode are: it is relatively conductive; it can be suppressed by increasing the pressure of the surrounding field. In addition, the appearance of a current pulse and the simultaneous occurrence of an emitted light pulse are common phenomena during this type of streamer growth.

The fact that the streamer is relatively conductive has been inferred from measurements in two ways. The first, and most direct, was the measurement of the electric-field distribution between the streamer and the electrode toward which it is propagating using the electro-optic Kerr effect (Kelley and Hebner, 1981b). In Fig. 7, typical 2nd cathode modes with Kerr fringes are redrawn from three frames of a high-speed photograph. As in Fig. 6, the dark bands, or fringes, between the streamer and the electrode toward which it is propagating indicate regions of equal electric field strength. This fringe pattern between the streamer

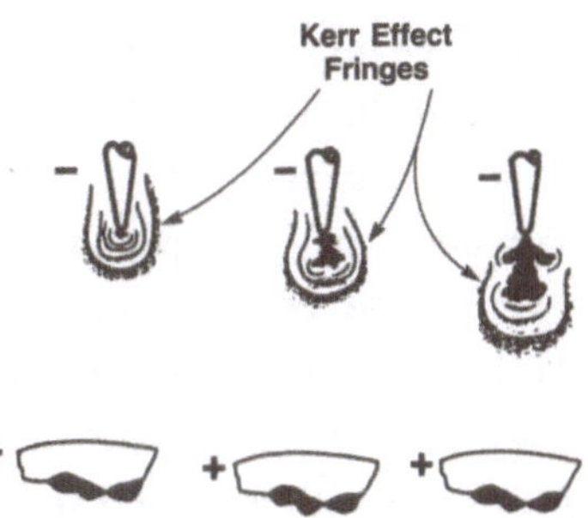

Fig. 7. The fringe pattern produced by the Kerr effect in the vicinity of a growing streamer. This type of data demonstrates that the field between the streamer and the electrode toward which it is propagating can be calculated by assuming that the streamer is a conductor at the same potential as the electrode from which it is propagating.

and the plane electrode was compared with the fringe pattern produced by a sphere over a plane (Kelley and Hebner, 1981b). In that case, the sphere was chosen to be about the same size as the streamer and the spacing between the sphere and the plane was chosen to be the same as the spacing between the streamer and the plane. If the sphere was chosen to have the same potential as the cathode, then the sphere produced the same fringe pattern as that measured in the vicinity of the streamer. If the size of the sphere, the spacing, or the voltage on the sphere were changed by 10%, then the resulting fringe pattern is clearly different from the pattern produced by the streamer. From measurements like these, it was concluded that the electric field distribution between the streamer and the electrode toward which it is propagating can be predicted by assuming that the streamer is a conductor at the potential of the electrode from which it is propagating.

It is, of course, true that the streamer is not a lossless conductor. As the streamer grows, work is done to produce the streamer. In modeling the voltage and current, for example, it was found to be useful to represent the streamer as a resistor and inductor in series (Sazama and Kenyon, 1979). The resistance is necessary to account for the fact that work is being done on the liquid. The Kerr effect data shows that the voltage drop across the streamer is a small fraction (less than 10%) of the total potential difference between the electrodes.

The second set of measurements (Forster et al., 1981) determines an upper limit of the conductivity by reversing the voltage on the electrodes as the streamer propagates from one electrode to the other. The experiment was configured so that the 2nd cathode mode was growing in a point-plane electrode system. While the growth was occurring, the polarity was reversed in approximately one microsecond. High-speed photographs showed that after reversal, an anode mode initiated from what had previously been a cathode mode of growth and bridged the gap leading to breakdown. Assuming that the streamer was a conductor yielded a capacitance between the streamer and the plane electrode of about 0.3 pF. For the polarity of the voltage to reverse in one microsecond, the resistance of the streamer must be less than three megohms. Because the rate of polarity reversal was determined by the external circuit and not the streamer properties, only an upper limit on the resistance could be determined from this measurement. These results. however, provide confirmation of the results of the more precise Kerr-effect measurements.

This mode of growth can be suppressed by increasing the rate of rise of the applied voltage pulse (Hebner, 1987) as shown in Fig. 8. It, therefore, follows that previously published results obtained with different waveshapes would be expected to differ.

The same suppression has been observed in hexane and toluene as the pressure is increased above about 10 MPa (Kelley et al., 1986). Similar behavior has been reported in cyclohexane for pressures as low as 0.15 MPa (Beroual and Tobazeon, 1986). In water, however, the suppression does not appear to occur at pressures up to 100 MPa. The reason for this difference in behavior between water and hydrocarbon liquids is the subject of present investigations.

The growth of this mode is accompanied by the emission of pulses of visible light. Studies have shown that the light pulses are time correlated with current pulses (Devins et al., 1981; Beroual and Tobazeon,

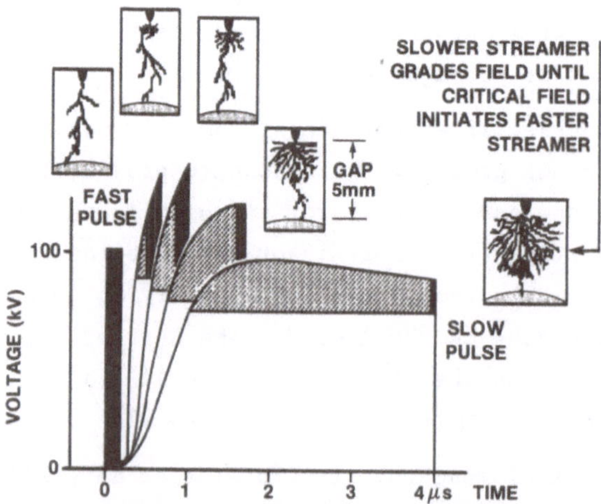

Fig. 8. Schematic diagram showing that the second cathode mode can be suppressed by increasing the rate of rise of the applied voltage. The crosshatched areas of the voltage waveforms show the region in which the second mode grows while the black areas denote the regions in which the third and fourth modes exist.

1986; Hanna et al., 1983; Nelson and McGrath, 1975). It has also been shown (Yamashita and Amano, 1985) that as the pressure is increased, the spatial distribution of the emitted light changes. Specifically, when the 2nd cathode mode is produced, the time integrated image of the emitted light is in the shape of the 2nd cathode mode. When the pressure is increased to suppress this mode, the time-integrated emitted light is in the shape of the 3rd cathode mode. To date, however, there have been no studies published which examine the correlation between streamer growth and the current of light pulses which go beyond the gross observation that the light is emitted as the streamer grows.

<u>3rd Cathode Mode and 4th Cathode Mode</u>

The 3rd cathode and 4th cathode mode, being much faster events than the 1st and 2nd cathode modes, have not been studied extensively. High-speed photography has been used to detect their existence. The addition of chemical impurities is a diagnostic probe which has been used in a number of studies to investigate these modes. The purpose of the additives is to modify the number of electrons and/or the electron energy distribution. Both here and, more particularly, in the study of positive streamers, such studies have led to contradictory results. There is evidence, for example, that the transition from the 2nd to the 3rd mode can be influenced by additives (Fitzpatrick et al., 1982). To obtain the data, chemicals were added to toluene so that the conductivity of the toluene was increased by four orders-of-magnitude from its purified value. It was found that the additives broadened the 2nd mode and induced the transition to the third mode to occur farther across the gap when the toluene was pure than when it was contaminated. The additives had a negligible effect on the velocity of either mode.

Other investigations, however, reported that the velocity of the streamers was increased by adding selected conducting impurities to the liquid (Devins et al., 1981; Beroual and Tobazeon, 1986). In these studies, single-frame photography was used and the particular mode of propagation which was studied was not identified. The fact that different behavior has been observed in different investigations suggests that the breakdown process is not yet well enough understood that the effect of adding a particular chemical impurity to a particular fluid can be predicted.

There is published evidence of acoustic emission from cathode streamers. The acoustic shock waves produced as the 3rd cathode mode propagates have been photographed (Kelley and Hebner, 1981a). In addition, a barium titanate ultrasound transducer has been used to measure the acoustic signals directly (Nelson and McGrath, 1975). This study showed that acoustic signals could be recorded as the cathode streamer propagated. Because no photographs were taken as a part of the investigation, it is not now possible to determine which mode or modes of propagation produced the acoustic signals. Subsequent investigations by other researchers now make it appear likely that the signals were associated with the 2nd and/or 3rd cathode mode. The 2nd cathode mode is a possible candidate because that study showed that, if the transit time between the transducer and the streamer were accounted for, there was good correlation between current pulses and acoustic pulses. Later studies have shown that there is current pulse activity during the growth of the 2nd cathode mode. The 3rd cathode mode is a possibility because photographs show that this mode produces acoustic signals.

<u>1st Anode Mode</u>

The 1st anode mode has been studied using high-speed photography. This mode of growth is similar in shape and in growth rate to the 2nd cathode mode. Data have been obtained (McKenny and McGrath, 1984) on the effect of voltage level, electrode geometry, liquid, and pressure on the production of this mode.

The probability of initiating the first anode mode appears to depend on the voltage level for the pulse rise time (300 ns) used in the experiments (McKenny and McGrath, 1984). When a potential difference of 20 kV was applied across a 1.5 mm gap, between one-third and one-half of the breakdowns involved the production of the 1st anode mode. If the peak voltage were raised to 24 kV, however, none of the breakdowns involved the 1st anode mode.

Measurements suggest that the appearance probability of the 1st anode mode is influenced by the point geometry in a point-sphere electrode system. The data were obtained by using a stainless steel sewing needle, a tungsten wire, and a stainless steel hypodermic needle as the point. Generally, reducing the radius of the point reduced the probability of observing the 1st anode mode. If the hypodermic needle were terminated with a square cut, multiple 1st anode modes could be generated.

To determine the effect of the liquid in which the breakdown occurs, measurements were made (McKenny and McGrath, 1984) in n-pentane, n-hexane, n-heptane, n-octane, n-nonane, n-decane, and toluene. The probability of observing a 1st anode event in n-pentane was 40% larger than n-hexane while the probability in n-heptane was about 90% smaller than in n-hexane. The 1st anode mode was rarely observed in n-octane and never observed in the other materials. Thus, that work suggested that the probability of observing the 1st anode mode decreased as the molecules became larger. Subsequent, high-magnification studies, however, have shown that this mode can occur in toluene and isooctane (Fitzpatrick et al., 1985).

Hydrostatic pressure was found to have a significant effect on the probability of initiating a prebreakdown streamer. When the pressure in n-hexane was increased to 0.3 MPa, the 1st anode event was suppressed. If the pressure is set to 0.25 MPa, the boiling point of n-hexane is nearly the same as the boiling point of n-heptane. Under this condition (McKenny and McGrath, 1984) the activity of the 1st anode mode in the two fluids was similar. These results suggest that boiling the fluid may contribute to the streamer growth processes under some conditions.

It should be noted that the apparent correlation of behavior of streamers with the boiling point of the fluid reported here and discussed in connection with the 1st cathode streamer has not been demonstrated definitively. In fact, estimates have been published that suggest that the energy deposited in the fluid is too small to raise the temperature by more than 10C (Chadband, 1980).

2nd Anode Mode and 3rd Anode Mode

The 2nd anode mode has been studied extensively. Like the faster cathode modes, however, the 3rd anode mode is less well studied because of its speed. Kerr effect measurements have been made of the 2nd anode mode (Kelley and Hebner, 1981) and it behaves similarly to the 2nd cathode mode in that the field between the streamer and the electrode toward which it is propagating can be calculated by assuming that the streamer is a conductor and that it is at the same voltage as the electrode from which it is propagating.

Considerable work has been done in determining the effect of chemical additives on the propagation of the 2nd anode mode. The seminal investigation (Devins et al., 1981) found that the addition of 0.05 M-n, n-dimethylaniline to a highly refined white mineral oil increased the velocity of the 2nd cathode mode by a factor of two. A subsequent investigation (Beroual and Tobazeon, 1986) showed a similar effect by adding tri-isoamylammonium to cyclohexane.

In an attempt to replicate the results of adding n,n-dimethylaniline to a white mineral oil by recording the streamer growth using multiple-frame high-speed photography, the change in the speed of the 2nd anode mode was not observed (Kelley et al., 1984). That work found instead that the effect of the additive was to increase the branching density of the 2nd anode mode and to suppress the formation of the 3rd anode mode. Similar results were reported for the effects of impurities in toluene (Hebner et al., 1985).

The disparity in the results of these three investigations may be due to differences in the rate-of-rise of the applied voltage. As reported above, increasing the rate-of-rise suppresses the 2nd cathode mode, and it may have the same effect on one or more of the anode modes. Thus, more information on the effect of the rate-of-rise of the voltage is necessary before the results from the various investigations can be compared.

As with cathode modes, the current and the light emission during streamer growth have been recorded for the 2nd anode mode (Devins et al., 1981; Beroual and Tobazeon, 1986). Light emission is similar for anode and cathode streamers, but the growth of the streamer is much smoother for the anode streamer than for the cathode streamer.

DEVELOPMENT OF THE ARC

When a streamer connects one electrode to the other, the arc begins to develop. This development is characterized by the rapid increase in the current, a rapid decrease in the voltage across the electrodes, and the emission of visible light. Studies have been made of the dynamic resistance of the arc as it is heated (Fuhr et al., 1986). These were done by measuring the voltage and current. Electrically, these are difficult measurements to perform. The relationship between the current through the arc, I, and the voltage across it, V, is (Fuhr et al., 1986).

$$V(t) = R(t)I(t) + d[I(t)L(t)]/dt \ , \tag{1}$$

where R is the arc resistance, L is the arc inductance, and t is time. Typical rates of change of the current are 1 to 1000 kA/μs. At these levels, significant voltages can be generated across small values of inductance which can lead to significant measurement errors. With careful attention to the measurement geometry and the operation of the measurement systems, however, reliable measurements can be made. As would be expected, these measurements show that the resistance of the arc decreases as the arc is heated (Fuhr and Schmidt, 1986). The data suggest that the growth of the arc is a result of at least three different processes.

The spectrum of the emitted light has also been measured (Wong and Forster, 1982). This measurement shows that the arc emits light at wavelengths characteristic of the atomic species which make up the surrounding liquid and the metal of the electrodes.

BEHAVIOR AFTER THE ARC IS INITIATED

Upon arc initiation, a shock wave is produced which propagates away from the arc at the sonic velocity. This shock wave (thunder) has been photographed using both single-frame and multiple-frame photographic systems and, from these photographs, its speed has been determined.

The second process which has been photographed is the growth of the gas column which surrounds the arc. The establishment of the arc deposits significant energy in a small volume in the fluid and thus raises that volume to a very high temperature. This temperature, which is related to the current passing through the arc, has not been measured directly. By analogy with arcs in gases, however, the temperature has been estimated to be in the range of 5000 K to 15,000 K. As this small volume cools, an expanding gas column is formed. The time to reach maximum size depends on a number of factors, but generally is of order 100 μs. The effect of pressure, between 0.1 and 100 MPa, has been measured photographically. In general, the effect of increasing pressure is to reduce the final size of the gas column.

After the gas column reaches its maximum size, it begins to recollapse. Photographs show that it then typically breaks up into a few bubbles which drift out of the interelectrode region and/or become absorbed by the liquid.

SUMMARY

In summary, the breakdown process in a liquid is complex, and important processes occur on widely different time scales. Useful modeling of the breakdown must take into account the various processes which occur and the factors which influence the transition from one process to another. Experimentally, we are learning which are the variables which are important to control to obtain results which are reproducible from laboratory to laboratory.

ACKNOWLEDGMENTS

The author has been fortunate to work with a number of respected investigators in attempting to understand the processes which contribute to electrical breakdown in a liquid. Particularly significant were discussions with E. Kelley, E. Forster, and G. Fitzpatrick on the various aspects of the breakdown process, and with M. Zahn and C. Fenimore on postbreakdown phenomena. In addition, P. McKenny, P. McGrath, J. Devins, S. Rzad, J. K. Nelson, and P. K. Watson have provided encouragement and criticism of the author's previous work in this field that have influenced the ideas outlined in this paper. Finally, special thanks are due to C. Fenimore for presenting this lecture at the Advanced Study Institute when the author was unable to attend.

REFERENCES

Beroual, A., and Tobazeon, R., 1986a, Effects de la pression hydrostatic sur les phenomenes de preclaquage dans les dielectriques liquides, C.R., Acad. Sc. Paris, Vol. 303, pp. 1081-1084.

Beroual, A., and Tobazeon, R., 1986b, Prebreakdown phenomena in liquid dielectrics, IEEE Trans. Elec. Insul., Vol. EI-21, pp. 613-627.

Chadband, W.G., 1980, On variations in the propagation of positive discharges between transformer oil and silicone fluids, J. Phys. D. Appl. Phys., Vol. 13, pp. 1299-1307.

Chadband, W.G., and Sufian, T.M., 1985, Experimental support for a model of positive streamer propagation in liquid insulation, IEEE Trans. Elec. Insul., Vol. EI-20, pp. 239-246.

Courtney-Pratt, J.S., 1986, Advances in high speed photography: 1972-1982, in: "Fast Electrical and Optical and Electrical Measurements," J.E. Thompson and L.H. Luessen, eds., Martinus Nijhoff, Boston, Vol. 2, pp. 595-607.

Devins, J.C., Rzad, S.J., and Schwabe, R.J., 1981, Breakdown and prebreakdown phenomena in liquids, J. Appl. Phys., Vol. 52, pp. 4531-4545.

Fenimore, C., and Hebner, R.E., The thermally induced growth of bubbles: an arc in a liquid, J. Appl. Phys., in press.

Fitzpatrick, G.J., Forster, E.O., Kelley, E.F., and Hebner, R.E., 1982, Effects of chemical impurities on prebreakdown events in toluene, 1982 annual rpt. Conf. Elec. Insul. Dielec. Phen., IEEE #82CH1773-1, pp. 464-472.

Fitzpatrick, G.J., Forster, E.O., Kelley, E.F., and Hebner, R.E., 1985, Streamer initiation in liquid hydrocarbons, 1985 Annual Rpt. Conf. Elec. Insul. Dielec. Phen., IEEE #85Ch2165-9, pp. 27-32.

Fitzpatrick, G.J., Forster, E.O., Hebner, R.E., and Kelley, E.F., 1987, Prebreakdown cathode processes in liquid hydrocarbons, IEEE Trans. Elec. Insul., Vol. EI-22, pp. 435-458.

Forster, E.O., and Wong, P., 1977, High speed laser schlieren studies of electrical breakdown in liquid hydrocarbons, IEEE Trans. Elec. Insul., Vol. EI-12, pp. 435-442.

Forster, E.O., Fitzpatrick, G.J., Hebner, R.E., and Kelley, E.F., 1981, Observations of prebreakdown and breakdown phenomena in liquid hydrocarbons II, point-plane geometry, 1981 Annual Rpt. Conf. Elec. Insul. Dielec. Phen., IEEE #81CH1668-3, pp. 377-389.

Fuhr, J., Schmidt, W.F., and Sato, S., 1986, Spark breakdown of liquid hydrocarbons I. Fast current and voltage measurements of the spark breakdown in liquid n-hexane, J. App. Phys., Vol. 59, pp. 3694-3701.

Fuhr, J., Schmidt, W.F., 1986, Spark breakdown of liquid hydrocarbons II. Temporal development of the electric spark resistance in n-pentane, n-hexane, 2,2 dimethylbutane, and n-decane, J. Appl. Phys., Vol. 59, pp. 3702-3708.

Hanna, M.C., Thompson, J.E., and Sudarshen, T.S., 1983, The simultaneous measurement of light and current pulses in liquid dielectrics, 1983 Annual Rpt. Conf. Elec. Insul. Dielec. Phen., IEEE #83Ch1902-6, pp. 245-250.

Hebner, R.E., Kelley, E.F., Forster, E.O., and Fitzpatrick, G.J., 1982, Observation of prebreakdown and breakdown phenomena in liquid hydrocarbons, J. Electrostatics, Vol. 12, pp. 265-283.

Hebner, R.E., Kelley, E.F., Forster, E.O., and Fitzpatrick, G.J., 1985, Observation of prebreakdown and breakdown phenomena in liquid hydrocarbons, II. Non-uniform field conditions, IEEE Trans. Elec. Insul., Vol. EI-20, pp. 281-292.

Hebner, R.E., 1986, Electro-optical measurement techniques, in: "Fast Electrical and Optical and Electrical Measurements," J.E., Thompson and L.H. Luessen, eds., (Martinus Nijhoff: Boston), Vol. 1, pp. 5-26.

Hebner, R.E., Ed. 1987, Research for electric energy systems -- An annual report, National Bureau of Stands., (USA) Report NBSIR 87-3643.

Kelley, E.F., and Hebner, R.E., 1981a, Prebreakdown phenomena between sphere-sphere electrodes in transformer oil, Appl. Phys. Lett., Vol. 38, pp. 231-233.

Kelley, E.F., and Hebner, R.E., 1981b, The electric field distribution associated with prebreakdown phenomena in nitrobenzene, J. Appl. Phys., Vol. 52, pp. 191-195.

Kelley, E.F., Hebner, R.E., Forster, E.O., and Fitzpatrick, G.J., 1982, Observations of pre- and post-breakdown events in polydimethylsilozanes, Proc. 1982 IEEE International Symp. Elec. Insul., IEEE #82CH1780-6, pp. 255-258.

Kelley, E.F., Hebner, R.E., Fitzpatrick, G.J., and Forster, E.O., 1984, The effect of aromatic impurities on the positive streamer growth in marcol 70, Conf. Record 1984 IEEE International Symp. on Elec. Insul., IEEE #84CH1964-6, pp. 284-287.

Kelley, E.F., and Hebner, R.E., 1986, Electro-optic field measurement at a needle tip and streamer initiation in nitrobenzene, 1986 Annual Rpt. Conf. Elec. Insul. Dielec. Phen., IEEE #86CH2315-0, pp. 272-277.

Kelley, E.F., Hebner, R.E., Fitzpatrick, G.J., and Forster, E.O., 1986, The effect of pressure on streamer initiation in n-hexane, Conf. Record 1986 IEEE Internat. Symp. Elec. Insul., IEEE #86CH2196-4, pp. 66-68.

McGrath, P.B., and Marsden, H.I., 1986, DC induced prebreakdown events in n-hexane, IEEE Trans. Elec. Insul., Vol. EI-21, pp. 669-672.

McKenny, P.J., and McGrath, P.B., 1984, Anomalous positive point prebreakdown behavior in dielectric liquids, IEEE Trans. Elec. Insul., Vol. EI-19, pp. 93-100.

Nelson, J.K., and McGrath, P.B., 1975, Evidence for transitions in the prebreakdown mechanism in liquid dielectrics, Proc. IEE, Vol. 122, pp. 1439-1442.

Sazama, F.J., and Kenyon, V.L., 1979, A streamer model for high voltage water switches, Digest of 2nd Internat. Pulsed Power Conf., IEEE #79CH1505-7, pp. 187-190.

Watson, P.K., 1981, EKD instability in the breakdown of point-plane gaps in dielectric liquids, 1981 Annual Report Conf. Elec. Insul. Dielec. Phen., IEEE #81Ch1668-3, pp. 370-376.

Watson, P.K., 1985, Electrostatic and hydrodynamic effects in the electrical breakdown of liquid dielectrics, IEEE Trans. Elec. Insul., Vol. EI-20, pp. 395-399.

Wong, P.P., and Forster, E.O., 1982, The dynamics of electrical breakdown in liquid hydrocarbons, IEEE Trans. Elec. Insul., Vol. EI-17, pp. 203220.

Yamashita, H., and Amano, H., 1985, Pre-breakdown current and light emission in transformer oil, IEEE Trans. Elec. Insul., Vol. EI-20, pp. 247-255.

Zahn, M., Forster, E.O., Kelley, E.F., and Hebner, R.E., 1982, Hydrodynamic shock wave propagation after electrical breakdown, J. Electrostatics, Vol. 12, pp. 535-546.

APPENDICES

APPENDIX A

ABSTRACTS OF POSTER PAPERS

ON THE DEVELOPMENT OF LIQUID IONIZATION DETECTORS AS SPECTROSCOPIC INSTRUMENTS

Elena Aprile

Physics Department
Columbia University
New York, NY 10027

A gridded ionization chamber has been used to study the maximum energy resolution achievable with liquid argon and liquid xenon. With its high density and atomic number, liquid xenon is a particularly interesting filling for a high resolution detector to be used in many different applications. Our specific interest is to develop a large volume high resolution imaging liquid xenon instrument for high energy gamma-ray astrophysics as well as to search for nuclear double beta decay of Xe-136 with much better sensitivity than existing instruments. The energy resolution of liquid argon or xenon ionization detectors is expected to be close to that achievable with Ge (Li) spectrometers, given the measured W values and the small Fano factors calculated by Doke. However, the best experimental results so far are nearly an order of magnitude worse than the theoretical values. In order to understand the reasons for this discrepancy, we have carried out several measurements with both liquids. The electric field dependence of conversion electrons has been measured up to 11 kV/cm with optimized grid geometry and liquid purity. In liquid argon, we obtain 2.7% fwhm for the resolution of the dominant 976 keV electron line in the Bi-207 spectrum. This value, the best reported so far in the literature, is still a factor of seven worse than the Fano limit. We find that our results can be explained if we take into account the additional statistical fluctuations associated with incomplete charge collection from delta-electron tracks produced in large number along the path of the primary ionizing particle. The strong recombination rate on these heavily ionizing delta-electrons is the limiting process to the ultimate energy resolution of noble liquid detectors, unless very high fields are used. Alternatively, one can increase the electron mobility.

The affects of photosensitive dopants added to the pure liquids have also been measured, both with electrons and alpha particles. We observe an improved energy resolution only in the case of alpha particles. In pure liquid argon, the 5.5 MeV alpha peak was measured with about 5% fwhm at a field of 38 kV/cm. In argon doped with 14 ppm of allene, the same resolution could be observed already with a field of less than 20 kV/cm.

THE BEHAVIOR OF ALKALI HALIDES NEAR THE MELTING POINTS

T. Armagan

Trakya University
Faculty of Science, Physics Department
Edirne, Turkey

S. Karagözlü and H. Gürbüz
Yildiz University
Faculty of Science, Physics Department
Istanbul, Turkey

In this work we have represented the variation of the first peaks of the partial pair distribution functions g(r) for the alkali halides which have various crystal radii versus the plasma parameter $\Gamma = e^2/a\,k_bT$ which specifies the coupling strength for Coulombic interactions with $a = (3/8\pi\, n)^{1/3}$ (n is the number density of ions). Our numerical calculations are also based analytically on the solution of the mean spherical approximation for a two-component plasma. The detailed analysis of the results has shown that the structural phenomena for the crystal-forming salts (e.g., liquid-gas and melting transitions) strongly depend on only the relative sizes of their ionic components.

CALCULATION OF THE MOBILITY OF ELECTRONS INJECTED IN SOME SIMPLE LIQUIDS

G. Ascarelli

Physics Department
Purdue University
W. Lafayette, IN 47907 USA

I wish to report the calculation of the mobility of electrons injected in liquid methane, argon, and xenon between their respective triple and critical points. The calculation was carried out as if all these molecules were rare gases.

Good agreement with the experimental time of flight data is obtained by considering only the scattering of electrons by acoustical phonons and by static density fluctuations. The only arbitrary parameter that was used in the case of argon and methane is the effective mass. In xenon instead, where this parameter is known, no adjustable parameters were used. The good agreement between experiment and the model calculation that is found even in the case of methane is an indication that the optical modes arising from both the vibrational and rotational modes of the isolated methane molecules do not affect very much the mobility of thermal electrons.

Finally some considerations are made regarding the parameters that are necessary to consider the mechanisms that will remove energy from the electron distribution when the latter does not correspond to thermal equilibrium.

QUANTUM STATISTICAL MECHANICS OF MOLECULAR LIQUIDS

Shalom Baer

Department of Physical Chemistry
The Hebrew University
91904 Jerusalem, Israel

Taking the intermolecular potential in a liquid as decomposable into a sum of a repulsive potential depending only on nuclear coordinates and the Coulomb potential between the electronic and nuclear charges of different molecules, the latter taken as a perturbation to a classical liquid with a purely repulsive intermolecular potential, one can perform a quantum perturbation expansion of the trace of the statistical operator $\exp(-\beta H)$ and similarly expansions of the thermally averaged, imaginary time displaced, molecular charge correlation functions $\langle\rho(\chi_1,\tau_1),\ldots,\rho(\chi_n,\tau_n)\rangle$. The individual terms in the expansions consist of multipolar interaction tensors and functionals of molecular polarizabilities of various orders. Summation of all terms depending only on linear molecular polarizabilities lead to:

(1) An explicit expression for an extended van der Waals equation of state including contributions of many molecule potentials to all orders.

(2) The liquid susceptibility, including excitonic contributions to the absorption spectrum of the liquid.

AN ELECTRIC-FIELD INDUCED DISPERSION AND COALESCENCE TECHNIQUE IN MULTIPHASE LIQUID SYSTEMS*

Charles H. Byers and Timothy C. Scott

Chemical Technology Division
Oak Ridge National Laboratory
Oak Ridge, TN 37831

In fluid-fluid separation processes, mass transfer rate frequently limits the efficiency of the operation. The fundamental physical problem is transport of a chemical species from a dispersed phase in the form of a bubble or drop to a continuous phase. Once one has selected a chemical system, the equilibrium distribution coefficient is established and one must turn to physical means to improve the transport efficiency. Promotion of convection in the region of the interface and increasing the interfacial area per unit volume of dispersed phase are the two possible approaches. The imposition of pulsed electric fields on the droplet with electrodes in the continuous phase has been studied with transfer enhancement by the two aforementioned mechanisms being the objective.

At low-to-moderate field strengths, droplets oscillate in the electric field. We have examined both the mass transfer and the fluid mechanics of oscillating drops. Our experimental and theoretical results indicated that we can expect a maximum mass transfer enhancement of approximately 50% through this type of operation over the conventional equipment which is currently in commercial use. To proceed beyond this level we must consider droplet rapture.

The natural oscillation frequency was found to be the point at which the minimum field intensity was needed to rapture droplets. A shallower relative minimum was noted at the first overtone frequency. The energy associated with droplet rapture with a pulsed electric field to form an emulsion was estimated using a simple conservative model and found to be

* Research sponsored by the office of Basic Energy Sciences, U.S. Department of Energy under contract DE-AC05-84OR21400 with Martin Marietta Energy Systems, Inc.

at least two orders of magnitude less than would be needed to perform the equivalent function with a mixer (the conventional means).

These findings led to the proposal of a fluidizing contactor in which an upflowing continuous phase (2-ethyl hexanol) in an expanding vertical column contacts an aqueous dispersed phase. The dispersed phase is introduced dropwise through a capillary partway up the expanding column in an area between parallel-plate electrodes. The pulsed field on these electrodes emulsifies the dispersed phase in the region between the electrodes. Because of the expanding channel flow, the emulsion disengages from the continuous phase before this phase leaves the system. Within the area of the column containing the emulsion phase, coalescence occurs, induced by the existence of the pulsed field. The coalesced dispersed phase collects at the bottom of the column and is continuously removed. There are indications that the mass transfer performance of the device is an order of magnitude more efficient than the best of the currently used conventional technology.

The fundamental implications of these findings and the applicability of this invention to separation problems are under investigation.

LASER MULTIPHOTON IONIZATION OF AROMATIC MOLECULES IN NONPOLAR LIQUIDS*

H. Faidas+ and L.G. Christophorou+

Atomic, Molecular, and High Voltage Physics Group
Health and Safety Research Division
Oak Ridge National Laboratory
Oak Ridge, TN 37831 USA

The results of a laser multiphoton ionization (MPI) study of fluoranthene in tetramethylsilane (TMS) and of azulene in n-tridecane (TRD), n-pentane (PNT), 2,2,4,4-tetramethylpentane (TMP), TMS and tetramethyltin (TMT) are reported. Three types of MPI mechanisms have been identified in all solute/solvent systems: 1) two-photon ionization occurring at laser wavelengths $\lambda < 400$-480 nm depending on the solute and the solvent; 2) stepwise three-photon ionization occurring for $\lambda >$ 400-480 nm; and 3) mixed two- and three-photon ionization occurring in an intermediate wavelength region $400 < \lambda < 480$ nm. The stepwise three-photon process consists of two-photon excitation, relaxation to a lower lying excited state with a lifetime comparable to the laser pulse duration (~5 ns) and subsequent ionization with the absorption of a third photon from this state. For azulene, this lower lying state is the second excited singlet state, S_2, while for fluoranthene both the first excited singlet state, S_1, and the S_2 participate in the ionization process. By monitoring the transition from the two- to the three-photon ionization mechanism, the two-photon ionization onset was located and thus the ionization threshold, I_L, for each solute/solvent system was found (Faidas and Christophorou, 1987, 1988). The I_L of fluoranthene in TMS was found to be 5.70 eV and that of azulene in TRD, PNT, TMP, TMS, and TMT 6.28, 6.12, 5.70, 5.45, and 5.33 eV, respectively. The accuracy of the I_L measurements was better than ~0.05 eV. The I_L of azulene

* Research sponsored by the Office of Health and Environmental Research, U.S. Department of Energy, under contract DE-AC05-84OR21400 with Martin Marietta Energy Systems, Inc.

+ Also, Department of Physics, The University of Tennessee, Knoxville, Tennessee 37996.

varied linearly with the V_o (the bottom of the electron conduc-tion band) of the liquid within the error of the present measurements.

References

Faidas, H., and Christophorou, L.G., 1987, J. Chem. Phys., 86:2505.

Faidas, H., and Christophorou, L.G., 1988, "Radiation Physics and Chemistry" (in press).

PRIX CHATEAU LAFITE - ROTHSCHILD: QUANTUM MECHANICS AND SINGLE ELECTRON DIFFRACTION

Gordon R. Freeman

Chemistry Department
University of Alberta
Edmonton, Canada

Every textbook that I have seen of quantum mechanics written for physics students states or implies that a sequence of single particles, separated in space and in time, gives the same diffraction pattern as does an intense beam of the particles.

An intense, coherent beam of photons or other particles gives a Fresnel pattern from a single slit and a Fraunhoffer pattern from a double slit (Feynman, 1965). Textbooks state or imply that a sequence of space-time separated single particles also give a Fraunhoffer pattern from a double slit.

Chateau Lafite was a wine preferred by King Louis XV of France, and is now probably the most famous wine in the world. I offer a bottle of this great wine (vintage 1982, the best since 1959) to the first person who gives me an article or reference, old or new, that contains an experimental proof that a sequence of single particles, separated in space and in time, that has been diffracted by an ensemble of two slits gives a pattern different from the sum of the two patterns of single particles diffracted by the same slits one at a time.

This offer is valid until 1997.

Reference

Feynman Lectures on Physics, 1965, Vol. 3.

WATER-BASED DIELECTRICS FOR HIGH-POWER PULSE FORMING LINES

Victor H. Gehman, Jr. and Ronald J. Gripshover

Naval Surface Weapons Center
Director Energy Branch
Mail Code F12
Dahlgren, VA 22448-5000

Water is often the dielectric of choice for energy storage and pulse forming lines (PFL) in pulsed power systems. This poster reviewed the properties needed by a dielectric for a PFL and/or an energy store and showed why water or a water-based mixture satisfies the requirements. Some Naval applications will require the energy store and the PFL to be combined into a single, compact unit that can be slowly charged (i.e., milliseconds) by rotating machines rather than quickly charged (i.e., microseconds) by Marx generators. The energy density capacity of the dielectric will limit how small the PFL can be made. The energy density of the dielectric is proportional to the square of the electric field. Thus, the electrical breakdown strength of the dielectric is an important material property that needs to be characterized and improved. This poster paper described the experimental apparatus needed to make measurements of the breakdown strength of water and the effects on breakdown strength of electrode materials. Theoretical explanations for the breakdown behavior of water were concerned with the properties of the electric double layer (i.e., high electric fields, ionic concentrations and mobilities) that could account for experimental observations. Finally, the paper described a full-scale, long-charging-time PFL demonstrator with a peak power capability of 2.5 gigawatts.

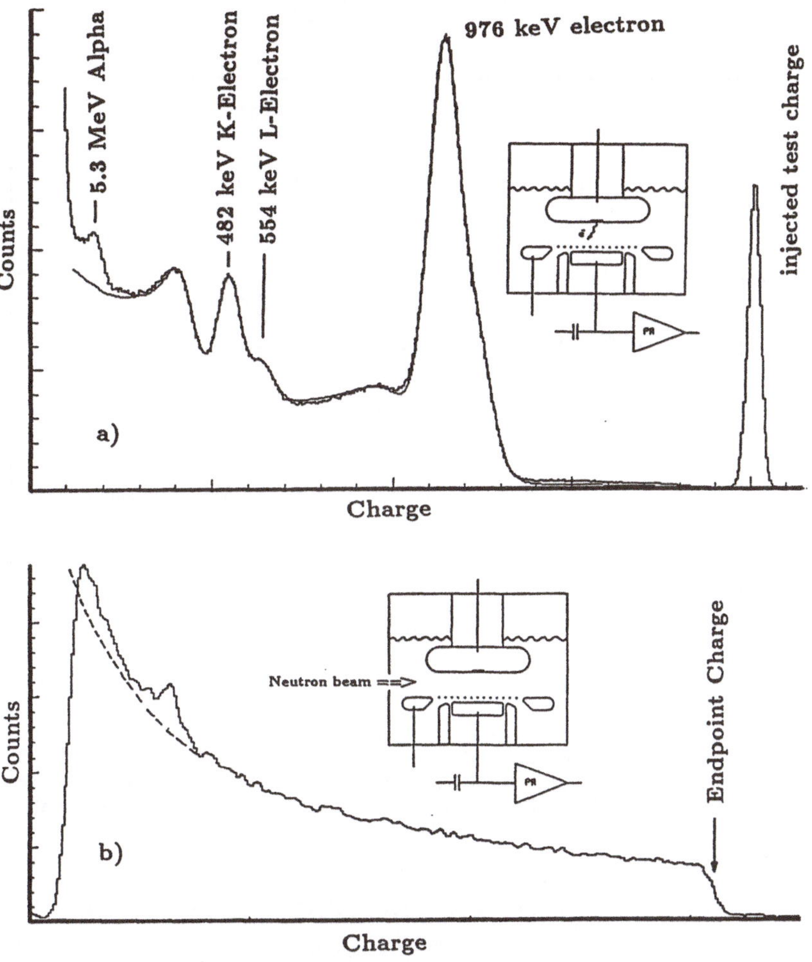

Fig. 1. Charge pulse height spectra for: (a) Implanted sources at 15 kV/cm. The fit shown is a superposition of electron double peaks (K- and L-shell conversion electrons), Compton background produced by Bi-207 γ- lines and estimated backscattering distributions; (b) 14.2 MeV neutrons at 5 kV/cm. The structure above the dashed line is due to accidental coincidences between neutron tag triggers and Bi-207 events in the cell.

Comparisons

Figure 2 shows charge yields versus mean energy loss rate for the particles and energies investigated. Included are results for X- and γ-ray induced ionization in 2,2,4,4-tetramethylpentane (TMP) from Holroyd and Sham (1985) and Am-241 data for TMS and TMP from Muñoz et al. (1986). A fit for 5 kV/cm X-ray yields in TMP, taken from Muñoz et al. (1986), is based on Jaffé's theory for columnar recombination (solid line). The model is in good qualitative agreement with TMS yields for electrons and

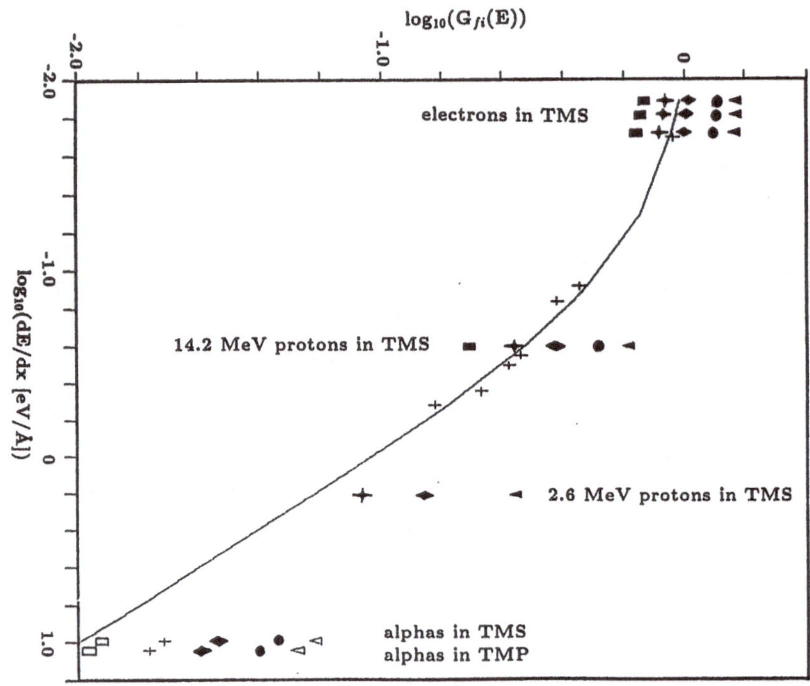

Fig. 2. Charge yields versus mean dE/dx in eV/Å for electrons, protons and alphas, and field strengths: ■, 2; +; 5; ◆, 10; ●, 20; ▼, 30 kV/cm. Included are X-ray yields for TMP 5 kV/cm + (Holroyd and Sham, 1985) and alpha yields for TMS and TMP □-∇ (Muñoz et al., 1986).

14.2 MeV protons but fails to reproduce results obtained with 2.6 MeV protons and alphas. Apparently a model is needed which takes into account the differences in track structure produced by different ionizing particles.

Acknowledgement

The neutron measurements were done at the University of Lausanne. We thank Prof. C. Joseph for letting us use his Van de Graaff accelerator. Very valuable help in operating the machine was contributed by Mr. L. Studer and Dr. T. Tran.

References

Holroyd, R.A., and Sham, T.K., 1985, J. Phys. Chem., 89:2909.
Miroshnicenko, V.P., Newskij, P.L., Rodionov, B.U., 1982, Elementary Particles and Cosmic Rays, Moscou, Energoizdat.
Muñoz, R.C., Cumming, J.B., and Holroyd, R.A., 1986, J. Chem. Phys., 85:1104.

ELECTROHYDRODYNAMICALLY-DRIVEN LIQUID ION SOURCE FOR INERTIAL CONFINEMENT FUSION

Arian L. Pregenzer
Sandia National Laboratories
Albuquerque, NM 87185 USA

At Sandia National Laboratories, an effort is underway to develop a two-dimensional electrohydrodynamically- (EHD) driven liquid lithium ion source for light ion fusion in the Particle Beam Fusion Accelerator II (PBFA II). PBFA II is designed to deliver a 30 MV voltage pulse of about 10 ns FWHM to a 2.0 cm anode-cathode gap. Ions generated at the inner surface of the barrell-shaped anode will be accelerated to an energy of 30 MeV and focussed on an inertial confinement fusion (ICF) target at the center. A necessary first step in the success of PBFA II is development of a Li^+ ion source which will begin to emit on a nanosecond time scale, have a beam divergence of less than 10 mrad, and produce a current density of about 5 kA/cm^2 from an active anode area of about 1000 cm^2. Theoretical work predicts that an EHD-driven liquid lithium or liquid lithium nitrate ($LiNO_3$) ion source can meet these requirements. The basic principles of EHD theory are that a liquid surface is deformed when subjected to an applied electric field E_o due to the competing effects of electric stress and surface tension. Unstable deformations grow into cusp-like structures. Field evaporation of ions occurs from the tips of these cusps when the <u>enhanced</u> electric field strength exceeds 100 MV/cm. The wavelength and growth time of the most rapidly growing mode are obtained by solving a dispersion relation resulting from a linear analysis of the EHD equations of motion (Pregenzer, 1985). Nonlinear analysis predicts that when the <u>applied</u> electric field is greater than 10 MV/cm, cusp formation and ion emission will occur in less than 3 ns (Pregenzer and Marder, 1986). Experiments with water and ethanol near the critical field (18 - 35 kV/cm) have been in good agreement with theoretical predictions of wavelength and are consistent with predictions of cusp formation time as a function of E_o (Pregenzer and Woodworth, 1988). Theory also indicates that the beam divergence will be less than

4 mrad if the applied voltage is greater than 10 MV and the initial applied electric field is greater than 10 MV/cm (Pregenzer and Kingham, 1986). For $E_o > 10$ MV/cm, the wavelength of the dominant instability is less than 0.5 μm. This means that the emitter density will be greater than 4.0×10^6 per cm^2. Calculations suggest that for the parameter range pertinent to PBFA II the effect of space-charge from adjacent emitters will not significantly reduce the current from an emitter in an array compared to that of a single emitter, and that current densities of 5 kA/cm^2 should be achievable (Pregenzer and Kingham, 1986). In summary, theory predicts that an EHD-driven liquid lithium ions source can meet the requirements for ICF on PBFA II.

Within the next year, we expect to field liquid anodes on PBFA II. A small-scale experiment with a liquid anode outside an ion diode is also planned. If successful, these experiments will represent the first two-dimensional EHD-driven ion source.

References

Pregenzer, A.L., 1985, J. Appl. Phys., 58:4509.

Pregenzer, A.L., and Marder B.M., 1986, J. Appl. Phys., 60:3821.

Pregenzer, A.L., and Kingham, D.R., 1986, Beam Divergence and Space Charge Effects for an Electrohydrodynamically-Driven Liquid Ion Source," Particle Beam Fusion Progress Report, July-December, (Albuquerque, NM, Sandia National Laboratories, to be published).

Pregenzer, A.L., and Woodworth, J.R., 1988, Submitted to J. Appl. Phys.

PHOTO-INJECTED ELECTRON CURRENT INTO LIQUID Ar, Ne, He, N_2, AND H_2

Yosuke Sakai, Hiroshi Kojima, and Hiroaki Tagashira

Department of Electrical Engineering
Hokkaido University
Sapporo, 060 Japan

The steady-state photoelectron currents J_L injected from the cathode irradiated by a deuterium lamp into liquid Ar, Ne, He, N_2, and H_2 near their boiling temperatures were measured for the external field strengths E_a up to 10^5 V/cm. The curve of the current attenuation in liquid Ar, Ne, and He from that available from the same light intensity into vacuum at the liquid temperature, J (= J_L/J_V, where J_V is the current in vacuum), against E_a was shown to be composed of two regions, i.e., $J \propto E_a^{1.3\sim2}$ for low E_a and $J \propto E_a^{0.5}$ for high E_a. However, the J in liquid N_2 and H_2 increased only in proportion to E_a^2 (i.e., space charge limited current). The J in liquid Ar was the largest among the liquids, and the J in the other liquids decreased in order of liquid Ne, He, N_2, and H_2. The magnitude of J was found to be arranged in reverse order of the dielectric strength of the liquids. This result may suggest that the facility of electron injection from the cathode plays an important role in determination of the dielectric strength.

IONIC PHOTODISSOCIATION AND PICOSECOND SOLVENT RELAXATION

Kenneth G. Spears, John Gong, and Martha Wach

Northwestern University
Chemistry Department
Evanston, IL 60208

We present rate data for an ionic photodissociation process as a function of temperature. The experiment uses picosecond ultraviolet excitation and fluorescence lifetime measurements to monitor the rate of ion pair formation from a neutral molecule. The molecule is malachite green leucocyanide, which ionizes to form a green colored carbonium ion and a cyanide anion. Prior work studied ion recombination and solvent stabilization of transition states (J. Phys. Chem., 1986, 90:779), and in this work we used temeprature dependent rate data to derive activation energies and entropies. The results for the lowest excited state process shows that dielectric constant changes ranging from 5 to 34 in ethyl acetate/acetonitrile mixtures affect the rate by changing the activation free energy. The rate changes are consistent with the dipole solvation model of Kirkwood and the assumption of an ionic dipole transition state. The activation energy for this process is very small (~1 kcal/mole) and independent of dielectric constant while the negative activation entropy is very dependent on dielectric constant. The electrostatic reduction of solvent motions probably is the dominant effect in reducing the entropy. The thermodynamic relation, $[\partial\Delta G / \partial T]_p = -\Delta S$, when applied to the Kirkwood formula, predicts that the dielectric constant, D, will correlate with ΔS by the function $[3D/(2D + 1)^2][\partial \ln D/\partial T]_p$. Our data correlate ΔS with $3D/(2D + 1)^2$, which tentatively supports the electrostatic model. Future measurements in dilute mixtures of molecular dipoles in nonpolar solvent should be amenable to analysis by molecular dynamics models in order to develop a microscopic understanding of the solvent entropy changes.

THEORY OF COMPLEXATION-SOLVATION

M. Szpakowska and O.B. Nagy

Universite Catholique de Louvain
Unite Cico, Laboratoire de Chimie
Organique Physique
Place Louis Pasteur, 1
B-1348 Louvain la Neuve
Belgium

The competitive preferential solvation (COPS) theory provides a general framework for interpreting, on the same footing, all complexation and solvation phenomena. Its applications to various physico-chemical techniques allowed to account for some anomalies (e.g., negative association constants) in a straightforward manner (Nagy et al., 1978, 1979). The basic idea of COPS theory is a microscopic partitioning in homogeneous media. This leads to the establishment of transferable solute-solvent affinity constants. The limiting case of partitioning in homogeneous media is coordination. Even in this case COPS theory is fully verified. Perusal of spectroscopic data obtained for Cu(II) complexes (Szpakowska et al., 1985) in non-aqueous binary solvent mixtures yielded transferable solute-solvent affinity constants and solvent-independent molar extinction coefficients. Detailed analysis shows that COPS theory and tranditional thermodynamics lead to the same results provided the latter takes into account the solvent explicitely (Parbhoo and Nagy, 1986). Ion fluxes (J) through liquid membranes obey also the laws of COPS theory. When K^+(A) is transported through binary mixed membranes (KCl in H_2O/membrane: CH_2Cl_2(S) and/or C_6H_5Cl(Z); carrier: dibenzocrown-18-6/pure H_2O; T = 20±0.1°C; mixing rate: 300 rpm) the following results are obtained:

Volume fraction of C_6H_5Cl:Y_Z	0.0	0.05	0.1	0.2	0.3	0.4	0.6	0.8	1.0
Ion flux:$J10^6$(M/h)	43.4 (J_S)	17.9	8.45	8.25	2.63	3.84	0.96	0.56	0.246 (J_Z)

The transport profile can be linearized using the equation:

$$(J_S-J)/C_Z = (J_S-J_Z)\kappa_{A(Z)}v_S/\kappa_{A(S)}-\left(\kappa_{A(Z)}v_S/\kappa_{A(S)}-v_Z\right)(J_S-J) ,$$

where C_Z is the concentration of C_6H_5Cl(M), $\kappa_{A(j)}$ is the solute A-solvent j affinity constant and v_S and v_Z are the molar volumes of CH_2Cl_2 and C_6H_5Cl, respectively. One obtains:

$$\kappa_{A(Z)}/\kappa_{A(S)} = 43.6; \; J_S-J_Z = 43.22 \times 10^{-6}(\text{M/h})(43.15 \text{ exp.});$$
$$v_Z = 0.106\ (\text{M}^{-1})(0.102 \text{ exp.}) .$$

The applicability of COPS theory to transport phenomena raises interesting questions about the solvation shell structure in the light of the irreversible thermodynamics.

References

Nagy, O.B., Muanda, M. wa, and Nagy, J.B., 1978, J. Chem. Soc., Faraday Trans. 1, 74:2210.
Nagy, O.B., Muanda, M. wa, and Nagy, J.B., 1961, J. Phys. Chem., 83:1979.
Parbhoo, B., and Nagy, O.B., 1986, J. Chem. Soc., Faraday Trans. 1, 82:1789.
Szpakowska, M., Uruska, I., and Zielkiewicz, J., 1985, J. Chem. Soc., Dalton Trans., p. 1849.

APPENDIX B

ORGANIZING COMMITTEE

ORGANIZING COMMITTEE

L. H. Luessen
Naval Surface Warfare Center
Code F12
Dahlgren, VA 22448-5000
USA

E. E. Kunhardt
Polytechnic University
Weber Research Institute
Route 110
Farmingdale, NY 11735
USA

L. G. Christophorou
Atomic, Molecular, and High Voltage Physics Group
Health and Safety Research Div.
Oak Ridge National Laboratory
Oak Ridge, TN 37831
USA

A. J. Policarpo
University of Coimbra
Departamento de Fisica
3000 Coimbra
Portugal

R. E. Hebner
Electrosystems Division
National Bureau of Standards
Gaithersburg, MD 20899
USA

W. F. Schmidt
Hahn-Meitner-Institut
Postface 390128
D-1000 Berlin 39
West Germany

E. A. Silva
Office of Naval Research
Code 1121
800 North Quincy Street
Arlington, VA 22217
USA

APPENDIX C

LECTURERS

LECTURERS

L. G. Christophorou
Atomic, Molecular, and High
Voltage Physics Group
Health and Safety Research
Division
Oak Ridge National Laboratory
Oak Ridge, TN 37831
USA

Brian Conway
University of Ottawa
Chemistry Department
32 George Glinski Street
Ottawa, Ontario KIN 6N5
Canada

K. B. Eisenthal
Columbia University
Department of Chemistry
New York, NY 10027
USA

Gordon R. Freeman
Department of Chemistry
E3-43 Chemistry Building East
University of Alberta
Edmonton, Alberta T6G 2G2
Canada

J. P. Gosse
CNRS
Laboratory for Electrostatics
25 Avenue des Martyrs
38042 Grenoble
France

R. E. Hebner
National Bureau of Standards
Electrosystems Division
Gaithersburg, MD 20899
USA

Richard Holroyd
Department of Chemistry
Building 555
Brookhaven National Laboratory
Upton, NY 11973
USA

Joseph Hubbard
National Bureau of Standards
Thermo-Physics Division
Building 221
Gaithersburg, MD 20899
USA

G. A. Kenney-Wallace
University of Toronto
Department of Chemistry
Toronto, Ontario MSS 1A1
Canada

Neil Kestner
Louisiana State University
Department of Chemistry
Baton Rouge, LA 70803
USA

T. J. Lewis
University College of North Wales
Bangor, Gwynedd LL57 1UT
United Kingdom

Donald McQuarrie
University of California at Davies
Chemistry Department
Davies, CA 95616
USA

J. C. Rasaiah
University of Maine
Department of Chemistry
Orono, ME 04469
USA

W. F. Schmidt
Hahn-Meitner-Institut
Postfach 390128
D-1000 Berlin 39
West Germany

Itzack T. Steinberger
Hebrew University of Jerusalem
The Racah Institute of Physics
Jerusalem 91904
Isreal

R. Tobazeon
CNRS
Laboratory for Electrostatics
25 Avenue des Martyrs
38042 Grenoble
France

Markus Zahn
Massachusetts Inst. of Technology
Building N-10
155 Massachusetts Avenue
Cambridge, MA 02139
USA

H. R. Zeller
BBC Research Center
CH-5405
Baden-Dattwil
Switzerland

APPENDIX D

PARTICIPANTS

PARTICIPANTS

Nikolaos Alexandropoulos
University of Ioannina
Department of Physics
453 32 Ioannina
Greece

Elena Aprile
Columbia University
Physics Department
538 W. 120th St.
New York, NY 10027
USA

Turgay Armagan
University of Trakya
Faculty of Science and Letters
Physics Department
Edirne
Turkey

Gianni Ascarelli
Purdue University
Physics Department
W. Lafayette, IN 47906
USA

Shalom Baer
Hebrew University of Jerusalem
Department of Physical Chemistry
91904 Jerusalem
Israel

Charles H. Byers
Oak Ridge National Laboratory
Chemical Technology Division
Box X
Building 4501
Oak Ridge, TN 37831-6224
USA

Eugenio Caponeti
Instituto di Chimica Fisica
Via Archirafi 26
90123 Palermo
Italy

Alexandros A. Christodoulides
University of Ioannina
Department of Physics
Ioannina 453 32
Greece

Marc Cuzin
CEA/CEN.Grenoble LETI/ESA
Avenue des Martyrs
85 X 38041
Grenoble Cedex
France

M. P. de Haas
Interuniversity Reactor Institute
Mekelweg 15
2629 JB Delft
The Netherlands

Ann Drury
Polytechnic University
Weber Research Institute
Route 110
Farmingdale, NY 11735
USA

Daniel A. Erwin
University of Southern California
Department of Aerospace Engineering
Los Angeles, CA 90089-1191
USA

Homer Faidas
University of Tennessee, Knoxville
Department of Physics
Knoxville, TN 37996
USA

Charles Fenimore
National Bureau of Standards
Metrology - B344
Gaithersburg, MD 20899
USA

Michel F. Frechette
Institut de Recherche D'Hydro-Quebec (IREQ)
P. O. Box 1000 or 1800, Montee
STE-Julie, Varennes
Quebec, J0L 2P0 (Room P-70)
Canada

Jitka Fuhr
Polythecnic University
Weber Research Institute
Route 110
Farmingdale, NY 11735
USA

Uno Gafvert
ASEA Research and Innovation
S-721 78 Vasteras
Sweden

Adolfas K. Gaigalas
National Bureau of Standards
Building 230, Room 105
Gaithersburg, MD 20849
USA

Victor H. Gehman, Jr.
Naval Surface Warfare Center
Code F12
Dahlgren, VA 22448-5000
USA

Ronald J. Gripshover
Naval Surface Weapons Center
Code F12
Dahlgren, VA 22448-5000
USA

Jean-Pierre Guelfucci
Universite Paul Sabatier
Centre de Physique Atomique
118 Route de Narbonne
31062 Toulouse Cedex
France

Linda C. Johnson
Naval Surface Warfare Center
Code F12
Dahlgren, VA 22448-5000
USA

Hugh Jones
Polytechnic University
Weber Research Institute
Route 110
Farmingdale, NY 11735
USA

B. R. Junker
Office of Naval Research
Code 111
800 N. Quincy Street
Arlington, VA 22217-5000
USA

Seyfeddin Karagozlu
Assistant Professor
Yildiz Universitesi
Physics Department
Fen-Edebiyat Fakultesi
Sisli - Instanbul
Turkey

Joseph Kunc
University of Southern California
Aerospace Engineering Department
Rapp Building
University Park-MC 1191
Los Angeles, CA 90089-1191
USA

Maria Salete Leite
University of Coimbra
Departmento de Fisica
3000 Coimbra
Portugal

Stephen Levy
Electronics Technology and Devices
Laboratory
ATTN: SLCET-ML
Ft. Monmouth, NJ 07703
USA

Maria Isabel Lopes
Universidade de Coimbra
Departmento de Fisica
3000 Coimbra
Portugal

N. Marcuvitz
Polytechnic University
Weber Research Institute
Route 110
Farmingdale, NY 11735
USA

Jane Messerschmitt
Polytechnic University
Electrical Engineering
Route 110
Farmingdale, NY 11735
USA

Paul C. Munoz
CERN
Group EP/VA1
CH-1211, Geneva 23
Switzerland

O. B. Nagy
Universite Catholique de Louvain
Unite Cico, Laboratoire de Chimie
Organique Physique
Place Louis Pasteur, 1
B-1348 Louvain La Veuve
Belgium

Stephan Ochsenbein
Swiss Inst. of Nuclear Research
(SIN)
5234 Villigen
Switzerland

Ian Penfold
University of East Anglia
School of Mathematics and Physics
Norwich NR4 7TJ
England

S. Petrucci
Polytechnic University
Weber Research institute
Route 110
Farmingdale, NY 11735
USA

PARTICIPANTS (Cont.)

Arian L. Pregenzer
Sandia National Laboratories
Simulation Physics Division 1231
P. O. Box 5800
Albuquerque, NM 87185
USA

Michael Reichling
Freie Universitat Berlin
Fachbereich Physik
Institut fur Atom- and
Festkorperphysik (WE 1)
Arnimallee 14, D-1000 Berlin 33
West Germany

Yosuke Sakai
Hokkaido University
Department of Electrical
Engineering
KIYA-13, NISHI-8
Sapporo 060
Japan

Kenneth G. Spears
Northwestern University
Chemistry Department
2145 Sheridan Road
Evanston, IL 60201
USA

Spyros Spyrou
Nat. Hellenic Research Foundation
Theoretical and Physical Chemistry
Institute
48 Vasileos Constatninou Avenue
Athens 116 - 35
Greece

Tangali S. Sudarshan
University of South Carolina
Electrical & Computer Engineering
Department
College of Engineering
Columbia, SC 29208
USA

Benson R. Sundheim
New York University
Department of Chemistry
4 Washington Place
Room 514
New York, NY 10003
USA

Bo R. Svensson
University of Lund
Physical Chemistry II
Chemical Center
P. O. B. 124
S-22100 Lund
Sweden

M. Szpakowska
Universite Catholique de Louvain
Unite CICO, Laboratoire de Chimie
Organique Physique, C/O Dr. Nagy
Place Louis Pasteur, 1
B-1348 Louvain La Neuve
Belgium

Mac A. Thompson
RTE Corporation
Corporate R&D Center
N25 W23131 Paul Road
Pewaukee, WI 53072-4025
USA

Aysen Turkman
Dokuz Eylul University
Faculty of Engineering and
Architecture
Dept. of Environmental Eng.
Bornova Izmir
Turkey

Isao Ueno
The University of Tokyo
Dept. of Electrical Engineering
7-3-1, Hongo, Bunkyo-Ku
Japan

Alan Watson
University of Windsor
Windsor, Ontario N9B 3P4
Canada

P. F. Williams
University of Nebraska
Dept. of Electrical Engineering
Lincoln, NE 68588-0511
USA

Clifford E. Woodward
University of Lund
Chemical Centre
Dept. of Physical Chemistry 2
S-22100 Lund
Sweden

INDEX